MW01343314

Modeling and Analysis Principles for Chemical and Biological Engineers

ISBN 978-0-9759377-1-6

Modeling and Analysis Principles for Chemical and Biological Engineers

Michael D. Graham and James B. Rawlings
Department of Chemical and Biological Engineering
University of Wisconsin-Madison
Madison, Wisconsin

Madison, Wisconsin

This book was set in Lucida using LaTeX, and printed and bound by Worzalla. It was printed on acid-free paper.

Cover design by Cheryl M. and James B. Rawlings, and Michael D. Graham

Nob Hill Publishing, LLC
Cheryl M. Rawlings, publisher
Madison, WI 53705
orders@nobhillpublishing.com
http://www.nobhillpublishing.com

Library of Congress Control Number: 2012956351

Graham, Michael D.
Modeling and Analysis Principles for
Chemical and Biological Engineers \
by Michael D. Graham and James B. Rawlings
p. cm.
Includes bibliographical references (p.) and index.
ISBN 978-0-9759377-1-6 (cloth)
1. Chemical engineering. 2. Mathematical modeling.
I. Rawlings, James B. II. Title.

Printed in the United States of America.

First Printing May 2013

To my father and the memory of my mother.

MDG

To my graduate students, who have been some of my best teachers.

JBR

Preface

Research undertaken by modern chemical and biological engineers incorporates a wide range of mathematical principles and methods. This book came about as the authors struggled to incorporate modern topics into a one- or two-semester course sequence for new graduate students, while not losing the essential aspects of traditional mathematical modeling syllabi. Topics that we decided are particularly important but not represented in traditional texts include: matrix factorizations such as the singular value decomposition, basic qualitative dynamics of nonlinear differential equations, integral representations of partial differential equations, probability and stochastic processes, and state estimation. The reader will find many more in the book. These topics are generally absent in many texts, which often have a bias toward the mathematics of 19th- through early 20th-century physics. We also believe that the book will be of substantial interest to active researchers, as it is in many respects a survey of the applied mathematics commonly encountered by chemical and biological engineering practitioners, and contains many topics that were almost certainly absent in their chemical engineering graduate coursework.

Due to the wide range of topics that we have incorporated, the level of discussion in the book ranges from very detailed to broadly descriptive, allowing us to focus on important core topics while also introducing the reader to more advanced or specialized ones. Some important but technical subjects such as convergence of power series have been treated only briefly, with references to more detailed sources. We encourage instructors and students to browse the exercises. Many of these illustrate applications of the chapter material, for example, the numerical stability of the Verlet algorithm used in molecular dynamics simulation. Others deepen, complement, and extend the discussion in the text.

During their undergraduate education in chemical and biological engineering, students become very accomplished at numerical examples and problem solving. This is not a book with lots of numerical examples. Engineering graduate students need to make the shift from applying mathematical tools to developing and understanding them. As such, substantial emphasis in this book is on derivations and sometimes short proofs. We believe the text contains a healthy mix of fundamental mathematics, analytical solution techniques, and numerical methods. Researchers in engineering must know mathematical structures, principles, and tools, because these guide analysis and understanding, and they also must be able to produce quantitative answers. We hope this text will enable them to do both.

MDG
Madison, Wisconsin

JBR
Madison, Wisconsin

Acknowledgments

This book grew out of the lecture notes for graduate level analysis courses taught by the authors in the Department of Chemical and Biological Engineering at the University of Wisconsin-Madison. We have benefited from the feedback of many graduate students taking these classes, and appreciate the enthusiasm with which they received some early and incomplete drafts of the notes. Especially Andres Merchan, Kushal Sinha, and Megan Zagrobelny provided helpful discussion and assistance.

We have had numerous helpful discussions with colleagues on many topics covered in the text. JBR would like to acknowledge especially Dave Anderson, David Mayne, Gabriele Pannocchia, and Joe Qin for their interest and helpful suggestions.

Several colleagues gave us helpful reviews of book chapters. We would like to thank Prodromos Daoutidis, Tunde Ogunnaike, Patrick Underhill, Venkat Ganesan, Dave Anderson, and Jean-Luc Thiffeault for their valuable feedback. We are also grateful to colleagues who responded to a survey that we conducted to gather information on mathematical modeling courses for chemical and biological engineering graduate students. Their valuable feedback had significant impact on the content of this book.

Several members of our research groups also reviewed chapters, and helped us typeset solutions to some of the exercises. Anubhav, Cuyler Bates, Ankur Gupta, Rafael Henriquez, Amit Kumar, Jae Sung Park, and Sung-Ning Wang deserve special mention.

John Eaton generously provided his usual invaluable computing and typesetting expertise. MDG is grateful to his family for their forbearance during the preparation of this book. Our special relationship with the staff at Nob Hill Publishing again made the book production process an enjoyable experience.

Contents

List of Figures

List of Tables

List of Examples and Statements

1

Linear Algebra

1.1 Vectors and Linear Spaces

A vector is defined in introductory physics courses as a quantity having magnitude and direction. For example, the position vector of an object in three dimensions is the triple of Cartesian coordinates that determine the position of the object relative to a chosen origin. Another way of thinking of the position vector is as a point in three-dimensional space, generally denoted $\mathbb{R}^3$. This view leads us to the more general and abstract definition of a vector: A VECTOR IS AN ELEMENT OF A LINEAR SPACE:

Definition 1.1 (Linear space). A linear space is a set V whose elements (vectors) satisfy the following properties: For all x, y, and z in V and for all scalars α and β

$$
\begin{aligned}
x + y &\in V && \text{closure under addition} \\
\alpha x &\in V && \text{closure under multiplication} \\
x + y &= y + x && \\
x + (y + z) &= (x + y) + z && \\
x + 0 &= x && \text{definition of the origin} \\
x + (-x) &= 0 && \text{definition of subtraction} \\
\alpha(\beta x) &= (\alpha\beta)x && \\
(\alpha + \beta)x &= \alpha x + \beta x && \\
\alpha(x + y) &= \alpha x + \alpha y && \\
1x &= x, 0x = 0 &&
\end{aligned}
$$

Naturally, these properties apply to vectors in normal 3-D space; but they also apply to vectors in any finite number of dimensions as well as to sets whose elements are, for example, 3 by 3 matrices or trigonometric functions. This latter case is an example of a *function space;* we will encounter these in Chapter 2. Not every set of vectors forms a linear space, however. For example, consider vectors pointing from the origin to a point on the unit sphere. The sum of two such vectors will no longer lie on the unit sphere—vectors defining points on the sphere do not form a linear space. Regarding notation, many readers will be familiar with vectors expressed in boldface type, **x**, **v**, etc. This notation is especially common in physics-based problems where these are vectors in three-dimensional physical space. In the applied mathematics literature, where a vector takes on a more general definition, one more commonly finds vectors written in italic type as we have done above and will do for most of the book.

1.1.1 Subspaces

Definition 1.2 (Subspace). A subspace S is a subset of a linear space V whose elements satisfy the following properties: For every $x, y \in S$ and for all scalars α

$$\begin{aligned} x + y &\in S \qquad \text{closure under addition} \\ \alpha x &\in S \qquad \text{closure under multiplication} \end{aligned} \tag{1.1}$$

For example, if V is the plane ($\mathbb{R}^2$), then any line through the origin on that plane is a subspace.

1.1.2 Length, Distance, and Alignment

The idea of a norm generalizes the concept of length.

Definition 1.3 (Norm). A norm of a vector x, denoted $\|x\|$, is a real number that satisfies

$$\begin{aligned} \|\alpha x\| &= |\alpha| \; \|x\| \\ \|x\| &> 0, \forall x \neq 0 \\ \|x\| &= 0 \text{ if } x = 0 \\ \|x + y\| &\leq \|x\| + \|y\| \qquad \text{triangle inequality} \end{aligned}$$

The Euclidean norm in $\mathbb{R}^n$ is our usual concept of length

$$\|x\|_2 = \sqrt{\sum_{i=1}^{n} |x_i|^2}$$

in which x_i is the ith component of the vector. Unless otherwise noted, this is the norm that will be used throughout this book, and will generally be denoted simply as $\|x\|$ rather than $\|x\|_2$. It should be noted, however, that this is not the only definition of a norm, nor is it always the most useful. For example, the so-called l_p norms for vectors in $\mathbb{R}^n$ are defined by the equation

$$\|x\|_p = \left(\sum_{i=1}^{n} |x_i|^p\right)^{1/p}$$

Particularly useful are the cases $p = 1$, sometimes called the "taxicab norm" (why?) and $p = \infty$: $\|x\|_\infty = \max_i |x_i|$.

The INNER PRODUCT generalizes the dot product of elementary algebra and measures the alignment of a pair of vectors: an inner product of two vectors, denoted (x, y) is a scalar that satisfies

$$\begin{aligned}(x + y, z) &= (x, z) + (y, z)\\ (\alpha x, y) &= \alpha(x, y)\\ (x, y) &= \overline{(y, x)}\\ (x, x) &> 0, \text{ if } x \neq 0\end{aligned}$$

The overbar denotes complex conjugate. Notice that the square root of the inner product $\sqrt{(x, x)}$ satisfies all the properties of a norm, so it is a measure of the length of x. The usual inner product in $\mathbb{R}^n$ is

$$(x, y) = \sum_{i=1}^{n} x_i y_i$$

in which case $\sqrt{(x, x)} = \|x\|_2$. This is a straightforward generalization of the formula for the dot product $\mathbf{x} \cdot \mathbf{y}$ in $\mathbb{R}^2$ or $\mathbb{R}^3$ and has the same geometric meaning

$$(x, y) = \|x\| \, \|y\| \cos\theta$$

where θ is the angle between the vectors. See Exercise 1.1 for a derivation. If we are considering a space of complex numbers rather than real numbers, the usual inner product becomes

$$(x, y) = \sum_{i=1}^{n} x_i \bar{y}_i$$

If $(x, y) = 0$, then x and y are said to be ORTHOGONAL.

Finally, we can represent a vector x in $\mathbb{R}^n$ as a single column of elements, a COLUMN VECTOR, and define its TRANSPOSE x^T as a ROW VECTOR

$$x = \begin{bmatrix} x_1 \\ x_2 \end{bmatrix} \qquad x^T = \begin{bmatrix} x_1 & x_2 \end{bmatrix}$$

Now the inner product (x, y) can be written $x^T y$ if x and y are real, and $x^T \bar{y}$ if they are complex.

1.1.3 Linear Independence and Bases

If we have a set of vectors, say $\{x_1, x_2, x_3\}$, in a space V, this set is said to be LINEARLY INDEPENDENT (LI) if the only solution to the equation

$$\alpha_1 x_1 + \alpha_2 x_2 + \alpha_3 x_3 = 0$$

is $\alpha_i = 0$ for all i. Otherwise the set is LINEARLY DEPENDENT. A space V is n-DIMENSIONAL if it contains a set of n linearly independent vectors, but no set of $n+1$ linearly independent vectors. If n LI vectors can be found for any n, no matter how large, then the space is INFINITE-DIMENSIONAL.

Everything said above holds independent of our choice of coordinate system for a space. To actually compute anything, however, we need a convenient way to represent vectors in a space. We define a BASIS $\{e_1, e_2, e_3 \ldots\}$ as a set of LI vectors that SPAN the space of interest, i.e., every vector x in the space can be represented

$$x = \alpha_1 e_1 + \alpha_2 e_2 + \alpha_3 e_3 + \cdots$$

If a space is n-dimensional, then a basis for it has exactly n vectors and vice versa. For example, in $\mathbb{R}^3$ the unit vectors in the x, y, and z directions form a basis. But more generally, any three LI vectors form a basis for $\mathbb{R}^3$.

Although any set of LI vectors that span a space form a basis, some bases are more convenient than others. The elements of an ORTHONORMAL (ON) basis satisfy these properties

$(e_i, e_i) = 1$	each basis vector has unit length
$(e_i, e_j) = 0, i \neq j$	the vectors are mutually orthogonal

These properties may be displayed more succinctly

$$(e_i, e_j) = \delta_{ij} \equiv \begin{cases} 1, & i = j \\ 0, & i \neq j \end{cases}$$

The symbol δ_{ij} is called the KRONECKER DELTA. In an orthonormal basis, any vector can be expressed

$$x = \sum_i (x, e_i) e_i$$

1.2 Linear Operators and Matrices

An OPERATOR transforms one vector into another. Operators appear everywhere in applied mathematics. For example, the operator d/dx transforms a function $f(x)$ into its derivative. More abstractly, an operator A is a mapping that takes elements of one set (the DOMAIN of A) and converts them into elements of another (the RANGE of A). LINEAR operators satisfy the following properties for all vectors u and v in their domain and all scalars α

$$\begin{aligned} A(u+v) &= Au + Av \\ A(\alpha u) &= \alpha(Au) \end{aligned} \tag{1.2}$$

We focus here on operators on finite-dimensional vector spaces $\mathbb{R}^n$; operators on spaces of complex numbers are similar. (In Chapter 2 we will look at an important class of operators in function spaces.) In these spaces, and having chosen a coordinate system in which to represent vectors, any linear operator can be expressed as multiplication by a MATRIX. A matrix is an array of numbers

$$A = \begin{bmatrix} A_{11} & A_{12} & \dots & A_{1n} \\ A_{21} & A_{22} & \dots & A_{2n} \\ \vdots & \vdots & \ddots & \vdots \\ A_{m1} & A_{m2} & \dots & A_{mn} \end{bmatrix}$$

The first subscript of each element denotes its row, while the second denotes its column. The transformation of a vector

$$x = \begin{bmatrix} x_1 \\ x_2 \\ \vdots \\ x_n \end{bmatrix} \qquad \text{into another} \qquad y = \begin{bmatrix} y_1 \\ y_2 \\ \vdots \\ y_n \end{bmatrix}$$

then occurs through matrix-vector multiplication. That is: $y = Ax$, which means

$$y_i = \sum_{j=1}^{n} A_{ij} x_j, i = 1, 2, \dots, m$$

In this example, the matrix A is m by n (rows by columns); it is an element of the linear space $\mathbb{R}^{m \times n}$ and multiplication by A maps vectors in $\mathbb{R}^n$ into vectors in $\mathbb{R}^m$. That is, for the function defined by matrix multiplication, $f(x) = Ax$, $f : \mathbb{R}^n \to \mathbb{R}^m$. Some readers will be familiar with matrices written in bold and the matrix-vector product between matrix $\mathbf{A}$ and vector $\mathbf{x}$ written as either $\mathbf{Ax}$ or $\mathbf{A} \cdot \mathbf{x}$.

One can also think of each row of A as a vector. In this case the ith component of y can be thought of as the dot product between the ith row of A and the vector x. This is probably the best way to remember the actual algebra of the matrix-vector multiplication formula. A more intuitive and general geometric interpretation (which will be used extensively as we proceed through the chapter) is allowed by considering each *column* of A as a vector, and thinking of the vector y as a linear combination of these vectors. That is, y is in the space spanned by the columns of A. If we let the ith column of A be the vector c_i, then

$$
\begin{aligned}
y &= x_1 c_1 + x_2 c_2 + x_3 c_3 + \dots \\
&= \sum_{j=1}^{n} x_j c_j
\end{aligned}
$$

(Note that in this equation x_j is a scalar component of the vector x, while c_j is a vector.) This equation implies that the number of columns of A must equal the length of x. That is, matrix-vector multiplication only makes sense if the vector x is in the domain of the operator A.

1.2.1 Addition and Multiplication of Matrices

The following terminology is used to describe important classes of matrices.

1. A is SQUARE if $m = n$.

2. A is DIAGONAL if A is square and $A_{ij} = 0$ for $i \neq j$. This is sometimes written as $A = \text{diag}(a_1, a_2, ..., a_n)$ in which a_i is the element A_{ii} for $i = 1, 2, \dots, n$.

3. A is UPPER (LOWER) TRIANGULAR if A is square and $A_{ij} = 0$ for $i > j$ $(i < j)$.

4. A is UPPER (LOWER) HESSENBERG if A is square and $A_{ij} = 0$ for $i > j + 1$ $(i < j - 1)$.

5. A is TRIDIAGONAL if it is both upper and lower Hessenberg.

6. A is SYMMETRIC if A is square and $A_{ij} = A_{ji}, i, j = 1, 2, \ldots, n$.

ADDITION of two matrices is straightforward: if A and B both have the same domain and range, then

$$(A + B)_{ij} = A_{ij} + B_{ij}$$

Otherwise, the matrices cannot be added.
SCALAR MULTIPLICATION is simple

$$(\alpha A)_{ij} = \alpha(A_{ij})$$

MATRIX-MATRIX MULTIPLICATION is not as simple. The product AB is the matrix of dot products of the rows of A with the columns of B. If $A \in \mathbb{R}^{m \times n}$ and $B \in \mathbb{R}^{p \times q}$, then AB only exists if $n = p$. Otherwise, the lengths of the rows of A are incompatible with the columns of B. If u_i^T represents the ith row of A, and v_j the jth column of B, then

$$(AB)_{ij} = u_i^T v_j, i = 1, \ldots, m, j = 1, \ldots q$$

Equivalently,

$$(AB)_{ij} = \sum_{k=1}^{n} A_{ik} B_{kj}, i = 1, \ldots, m, j = 1, \ldots q$$

So AB is an m by q matrix. Note that the existence of AB does not imply the existence of BA. Both exist if and only if $n = p$ and $m = q$. Even when both products exist, AB is not generally equal to BA. In other words, the final result of a sequence of operations on a vector generally depends on the order of the operations, i.e., $A(Bx) \neq B(Ax)$. One important exception to this rule is when one of the matrices is the IDENTITY MATRIX I. The elements of I are given by $I_{ij} = \delta_{ij}$, so for example, in $\mathbb{R}^{3 \times 3}$,

$$I = \begin{bmatrix} 1 & 0 & 0 \\ 0 & 1 & 0 \\ 0 & 0 & 1 \end{bmatrix}$$

For any vector x and matrix A, $Ix = x$ and $AI = IA$.

Example 1.4: Common transformations do not commute

Let matrices A and B be given by

$$A = \begin{bmatrix} \frac{\sqrt{2}}{2} & -\frac{\sqrt{2}}{2} \\ -\frac{\sqrt{2}}{2} & \frac{\sqrt{2}}{2} \end{bmatrix} \qquad B = \begin{bmatrix} 2 & 0 \\ 0 & \frac{1}{2} \end{bmatrix}$$

The matrix A rotates a vector counterclockwise by $\pi/4$, while B stretches it by a factor of 2 in the "1" direction while compressing it by the same factor in the "2" direction. Show that the operations of stretching and rotating a vector do not commute.

Solution

The matrices AB and BA are

$$AB = \begin{bmatrix} \sqrt{2} & -\frac{1}{2\sqrt{2}} \\ -\sqrt{2} & \frac{1}{2\sqrt{2}} \end{bmatrix} \qquad BA = \begin{bmatrix} \sqrt{2} & -\sqrt{2} \\ -\frac{1}{2\sqrt{2}} & \frac{1}{2\sqrt{2}} \end{bmatrix}$$

Since these are not equal, we conclude that the two vector operations do not commute. □

1.2.2 Transpose and Adjoint

For every matrix A there exists another matrix, called the TRANSPOSE of A and denoted A^T, such that $(A^T)_{ij} = A_{ji}$. The rows of A become the columns of A^T and vice versa. (We already saw this notion in the context of vectors: viewing x as a matrix with one column, then x^T is a matrix with one row.) A matrix that equals its transpose satisfies $A_{ji} = A_{ij}$ and is said to be SYMMETRIC; this can occur only for square matrices. Some properties of the transpose of a matrix are

$$\begin{aligned} (A^T)^T &= A \\ (A+B)^T &= A^T + B^T \\ (AB)^T &= B^T A^T \\ (ABC)^T &= C^T B^T A^T \end{aligned}$$

Properties involving matrix-vector products follow from the treatment of a vector x as a matrix with only one column. For example

$$(Ax)^T = x^T A^T$$

If A, x, and y are real, then the inner product between the vector Ax and the vector y is given by

$$(Ax)^T y = x^T A^T y \tag{1.3}$$

One can generalize the idea of a transpose to more general operators. The ADJOINT of an operator L (not necessarily a matrix) is denoted L^* and is defined by this equation

$$(Lx, y) = (x, L^* y) \tag{1.4}$$

If L is a real matrix A, then (Lx, y) becomes $(Ax)^T y$ and comparison of (1.3) and (1.4) shows that

$$A^* = A^T$$

Similarly, if L is a complex matrix A then we show in the following section that

$$A^* = \bar{A}^T$$

By analogy with this expression for matrices, we will use the notation $x^* = \bar{x}^T$ for vectors as well. Some general properties of the adjoint of an operator are

$$\begin{aligned}
(L^*)^* &= L \\
(L_1 + L_2)^* &= L_1^* + L_2^* \\
(L_1 L_2)^* &= L_2^* L_1^* \\
(L_1 L_2 L_3)^* &= L_3^* L_2^* L_1^*
\end{aligned}$$

If $L = L^*$, then L is said to be SELF-ADJOINT or HERMITIAN. Self-adjoint operators have special properties, as we shall see shortly, and show up in many applications.

1.2.3 Einstein Summation Convention

Notice that when performing matrix-matrix or matrix-vector multiplications, the index over which the sum is taken appears twice in the formula, while the unsummed indices appear only once. For example, in the formula

$$(ABC)_{ij} = \sum_{k=1}^{N} \sum_{l=1}^{N} A_{ik} B_{kl} C_{lj}$$

the indices k and l appear twice in the summations, while the indices i and j only appear once. This observation suggests a simplified notation for products, in which the presence of the repeated indices implies summation, so that the explicit summation symbols do not need to be written. Using this EINSTEIN SUMMATION CONVENTION, the inner product $x^T y$ is simply $x_i y_i$ and the matrix-vector product $y = Ax$ is $y_i = A_{ij} x_j$. This convention allows us to concisely derive many key results.

Example 1.5: Matrix identities derived with index notation

Establish the following matrix identities using index notation

(a) $(Ax, y) = (x, \bar{A}^T y)$ (b) $(AB)^T = B^T A^T$ (c) $AA^T = (AA^T)^T$

(d) $A + A^T = (A + A^T)^T$ (e) $A^T A = (A^T A)^T$

Solution

(a) $(Ax, y) = (x, \bar{A}^T y)$

$$\begin{aligned}(Ax, y) &= A_{ij}x_j\bar{y}_i \\ &= x_j A_{ij}\bar{y}_i \\ &= x_j A^T_{ji}\bar{y}_i \\ &= x_j \overline{\bar{A}^T_{ji} y_i} \\ &= (x, \bar{A}^T y)\end{aligned}$$

(b) $(AB)^T = B^T A^T$

$$\begin{aligned}(AB)^T_{ij} &= (A_{ik}B_{kj})^T \\ &= A_{jk}B_{ki} \\ &= B_{ki}A_{jk} \\ &= B^T_{ik}A^T_{kj} \\ &= (B^T A^T)_{ij}\end{aligned}$$

(c) $AA^T = (AA^T)^T$

$$\begin{aligned}(AA^T)_{ij} &= A_{ik}A^T_{kj} \\ &= A_{ik}A_{jk} \\ &= A_{jk}A_{ik} \\ &= A_{jk}A^T_{ki} \\ &= (AA^T)_{ji} \\ &= (AA^T)^T_{ij}\end{aligned}$$

(d) $A + A^T = (A + A^T)^T$

$$\begin{aligned}(A + A^T)_{ij} &= A_{ij} + A_{ji} \\ &= A_{ji} + A_{ij} \\ &= (A + A^T)_{ji} \\ &= (A + A^T)^T_{ij}\end{aligned}$$

(e) $A^T A = (A^T A)^T$

$$\begin{aligned}(A^T A)_{ij} &= A^T_{ik}A_{kj} \\ &= A_{ki}A_{kj} \\ &= A^T_{jk}A_{ki} \\ &= (A^T A)_{ji} \\ &= (A^T A)^T_{ij}\end{aligned}$$

□

1.2.4 Gram-Schmidt Orthogonalization and the QR Decomposition

We will encounter a number of situations where a linearly independent set of vectors are available and it will be useful to construct from them a set of orthogonal vectors. The classical approach to doing this is called GRAM-SCHMIDT orthogonalization. As a simple example, consider LI vectors v_1 and v_2, from which we wish to find an orthogonal pair u_1 and u_2. Without loss of generality we can set

$$u_1 = v_1$$

It is straightforward to find the component of v_2 that is orthogonal to u_1—we just subtract from v_2 the component that is parallel to u_1 (the PROJECTION of v_2 onto u_1)

$$u_2 = v_2 - \frac{(v_2, u_1)}{\|u_1\|} \frac{u_1}{\|u_1\|}$$

In higher dimensions, where we have v_3, v_4, etc., we continue the process, subtracting off the components parallel to the previously determined orthogonal vectors

$$u_3 = v_3 - \frac{(v_3, u_1)}{\|u_1\|^2} u_1 - \frac{(v_3, u_2)}{\|u_2\|^2} u_2$$

and so on.

We can apply Gram-Schmidt orthogonalization to the columns of any $m \times n$ matrix A whose columns are linearly independent (which implies that $m \geq n$). Specifically, we can write

$$A = QR$$

where Q is an $m \times n$ matrix of orthonormal vectors formed from the columns of A and R is an $n \times n$ upper triangular matrix. This result is known as the QR DECOMPOSITION. We have the following theorem.

Theorem 1.6 (QR decomposition). *If $A \in \mathbb{R}^{m \times n}$ has linearly independent columns, then there exists $Q \in \mathbb{R}^{m \times n}$ with orthonormal columns, and upper triangular R such that*

$$A = QR$$

See Exercise 1.38 for the proof. Because the columns of Q are orthonormal, $Q^T Q = I$.

1.2.5 The Outer Product, Dyads, and Projection Operators

Given two LI vectors v_1 and v_2 in $\mathbb{R}^n$, Gram-Schmidt uses projection to construct an orthogonal pair

$$u_1 = v_1$$
$$u_2 = v_2 - (v_2^T \hat{u}_1)\hat{u}_1$$

where $\hat{u}_1 = u_1/\|u_1\|$ is a unit vector in the u_1 direction, and now we have used the inner product definition $(u, v) = u^T v$. Observe that the right-hand side of the second equation is linear in v_2, so we should be able to put this equation in the form $u_2 = Av_2$, where A is a matrix. The form of A illustrates some important concepts so we explicitly construct it here. We can write $A = I - P$, where

$$Pv_2 = (v_2^T \hat{u}_1)\hat{u}_1$$

Noting that $a^T b = b^T a$ for vectors a and b, this rearranges to

$$Pv_2 = \hat{u}_1(\hat{u}_1^T v_2)$$

which has the form we seek if we move the parentheses to have

$$Pv_2 = (\hat{u}_1\hat{u}_1^T)v_2$$

That is, P is given by what we will call the OUTER PRODUCT between $\hat{u}_1$ and itself: $\hat{u}_1\hat{u}_1^T$. More generally, the outer project uv^T between vectors u and v is a matrix, called a DYAD, that satisfies the following properties

$$\begin{aligned}(uv^T)_{i,j} &= u_i v_j \\ (uv^T)w &= u(v^T w) \\ w^T(uv^T) &= (w^T u)v^T\end{aligned}$$

where w is any vector. The outer product is sometimes denoted $u \otimes v$. When the notation $\mathbf{u} \cdot \mathbf{v}$ is used to represent the inner product, $\mathbf{u} \otimes \mathbf{v}$ or $\mathbf{uv}$ is used to represent the outer.

Finally, returning to the specific case $P = \hat{u}_1\hat{u}_1^T$, we can observe that $Pw = \hat{u}_1(\hat{u}_1^T w)$; the operation of P on w results in a vector that is the projection of w in the $\hat{u}_1$ direction: P is a PROJECTION OPERATOR. The operator $I - P$ is also a projection—it takes a vector and produces the projection of that vector in the direction(s) orthogonal to $\hat{u}_1$. We can check that both $\hat{u}_1\hat{u}_1^T$ and $I - \hat{u}_1\hat{u}_1^T$ satisfy the general definition of a projection operator

$$P^2 = P$$

1.2.6 Partitioned Matrices and Matrix Operations

It is often convenient to consider a large matrix to be composed of other *matrices*, rather than its scalar elements. We say the matrix is *partitioned* into other smaller dimensional matrices. To make this explicit, first we define a submatrix as follows. Let matrix $A \in \mathbb{R}^{m\times n}$, and define indices $1 \leq i_1 < i_2 < \cdots < i_k \leq m$, and $1 \leq j_1 < j_2 < \cdots < j_\ell \leq m$, then the $k \times \ell$ matrix S, whose (a,b) element is

$$S_{ab} = A_{i_a,j_b}$$

is called a submatrix of A.

A matrix $A \in \mathbb{R}^{m\times n}$ is partitioned when it is written as

$$A = \begin{bmatrix} A_{11} & A_{12} & \cdots & A_{1\ell} \\ A_{21} & A_{22} & \cdots & A_{2\ell} \\ \vdots & \vdots & \ddots & \vdots \\ A_{k1} & A_{k2} & \cdots & A_{k\ell} \end{bmatrix}$$

where each A_{ij} is an $m_i \times n_j$ submatrix of A. Note that $\sum_{i=1}^{k} m_i = m$ and $\sum_{j=1}^{\ell} n_j = n$. Two of the more useful matrix partitions are column partitioning and row partitioning. If we let the m-vectors $a_i, i = 1, 2, \ldots n$ denote the n column vectors of A, then the column partitioning of A is

$$A = \begin{bmatrix} a_1 & a_2 & \cdots & a_n \end{bmatrix}$$

If we let the row vectors ($1 \times n$ matrices) $\overline{a}_j, j = 1, 2, \ldots, m$ denote the m row vectors of A, then the row partitioning of A is

$$A = \begin{bmatrix} \overline{a}_1 \\ \overline{a}_2 \\ \vdots \\ \overline{a}_m \end{bmatrix}$$

The operations of matrix transpose, addition, and multiplication become even more useful when we apply them to partitioned matrices. Consider the two partitioned matrices

$$A = \begin{bmatrix} A_{11} & A_{12} & \cdots & A_{1\ell} \\ \vdots & \vdots & \ddots & \vdots \\ A_{k1} & A_{k2} & \cdots & A_{k\ell} \end{bmatrix} \qquad B = \begin{bmatrix} B_{11} & B_{12} & \cdots & B_{1n} \\ \vdots & \vdots & \ddots & \vdots \\ B_{m1} & B_{m2} & \cdots & B_{mn} \end{bmatrix}$$

in which A_{ij} has dimension $p_i \times q_j$ and B_{ij} has dimension $r_i \times s_j$. We then have the following formulas for scalar multiplication, transpose, matrix addition, and matrix multiplication of partitioned matrices.

1. Scalar multiplication.

$$\lambda A = \begin{bmatrix} \lambda A_{11} & \lambda A_{12} & \cdots & \lambda A_{1\ell} \\ \vdots & \vdots & \ddots & \vdots \\ \lambda A_{k1} & \lambda A_{k2} & \cdots & \lambda A_{k\ell} \end{bmatrix}$$

2. Transpose.

$$A^T = \begin{bmatrix} A_{11}^T & A_{21}^T & \cdots & A_{k1}^T \\ \vdots & \vdots & \ddots & \vdots \\ A_{1\ell}^T & A_{2\ell}^T & \cdots & A_{k\ell}^T \end{bmatrix}$$

3. Matrix addition. If $p_i = r_i$ and $q_j = s_j$ for $i = 1, \ldots, k$ and $j = 1, \ldots, \ell$; and $k = m$ and $\ell = n$; then the partitioned matrices can be added

$$A + B = \begin{bmatrix} C_{11} & \cdots & C_{1\ell} \\ \vdots & \ddots & \vdots \\ C_{k1} & \cdots & C_{k\ell} \end{bmatrix} \qquad C_{ij} = A_{ij} + B_{ij}$$

4. Matrix multiplication. If $q_i = r_i$ for $i = 1, \ldots, \ell$, then we say the partitioned matrices conform, and the matrices can be multiplied

$$AB = \begin{bmatrix} C_{11} & \cdots & C_{1\ell} \\ \vdots & \ddots & \vdots \\ C_{k1} & \cdots & C_{k\ell} \end{bmatrix} \qquad C_{ij} = \sum_{t=1}^{\ell} A_{it} B_{tj}$$

These formulas are all easily verified by reducing all the partitioned matrices back to their scalar elements. Notice that we do not have to remember any *new* formulas. These are the same formulas that we learned for matrix operations when the submatrices A_{ij} and B_{ij} were scalar elements (except we normally do not write the transpose for scalars in the transpose formula). The conclusion is that all the usual rules apply provided that the matrices are partitioned so that all the implied operations are defined.

1.3 Systems of Linear Algebraic Equations

1.3.1 Introduction to Existence and Uniqueness

Any set of m linear algebraic equations for n unknowns can be written in the form

$$Ax = b$$

where $A \in \mathbb{R}^{m \times n}, b \in \mathbb{R}^n$ and x ($\in \mathbb{R}^n$) is the vector of unknowns. Consider the vectors c_i that form the columns of A. The solution x (if it exists) is the linear combination of these columns that equals b

$$b = x_1 c_1 + x_2 c_2 + x_3 c_3 + \ldots + x_n c_n$$

This view of $Ax = b$ leads naturally to the following result. The system of equations

$$Ax = b, \qquad A \in \mathbb{R}^{m \times n}, x \in \mathbb{R}^n, b \in \mathbb{R}^m$$

has at least one solution x if and only if the columns of A are *not* linearly independent from b.

For example, if $m = n = 3$ and the columns of A form an LI set, then they span $\mathbb{R}^3$. Therefore, no vector $b \in \mathbb{R}^3$ can be linearly independent from the columns of A and therefore $Ax = b$ has a solution for all $b \in \mathbb{R}^3$. Conversely, if the column vectors of A are not LI, then they do not span $\mathbb{R}^3$ so there will be some vectors b for which no solution x exists.

Consider the case where there are the same number of equations as unknowns: $n = m$. Here the above result leads to this general theorem.

Theorem 1.7 (Existence and uniqueness of solutions for square systems). *If $A \in \mathbb{R}^{n \times n}$, then*

(a) If the columns of A are LI, then the matrix is INVERTIBLE. *The problem $Ax = b$ has the following properties:*

(1) $Ax = 0$ (the homogeneous problem) has only the trivial solution $x = 0$,

(2) $Ax = b$ (the inhomogeneous problem) has a unique nonzero solution for all $b \neq 0$.

(b) If the columns of A are NOT LI, then the matrix is SINGULAR *or* NONINVERTIBLE. *In this case:*

(1) $Ax = 0$ has an infinite number of nonzero solutions. These solutions comprise the NULL SPACE *of A.*

(2) For $b \neq 0$, $Ax = b$ has either:

(i) No solution, if b is LI of the columns of A. That is, b is not in the RANGE *of A, or*

(ii) *An infinite number of solutions, if b is in the range of A. These solutions correspond to the superposition of a particular solution to $Ax = b$ and any combination of the solutions of $Ax = 0$, i.e., $x = x_H + x_P$ where $Ax_P = b$ and $Ax_H = 0$.*

1.3.2 Solving $Ax = b$: LU Decomposition

We now turn to the issue of explicitly constructing solutions. For the present, we restrict attention to $n = m$ and to case (a) of the above theorem. In this case, we can define the INVERSE of A, denoted A^{-1}. This is a matrix operator that satisfies

1. $A^{-1}A = I$ (definition of A^{-1})
2. $AA^{-1} = I$
3. $(AB)^{-1} = B^{-1}A^{-1}$

The first property implies that $A^{-1}Ax = A^{-1}b$ reduces to $x = A^{-1}b$, so $Ax = b$ can be solved by finding A^{-1}. Finding A^{-1} is not necessary, however, to solve $Ax = b$, nor is it particularly efficient. We describe a widely used approach called LU decomposition.

LU decomposition is essentially a modification of Gaussian elimination, with which everyone should be familiar. It is based on the fact that triangular systems of equations are easy to solve. For example, this matrix is upper triangular

$$\begin{bmatrix} 1 & 2 & 3 \\ 0 & 4 & 8 \\ 0 & 0 & 7 \end{bmatrix}$$

All the elements below the diagonal are zero. Since the third row has only one nonzero element, it corresponds to a single equation with a single unknown. Once this equation is solved, the equation above it has only a single unknown and is therefore easy to solve, and so on. LU decomposition depends on the fact that a square matrix A can be written $A = LU$, where L is lower triangular and U is upper triangular. Using this fact, solving $Ax = b$ consists of three steps, the first of which takes the most computation:

1. Find L and U from A: LU factorization.
2. Solve $Lc = b$ for c: forward substitution.

3. Solve $Ux = c$ for x: back substitution.

The latter two steps are simple operations, because L and U are triangular. Note that L and U are independent of b, so to solve $Ax = b$ for many different values of b, then once A is factored, only the inexpensive steps 2 and 3 of the above process need be repeated. The LU decomposition procedure (first step above) is illustrated on the matrix

$$A = \begin{bmatrix} 3 & 5 & 2 \\ 0 & 8 & 2 \\ 6 & 2 & 8 \end{bmatrix}$$

Step a. Replace row 2 with a linear combination of row 1 and row 2 that makes the first element zero. That is, r_2 is replaced by $r_2 - L_{21}r_1$, where $L_{21} = A_{21}/A_{11}$. For this example, A_{21} is already zero, so $L_{21} = 0$ and r_2 is unchanged.

Step b. Replace row 3 with a linear combination of row 1 and row 3 that makes the first element zero. That is, r_3 is replaced by $r_3 - L_{31}r_1$, where $L_{31} = A_{31}/A_{11}$. So $L_{31} = 6/3 = 2$ and A is modified to

$$\begin{bmatrix} 3 & 5 & 2 \\ 0 & 8 & 2 \\ 0 & -8 & 4 \end{bmatrix}$$

Step c. Now the first column of the matrix is zero below the diagonal. We move to the second column. Replace row 3 with a linear combination of row 2 and row 3 that makes the second element zero. That is, $r_{(3)}$ is replaced by $r_3 - L_{32}r_2$, where $L_{32} = A_{32}/A_{22}$. So $L_{32} = -1$ and A is modified to

$$\begin{bmatrix} 3 & 5 & 2 \\ 0 & 8 & 2 \\ 0 & 0 & 6 \end{bmatrix} = U$$

This matrix is now the upper triangular matrix U. For a matrix in higher dimensions, the procedure would be continued until all of the elements below the diagonal were zero. The matrix L is simply composed of the multipliers L_{ij} that were computed at each step

$$L = \begin{bmatrix} 1 & 0 & 0 \\ L_{21} & 1 & 0 \\ L_{31} & L_{32} & 1 \end{bmatrix} = \begin{bmatrix} 1 & 0 & 0 \\ 0 & 1 & 0 \\ 2 & -1 & 1 \end{bmatrix}$$

> Note that all the diagonal elements of L are 1 and all above-diagonal elements are zero. The elements on the diagonal of U are called the PIVOTS.

Now, for any vector b the simple systems $Lc = b$ and then $Ux = c$ can be solved to yield x. Notice that as written, the method will fail if $A_{ii} = 0$ at any step of the procedure. Modern computational routines actually compute a slightly different factorization $PA = LU$ where P is a permutation matrix that exchanges rows to avoid the case $A_{ii} = 0$ (see Exercise 1.9). With this modification, known as PARTIAL PIVOTING, even singular or nonsquare ($m > n$) matrices can be factored. However, the substitution steps will fail except for values of b in the range of A. To see this, try to perform the back substitution step with a matrix U that has a zero pivot.

1.3.3 The Determinant

In elementary discussions of the solution to $Ax = b$ that are based on CRAMER'S RULE, the DETERMINANT of the matrix A, denoted $\det A$, arises. One often finds a complicated definition based on submatrices, but having the LU decomposition in hand a much simpler formula emerges (Strang, 1980). For a square matrix A that can be decomposed into LU, the determinant is the product of the pivots

$$\det A = \prod_{i=1}^{n} U_{ii}$$

If m permutations of rows must be performed to complete the decomposition, then the decomposition has the form $PA = LU$, and

$$\det A = (-1)^m \prod_{i=1}^{n} U_{ii}$$

The matrix A^{-1} exists if and only if $\det A \neq 0$, in which case $\det A^{-1} = (\det A)^{-1}$. Another key property of the determinant is that

$$\det AB = \det A \ \det B$$

The most important use of the determinant that we will encounter in this book is its use in the ALGEBRAIC EIGENVALUE PROBLEM that appears in Section 1.4.

1.3.4 Rank of a Matrix

Before we define the rank of a matrix, it is useful to establish the following property of matrices: the number of linearly independent columns of a matrix is equal to the number of linearly independent rows.

Example 1.8: Linearly independent columns, rows of a matrix

Given $A \in \mathbb{R}^{m \times n}$. Assume A has c linearly independent columns and r linearly independent rows. Show $c = r$.

Solution

Let $\{v_i\}_{i=1}^{c}$ be the set of A's linearly independent column vectors. Let the a_i be all of A's column vectors and $\overline{a}_i$ be all of A's row vectors, so the A matrix can be partitioned by its columns or rows as

$$A = m \overset{n}{\begin{bmatrix} a_1 & a_2 & \cdots & a_n \end{bmatrix}} \qquad A = m \overset{n}{\begin{bmatrix} \overline{a}_1 \\ \overline{a}_2 \\ \vdots \\ \overline{a}_m \end{bmatrix}}$$

Each column of the A matrix can be expressed as a linear combination of the c linearly independent v_i vectors. We denote this statement as follows

$$\underbrace{\begin{bmatrix} \cdots & \Big| \; a_j \; \Big| & \cdots \end{bmatrix}}_{A:m \times n} = \underbrace{\begin{bmatrix} v_1 \; \Big| & \cdots & \Big| \; v_c \end{bmatrix}}_{V:m \times c} \underbrace{\begin{bmatrix} \cdots & \Big| \; \delta_j \; \Big| & \cdots \end{bmatrix}}_{\Delta:c \times n}$$

in which the column vector $\delta_j \in \mathbb{R}^c$ contains the coefficients of the linear combination of the v_i representing the jth column vector of matrix A. If we place all the $\delta_j, j = 1, \ldots, n$ next to each other, we have matrix Δ. Next comes the key step. Repartition the relationship above as follows

$$\underbrace{\begin{bmatrix} \vdots \\ \hline \overline{a}_i \\ \hline \vdots \end{bmatrix}}_{A:m \times n} = \underbrace{\begin{bmatrix} \vdots \\ \hline \overline{v}_i \\ \hline \vdots \end{bmatrix}}_{V:m \times c} \underbrace{\begin{bmatrix} \overline{\delta}_1 \\ \hline \vdots \\ \hline \overline{\delta}_c \end{bmatrix}}_{\Delta:c \times n}$$

and we see that the rows of A can be expressed as linear combinations of the rows of Δ. The multipliers of the ith row of A are given by the elements of the ith row of V, written as the row vector $\overline{v}_i$. We know that all rows of A are expressible as linear combinations of the c rows of Δ, but we do not know if the rows of Δ are independent. So we can conclude that the number of linearly independent rows of A is less than or equal to the number of rows of Δ or

$$r \leq c$$

Thus, for any A, the number of linearly independent rows of A is less than or equal to the number of linearly independent columns of A. But if we apply that result to A^T, we obtain $c \leq r$ because the linearly independent row (column) vectors of A are also the linearly independent column (row) vectors of A^T. Combining $c \leq r$ with $r \leq c$, we conclude that

$$c = r$$

and the result is established. The number of linearly independent columns of a matrix is equal to the number of linearly independent rows, and this number is called the rank of the matrix. □

Definition 1.9 (Rank of a matrix)**.** The rank of a matrix is the number of linearly independent rows, equivalently, columns, of the matrix.

We also see clearly why partitioned matrices are so useful. The proof that the number of linearly independent rows of a matrix is equal to the number of linearly independent columns consisted of little more than partitioning a matrix by its columns and then repartitioning the same matrix by its rows. For another example of why partitioned matrices are useful, see Exercise 1.17 on deriving the partitioned matrix inversion formula, which often arises in applications.

1.3.5 Range Space and Null Space of a Matrix

Given $A \in \mathbb{R}^{m \times n}$, we define the range of A as

$$R(A) = \{y \in \mathbb{R}^m \mid y = Ax, \quad x \in \mathbb{R}^n\}$$

The range of a matrix is the set of all vectors that can be generated with the product Ax for all $x \in \mathbb{R}^n$. Equivalently, if $v_i \in \mathbb{R}^n$ are the linearly independent columns of A, then the range of A is the span of the v_i.

The v_i are a basis for the range of A. Given $A \in \mathbb{R}^{m \times n}$, we define the null space of A as follows

$$N(A) = \{x \in \mathbb{R}^n \mid Ax = 0\}$$

Similarly the range and null spaces of A^T are defined to be

$$R(A^T) = \{x \in \mathbb{R}^n \mid x = A^T y, \quad y \in \mathbb{R}^m\}$$
$$N(A^T) = \{y \in \mathbb{R}^m \mid A^T y = 0\}$$

A basis for $R(A^T)$ is the set of linearly independent rows of A, transposed to make column vectors. We can show that these four sets also satisfy the two properties of a subspace, so they are also subspaces (see Exercise 1.14).

Let r be the rank of matrix A. We know from the previous example that r is equal to the number of linearly independent rows of A and is also equal to the number of linearly independent columns of A. Equivalently, the dimension of $R(A)$ and $R(A^T)$ is also r

$$\dim(R(A)) = \dim(R(A^T)) = r = \text{rank}(A)$$

We also can demonstrate the following pair of orthogonality relations among these four fundamental subspaces

$$R(A) \perp N(A^T) \qquad R(A^T) \perp N(A)$$

Consider the first orthogonality relationship. Let y be any element of $N(A^T)$. We know $N(A^T) = \{y \in \mathbb{R}^m \mid A^T y = 0\}$. Transposing this relation and using column partitioning for A gives

$$\begin{aligned} y^T A &= 0 \\ y^T \begin{bmatrix} a_1 & a_2 & \cdots a_n \end{bmatrix} &= 0 \\ \begin{bmatrix} y^T a_1 & y^T a_2 & \cdots & y^T a_n \end{bmatrix} &= 0 \end{aligned}$$

The last equation gives $y^T a_i = 0, i = 1, \ldots, n$, or y is orthogonal to every column of A. Since every element of the range of A is a linear combination of the columns of A, y is orthogonal to every element of $R(A)$, which gives $N(A^T) \perp R(A)$. The second orthogonality relationship follows by switching the roles of A and A^T in the preceding argument (see Exercise 1.15). Note that the range of a matrix is sometimes called the image, and the null space is sometimes called the kernel.

1.3.6 Existence and Uniqueness in Terms of Rank and Null Space

We return now to the general case where $A \in \mathbb{R}^{m \times n}$. The FUNDAMENTAL THEOREM OF LINEAR ALGEBRA gives a complete characterization of the existence and uniqueness of solutions to $Ax = b$ (Strang, 1980): every matrix A decomposes the spaces $\mathbb{R}^n$ and $\mathbb{R}^m$ into the four fundamental subspaces depicted in Figure 1.1. The answer to the question of existence and uniqueness of solutions to $Ax = b$ can be summarized as follows.

1. **Existence.** Solutions to $Ax = b$ exist for all b if and only if the *rows* of A are linearly independent ($m = r$).

2. **Uniqueness.** A solution to $Ax = b$ is unique if and only if the *columns* of A are linearly independent ($n = r$).

We can also state this result in terms of the null spaces. A solution to $Ax = b$ exists for all b if and only if $N(A^T) = \{0\}$ and a solution to $Ax = b$ is unique if and only if $N(A) = \{0\}$. More generally, a solution to $Ax = b$ exists *for a particular* b if and only if $b \in R(A)$, by the definition of the range of A. From the fundamental theorem, that means $y^T b = 0$ for all $y \in N(A^T)$. And if $N(A^T) = \{0\}$ we recover the existence condition 1 stated above. These statements provide a succinct generalization of the results described in Section 1.3.1.

1.3.7 Least-Squares Solution for Overdetermined Systems

Now consider the OVERDETERMINED problem, $Ax = b$ where $A \in \mathbb{R}^{m \times n}$ with $m > n$. In general, this problem has no exact solution, because the n columns of A cannot span $\mathbb{R}^m$, the space where b exists. This problem arises naturally in fitting models to data. In general, the best we can hope for is an approximate solution x that minimizes the residual (or error) $r = Ax - b$. In particular, the "least squares" method attempts to minimize the square of the Euclidean norm of the residual, $\|r\|^2 = r^T r$. Replacing r by $Ax - b$, this quantity (divided by 2) reduces to the function

$$P(x) = \frac{1}{2} x^T A^T A x - x^T A^T b + \frac{1}{2} b^T b$$

P is a scalar function of x and the value of the vector x that minimizes P is the solution we seek. That is, we now want to solve $\partial P / \partial x_l = 0, l =$

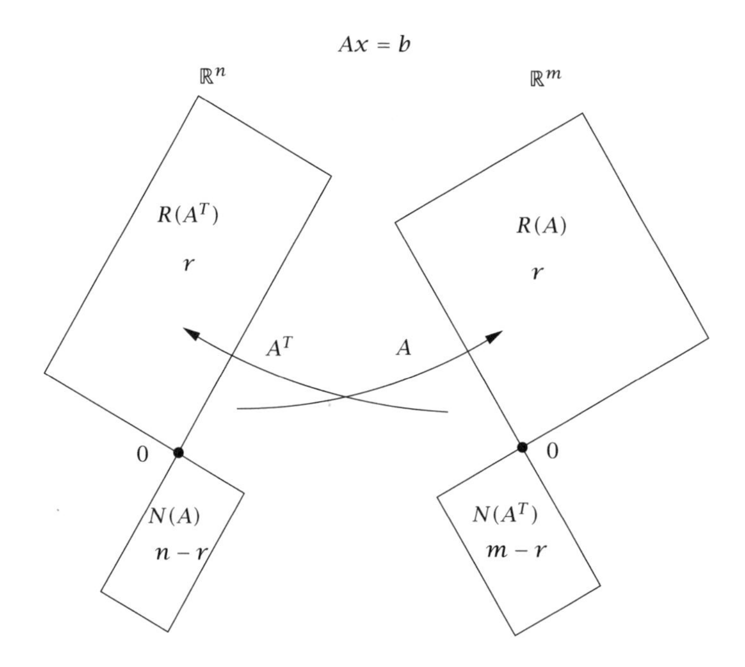

Figure 1.1: The four fundamental subspaces of matrix A (after (Strang, 1980, p.88)). The dimension of the range of A and A^T is r, the rank of matrix A. The nullspace of A and range of A^T are orthogonal as are the nullspace of A^T and range of A. Solutions to $Ax = b$ exist for all b if and only if $m = r$ (rows independent). A solution to $Ax = b$ is unique if and only if $n = r$ (columns independent).

$1, \ldots, n$, or in different notation, $\nabla P(x) = 0$. Performing the gradient operation yields

$$\frac{\partial P}{\partial x_l} = A^T_{lj} A_{jk} x_k - A^T_{lj} b_j = 0$$

or in matrix form

$$\frac{dP}{dx} = A^T A x - A^T b = 0$$

Therefore, the condition that P be minimized is equivalent to solving

$$A^T A x = A^T b$$

These are called the NORMAL EQUATIONS. Notice that A^TA is a square matrix, so that we can solve this problem with LU decomposition, provided A^TA has full rank. QR decomposition also can be used: $A^TA = R^TQ^TQR = R^TR$. Since R is triangular, $R^TRx = A^Tb$ is as easy to solve as $LUx = A^Tb$. In Exercise 1.41 you are asked to prove that A^TA has full rank if and only if the columns of A are linearly independent.[1]

If A^TA has full rank, the inverse is uniquely defined, and we can write the least-squares solution to the normal equations as

$$x_{\text{ls}} = (A^TA)^{-1}A^Tb \tag{1.5}$$

The matrix on the right-hand side is ubiquitous in least-squares problems; it is known as the pseudoinverse of A (or Moore-Penrose pseudoinverse in honor of mathematician E. H. Moore and mathematical physicist Roger Penrose) and given the symbol $A^\dagger$. The least-squares solution is then denoted

$$x_{\text{ls}} = A^\dagger b \qquad A^\dagger = (A^TA)^{-1}A^T$$

The normal equations have a compelling geometric interpretation that illustrates the origin of their name. Substituting r into the normal equations gives the condition $A^Tr = 0$. That is, the residual $r = Ax - b$ is an element of the null space of A^T, $N(A^T)$, which means r is orthogonal, i.e., normal, to the range of A, $R(A)$ (right side of Figure 1.1). This is just a generalization of the fact that the shortest path (minimum $\|r\|$) connecting a plane and a point b not on that plane is perpendicular to the plane. Note that this geometric insight is our second use of the fundamental theorem of linear algebra. This geometric interpretation is perhaps best reinforced by a simple example.

Example 1.10: The geometry of least squares

We are interested in solving $Ax = b$ for the following A and b.

$$A = \begin{bmatrix} 1 & 1 \\ 2 & 1 \\ 0 & 0 \end{bmatrix} \qquad b = \begin{bmatrix} 1 \\ 1 \\ 1 \end{bmatrix}$$

[1]Putting proof aside for a moment, the condition is at least easy to remember. The A in the overdetermined system for which we apply least squares has more rows than columns. So the rank of A is at *most* the number of columns. The least-squares solution is unique if and only if the rank is *equal* to this largest value, i.e., rank of A equals the number of columns.

(a) What is the rank of A? Justify your answer.

(b) Draw a sketch of the subspace $R(A)$.

(c) Draw a sketch of the subspace $R(A^T)$.

(d) Draw a sketch of the subspace $N(A)$.

(e) Draw a sketch of the subspace $N(A^T)$.

(f) Is there a solution to $Ax = b$ for all b? Justify your answer.

(g) Is there a solution for the particular b given above? Justify your answer.

(h) Assume we give up on solving $Ax = b$ and decide to solve instead the least-squares problem

$$\min_x (Ax - b)^T (Ax - b)$$

What is the solution to this problem, x^0?

(i) Is this solution unique? Justify your answer.

(j) Sketch the location of the b^0 for which this x^0 does solve $Ax = b$. In particular, sketch the relationship between this b^0 and one of the subspaces you sketched previously. Also on this same drawing, sketch the residual $r = Ax^0 - b$.

Solution

(a) The rank of A is 2. The two columns are linearly independent.

(b) $R(A)$ is the xy plane in $\mathbb{R}^3$.

(c) $R(A^T)$ is $\mathbb{R}^2$. Notice these are not the same subspaces, even though they have the same dimension 2.

(d) $N(A)$ is the zero element in $\mathbb{R}^2$.

(e) $N(A^T)$ is the z axis in $\mathbb{R}^3$.

(f) No. The rows are not independent.

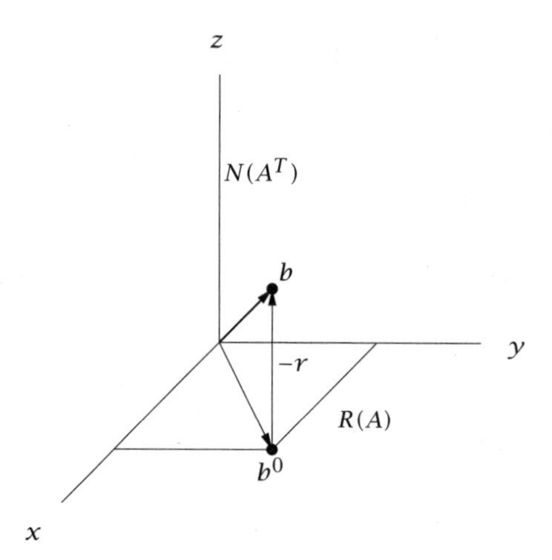

Figure 1.2: Least-squares solution of $Ax = b$; projection of b into $R(A)$ and residual $r = Ax^0 - b$ in $N(A^T)$.

(g) No. The range of A does not have a nonzero third element and this b does.

(h) The solution is

$$x^0 = (A^T A)^{-1} A^T b \qquad x^0 = \begin{bmatrix} 0 \\ 1 \end{bmatrix}$$

(i) Yes, the least-squares solution is unique because the columns of A are linearly independent.

(j) The vector b is decomposed into $b^0 \in R(A)$ and $r = Ax^0 - b \in N(A^T)$.

$$b = b^0 + (-r) \qquad r \in N(A^T) \quad b^0 \in R(A)$$

We want $Ax^0 = b^0$, so $b^0 = A(A^T A)^{-1} A^T b = Pb$ and the projection operator is $P = A(A^T A)^{-1} A^T$. The residual is $r = Ax^0 - b = (P - I)b$ and we have for this problem

$$P = A(A^T A)^{-1} A^T = \begin{bmatrix} 1 & 0 & 0 \\ 0 & 1 & 0 \\ 0 & 0 & 0 \end{bmatrix} \qquad P - I = \begin{bmatrix} 0 & 0 & 0 \\ 0 & 0 & 0 \\ 0 & 0 & -1 \end{bmatrix}$$

Substituting in the value for b gives

$$b^0 = \begin{bmatrix} 1 \\ 1 \\ 0 \end{bmatrix} \qquad r = \begin{bmatrix} 0 \\ 0 \\ -1 \end{bmatrix}$$

The spaces $R(A)$ and $N(A^T)$ are orthogonal, and therefore so are b^0 and r. The method of least squares projects b into the range of A, giving b^0, and then solves exactly $Ax = b^0$ to obtain x^0. These relationships are shown in Figure 1.2. □

The above analysis is only the beginning of the story for parameter estimation. We have not dealt with important issues such as errors in the measurements, quantifying the uncertainty in parameters, choice of model form, etc. Many of these issues will be studied in Chapter 4 as part of maximum-likelihood estimation.

1.3.8 Minimum Norm Solution of the Underdetermined Problem

Consider the case of solving $Ax = b$ with fewer equations than unknowns, the so-called UNDERDETERMINED problem. Assume that the rows of A are linearly independent, so a solution exists for all b. But we also know immediately that $N(A) \neq \{0\}$, and there are infinitely many solutions. One natural way to choose a specific solution from the infinite number of possibilities is to seek the MINIMUM-NORM solution. That is, we minimize $\|x\|^2$ subject to the constraint that $Ax = b$. By analogy with the approach taken above in constructing the least-squares solution, we define an objective function

$$P(x) = \frac{1}{2}x^T x - z^T(Ax - b) = \frac{1}{2}x_i x_i - z_i(A_{ij}x_j - b_i)$$

where now z is a vector of Lagrange multipliers. The minimization condition $\partial P / \partial x_k = 0$ is thus

$$x_k = z_j A_{jk}$$

or $x = A^T z$. Inserting this into the equation $Ax = b$ yields

$$AA^T z = b$$

Since the rows of A are linearly independent, AA^T is full rank.[2] We can solve this equation for z and insert into the equation $x = A^T z$ that we

[2]Transpose the result of Exercise 1.41.

found above to deduce the minimum-norm solution

$$x_{\mathrm{mn}} = A^T(AA^T)^{-1}b \tag{1.6}$$

Note the similarity in the solution structure of the underdetermined, minimum-norm problem to the overdetermined, least-squares problem given in (1.5). The singular value decomposition, which we introduce in Section 1.4.7, allows for a unified and general treatment of both the underdetermined and overdetermined problems.

1.3.9 Rank, Nullity, and the Buckingham Pi Theorem

As engineers, we often encounter situations where we have a number of measurements or other quantities d_i and we expect there to be a functional relationship Φ_d between them

$$\Phi_d(d_1, d_2, \ldots, d_n) = 0$$

In general, we would like to have a *dimensionless* representation of this relation, one that does not depend on the units of measurement, i.e.,

$$\Phi_\Pi(\Pi_1, \Pi_2, \ldots, \Pi_l) = 0$$

where each Π_i has the form

$$\Pi = d_1^{a_1} d_2^{a_2} \times \ldots \times d_n^{a_n}$$

and the exponents a_i are chosen so that each Π_i is dimensionless. If the set of n quantities d_i depend on m units (kilograms, meters, seconds, amperes, ...), the key question is: what is the relationship between n, m, and the number l of dimensionless variables Π_i that is required to characterize the relationship between the variables?

We will address this issue with a specific example. Consider fluid flow through a tube. The fluid has density ρ and viscosity η, and flows with average velocity U through a tube with radius R and length L, driven by a pressure drop Δp. Defining $[=]$ to mean "has dimensions of," we seek dimensionless quantities of the form

$$\begin{aligned}\Pi &= \Delta p^{a_1} U^{a_2} \rho^{a_3} \eta^{a_4} R^{a_5} L^{a_6} \\ &[=] \left(\frac{\text{kg m}}{\text{s}^2\text{m}^2}\right)^{a_1} \left(\frac{\text{m}}{\text{s}}\right)^{a_2} \left(\frac{\text{kg}}{\text{m}^3}\right)^{a_3} \left(\frac{\text{kg m s}}{\text{m}^2\text{s}^2}\right)^{a_4} (\text{m})^{a_5}\,(\text{m})^{a_6}\end{aligned}$$

All the units must cancel, so we require that

$$\begin{aligned} \text{kg}: \quad & a_1 + a_3 + a_4 = 0 \\ \text{m}: \quad & -a_1 + a_2 - 3a_3 - a_4 + a_5 + a_6 = 0 \\ \text{s}: \quad & -2a_1 - a_2 - a_4 = 0 \end{aligned}$$

This is a system of three equations with six unknowns and has the form $Ax = 0$, where $A \in \mathbb{R}^{m \times n}$, $m = 3, n = 6$, and $x = (a_1, \ldots, a_6)^T$. We know that A has at most three LI columns, so in six dimensions there must be at least three dimensions that cannot be spanned by these three columns. In this case it is easy show that A does have three LI columns, which means that there are $6 - 3 = 3$ families of solutions a_i that will yield proper dimensionless quantities. By inspection, we can find the solutions $x = (0, 1, 1, -1, 1, 0)^T$, $(1, -2, -1, 0, 0, 0)^T$, and $(0, 0, 0, 0, 1, -1)^T$, yielding the three dimensionless groups

$$\Pi_1 = \frac{\rho U R}{\eta} \qquad \Pi_2 = \frac{\Delta p}{\rho U^2} \qquad \Pi_3 = \frac{R}{L}$$

Readers with a background in fluid mechanics will recognize Π_1 as the REYNOLDS NUMBER (Bird, Stewart, and Lightfoot, 2002).

Because the solution to $Ax = 0$ is not unique, this choice of dimensionless groups is not unique: each Π_i can be replaced by any nonzero power of it, and the Π_is can be multiplied by one another and by any constant to yield other equally valid dimensionless groups. For example, Π_2 can be replaced in this set by $\Pi_2\Pi_3 = \frac{\Delta p R}{\rho U^2 L}$; fluid mechanicians will recognize this quantity as the FRICTION FACTOR.

Now we return to the general case where we have n quantities and m units. Because A has m LI rows (and thus m LI columns—see Example 1.8), it has a nullspace of $n - m$ dimensions, and therefore there is an $n - m$ dimensional subspace of vectors x that will solve $Ax = 0$. This result gives us the BUCKINGHAM PI THEOREM: given a problem with n dimensional parameters containing m units, the problem can be recast in terms of $l = n - m$ dimensionless groups (Lin and Segel, 1974). This theorem holds under the condition that $\text{rank}(A) = m$; in principle it is possible for the rank of A to be less than m. One somewhat artificial example where this issue arises is the following: if all units of length are represented as hectares per meter, then the equations corresponding to those two units would differ only by a sign. They would thus be redundant and the rank of A would be one less than the number of units. If m were replaced by $\text{rank}(A)$, then the Pi theorem would still hold.

A less trivial example in which the Buckingham Pi theorem can cause confusion is the case of problems involving mixtures. One might expect that moles (or masses) of chemical species A and moles of chemical species B (or mole or mass fractions of these species) would be independent units, but they are not. Unlike kilograms and meters, which cannot be added to one another, moles of A and moles of B *can* be added to one another so they do not yield separate equations for exponents the way that kilograms and meters do.

1.3.10 Nonlinear Algebraic Equations: the Newton-Raphson Method

Many if not most of the mathematical problems encountered by engineers are nonlinear: second-order reactions, fluid dynamics at finite Reynolds number, and phase equilibrium are a few examples. We will write a general nonlinear system of n equations and n unknowns as

$$f(x) = 0 \tag{1.7}$$

where $x \in \mathbb{R}^n$ and $f \in \mathbb{R}^n$. In contrast to the case with linear equations, where LU decomposition will lead to an exact and unique solution (if the problem is not singular), there is no general theory of existence and uniqueness for nonlinear equations. In general, many solutions can exist and there is no way of knowing *a priori* where they are or how many there are. To find solutions to nonlinear equations, one almost always needs to make an initial guess and use an iterative method to find a solution. A powerful and general method for doing this is called NEWTON-RAPHSON iteration.

Consider an initial guess x and assume for the moment that the exact solution x_e is given by $x + d$, where d is as yet unknown, but is assumed to be small, i.e., the initial guess is good. In this case

$$f(x_e) = f(x + d) = 0$$

We next expanding the right-hand side in a Taylor series around x. It is now convenient to switch to component notation to express the second-order Taylor series approximation for *vector* f

$$f_i(x + d) = f_i(x) + \left.\frac{\partial f_i}{\partial x_j}\right|_x d_j + \frac{1}{2}\left.\frac{\partial^2 f_i}{\partial x_j \partial x_l}\right|_x d_j d_l + O\left(\|d\|^3\right)$$

where the notation $O(\delta^p)$ denotes terms that are "of order δ^p," which means that they decay to zero at least as fast as δ^p in the limit $\delta \to 0$.

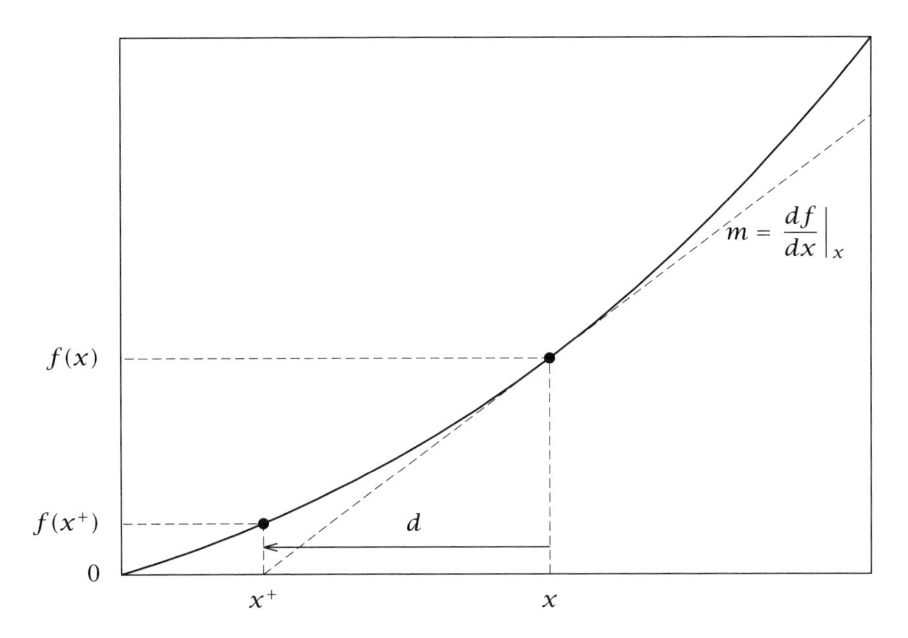

Figure 1.3: An iteration of the Newton-Raphson method for solving $f(x) = 0$ in the scalar case.

An approximate solution to this equation can be found if the terms that are quadratic and higher degree in d are neglected, yielding the *linearized* problem

$$f(x + d) = f(x) + \left.\frac{\partial f}{\partial x}\right|_x d$$

Setting $f(x+d) = 0$ and defining the JACOBIAN matrix $J_{ij}(x) = \partial f_i / \partial x_j \big|_x$, this can be rearranged into the linear system

$$J(x)d = -f(x)$$

This equation can be solved for d (e.g., by LU decomposition) to yield a new guess for the solution $x^+ = x + d$ in which we use the superscript x^+ to denote the variable x at the *next* iterate. Denoting the solution by $d = -J^{-1}(x)f(x)$, the process can be summarized as

$$x^+ = x - J^{-1}(x)f(x) \tag{1.8}$$

This equation is *iterated* until $\|x^+ - x\|$ or $||f(x)||$ reaches a prescribed error tolerance. One iteration of (1.8) is depicted for a scalar function in Figure 1.3.

An important question for any iterative method is how rapidly it converges. To address this issue for the Newton-Raphson method, let $\epsilon = x - x_e$ be the difference between the approximate solution and the exact solution. Similarly, $\epsilon^+ = x^+ - x_e$ and therefore $\epsilon^+ - \epsilon = x^+ - x$. Using this result and (1.8), the evolution equation for the error is

$$\epsilon^+ = \epsilon - J^{-1}(x_e + \epsilon) f(x_e + \epsilon)$$

Taylor expanding this equation around x_e yields, again in index notation due to the Taylor series,

$$\begin{aligned}
\epsilon_i^+ &= \epsilon_i - \left(J_{ij}^{-1}|_{x_e} + \left.\frac{\partial J_{ij}^{-1}}{\partial x_l}\right|_{x_e} \epsilon_l + O\left(\|\epsilon\|^2\right) \right) \cdot \\
&\qquad \left(0 + J_{jk}|_{x_e} \epsilon_k + \frac{1}{2} \left.\frac{\partial J_{jk}}{\partial x_l}\right|_{x_e} \epsilon_k \epsilon_l + O\left(\|\epsilon\|^3\right) \right) \\
&= \epsilon_i - \left(J_{ij}^{-1} J_{jk}\Big|_{x_e} \epsilon_k + \left.\left(\frac{\partial J_{ij}^{-1}}{\partial x_l} J_{jk} + \frac{1}{2} J_{ij}^{-1} \frac{\partial J_{jk}}{\partial x_l} \right)\right|_{x_e} \epsilon_l \epsilon_k + O\left(\|\epsilon\|^3\right) \right) \\
&= \epsilon_i - \left(\delta_{ik} \epsilon_k + \left.\left(\frac{\partial J_{ij}^{-1}}{\partial x_l} J_{jk} + \frac{1}{2} J_{ij}^{-1} \frac{\partial J_{jk}}{\partial x_l} \right)\right|_{x_e} \epsilon_l \epsilon_k + O\left(\|\epsilon\|^3\right) \right) \\
&= -\left.\left(\frac{\partial J_{ij}^{-1}}{\partial x_l} J_{jk} + \frac{1}{2} J_{ij}^{-1} \frac{\partial J_{jk}}{\partial x_l} \right)\right|_{x_e} \epsilon_l \epsilon_k + O\left(\|\epsilon\|^3\right) \\
&= -\left.\left(\frac{\partial}{\partial x_l} (J_{ij}^{-1} J_{jk}) - \frac{1}{2} J_{ij}^{-1} \frac{\partial J_{jk}}{\partial x_l} \right)\right|_{x_e} \epsilon_l \epsilon_k + O\left(\|\epsilon\|^3\right) \\
&= -\left.\left(\frac{\partial}{\partial x_l} \delta_{ik} - \frac{1}{2} J_{ij}^{-1} \frac{\partial J_{jk}}{\partial x_l} \right)\right|_{x_e} \epsilon_l \epsilon_k + O\left(\|\epsilon\|^3\right) \\
\epsilon_i^+ &= \frac{1}{2} J_{ij}^{-1} \left.\frac{\partial J_{jk}}{\partial x_l}\right|_{x_e} \epsilon_l \epsilon_k + O\left(\|\epsilon\|^3\right)
\end{aligned}$$

This result, which we can summarize as $\|\epsilon^+\| = O\left(\|\epsilon\|^2\right)$, illustrates that given a sufficiently good guess, the Newton-Raphson iteration converges rapidly, specifically *quadratically*, to the exact solution.

For example, if the error in iteration (1.8) after step k is 10^{-2}, the error after step $k + 1$ is $\sim 10^{-4}$ and after step $k + 2$ is $\sim 10^{-8}$. Indeed, a good check of whether a code for implementing Newton-Raphson is correct is to verify the quadratic convergence. Of course, this result only holds if a sufficiently good guess is given. If the initial guess is poor, the iteration may not converge, or alternately may converge to a solution far from the initial guess.

1.3.11 Linear Coordinate Transformations

As noted above, the components of a matrix operator depend on the coordinate system in which it is expressed. Here we illustrate how the components of a matrix operator change upon a change in coordinate system. Consider two vectors x and y and a matrix operator A, where $y = Ax$. For example, we can take x and y to be two-dimensional, in which case

$$x = \begin{bmatrix} x_1 \\ x_2 \end{bmatrix}$$

Now consider new variables x_1' and x_2', where

$$\begin{aligned} x_1' &= T_{11}x_1 + T_{12}x_2 \\ x_2' &= T_{21}x_1 + T_{22}x_2 \end{aligned}$$

This can be written $x' = Tx$. Here x and x' are the same vector, but represented in the original (unprimed) and new (primed) coordinate systems, and T is the operator that generates the new coordinate values from the original ones. It must be invertible—otherwise there is not a unique mapping between the coordinate systems. Therefore, we can write $x = T^{-1}x'$ and $y = AT^{-1}x'$; the matrix AT^{-1} yields the mapping between x' and y. If we also consider a coordinate transformation of the vector y of the form $y' = Wy$, then $y' = WAT^{-1}x'$. The matrix WAT^{-1} provides the mapping from x' to y'. Some important coordinate transformations that take advantage of the properties of the operator A are described in Section 1.4.

1.4 The Algebraic Eigenvalue Problem

1.4.1 Introduction

Eigenvalue problems arise in a variety of contexts. One of the most important is in the solution of systems of linear ordinary differential equations. Consider the system of two ordinary differential equations

$$\frac{dz}{dt} = Az \tag{1.9}$$

Here $z \in \mathbb{R}^2$ and $A \in \mathbb{R}^{2\times 2}$. If we guess, based on what we know about the scalar case, that solutions will have the form

$$z(t) = xe^{\lambda t}$$

then we have that

$$Ax = \lambda x \tag{1.10}$$

If we can find a solution to this equation, then we have a solution to (1.9). (To obtain the general solution to (1.9) we must find two solutions to this problem.) This is the algebraic version of the EIGENVALUE PROBLEM.

The eigenvalue problem can be rewritten as the homogeneous system of equations

$$(A - \lambda I)x = 0$$

As with any homogeneous system, this generally has only the trivial solution $x = 0$. For special values of λ, known as the EIGENVALUES of A, the equation has a nontrivial solution, however. The solutions corresponding to these eigenvalues, which can be real or complex, are the EIGENVECTORS of A. Geometrically, the eigenvectors of A are those vectors that change only a scalar multiple when operated on by A. This property is of great importance because for the eigenvectors, matrix multiplication reduces to simple scalar multiplication; Ax can be replaced by λx. Because of this property, the eigenvectors of a matrix provide a natural coordinate system for working with that matrix. This fact is used extensively in applied mathematics.

From the existence and uniqueness results for linear systems of equations that we saw in Section 1.3.1, we know that the above homogeneous problem has a nontrivial solution if an only if $A - \lambda I$ is noninvertible: that is, when

$$\det(A - \lambda I) = 0$$

This equation is called the CHARACTERISTIC EQUATION for A, and $\det(A-\lambda I)$ is the CHARACTERISTIC POLYNOMIAL. For an $n\times n$ matrix, this polynomial is always nth degree in λ; this can be seen by performing LU decomposition on $A-\lambda I$; therefore, the characteristic polynomial has n roots (not necessarily all real or distinct). Each root is an eigenvalue, so an $n \times n$ matrix has exactly n eigenvalues. Each distinct eigenvalue has a distinct (i.e. linearly independent) eigenvector. Each set of repeated roots will have at least one distinct eigenvector, but may have fewer than the multiplicity of the root. So a matrix may have fewer than n

eigenvectors. The nature of the eigenvectors depends on the structure of the matrix.

In principle, the eigenvalues of a matrix may be found by finding the roots of its characteristic polynomial. Since polynomials of degree greater than four cannot be factored analytically, approximate numerical methods must be used for virtually all matrix eigenvalue problems. There are numerical methods for finding the roots of a polynomial, but in practice, this procedure is difficult and inefficient. An extremely robust iterative method, based on the QR factorization of a matrix (Exercise 1.38), is the most commonly used technique for general matrices. In some cases, only the "dominant" eigenvalue (the eigenvalue with the largest magnitude) needs to be found. The POWER METHOD (Exercise 1.57) is a rapid iterative technique for this problem. Generalizations of the idea behind the power method form the basis of powerful KRYLOV SUBSPACE methods for iterative solutions of many computational linear algebra problems (Trefethen and Bau III, 1997).

1.4.2 Self-Adjoint Matrices

Consider the real symmetric (thus self-adjoint) matrix

$$A = \begin{bmatrix} 2 & 1 \\ 1 & 2 \end{bmatrix}$$

The characteristic equation for A is $\lambda^2 - 4\lambda + 3 = 0$ and its solutions are $\lambda_1 = 1, \lambda_2 = 3$. The corresponding eigenvectors $x = v_1$ and $x = v_2$ are solutions to

$$\begin{bmatrix} 2-\lambda & 1 \\ 1 & 2-\lambda \end{bmatrix} x = 0$$

These solutions are (to within an arbitrary multiplicative constant)

$$v_1 = \begin{bmatrix} 1 \\ -1 \end{bmatrix} \qquad v_2 = \begin{bmatrix} 1 \\ 1 \end{bmatrix}$$

Note that these vectors, when normalized to have unit length, form an ON basis for $\mathbb{R}^2$. Now let

$$Q = \begin{bmatrix} v_1 & v_2 \end{bmatrix} = \frac{1}{\sqrt{2}} \begin{bmatrix} 1 & 1 \\ -1 & 1 \end{bmatrix}$$

A vector x in $\mathbb{R}^2$ can now be represented in two coordinate systems, either the original basis or the eigenvector basis. A representation in the eigenvector basis will be indicated by a $'$, so $x' = [x_1' \; x_2']^T$ is the vector containing the coordinates of x expressed in the eigenvector basis. It can be shown that the coordinate transformation between these bases is defined by Q, so that $x = Qx'$ and $x' = Q^{-1}x$. Remember that A is defined in the original basis so Ax makes sense, but Ax' does not. However, we can write

$$\begin{aligned} Ax &= A(x_1'v_1 + x_2'v_2) \\ &= x_1'Av_1 + x_2'Av_2 \\ &= x_1'\lambda_1 v_1 + x_2'\lambda_2 v_2 \\ &= Q\Lambda x' \end{aligned}$$

where

$$\Lambda = \begin{bmatrix} \lambda_1 & 0 \\ 0 & \lambda_2 \end{bmatrix} = \begin{bmatrix} 1 & 0 \\ 0 & 3 \end{bmatrix}$$

Therefore, $Ax = Q\Lambda x'$. Using the transformation $x' = Q^{-1}x$ gives that $Ax = Q\Lambda Q^{-1}x$, or $A = Q\Lambda Q^{-1}$. This expression can be reduced further by noting that since the columns of Q form an orthonormal basis, $Q_{ki}Q_{kj} = \delta_{ij}$, or $Q^TQ = I$. Since $Q^{-1}Q = I$ by definition, it follows that $Q^{-1} = Q^T$. Matrices for which this property holds are called ORTHOGONAL. In the complex case, the property becomes $Q^{-1} = \bar{Q}^T$ and Q is said to be UNITARY. Returning to the example, the property means that A can be expressed

$$A = Q\Lambda Q^T$$

As an example of the usefulness of this result, consider the system of equations

$$\frac{dx}{dt} = Ax = Q\Lambda Q^T x$$

By multiplying both sides of the equation by Q^T and using the facts that $Q^TQ = I$ and $x' = Q^Tx$, the equation can be rewritten

$$\frac{dx'}{dt} = \Lambda x' = \begin{bmatrix} 1 & 0 \\ 0 & 3 \end{bmatrix} x'$$

or $dx_1/dt = x_1$, $dx_2/dt = 3x_2$. In the eigenvector basis, the differential equations are *decoupled*. They can be solved separately.

The above representation of A can be found for *any* matrix A that satisfies the self-adjointness condition $A = \bar{A}^T$. We have the following theorem.

Theorem 1.11 (Self-adjoint matrix decomposition). *If $A \in \mathbb{C}^{n\times n}$ is self-adjoint, then there exists a unitary $Q \in \mathbb{C}^{n\times n}$ and real, diagonal $\Lambda \in \mathbb{R}^{n\times n}$ such that*

$$A = Q\Lambda Q^*$$

The diagonal elements of Λ, Λ_{ii}, are the eigenvalues λ_i of A. The eigenvalues are all real, even if A is not. The columns of the matrix Q are the (normalized) eigenvectors v_i corresponding to the eigenvalues. The eigenvectors are orthonormal and form a basis for $\mathbb{C}^n$.

This result shows that for every self-adjoint matrix operator, there is a natural orthogonal basis, in which the matrix becomes diagonal. That is, the transformation DIAGONALIZES the matrix. Since the eigenvalues are all real, matrix multiplication reduces to simple contraction or stretching along the (eigenvector) coordinate axes. In this basis, any linear systems of algebraic or differential equations containing Ax reduce to n decoupled equations.

That the eigenvalues are real can be established as follows. We have $Av = \lambda v$ and, by taking adjoints, $v^*A = \bar{\lambda}v^*$, after noting that $A^* = A$. Multiply the first on the left by v^* and the second on the right by v and subtract to obtain $0 = (\lambda - \bar{\lambda})v^*v$. We have that v^*v is not zero since $v \neq 0$ is an eigenvector, and therefore $\lambda = \bar{\lambda}$ and λ is real.

If A has distinct eigenvalues, the eigenvectors are orthogonal, which is also readily established. Given an eigenvalue λ_i and corresponding eigenvector v_i, we have that $Av_i = \lambda_i v_i$. Let (λ_j, v_j) be another eigenpair so that $Av_j = \lambda_j v_j$. Multiplying $Av_i = \lambda_i v_i$ on the left by v_j^*, and $Av_j = \lambda_j$ on the left by v_i^* and subtracting gives $(\lambda_i - \lambda_j)(v_i^* v_j) = 0$. If the eigenvalues are distinct, $\lambda_i \neq \lambda_j$ this equation can hold only if $v_i^* v_j = 0$, and therefore v_i and v_j are orthogonal.

For the case of repeated eigenvalues, since orthogonality holds for eigenvalues that are arbitrarily close together but unequal, we might expect intuitively that it continues to hold when the eigenvalues become equal. This turns out to be true, and we delay the proof until we have introduced the Schur decomposition in Section 1.4.6.

1.4.3 General (Square) Matrices

Although many matrices arising in applications are self-adjoint, many others are not, so it is important to include the results for these cases. Now the eigenvectors do not necessarily form an ON basis, nor can the matrix always be diagonalized. But it is possible to come fairly close. There are three cases:

1. If A is not self-adjoint, but has distinct eigenvalues ($\lambda_i \neq \lambda_j, i \neq j$), then A can be diagonalized

$$A = S\Lambda S^{-1} \tag{1.11}$$

As before, $\Lambda = S^{-1}AS$ is diagonal, and contains the eigenvalues (not necessarily real) of A. The columns of S contain the corresponding eigenvectors. The eigenvectors are LI, so they form a basis, but are not orthogonal.

2. If A is not self-adjoint and has repeated eigenvalues, it may still be the case that the repeated eigenvalues have distinct eigenvectors – e.g. a root with multiplicity two that has two linearly independent eigenvectors. Here A can be diagonalized as above.

3. If A is not self-adjoint and has repeated eigenvalues that do not yield distinct eigenvectors, it cannot be completely diagonalized; a matrix of this type is called DEFECTIVE. Nevertheless, it can always be put into *Jordan form* J

$$A = MJM^{-1} \tag{1.12}$$

where $J = M^{-1}AM$ is organized as follows: each distinct eigenvalue appears on the diagonal with the nondiagonal elements of the corresponding row and column being zero, just as above. However, repeated eigenvalues appear in JORDAN BLOCKS with this structure (shown here for an eigenvalue of multiplicity three)

$$\begin{bmatrix} \lambda & 1 & 0 \\ 0 & \lambda & 1 \\ 0 & 0 & \lambda \end{bmatrix}$$

In the case of repeated eigenvalues, we can distinguish between ALGEBRAIC multiplicity and GEOMETRIC multiplicity. Algebraic multiplicity of an eigenvalue is simply its multiplicity as a root of the characteristic equation. Geometric multiplicity is the number of distinct eigenvectors that correspond to the repeated eigenvalue. In case 2 above, the geometric multiplicity of each repeated eigenvalue is equal to its algebraic multiplicity. In case 3, the algebraic multiplicity exceeds the geometric multiplicity.

For a non-self-adjoint 5 by 5 matrix with repeated eigenvalues

$\lambda_2 = \lambda_3 = \lambda_4$,

$$J = \begin{bmatrix} \lambda_1 & 0 & 0 & 0 & 0 \\ 0 & \lambda_2 & 1 & 0 & 0 \\ 0 & 0 & \lambda_3 & 1 & 0 \\ 0 & 0 & 0 & \lambda_4 & 0 \\ 0 & 0 & 0 & 0 & \lambda_5 \end{bmatrix}$$

The eigenvectors corresponding to the distinct eigenvalues are the corresponding columns of M. A distinct eigenvector does not exist for each of the repeated eigenvalues, but a GENERALIZED EIGENVECTOR can be found for each occurrence of the eigenvalue. These vectors, along with the eigenvectors, form a basis for $\mathbb{R}^n$.

Example 1.12: A nonsymmetric matrix

Find the eigenvalues and eigenvectors of the nonsymmetric matrix

$$A = \begin{bmatrix} 1 & 2 \\ 0 & 3 \end{bmatrix}$$

and show that it can be put in the form of (1.11).

Solution

This matrix has characteristic equation $(1-\lambda)(3-\lambda) = 0$ and thus has eigenvalues $\lambda = 1, \lambda = 3$. For $\lambda = 1$, the eigenvector solves

$$\begin{bmatrix} 1-\lambda & 2 \\ 0 & 3-\lambda \end{bmatrix} \begin{bmatrix} x_1 \\ x_2 \end{bmatrix} = \begin{bmatrix} 0 & 2 \\ 0 & 2 \end{bmatrix} \begin{bmatrix} x_1 \\ x_2 \end{bmatrix} = \begin{bmatrix} 0 \\ 0 \end{bmatrix}$$

and it is straightforward to see that this is satisfied by $[x_1, x_2]^T = v_1 = [1, 0]^T$. For $\lambda = 3$ we have

$$\begin{bmatrix} -2 & 2 \\ 0 & 0 \end{bmatrix} \begin{bmatrix} x_1 \\ x_2 \end{bmatrix} = \begin{bmatrix} 0 \\ 0 \end{bmatrix}$$

which has solution $v_2 = [1, 1]^T$. Here v_1 and v_2 are not orthogonal, but they are LI, so they still form a basis. Letting

$$S = \begin{bmatrix} v_1 & v_2 \end{bmatrix} = \begin{bmatrix} 1 & 1 \\ 0 & 1 \end{bmatrix}$$

one can determine that

$$S^{-1} = \begin{bmatrix} 1 & -1 \\ 0 & 1 \end{bmatrix}$$

Since the columns of S are not orthogonal, they cannot be normalized to form a matrix that satisfies $S^{-1} = S^T$. Nevertheless, A can be diagonalized

$$S^{-1}AS = \Lambda = \begin{bmatrix} 1 & 0 \\ 0 & 3 \end{bmatrix}$$

□

Example 1.13: A defective matrix

Find the eigenvalues and eigenvectors of the nonsymmetric matrix

$$A = \begin{bmatrix} 3 & 2 \\ 0 & 3 \end{bmatrix}$$

and show that it cannot be put in the form of (1.11), but can be put in the form of (1.12).

Solution

The characteristic equation for A is $(3 - \lambda)^2 = 0$, so A has the repeated eigenvalue $\lambda = 3$. The eigenvector is determined from

$$\begin{bmatrix} 0 & 2 \\ 0 & 0 \end{bmatrix} x = \begin{bmatrix} 0 \\ 0 \end{bmatrix}$$

which has solution $x = v_1 = [1, 0]^T$. There is not another nontrivial solution to this equation so the repeated eigenvalue $\lambda = 3$ has only one eigenvector. We cannot diagonalize this A.

Nevertheless, we will seek to nearly diagonalize it, by finding a generalized eigenvector v_2 that allows us to construct a matrix

$$M = \begin{bmatrix} v_1 & v_2 \end{bmatrix}$$

satisfying

$$M^{-1}AM = J = \begin{bmatrix} \lambda & 1 \\ 0 & \lambda \end{bmatrix}$$

Multiplying both sides of this equation by M yields that

$$AM = MJ = \begin{bmatrix} v_1 & v_2 \end{bmatrix} \begin{bmatrix} \lambda & 1 \\ 0 & \lambda \end{bmatrix}$$

which can be rearranged to

$$(A - \lambda I) \begin{bmatrix} v_1 & v_2 \end{bmatrix} = \begin{bmatrix} 0 & v_1 \end{bmatrix}$$

This equation can be rewritten as the pair of equations

$$(A - \lambda I)v_1 = 0,$$
$$(A - \lambda I)v_2 = v_1$$

The first of these is simply the equation determining the true eigenvector v_1, while the second will give us the generalized eigenvector v_2. For the present problem this equation is

$$\begin{bmatrix} 0 & 2 \\ 0 & 0 \end{bmatrix} v_2 = \begin{bmatrix} 1 \\ 0 \end{bmatrix}$$

A solution to this equation is $v_2 = [0, 1/2]^T$. (Any solution v_2 must be LI from v_1. Why?) Constructing the matrix

$$M = \begin{bmatrix} v_1 & v_2 \end{bmatrix} = \begin{bmatrix} 1 & 0 \\ 0 & 1/2 \end{bmatrix}$$

one can show that

$$M^{-1} = \begin{bmatrix} 1 & 0 \\ 0 & 2 \end{bmatrix}$$

and that

$$J = M^{-1}AM = \begin{bmatrix} 3 & 1 \\ 0 & 3 \end{bmatrix}$$

Note that we can replace v_2 by $v_2 + \alpha v_1$ for any α and still obtain this result. □

1.4.4 Positive Definite Matrices

Positive definite and positive semidefinite matrices show up often in applications. Here are some basic facts about them. In the following, A is real and symmetric and B is real. The matrix A is POSITIVE DEFINITE (denoted $A > 0$), if

$$x^T A x > 0, \quad \forall \text{ nonzero } x \in \mathbb{R}^n$$

The matrix A is POSITIVE SEMIDEFINITE (denoted $A \geq 0$), if

$$x^T A x \geq 0, \quad \forall x \in \mathbb{R}^n$$

You should be able to prove the following facts.

1. $A > 0 \iff \lambda > 0, \quad \lambda \in \text{eig}(A)$

2. $A \geq 0 \iff \lambda \geq 0, \quad \lambda \in \text{eig}(A)$

3. $A \geq 0 \iff B^T AB \geq 0 \quad \forall B$

4. $A > 0$ and B nonsingular $\iff B^T AB > 0$

5. $A > 0$ and B full column rank $\iff B^T AB > 0$

6. $A_1 > 0, A_2 \geq 0 \implies A = A_1 + A_2 > 0$

7. $A > 0 \iff z^* Az > 0 \quad \forall$ nonzero $z \in \mathbb{C}^n$

8. For $A \geq 0$, $x^T Ax = 0 \iff Ax = 0$

If symmetric matrix A is not positive semidefinite nor negative semidefinite, then it is termed indefinite. In this case A has both positive and negative eigenvalues.

1.4.5 Eigenvalues, Eigenvectors, and Coordinate Transformations

Under the general linear transformation

$$y = Ax \tag{1.13}$$

all the components of the vector y are coupled to all the components of the vector x via the elements of A, all of which are generally nonzero. We can always rewrite this transformation using the eigenvalue decomposition as

$$y = MJM^{-1}x$$

Now consider the coordinate transformation $x' = M^{-1}x$ and $y' = M^{-1}y$. In this new coordinate system, the linear transformation, (1.13) becomes

$$y' = Jx'$$

In the "worst case scenario," J has eigenvalues on the diagonal, some values of 1 just above the diagonal and is otherwise zero. In the more usual scenario $J = \Lambda$ and each component of y' is coupled only to one component of x'—the coordinate transformation associated with the eigenvectors of A provides a coordinate system in which the different components are decoupled. This result is powerful and is used in a wide variety of applications.

Further considering the idea of coordinate transformations leads naturally to the question of the dependence of the eigenvalue problem on the coordinate system that is used to set up the problem. Given that

$$Ax = \lambda x$$

let us take $x' = Tx$, where T is invertible but otherwise arbitrary; this expression represents a coordinate transformation between unprimed and primed coordinates, as we have already described in Section 1.3.11. Now $x = T^{-1}x'$, and thus

$$AT^{-1}x' = \lambda T^{-1}x'$$

Multiplying both sides by T to eliminate T^{-1} on the right-hand side yields

$$TAT^{-1}x' = \lambda x'$$

Recall that we have done nothing to the eigenvalues λ—they are the same in the last equation of this sequence as the first. Thus the eigenvalues of TAT^{-1} are the same as the eigenvalues of A. Therefore, if two matrices are related by a transformation $B = TAT^{-1}$, which is called a SIMILARITY TRANSFORMATION, their eigenvalues are the same. In other words, eigenvalues of a matrix are INVARIANT under similarity transformations.

In many situations, invariants other than the eigenvalues are used. These can be expressed in terms of the eigenvalues. The two most common are the TRACE of a matrix A

$$\operatorname{tr} A = \sum_{i=1}^{n} A_{ii} = \sum_{i=1}^{n} \lambda_i$$

and the determinant

$$\det A = (-1)^m \prod_{i=1}^{n} U_{ii} = \prod_{i=1}^{n} \lambda_i$$

Example 1.14: Vibrational modes of a molecule

The individual atoms that make up a molecule vibrate around their equilibrium positions and orientations. These vibrations can be used to characterize the molecule by spectroscopy and are important in determining many of its properties, such as heat capacity and reactivity. We examine here a simple model of a molecule to illustrate the origin and nature of these vibrations.

Let the αth atom of a molecule be at position $\mathbf{x}_\alpha = [x_\alpha, y_\alpha, z_\alpha]^T$ and have mass m_α. The bond energy of the molecule is $U(\mathbf{x}_1, \mathbf{x}_2, \mathbf{x}_3, \ldots, \mathbf{x}_N)$ where N is the number of atoms in the molecule. Newton's second law for each atom is

$$m_\alpha \frac{d^2\mathbf{x}_\alpha}{dt^2} = -\frac{\partial U(\mathbf{x}_1, \ldots, \mathbf{x}_N)}{\partial \mathbf{x}_\alpha}$$

Let $X = [x_1, y_1, z_1, x_2, y_2, z_2, \ldots, x_N, y_N, z_N]^T$ and M be a $3N \times 3N$ diagonal matrix with the masses of each atom on the diagonals. That is, $M_{11} = M_{22} = M_{33} = m_1$, $M_{44} = M_{55} = M_{66} = m_2$, $M_{3N-2,3N-2} = M_{3N-1,3N-1} = M_{3N,3N} = m_N$. Now the equations of motion for the coordinates of the atom become

$$M_{ij}\frac{d^2 X_j}{dt^2} = -\frac{\partial U(X)}{\partial X_i}$$

An equilibrium shape X_{eq} of the molecule is a minimum of the bond energy U, and can be found by Newton-Raphson iteration on the problem $\frac{\partial U}{\partial X_i} = 0$. Assume X_{eq} is known and characterize small-amplitude vibrations around that shape.

Solution

Let $\hat{X} = X - X_{\text{eq}}$ be a small perturbation away from the equilibrium shape. Because $\frac{d^2 X_{\text{eq}}}{dt^2} = 0$, this perturbation satisfies the equation

$$M_{ij}\frac{d^2 \hat{X}_j}{dt^2} = -\frac{\partial U(X_{\text{eq}} + \hat{X})}{\partial X_i}$$

Taylor expanding the right-hand side of this equation, using the fact that $\left.\frac{\partial U}{\partial X_i}\right|_{X_{\text{eq}}} = 0$, and neglecting terms of $O(||\hat{X}||^2)$ yield

$$\frac{\partial U(X_{\text{eq}} + \hat{X})}{\partial X_i} \approx H_{ik}\hat{X}_k$$

where

$$H_{ik} = \left.\frac{\partial^2 U}{\partial X_i \partial X_k}\right|_{X_{\text{eq}}}$$

is called the HESSIAN matrix for the function U. Thus the governing equation for the vibrations is given by

$$M_{ij}\frac{d^2 \hat{X}_j}{dt^2} = -H_{ik}\hat{X}_k$$

By definition, H is symmetric. Furthermore, rigidly translating the entire molecule does not change its bond energy, so H has three zero eigenvalues, with eigenvectors

$$V = [1, 0, 0, \ldots, 1, 0, 0]^T \qquad V = [0, 1, 0, \ldots, 0, 1, 0]^T$$
$$V = [0, 0, 1, \ldots, 0, 0, 1]^T$$

These correspond to moving the whole molecule in the x, y, and z directions, respectively. Furthermore, because X_{eq} is a minimum of the bond energy, H is also positive semidefinite.

We expect the molecule to vibrate, so we will seek oscillatory solutions. A convenient way to do so is to let

$$\hat{X}(t) = Ze^{i\omega t} + \bar{Z}e^{-i\omega t}$$

recalling that for real ω, $e^{i\omega t} = \cos\omega t + i\sin\omega t$. Substituting into the governing equation yields

$$-\omega^2 M_{ij}(Z_j e^{i\omega t} + \bar{Z}_j e^{-i\omega t}) = -H_{ik}(Z_k e^{i\omega t} + \bar{Z}_k e^{-\omega t})$$

Gathering terms proportional to $e^{i\omega t}$ and $e^{-i\omega t}$, we can see that this equation will be satisfied at all times if and only if

$$\omega^2 M_{ij} Z_j = H_{ik} Z_k \qquad (1.14)$$

This looks similar to the linear eigenvalue problem, (1.10), and reduces exactly to one if all atoms have the same mass m (in which case $M = mI$).

We can learn more about this problem by considering the properties of M and H. Since M is diagonal and the atomic masses are positive, M is clearly positive definite. Also recall that H is symmetric positive semidefinite. Writing $M = L^2$, where L is diagonal and its diagonal entries are the square roots of the masses, we can write that $\omega^2 L^2 Z = HZ$. Multiplying by L^{-1} on the left yields $\omega^2 LZ = L^{-1}HZ$ and letting $\tilde{Z} = LZ$ results in $\omega^2\tilde{Z} = L^{-1}HL^{-1}\tilde{Z}$. This has the form of an eigenvalue problem $\tilde{H}\tilde{Z} = \omega^2\tilde{Z}$, where $\tilde{H} = L^{-1}HL^{-1}$. Solving this eigenvalue problem gives the frequencies ω at which the molecule vibrates. The corresponding eigenvectors $\tilde{Z}$, when transformed back into the original coordinates via $Z = L^{-1}\tilde{Z}$, give the so-called "normal modes." Each frequency is associated with a mode of vibration that in general involves different atoms of the molecule in different ways. Because $\tilde{H}$ is symmetric, these modes form an orthogonal basis in which to describe the motions of the molecule. A further result can be obtained by multiplying (1.14) on the left by Z^T, yielding

$$\omega^2 Z^T M Z = Z^T H Z$$

Because $Z^T M Z > 0$ and $Z^T H Z \geq 0$, we can conclude that $\omega^2 \geq 0$ with equality only when Z is a zero eigenvector of H. This result shows that the frequencies ω are real and thus that the dynamics are purely oscillatory.

Observe that the quantity $Z^T MZ$ arises naturally in this problem: via the transformation $\tilde{Z} = LZ$ it is equivalent to the inner product $\tilde{Z}^T\tilde{Z}$. It is straightforward to show that for any symmetric positive definite W, the quantity $x^T Wy$ satisfies all the conditions of an inner product between real vectors x and y; it is called a WEIGHTED inner product. In the current case, the eigenvectors $\tilde{Z}$ are orthogonal under the usual "unweighted" inner product, in which case the vectors $Z = L^{-1}\tilde{Z}$ are orthogonal under the weighted inner product with $W = M$. □

1.4.6 Schur Decomposition

A major problem with using the Jordan form when doing calculations on matrices that have repeated eigenvalues is that the Jordan form is numerically unstable. For matrices with repeated eigenvalues, if diagonalization is not possible, it is usually better computationally to use the Schur form instead of the Jordan form. The Schur form only triangularizes the matrix. Triangularizing a matrix, even one with repeated eigenvalues, is numerically well conditioned. Golub and Van Loan (1996, p.313) provide the following theorem.

Theorem 1.15 (Schur decomposition). *If $A \in \mathbb{C}^{n\times n}$ then there exists a unitary $Q \in \mathbb{C}^{n\times n}$ such that*

$$Q^*AQ = T$$

in which T is upper triangular.

The proof of this theorem is discussed in Exercise 1.43. Note that even though T is upper triangular instead of diagonal, its diagonal elements are still its eigenvalues. The eigenvalues of T are also equal to the eigenvalues of A because T is a the result of a similarity transformation of A. Even if A is a real matrix, T can be complex because the eigenvalues of a real matrix may come in complex conjugate pairs. Recall a matrix Q is unitary if $Q^*Q = I$. You should also be able to prove the following facts (Horn and Johnson, 1985).

1. If $A \in \mathbb{C}^{n\times n}$ and $BA = I$ for some $B \in \mathbb{C}^{n\times n}$, then

   (a) A is nonsingular

   (b) B is unique

(c) $AB = I$

2. The matrix Q is unitary if and only if

   (a) Q is nonsingular and $Q^* = Q^{-1}$

   (b) $QQ^* = I$

   (c) Q^* is unitary

   (d) The rows of Q form an orthonormal set

   (e) The columns of Q form an orthonormal set

If A is self-adjoint, then by taking adjoints of both sides of the Schur decomposition equality, we have that T is real and diagonal, and the columns of Q are the eigenvectors of A, which is one way to show that the eigenvectors of a self-adjoint matrix are orthogonal, regardless of whether the eigenvalues are distinct. Recall that we delayed the proof of this assertion in Section 1.4.2 until we had introduced the Schur decomposition.

If A is real and symmetric, then not only is T real and diagonal, but Q can be chosen real and orthogonal. This fact can be established by noting that if complex-valued $q = a + bi$ is an eigenvector of A, then so are both real-valued vectors a and b. And if complex eigenvector q_j is orthogonal to q_k, then real eigenvectors a_j and b_j are orthogonal to real eigenvectors a_k and b_k, respectively. The theorem summarizing this case is the following (Golub and Van Loan, 1996, p.393), where, again, it does not matter if the eigenvalues of A are repeated.

Theorem 1.16 (Symmetric Schur decomposition). *If $A \in \mathbb{R}^{n \times n}$ is symmetric, then there exists a real, orthogonal Q and a real, diagonal Λ such that*

$$Q^T A Q = \Lambda = \operatorname{diag}(\lambda_1, \lambda_2, \ldots, \lambda_n)$$

where $\operatorname{diag}(a, b, c, \ldots)$ *denotes a diagonal matrix with elements $a, b, c, \ldots$ on the diagonal.*

Note that the $\{\lambda_i\}$ are the eigenvalues of A and the columns of Q, $\{q_i\}$, are the corresponding eigenvectors.

For real but not necessarily symmetric A, you can restrict yourself to real matrices by using the real Schur decomposition (Golub and Van Loan, 1996, p.341). But the price you pay is that you can achieve only block upper triangular T, rather than strictly upper triangular T.

Theorem 1.17 (Real Schur decomposition). *If $A \in \mathbb{R}^{n\times n}$ then there exists a real, orthogonal Q such that*

$$Q^T A Q = \begin{bmatrix} R_{11} & R_{12} & \cdots & R_{1m} \\ 0 & R_{22} & \cdots & R_{2m} \\ \vdots & \vdots & \ddots & \vdots \\ 0 & 0 & \cdots & R_{mm} \end{bmatrix}$$

in which each R_{ii} is either a real scalar or a 2×2 real matrix having complex conjugate eigenvalues; the eigenvalues of R_{ii} are the eigenvalues of A.

1.4.7 Singular Value Decomposition

Another highly useful matrix decomposition that can be applied to nonsquare in addition to square matrices is the singular value decomposition (SVD). Any matrix $A \in \mathbb{C}^{m\times n}$ has an SVD

$$A = USV^*$$

in which $U \in \mathbb{C}^{m\times m}$ and $V \in \mathbb{C}^{n\times n}$ are square and unitary

$$U^*U = UU^* = I_m \qquad V^*V = VV^* = I_n$$

and $S \in \mathbb{R}^{m\times n}$ is partitioned as

$$S = \begin{bmatrix} \Sigma_{r\times r} & 0_{r\times(n-r)} \\ 0_{(m-r)\times r} & 0_{(m-r)\times(n-r)} \end{bmatrix}$$

in which r is the rank of the A matrix. The matrix Σ is diagonal and real

$$\Sigma = \begin{bmatrix} \sigma_1 & & \\ & \ddots & \\ & & \sigma_r \end{bmatrix} \qquad \sigma_1 \geq \sigma_2 \geq \cdots \geq \sigma_r > 0$$

in which the diagonal elements, σ_i are known as the singular values of matrix A. The singular values are real and positive and can be ordered from largest to smallest as indicated above.

Connection of SVD and eigenvalue decomposition. Given $A \in \mathbb{C}^{m\times n}$ with rank r, consider the Hermitian matrix $AA^* \in \mathbb{R}^{m\times m}$, also of rank r. We can deduce that the eigenvalues of AA^* are real and nonnegative as follows. Given (λ, v) are an eigenpair of AA^*, we have

$AA^*v = \lambda v, v \neq 0$. Taking inner products of both sides with respect to v and solving for λ gives $\lambda = v^*AA^*v/v^*v$. We know v^*v is a real, positive scalar since $v \neq 0$. Let $y = A^*v$ and we have that $\lambda = y^*y/v^*v$ and we know that y^*y is a real scalar and $y^*y \geq 0$. Therefore λ is real and $\lambda \geq 0$. And we can connect the eigenvalues and eigenvectors of AA^* to the singular values and vectors of A. The r nonzero eigenvalues of AA^* (λ_i) are the squares of the singular values (σ_i) and the eigenvectors of AA^* (q_i) are the columns of U (u_i)

$$\lambda_i(AA^*) = \sigma_i^2(A) \quad i = 1, \dots r$$
$$q_i = u_i, \quad i = 1, \dots m$$

Next consider the Hermitian matrix $A^*A \in \mathbb{R}^{n\times n}$, also of rank r. The r nonzero eigenvalues of A^*A (λ_i) are also the squares of the singular values (σ_i) and the eigenvectors of A^*A ($\overline{q}_i$) are the columns of V (v_i)

$$\lambda_i(A^*A) = \sigma_i^2(A) \quad i = 1, \dots r$$
$$\overline{q}_i = v_i, \quad i = 1, \dots n$$

These results follow from substituting the SVD into both products and comparing with the eigenvalue decomposition

$$AA^* = (USV^*)(VSU^*) = US^2U^* = Q\Lambda Q^{-1}$$
$$A^*A = (VSU^*)(USV^*) = VS^2V^* = \overline{Q}\Lambda\overline{Q}^{-1}$$

Real matrix with full row rank. Consider a real matrix A with more columns than rows (wide matrix, $m < n$) and full row rank, $r = m$. In this case both U and V are real and orthogonal, and the SVD takes the form

$$A = U\begin{bmatrix}\Sigma & 0\end{bmatrix}\begin{bmatrix}V_1^T \\ V_2^T\end{bmatrix}$$

in which V_1 contains the first m columns of V, and V_2 contains the remaining $m - n$ columns. Multiplying the partitioned matrices gives

$$A = U\Sigma V_1^T$$

and notice that we do not need to store the V_2 matrix if we wish to represent A. This fact is handy if A has many more columns than rows, $n \gg m$ because $V_2 \in \mathbb{R}^{n-m\times n}$ requires a large amount of storage compared to A.

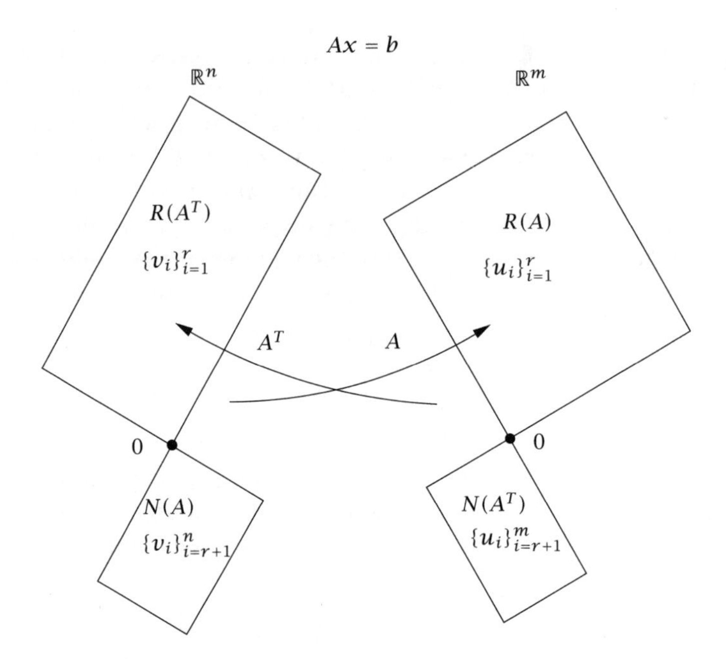

Figure 1.4: The four fundamental subspaces of matrix $A = USV^T$. The range of A is spanned by the first r columns of U, $\{u_1, \ldots, u_r\}$. The range A^T is spanned by the first r columns of V, $\{v_1, \ldots, v_r\}$. The null space of A is spanned by $\{v_{r+1}, \ldots, v_n\}$, and the null space of A^T is spanned by $\{u_{r+1}, \ldots, u_m\}$.

Real matrix with full column rank. Next consider the case in which real matrix A has more rows than columns (tall matrix, $m > n$) and full column rank. In this case the SVD takes the form

$$A = \begin{bmatrix} U_1 & U_2 \end{bmatrix} \begin{bmatrix} \Sigma \\ 0 \end{bmatrix} V^T$$

in which U_1 contains the first n columns of U, and U_2 contains the remaining $n - m$ columns. Multiplying the partitioned matrices gives

$$A = U_1 \Sigma V^T$$

and notice that we do not need to store the U_2 matrix if we wish to represent A.

SVD and fundamental theorem of linear algebra. The SVD provides an orthogonal decomposition of all four of the fundamental subspaces of matrix A. Consider first the partitioned SVD for real-valued A

$$A = \begin{bmatrix} U_1 & U_2 \end{bmatrix} \begin{bmatrix} \Sigma & 0 \\ 0 & 0 \end{bmatrix} \begin{bmatrix} V_1^T \\ V_2^T \end{bmatrix}$$
$$A = U_1 \Sigma V_1^T$$

Now consider Av_k in which $k \geq r + 1$. Because v_k is orthogonal to $V_1 = \{v_1, \ldots, v_r\}$, we have $Av_k = 0$, and these $n - r$ orthogonal v_k span the null space of A. Because the columns of V_1 are orthogonal to this set, they span the range of A^T. Transposing the previous equation gives

$$A^T = V_1 \Sigma U_1^T$$

and we have $\{u_{r+1}, \ldots, u_m\}$ span the null space of A^T. Because the columns of U_1 are orthogonal to this set, they span the range of A. These results are summarized in Figure 1.4.

SVD and least-squares problems. We already have shown that if A has independent columns, the unique least-squares solution to the overdetermined problem

$$\min_x \|Ax - b\|^2$$

is given by

$$x_{\text{ls}} = (A^T A)^{-1} A^T b$$
$$x_{\text{ls}} = A^\dagger b$$

The SVD also provides a means to compute x_{ls}. For real A, the SVD satisfies

$$A = U_1 \Sigma V^T \qquad A^T = V \Sigma U_1^T$$
$$A^T A = V \Sigma U_1^T U_1 \Sigma V^T = V \Sigma^2 V^T$$

The pseudoinverse is therefore given by

$$A^\dagger = V \Sigma^{-2} V^T V \Sigma U_1^T$$
$$A^\dagger = V \Sigma^{-1} U_1^T$$

and the least-squares solution is

$$x_{\text{ls}} = V \Sigma^{-1} U_1^T b$$

SVD and underdetermined problems. We already have shown that if A has independent rows, the unique minimum-norm solution to the underdetermined problem

$$\min_x \|x\|^2 \quad \text{subject to } Ax = b$$

is given by

$$x_{\text{mn}} = A^T(AA^T)^{-1}b$$

The SVD also provides a means to compute x_{mn}. In this case we have $A = U\Sigma V_1^T$ and substituting this into the minimum-norm solution gives

$$x_{\text{mn}} = V_1\Sigma^{-1}U^Tb$$

Note the similarity to the least-squares solution above.

1.5 Functions of Matrices

1.5.1 Polynomial and Exponential

We have already defined some functions of square matrices using matrix multiplication and addition. These operations create the class of polynomial functions

$$p(A) - \alpha_n A^n + \alpha_{n-1}A^{n-1} + \cdots + \alpha_0 I$$

with $A \in \mathbb{C}^{n\times n}$, $\alpha_i \in \mathbb{C}, i = 1, \ldots n$. We wish to expand this set of functions so that we have convenient ways to express solutions to coupled sets of differential equations, for example. Probably the most important function for use in applications is the matrix exponential. The standard exponential of a scalar can be defined in terms of its Taylor series

$$e^a = 1 + a + \frac{1}{2!}a^2 + \frac{1}{3!}a^3 + \cdots \qquad a \in \mathbb{C}$$

This series converges for all $a \in \mathbb{C}$. Notice that this expression is an infinite-order series and therefore not a polynomial function. We can proceed to define the matrix exponential analogously

$$e^A = I + A + \frac{1}{2!}A^2 + \frac{1}{3!}A^3 + \cdots \qquad A \in \mathbb{C}^{n\times n}$$

and this series converges for all $A \in \mathbb{C}^{n\times n}$. Let's see why the matrix exponential is so useful. Consider first the scalar first-order linear differential equation

$$\frac{dx}{dt} = ax \qquad x(0) = x_0 \qquad x \in \mathbb{R}, a \in \mathbb{R}$$

which arises in the simplest chemical kinetics models. The solution is given by

$$x(t) = x_0 e^{at}$$

and this is probably the first and most important differential equation that is discussed in the introductory differential equations course. By defining the *matrix* exponential we have the solution to *all coupled sets* of linear first-order differential equations. Consider the coupled set of linear first-order differential equations

$$\frac{d}{dt}\begin{bmatrix} x_1 \\ x_2 \\ \vdots \\ x_n \end{bmatrix} = \begin{bmatrix} a_{11} & a_{12} & \cdots & a_{1n} \\ a_{21} & a_{22} & \cdots & a_{2n} \\ \vdots & \vdots & \ddots & \vdots \\ a_{n1} & a_{n2} & \cdots & a_{nn} \end{bmatrix} \begin{bmatrix} x_1 \\ x_2 \\ \vdots \\ x_n \end{bmatrix}$$

with initial condition

$$\begin{bmatrix} x_1(0) \\ x_2(0) \\ \vdots \\ x_n(0) \end{bmatrix} = \begin{bmatrix} x_{10} \\ x_{20} \\ \vdots \\ x_{n0} \end{bmatrix}$$

which we express compactly as

$$\frac{dx}{dt} = Ax \qquad x(0) = x_0 \qquad x \in \mathbb{R}^n, A \in \mathbb{R}^{n \times n} \tag{1.15}$$

The payoff for knowing the solution to the scalar version is that we also know the solution to the matrix version. We propose as the solution

$$x(t) = e^{At} x_0 \tag{1.16}$$

Notice that we must put the x_0 after the e^{At} so that the matrix multiplication on the right-hand side is defined and gives the required $n \times 1$ column vector for $x(t)$. Let's establish that this proposed solution is indeed the solution to (1.15). Substituting $t = 0$ to check the initial condition gives

$$x(0) = e^{A0} x_0 = e^0 x_0 = I x_0 = x_0$$

and the initial condition is satisfied. Next differentiating the matrix exponential with respect to scalar time gives

$$\begin{aligned}\frac{d}{dt}e^{At} &= \frac{d}{dt}(I + tA + \frac{t^2}{2!}A^2 + \frac{t^3}{3!}A^3 + \cdots) \\ &= 0 + A + \frac{t}{1!}A^2 + \frac{t^2}{2!}A^3 + \cdots \\ &= A(I + \frac{t}{1!}A^1 + \frac{t^2}{2!}A^2 + \cdots) \\ &= Ae^{At}\end{aligned}$$

We have shown that the scalar derivative formula $d/dt(e^{at}) = ae^{at}$ also holds for the matrix case, $d/dt(e^{At}) = Ae^{At}$. We also could have factored the A to the right instead of the left side in the derivation above to obtain $d/dt(e^{At}) = e^{At}A$. Note that although matrix multiplication does not commute in general, it does commute for certain matrices, such as e^{At} and powers of A. Finally, substituting the derivative result into (1.15) gives

$$\frac{dx}{dt} - \frac{d}{dt}(e^{At}x_0) = (Ae^{At})x_0 = A(e^{At}x_0) = Ax$$

and we see that the differential equation also is satisfied.

Another insight into functions of matrices is obtained when we consider their eigenvalue decomposition. Let $A = S\Lambda S^{-1}$ in which we assume for simplicity that the eigenvalues of A are not repeated so that Λ is diagonal. First we see that powers of A can be written as follows for $p \geq 1$

$$\begin{aligned}A^p &= \underbrace{AA \cdots A}_{p\text{times}} \\ &= (S\Lambda S^{-1})(S\Lambda S^{-1}) \cdots (S\Lambda S^{-1}) \\ &= S(\underbrace{\Lambda\Lambda \cdots \Lambda}_{p\text{times}})S^{-1} \\ &= S\Lambda^p S^{-1}\end{aligned}$$

Substituting the eigenvalue decomposition into the definition of the matrix exponential gives

$$\begin{aligned}
e^{At} &= I + tA + \frac{t^2}{2!}A^2 + \frac{t^3}{3!}A^3 + \cdots \\
&= SS^{-1} + tS\Lambda S^{-1} + \frac{t^2}{2!}S\Lambda^2 S^{-1} + \frac{t^3}{3!}S\Lambda^3 S^{-1} + \cdots \\
&= S(I + t\Lambda + \frac{t^2}{2!}\Lambda^2 + \frac{t^3}{3!}\Lambda^3 + \cdots)S^{-1} \\
e^{At} &= Se^{\Lambda t}S^{-1}
\end{aligned}$$

Therefore, we can determine the time behavior of e^{At} by examining the behavior of

$$e^{\Lambda t} = \begin{bmatrix} e^{\lambda_1 t} & & & \\ & e^{\lambda_2 t} & & \\ & & \ddots & \\ & & & e^{\lambda_n t} \end{bmatrix}$$

and we deduce that e^{At} asymptotically approaches zero as $t \to \infty$ if and only if $\mathrm{Re}(\lambda_i) < 0, i = 1, \ldots, n$. We also know that e^{At} is oscillatory if any eigenvalue has a nonzero imaginary part, and so on.

The matrix exponential is just one example of expanding scalar functions to matrix functions. Any of the transcendental functions (trigonometric functions, hyperbolic trigonometric functions, logarithm, square root, etc.) can be extended to matrix arguments as was shown here for the matrix exponential (Higham, 2008). For example, a square root of a matrix A is any matrix B that satisfies $B^2 = A$. If $A = S\Lambda S^{-1}$, then one solution is $B = S\Lambda^{1/2}S^{-1}$, where $\Lambda^{1/2} = \mathrm{diag}(\lambda_1^{1/2}, \lambda_2^{1/2}, \ldots)$. More generally, $\Lambda^{1/2}$ can be replaced by $Q^*\Lambda^{1/2}Q$ for any unitary matrix Q. Moreover, for any linear scalar differential equation having solutions consisting of these scalar functions, coupled sets of the corresponding linear differential equations are solved by the matrix version of the function.

Bound on e^{At}. When analyzing solutions to dynamic models, we often wish to bound the asymptotic behavior as time increases to infinity. For linear differential equations, this means we wish to bound the asymptotic behavior of e^{At} as $t \to \infty$. We build up to a convenient bound in a few steps. First, for scalar $z \in \mathbb{C}$, we know that

$$|e^z| = \left|e^{\mathrm{Re}(z)+\mathrm{Im}(z)i}\right| = \left|e^{\mathrm{Re}(z)}\right| \left|e^{\mathrm{Im}(z)i}\right| = e^{\mathrm{Re}(z)}$$

Similarly, if we have a diagonal matrix $D \in \mathbb{C}^{n\times n}$, $D = \text{diag}(d_1, d_2, \ldots, d_n)$, then the matrix norm of e^D is

$$\left\|e^D\right\| = \max_{x\neq 0} \frac{\|e^D x\|}{\|x\|} = e^{\lambda}$$

in which $\lambda = \max_i(\text{Re}(d_i))$. In fact, if this max over the real parts of the eigenvalues occurs for index i^*, then $x = e_{i^*}$ achieves the maximum in $\|e^D x\| / \|x\|$. Given a real, nonnegative time argument $t \geq 0$, we also have that

$$\left\|e^{Dt}\right\| = e^{\lambda t} \qquad t \geq 0$$

Next, if the matrix A is diagonalizable, we can use $A = S\Lambda S^{-1}$ to obtain

$$e^{At} = Se^{\Lambda t}S^{-1}$$

and we can obtain a bound by taking norms of both sides

$$\left\|e^{At}\right\| = \left\|Se^{\Lambda t}S^{-1}\right\| \leq \|S\| \left\|e^{\Lambda t}\right\| \left\|S^{-1}\right\|$$

For any nonsingular S, the product $\|S\| \left\|S^{-1}\right\|$ is defined as the condition number of S, denoted $\kappa(S)$. A bound on the norm of e^{At} is therefore

$$\left\|e^{At}\right\| \leq \kappa(S)e^{\lambda t} \qquad t \geq 0 \qquad A \text{ diagonalizable}$$

in which $\lambda = \max_i \text{Re}(\lambda_i) = \max(\text{Re}(\text{eig}(A)))$. So this leaves only the case in which A is not diagonalizable. In the general case we use the Schur form $A = QTQ^*$, with T upper triangular. Van Loan (1977) shows that[3]

$$\left\|e^{At}\right\| \leq e^{\lambda t} \sum_{k=0}^{n-1} \frac{\|Nt\|^k}{k!} \qquad t \geq 0 \tag{1.17}$$

in which $N = T - \Lambda$ where Λ is the diagonal matrix of eigenvalues and N is strictly upper triangular, i.e., has zeros on as well as below the diagonal. Note that this bound holds for any $A \in \mathbb{C}^{n\times n}$. Van Loan (1977) also shows that this is a fairly tight bound compared to some popular alternatives. If we increase the value of λ by an arbitrarily small amount, we can obtain a looser bound, but one that is more convenient for analysis. For any λ' satisfying

$$\lambda' > \max(\text{Re}(\text{eig}(A)))$$

[3]Note that there is a typo in (2.11) in Van Loan (1977), which is corrected here.

there is a constant $c > 0$ such that

$$\left\| e^{At} \right\| \leq c e^{\lambda' t} \qquad t \geq 0 \tag{1.18}$$

This result holds also for any $A \in \mathbb{C}^{n \times n}$. Note that the constant c depends on the matrix A. Establishing this result is discussed in Exercise 1.71. To demonstrate one useful consequence of this bound, consider the case in which all eigenvalues of A have strictly negative real parts. Then there exists λ' such that $\text{Re}(\text{eig}(A)) < \lambda' < 0$, and (1.18) tells us that $e^{At} \to 0$ exponentially fast as $t \to \infty$ for the entire class of "stable" A matrices. We do not need to assume that A has distinct eigenvalues or is diagonalizable, for example, to reach this conclusion.

1.5.2 Optimizing Quadratic Functions

Optimization is a large topic of fundamental importance in many engineering activities such as process design, process control, and process operations. Here we would like to introduce some of the important concepts of optimization in the simple setting of quadratic functions. You now have the required linear algebra tools to make this discussion accessible.

Scalar argument. The reader is undoubtedly familiar with finding the maximum and minimum of scalar functions by taking the first derivative and setting it to zero. For conciseness, we restrict attention to (unconstrained) minimization, and we are interested in the problem[4]

$$\min_x f(x)$$

What do we expect of a solution to this problem? A point x^0 is termed a minimizer if $f(x^0) \leq f(p)$ for all p. A minimizer x^0 is unique if no other point has this property. In other words, the minimizer x^0 is unique provided $f(x^0) < f(p)$ for all $p \neq x^0$. We call x^0 the minimizer and $f^0 = f(x^0)$ the (optimal) value function. Note that to avoid confusion $f^0 = \min_x f(x)$ is called the "solution" to the problem, even though x^0 is usually the item of most interest, and $x^0 = \arg\min_x f(x)$ is called the "argument of the solution."

We wish to know when the minimizer exists and is unique, and how to compute it. We consider first the real, scalar-valued quadratic function of the real, scalar argument x, $f(x) = (1/2)ax^2 + bx + c$, with

[4]We do not lose generality with this choice; if the problem of interest is instead maximization, use the following identity to translate: $\max_x f(x) = \min_x -f(x)$.

$a, b, c, x \in \mathbb{R}$. Putting the factor of $1/2$ in front of the quadratic term is a convention to simplify various formulas to be derived next. If we take the derivative and set it to zero we obtain

$$f(x) = (1/2)ax^2 + bx + c$$
$$\frac{d}{dx}f(x) = ax + b = 0$$
$$x^0 = -b/a$$

This last result for x^0 is at least well defined provided $a \neq 0$. But if we are interested in minimization, we require more: $a > 0$ is required for a unique solution to the problem $\min_x f(x)$. Indeed, taking a second derivative of $f(\cdot)$ gives $d^2/dx^2 f(x) = a$. The condition $a > 0$ is usually stated in beginning calculus courses as: the function is concave upward. This idea is generalized to the condition that the function is *strictly convex*, which we define next. Evaluating $f(x)$ at the proposed minimizer gives $f^0 = f(x^0) = -(1/2)b^2/a + c$.

Convex functions. Generalizing the simple notion of a function having positive curvature (or being concave upward) to obtain existence and uniqueness of the minimizer leads to the concept of a convex function, which is defined as follows (Rockafellar and Wets, 1998, p. 38).

Definition 1.18 (Convex function). Let function $f(\cdot)$ be defined on all reals. Consider two points x, y and a scalar α that satisfy $0 \leq \alpha \leq 1$. The function f is convex if the following inequality holds for all x, y and $\alpha \in [0, 1]$

$$f(\alpha x + (1 - \alpha)y) \leq \alpha f(x) + (1 - \alpha)f(y)$$

Figure 1.5 shows a convex function. Notice that if you draw a straight line connecting any two points on the function curve, if the function is convex the straight line lies *above* the function. We say the function $f(\cdot)$ is STRICTLY CONVEX if the inequality is strict for all $x \neq y$ and $\alpha \in (0, 1)$

$$f(\alpha x + (1 - \alpha)y) < \alpha f(x) + (1 - \alpha)f(y)$$

Notice that x and y are restricted to be nonequal and α is restricted to lie in the *open* interval $(0, 1)$ in the definition of strict convexity (or no function would be strictly convex).

That strict convexity is sufficient for uniqueness of the minimizer is established readily as follows. Assume that one has found a (possibly nonunique) minimizer of $f(\cdot)$, denoted x^0, and consider another point

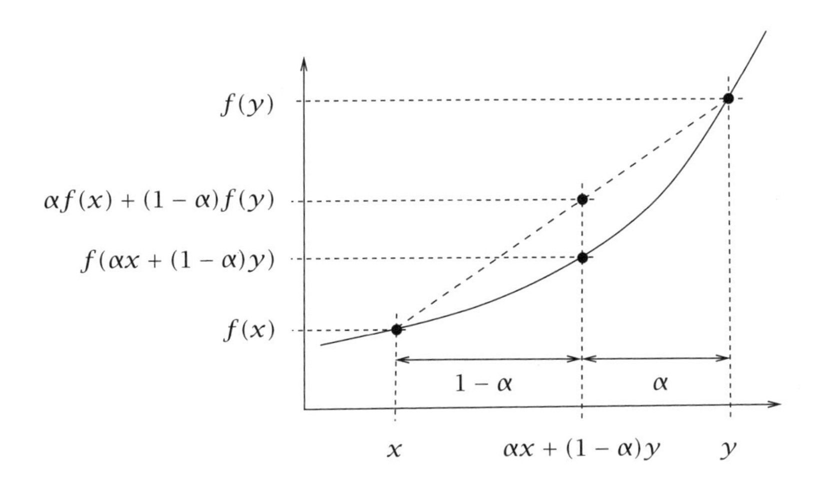

Figure 1.5: Convex function. The straight line connecting two points on the function curve lies above the function; $\alpha f(x) + (1-\alpha)f(y) \geq f(\alpha x + (1-\alpha)y)$ for all x, y.

$p \neq x^0$. We know that $f(p)$ cannot be less than $f(x^0)$ or we contradict optimality of x^0. We wish to rule out $f(p) = f(x^0)$, also, because equality implies that the minimizer is not unique. If $f(\cdot)$ is strictly convex and $f(p) = f(x^0)$, we have that

$$f(\alpha x^0 + (1-\alpha)p) < \alpha f(x^0) + (1-\alpha)f(p) = f(x^0)$$

So for all $z = \alpha x^0 + (1-\alpha)p$ with $\alpha \in (0,1)$, $z \neq x^0$ and $f(z) < f(x^0)$, which also contradicts optimality of x^0. Therefore $f(x^0) < f(p)$ for all $p \neq x^0$, and the minimizer is unique. Notice that the definition of convexity does not require $f(\cdot)$ to have even a first derivative, let alone a second derivative as required when using a curvature condition for uniqueness. But if $f(\cdot)$ is quadratic, then strict convexity is equivalent to positive curvature as discussed in Exercise 1.72.

Vector argument. We next take real-valued vector $x \in \mathbb{R}^n$, and the general real, scalar-valued, quadratic function is $f(x) = (1/2)x^T Ax + b^T x + c$ with $A \in \mathbb{R}^{n \times n}$, $b \in \mathbb{R}^n$, and $c \in \mathbb{R}$. Without loss of generality, we can assume A is symmetric.[5] We know that the eigenvalues of a

[5]If A is not symmetric, show that replacing A with the symmetric $\widetilde{A} = (1/2)(A + A^T)$ does not change the function $f(\cdot)$.

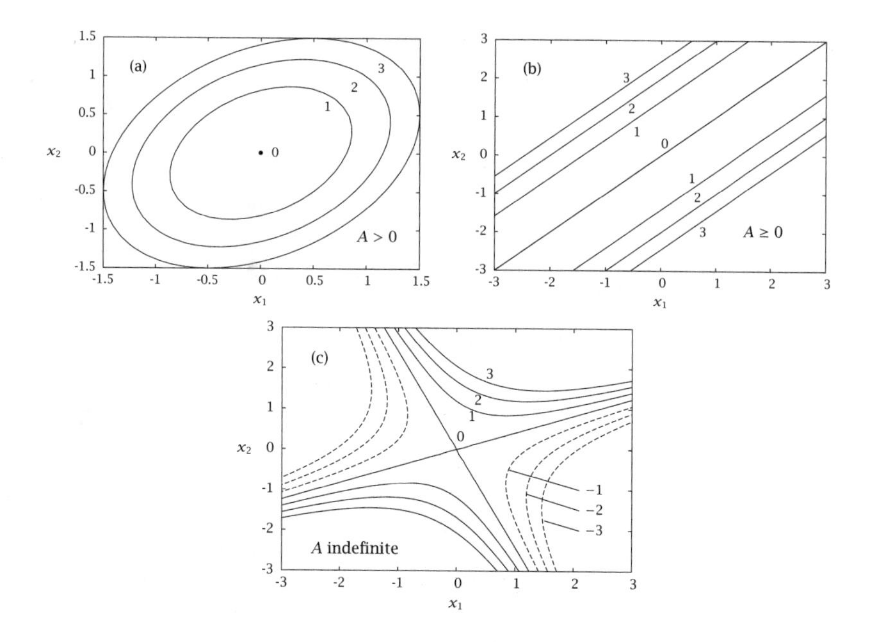

Figure 1.6: Contours of constant $f(x) = x^T Ax$; (a) $A > 0$ (or $A < 0$), ellipses; (b) $A \geq 0$ (or $A < 0$), straight lines; (c) A indefinite, hyperbolas. The coordinate axes are aligned with the contours if and only if A is diagonal.

symmetric matrix are real (see Theorem 1.16); the following cases are of interest and cover all possibilities for symmetric A: (a) $A > 0$ (or $A < 0$), (b) $A \geq 0$ (or $A \leq 0$), and (c) A indefinite. Figure 1.6 shows contours of the quadratic functions that these A generate for $x \in \mathbb{R}^2$. Since A is the parameter of interest here, we set $b = 0$, $c = 0$.[6] The positive cost contours are concentric ellipses for the case $A > 0$. The contour for $f = 0$ is the origin, and minimizing f for $A > 0$ has the origin as a unique minimizer. This problem corresponds to finding the bottom of a bowl. If $A < 0$, contours remain ellipses, but the sign of the contour value changes. The case $A > 0$ has the origin as a unique *maximizer*. This problem corresponds to finding the top of a mountain.

For the case $A \geq 0$, the positive contours are straight lines. The line through the origin corresponds to $f = 0$. All points on this line

[6]Note that c merely shifts the function $f(\cdot)$ up and down, and b merely shifts the origin, so they are not important to the *shape* of the contours of the quadratic function.

are minimizers for $f(\cdot)$ in this case, and the minimizer is nonunique. The quadratic function is convex but not strictly convex. The function corresponds to a long valley. As before, if $A \leq 0$, contours remain straight lines, but the sign of the contour value changes. For $A \leq 0$, the maximizer exists, but is not unique. The function is now a ridge. And some specialized techniques for numerically finding optima with badly conditioned functions approaching this case are known as "ridge regression."

For indefinite A, Figure 1.6 shows that the contours are hyperbolas. The origin is termed a saddle point in this case, because the function resembles a horse's saddle, or a mountain pass if one prefers to maintain the topography metaphor. Note that $f(\cdot)$ increases without bound in the northeast and southwest directions, but decreases without bound in the southeast and northwest directions. So *neither* a minimizer nor a maximizer exists for the indefinite case. But there is an important class of problems for which the origin is the solution. These are the minmax or maxmin problems. Consider the two problems

$$\max_{x_1} \min_{x_2} f(x) \qquad \min_{x_2} \max_{x_1} f(x) \tag{1.19}$$

These kinds of problems are called noncooperative *games*, and *players* one and two are optimizing over decision variables x_1 and x_2, respectively. In this type of noncooperative problem, player one strives to maximize function f while player two strives to minimize it. Noncooperative games arise in many fields, especially as models of economic behavior. In fact, von Neumann and Morgenstern (1944) originally developed game theory for understanding economic behavior in addition to other features of classical games such as bluffing in poker. These kinds of problems also are useful in worst-cases analysis and design. For example, the outer problem can represent a standard design optimization while the inner problem finds the worst-case scenario over some set of uncertain model parameters.

Another important engineering application of noncooperative games arises in the introduction of Lagrange multipliers and the Lagrangian function when solving *constrained* optimization problems. For the quadratic function shown in Figure 1.6 (c), Exercise 1.74 asks you to establish that the origin is the unique solution to both problems in (1.19). The solution to a noncooperative game is known as a Nash equilibrium or Nash point in honor of the mathematician John Nash who established some of the early fundamental results of game theory (Nash, 1951).

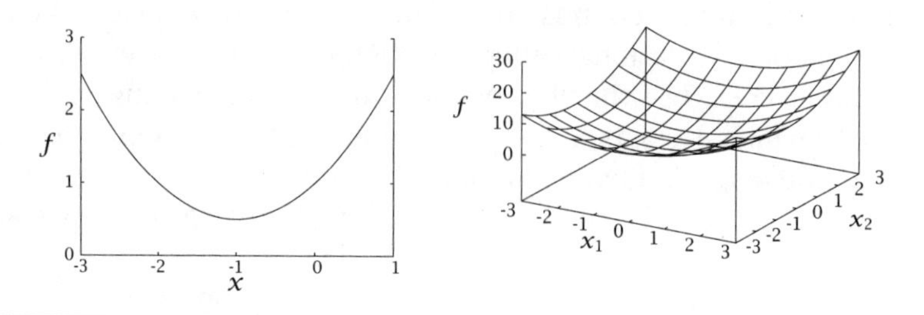

	Scalar	Vector
$f(x)$	$(1/2)ax^2 + bx + c$	$(1/2)x^TAx + b^Tx + c$
$\dfrac{d}{dx}f$	$ax + b$	$Ax + b$
$\dfrac{d^2}{dx^2}f$	a	A
x^0	$-b/a$	$-A^{-1}b$
f^0	$-(1/2)b^2/a + c$	$-(1/2)b^TA^{-1}b + c$
$f(x)$	$(1/2)a(x - x^0)^2 + f^0$	$(1/2)(x - x^0)^TA(x - x^0) + f^0$

Table 1.1: Quadratic function of scalar and vector argument; $a > 0$, A positive definite.

Finally, to complete the vector *minimization* problem, we restrict attention to the case $A > 0$. Taking two derivatives in this case produces

$$\begin{aligned}
f(x) &= (1/2)x^TAx + b^Tx + c \\
\frac{d}{dx}f(x) &= (1/2)(Ax + A^Tx) + b = Ax + b \\
\frac{d^2}{dx^2}f(x) &= A
\end{aligned}$$

Setting $df/dx = 0$ and solving for x^0, and then evaluating $f(x^0)$ gives

$$x^0 = -A^{-1}b \qquad f^0 = -(1/2)b^TA^{-1}b + c$$

These results for the scalar and vector cases are summarized in Table 1.1. Notice also in the last line of the table that one can reparameterize the function $f(\cdot)$ in terms of x^0 and f^0, in place of b and c, which is often useful.

Revisiting linear least squares. Consider again the linear least-squares problem of Section 1.3.7

$$\min_x (1/2) \left\| \widetilde{A}x - \widetilde{b} \right\|^2$$

where we have changed $Ax - b$ to $\widetilde{A}x - \widetilde{b}$ to not conflict with the notation of this section. We see that least squares is the special case of a quadratic function with the parameters

$$A = \widetilde{A}^T \widetilde{A} \qquad b = -\widetilde{A}^T \widetilde{b} \qquad c = (1/2)\widetilde{b}^T \widetilde{b}$$

Obviously A is symmetric in the least-squares problem. We have already derived the fact that $\widetilde{A}^T \widetilde{A} > 0$ if the columns of $\widetilde{A}$ are independent in the discussion of the SVD in Section 1.4.7. So independent columns of $\widetilde{A}$ correspond to case (a) in Figure 1.6. If the columns of $\widetilde{A}$ are not independent, then $A \geq 0$, and we are in case (b) and lose uniqueness of the least-squares solution. See Exercise 1.64 for a discussion of this case. It is not possible for a least-squares problem to be in case (c), which is good, because we are posing a minimization problem in least squares. So the solution to a least-squares problem always exists.

1.5.3 Vec Operator and Kronecker Product of Matrices

We introduce two final matrix operations that prove highly useful in applications, but often are neglected in an introductory linear algebra course. These are the vec operator and the Kronecker product of two matrices.

The vec operator. For $A \in \mathbb{R}^{m \times n}$, the vec operator is defined as the restacking of the matrix by its columns into a single large column vector

$$A = \begin{bmatrix} A_{11} & A_{12} & \cdots & A_{1n} \\ A_{21} & A_{22} & \cdots & A_{2n} \\ & & \ddots & \\ A_{m1} & A_{m2} & \cdots & A_{mn} \end{bmatrix} \qquad \text{vec}A = \begin{bmatrix} A_{11} \\ A_{21} \\ \vdots \\ A_{m1} \\ \hline A_{12} \\ A_{22} \\ \vdots \\ A_{m2} \\ \hline \vdots \\ \hline A_{1n} \\ A_{2n} \\ \vdots \\ A_{mn} \end{bmatrix}$$

Note that vec$A$ is a column vector in $\mathbb{R}^{mn}$. If we denote the n column vectors of A as a_i, we can express the vec operator more compactly using column partitioning as

$$A = \begin{bmatrix} a_1 & a_2 & \cdots & a_n \end{bmatrix} \qquad \text{vec}A = \begin{bmatrix} a_1 \\ a_2 \\ \vdots \\ a_n \end{bmatrix}$$

Matrix Kronecker product. For $A \in \mathbb{R}^{m \times n}$ and $B \in \mathbb{R}^{p \times q}$, the Kronecker product of A and B, denoted $A \otimes B$, is defined as

$$A \otimes B = \begin{bmatrix} A_{11}B & A_{12}B & \cdots & A_{1n}B \\ A_{21}B & A_{22}B & \cdots & A_{2n}B \\ & & \ddots & \\ A_{m1}B & A_{m2}B & \cdots & A_{mn}B \end{bmatrix} \tag{1.20}$$

Note that the Kronecker product is defined for all matrices A and B, and the matrices do not have to conform as in normal matrix multiplication. By counting the number of rows and columns in the definition above, we see that matrix $A \otimes B \in \mathbb{R}^{mp \times nq}$. Notice also that the vector outer product, defined in Section 1.2.5, is a special case of this more general matrix Kronecker product.

Some useful identities. We next establish four useful identities involving the vec operator and Kronecker product.

$$\operatorname{vec}(ABC) = (C^T \otimes A)\operatorname{vec}B \tag{1.21}$$
$$(A \otimes B)(C \otimes D) = (AC) \otimes (BD) \quad A, C \text{ conform}, B, D \text{ conform} \tag{1.22}$$
$$(A \otimes B)^T = (A^T \otimes B^T) \tag{1.23}$$
$$(A \otimes B)^{-1} = A^{-1} \otimes B^{-1} \qquad A \text{ and } B \text{ invertible} \tag{1.24}$$

Establishing (1.21). Let $A \in \mathbb{R}^{m\times n}$, $B \in \mathbb{R}^{n\times p}$, and $C \in \mathbb{R}^{p\times q}$. Let the column partitions of matrices B and C be given by

$$B = \begin{bmatrix} b_1 & b_2 & \dots & b_p \end{bmatrix} \qquad C = \begin{bmatrix} c_1 & c_2 & \dots & c_q \end{bmatrix}$$

We know from the rules of matrix multiplication that the jth column of the product $ABC = (AB)C$ is given by ABc_j. So when we stack these columns we obtain

$$\operatorname{vec}(ABC) = \begin{bmatrix} ABc_1 \\ ABc_2 \\ \vdots \\ ABc_q \end{bmatrix}$$

Now we examine the right-hand side of (1.21). We have from the definitions of vec operator and Kronecker product

$$(C^T \otimes A)\operatorname{vec}B = \begin{bmatrix} c_{11}A & c_{21}A & \cdots & c_{p1}A \\ c_{12}A & c_{22}A & \cdots & c_{p2}A \\ & & \ddots & \\ c_{1q}A & c_{2q}A & \cdots & c_{pq}A \end{bmatrix} \begin{bmatrix} b_1 \\ b_2 \\ \vdots \\ b_p \end{bmatrix}$$

The jth row of this partitioned matrix multiplication can be rearranged as follows

$$\begin{aligned} c_{1j}Ab_1 + c_{2j}Ab_2 + \cdots + c_{pj}Ab_p &= \begin{bmatrix} Ab_1 & Ab_2 & \cdots & Ab_p \end{bmatrix} \begin{bmatrix} c_{1j} \\ c_{2j} \\ \vdots \\ c_{pj} \end{bmatrix} \\ &= A\begin{bmatrix} b_1 & b_2 & \cdots & b_p \end{bmatrix} c_j \\ &= ABc_j \end{aligned}$$

Inserting this result into the previous equation gives

$$(C^T \otimes A)\text{vec}B = \begin{bmatrix} ABc_1 \\ ABc_2 \\ \vdots \\ ABc_q \end{bmatrix}$$

which agrees with the expression for $\text{vec}(ABC)$.

Establishing (1.22). Here we let $A \in \mathbb{R}^{m\times n}$, $B \in \mathbb{R}^{p\times q}$, $C \in \mathbb{R}^{n\times r}$, $D \in \mathbb{R}^{q\times s}$ so that A, C conform and B, D conform. Let $c_1, c_2, \ldots c_r$ be the column vectors of matrix C. We know from the rules of matrix multiplication that the jth (block) column of the product $(A\otimes B)(C\otimes D)$ is given by $A \otimes B$ times the jth (block) column of $C \otimes D$, which is

$$\begin{aligned}
\begin{bmatrix} A_{11}B & \cdots & A_{1n}B \\ \vdots & \ddots & \vdots \\ A_{m1}B & \cdots & A_{mn}B \end{bmatrix} \begin{bmatrix} C_{1j}D \\ \vdots \\ C_{nj}D \end{bmatrix} &= \begin{bmatrix} A_{11}B & \cdots & A_{1n}B \\ \vdots & \ddots & \vdots \\ A_{m1}B & \cdots & A_{mn}B \end{bmatrix} \begin{bmatrix} C_{1j} \\ \vdots \\ C_{nj} \end{bmatrix} D \\
&= (Ac_j \otimes B)D \\
&= (Ac_j) \otimes (BD)
\end{aligned}$$

Since this is the jth (block) column of $(A\otimes B)(C\otimes D)$, the entire matrix is

$$\begin{aligned}
(A \otimes B)(C \otimes D) &= \begin{bmatrix} (Ac_1) \otimes (BD) & (Ac_2) \otimes (BD) & \cdots & (Ac_r) \otimes (BD) \end{bmatrix} \\
&= \begin{bmatrix} Ac_1 & Ac_2 & \cdots & Ac_r \end{bmatrix} \otimes (BD) \\
&= A\begin{bmatrix} c_1 & c_2 & \cdots & c_r \end{bmatrix} \otimes (BD) \\
&= (AC) \otimes (BD)
\end{aligned}$$

and the result is established.

Establishing (1.23). Given $A \in \mathbb{R}^{m\times n}$ and $B \in \mathbb{R}^{p\times q}$, from the definition of transpose and cross product we have that

$$(A \otimes B)^T = \begin{bmatrix} A_{11}B & A_{12}B & \cdots & A_{1n}B \\ A_{21}B & A_{22}B & \cdots & A_{2n}B \\ & & \ddots & \\ A_{m1}B & A_{m2}B & \cdots & A_{mn}B \end{bmatrix}^T$$
$$= \begin{bmatrix} A_{11}B^T & A_{21}B^T & \cdots & A_{m1}B^T \\ A_{12}B^T & A_{22}B^T & \cdots & A_{m2}B^T \\ & & \ddots & \\ A_{1n}B^T & A_{2n}B^T & \cdots & A_{mn}B^T \end{bmatrix}$$
$$= A^T \otimes B^T$$

Establishing (1.24). Apply (1.22) to the following product and obtain

$$(A \otimes B)(A^{-1} \otimes B^{-1}) = (AA^{-1}) \otimes (BB^{-1}) = I \otimes I = I$$

and therefore

$$(A \otimes B)^{-1} = A^{-1} \otimes B^{-1}$$

Eigenvalues, singular values, and rank of the Kronecker product. When solving matrix equations, we will want to know about the rank of the Kronecker product $A \otimes B$. Since rank is closely tied to the singular values, and these are closely tied to the eigenvalues, the following identities prove highly useful.

$$\operatorname{eig}(A \otimes B) = \operatorname{eig}(A)\operatorname{eig}(B) \qquad A \text{ and } B \text{ square} \tag{1.25}$$
$$\sigma(A \otimes B) = \sigma(A)\sigma(B) \qquad \textit{nonzero} \text{ singular values} \tag{1.26}$$
$$\operatorname{rank}(A \otimes B) = \operatorname{rank}(A)\operatorname{rank}(B) \tag{1.27}$$

Given our previous identities, these three properties are readily established. Let A and B be square of order m and n, respectively. Let (λ, v) and (μ, w) be eigenpairs of A and B, respectively. We have that $Av \otimes Bw = (\lambda v) \otimes (\mu w) = (\lambda\mu)(v \otimes w)$. Using (1.22) on $Av \otimes Bw$ then gives

$$(A \otimes B)(v \otimes w) = (\lambda\mu)(v \otimes w)$$

and we conclude that (nonzero) mn-vector $(v \otimes w)$ is an eigenvector of $A \otimes B$ with product $\lambda\mu$ as the corresponding eigenvalue. This establishes (1.25). For the nonzero singular values, recall that the nonzero singular values of real, nonsquare matrix A are the nonzero eigenvalues of AA^T

(and A^TA). We then have for $\sigma(A)$ and $\lambda(A)$ denoting nonzero singular and eigenvalues, respectively

$$\begin{aligned}
\sigma(A \otimes B) &= \lambda((A \otimes B)(A \otimes B)^T) \\
&= \lambda((A \otimes B)(A^T \otimes B^T)) \\
&= \lambda((AA^T) \otimes (BB^T)) \\
&= \lambda(AA^T)\lambda(BB^T) \\
&= \sigma(A)\sigma(B)
\end{aligned}$$

which establishes (1.26). Since the number of nonzero singular values of a matrix is equal to the rank of the matrix, we then also have (1.27).

Properties (1.21)–(1.27) are all that we require for the material in this text, but the interested reader may wish to consult Magnus and Neudecker (1999) for a more detailed discussion of Kronecker products.

Solving linear matrix equations. We shall find the properties (1.21)–(1.27) highly useful when dealing with complex maximum-likelihood estimation problems in Chapter 4. But to provide here a small illustration of their utility, consider the following linear matrix equation for the unknown matrix X

$$AXB = C$$

in which neither A nor B is invertible. The equations are *linear* in X, so they should be solvable as some form of linear algebra problem. But since we cannot operate with A^{-1} from the left, nor B^{-1} from the right, it seems difficult to isolate X and solve for it. This is an example where the linear equations are simply not packaged in a convenient form for solving them. But if we apply the vec operator and use (1.21) we have

$$(B^T \otimes A)\text{vec}X = \text{vec}C$$

Note that this is now a standard linear algebra problem for the unknown *vector* vec$X$. We can examine the rank, and linear independence of the rows and columns of $B^T \otimes A$, to determine the existence and uniqueness of the solution vec$X$, and whether we should solve a least-squares problem or minimum-norm problem. After solution, the vec$X$ column vector can then be restacked into its original matrix form X if desired. Exercise 1.77 provides further discussion of solving $AXB = C$.

As a final example, in Chapter 2 we will derive the matrix Lyapunov equation, which tells us about the stability of a linear dynamic system,

$$A^TS + SA = -Q$$

in which matrices A and Q are given, and S is the unknown. One way to think about solving the matrix Lyapunov equation is to apply the vec operator to obtain

$$((I \otimes A^T) + (A^T \otimes I))\text{vec}S = -\text{vec}Q$$

and then solve this linear algebra problem for vec$S$. Although this approach is useful for characterizing the solution, given the special structure of the Lyapunov equation, more efficient numerical solution methods are available and coded in standard software. See the function `lyap(A',Q)` in Octave or MATLAB, for example. Exercise 1.78 asks you to solve a numerical example using the Kronecker product approach and compare your result to the `lyap` function.

1.6 Exercises

Exercise 1.1: Inner product and angle in $\mathbb{R}^2$

Consider the two vectors $a, b \in \mathbb{R}^2$ shown in Figure 1.7 and let θ denote the angle between them. Show the usual inner product and norm formulas

$$(a, b) = \sum_i a_i b_i \qquad \|a\| = \sqrt{(a, a)}$$

satisfy the following relationship with the angle

$$\cos\theta = \frac{(a, b)}{\|a\| \, \|b\|}$$

This relationship allows us to generalize the concept of an angle between vectors to any inner product space.

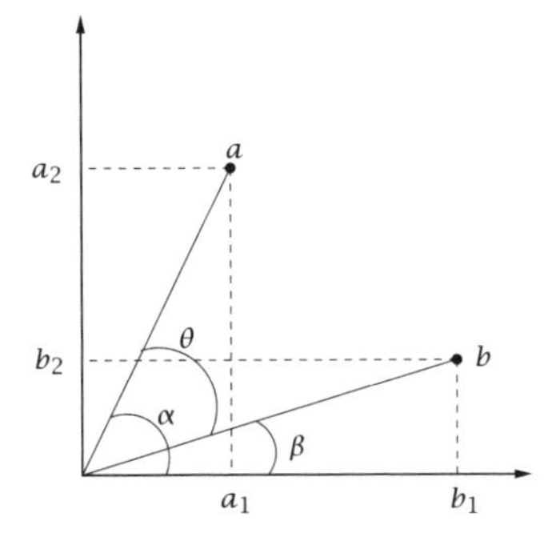

Figure 1.7: Two vectors in $\mathbb{R}^2$ and the angle between them.

Exercise 1.2: Scaling and vector norm

Consider the vector $x \in \mathbb{R}^2$, whose elements are the temperature (in K) and pressure (in Pa) in a reactor. A typical value of x would be $\begin{bmatrix} 300 \\ 1.0 \times 10^6 \end{bmatrix}$.

(a) Let $y_1 = \begin{bmatrix} 310 \\ 1.0 \times 10^6 \end{bmatrix}$ and $y_2 = \begin{bmatrix} 300 \\ 1.2 \times 10^6 \end{bmatrix}$ be two measurements of the state of the reactor. Use the Euclidean norm to calculate the error $\|y - x\|$ for the two values of y. Do you think that the calculated errors give a meaningful idea of the difference between y_1 and y_2?

(b) A problem with the Euclidean norm is that it takes no account of different scales for different elements of a vector. Consider the following formula

$$\|x\|_w = \sqrt{\sum_{i=1}^{n} |x_i|^2 w_i}$$

where x_i is the ith component of the vector x. Show that this formula is a norm if and only if $w_i > 0$ for all i. This is known as a weighted norm, with weight vector w.

(c) Propose a weight vector that is appropriate for the example in part (a). Justify your choice.

Exercise 1.3: Linear independence

Verify that the following sets are LI.

(a) $e_1 = [0, 1, 0]^T, e_2 = [2, 0, 1]^T, e_3 = [1, 1, 1]^T$.

(b) $e_1 = [1 + i, 1 - i, 0]^T, e_2 = [1 + 2i, 1 - i, 0]^T, e_3 = [0, 0, 2]^T$.

Hint: Express $\alpha_1 e_1 + \alpha_2 e_2 + \alpha_3 e_3 = 0$. Taking inner products of this equation with e_1, e_2, and e_3 yields three linear equations for the α_i.

Exercise 1.4: Gram-Schmidt procedure

Using Gram-Schmidt orthogonalization, obtain ON sets from the LI sets given in the previous problem.

Exercise 1.5: Failure of Gram-Schmidt

The Gram-Schmidt process will fail if the initial set of vectors is not LI.

(a) Construct a set of three vectors in $\mathbb{R}^3$ that are not LI and apply Gram-Schmidt, pinpointing where it fails.

(b) Similarly, in an n-dimensional space, no more than n LI vectors can be found. Construct a set of four vectors in $\mathbb{R}^3$ and use Gram-Schmidt to show that if three of the vectors are LI, then a fourth orthogonal vector cannot be found.

Exercise 1.6: Linear independence and expressing one vector as a linear combination of others

We often hear that a set of vectors is linearly independent if none of the vectors in the set can be expressed as a linear combination of the remaining vectors. Although the statement is correct, as a *definition* of linear independence, this idea is a bit unwieldy because we do not know a priori which vector(s) in a linearly dependent set is(are) expressible as a linear combination of the others.

The following statement is a more precise variation on this theme. Given the vectors $\{x_i\}, i = 1, \ldots, k, x_i \in \mathbb{R}^n$ are linearly independent and the vectors $\{x_i, a\}$ are linearly dependent, show that a can be expressed as a linear combination of the x_i.

Using the definition of linear independence provided in the text, prove this statement.

Exercise 1.7: Some properties of subspaces

Establish the following properties

(a) The zero element is an element of every subspace.

(b) The span of any set of j elements of $\mathbb{R}^n$ is a subspace (of the linear space $\mathbb{R}^n$).

(c) Except for the zero subspace, a subspace cannot have a finite, largest element. Hence, every subspace, except the zero subspace, is unbounded.

Exercise 1.8: Some subspaces in 2-D and 3-D

(a) Consider the line in $\mathbb{R}^2$

$$S = \left\{ y \mid y = \alpha \begin{bmatrix} 1 \\ 1 \end{bmatrix}, \quad \alpha \in \mathbb{R} \right\}$$

Draw a sketch of S. Show that S is a subspace.

(b) Next consider the shifted line

$$S' = \left\{ y \mid y = \begin{bmatrix} 0 \\ 1 \end{bmatrix} + \alpha \begin{bmatrix} 1 \\ 1 \end{bmatrix}, \quad \alpha \in \mathbb{R} \right\}$$

Draw a sketch of S'. Show that S' is *not* a subspace.

(c) Describe all of the subspaces of $\mathbb{R}^3$.

Exercise 1.9: Permutation matrices

(a) Given the matrix

$$P = \begin{bmatrix} 1 & 0 & 0 \\ 0 & 0 & 1 \\ 0 & 1 & 0 \end{bmatrix}$$

show that PA interchanges the second and third rows of A for any 3×3 matrix. What does AP do?

(b) A general permutation matrix involving p row exchanges can be written $P = P_p P_{p-1} \ldots P_2 P_1$ where P_i corresponds to a simple row exchange as above. Show that P is orthogonal.

Exercise 1.10: Special matrices

Consider operations on vectors in $\mathbb{R}^2$.

(a) Construct a matrix operator A that multiplies the horizontal (x_1) component of a vector by 2, but leaves its vertical component (x_2) unchanged.

(b) Construct a matrix operator B that rotates a vector counterclockwise by an angle of $2\pi/3$.

(c) Compute and draw ABx and BAx for $x = \begin{bmatrix} 1 \\ 2 \end{bmatrix}$.

(d) Show that $B^3 = I$. With drawings, show how this makes geometric sense.

Exercise 1.11: Integral operators and matrices

Many problems can be posed with the use of *integral* operators. Consider the integral operator K defined by its action on the function x by the following equation:

$$K\{x(s)\} = \int_0^1 k(t,s)x(s)ds$$

where $k(t,s)$ is a known function called the KERNEL of the operator.

(a) Show that K is a linear operator.

(b) Read Section 2.4.1. Use the usual (i.e., unweighted) inner product on the interval $(0,1)$ and show that if $k(t,s) = \overline{k(s,t)}$, then K is self-adjoint.

(c) An integral can be approximated as a sum, so the above integral operator can be approximated like this:

$$K_a\{x(i\Delta t)\} = \sum_{j=1}^{N} k(i\Delta t, j\Delta t)x(j\Delta t)\Delta t, i = 1, N$$

where $\Delta t = 1/N$. Show how this approximation can be rewritten as a standard matrix-vector product. What is the matrix approximation to the integral operator?

Exercise 1.12: Projections and matrices

Given a unit vector n, use index notation (and the summation convention) to simplify the following expressions:

(a) $(nn^T)(nn^T)u$ for any vector u. Recalling that nn^T is the projection operator, what is the geometric interpretation of this result?

(b) $(I - 2nn^T)^2$. What is the geometric interpretation of this result?

Exercise 1.13: Use the source, Luke

Someone in your research group wrote a computer program that takes an n-vector input, $x \in \mathbb{R}^n$ and returns an m-vector output, $y \in \mathbb{R}^m$.

$$y = f(x)$$

All we know about the function f is that it is *linear*.

The code was compiled and now the source code has been lost; the author has graduated and won't respond to our email. We need to create the source code for function f so we can compile it for our newly purchased hardware, which no longer runs the old compiled code. To help us accomplish this task, all we can do is execute the function on the old hardware.

(a) How many function calls do you need to make before you can write the source code for this function?

(b) What inputs do you choose, and how do you construct the linear function f from the resulting outputs?

(c) To make matters worse, your advisor has a hot new project idea that requires you to write a program to evaluate the inverse of this linear function,

$$x = f^{-1}(y)$$

and has asked you if this is possible. How do you respond? Give a complete answer about the existence and uniqueness of x given y.

Exercise 1.14: The range and null space of a matrix are subspaces

Given $A \in \mathbb{R}^{m\times n}$, show that the sets $R(A)$ and $N(A)$ satisfy the properties of a subspace, (1.1), and therefore $R(A)$ and $N(A)$ are subspaces.

Exercise 1.15: Null space of A is orthogonal to range of A^T

Given $A \in \mathbb{R}^{m\times n}$ show that $N(A) \perp R(A^T)$.

Exercise 1.16: Rank of a dyad

What is the rank of the $n \times n$ dyad uv^T?

Exercise 1.17: Partitioned matrix inversion formula

(a) Let the matrix A be partitioned as

$$A = \begin{bmatrix} B & C \\ D & E \end{bmatrix}$$

in which B, C, D, E are suitably dimensioned matrices and B and E are square. Derive a formula for A^{-1} in terms of B, C, D, E by block Gaussian elimination. Check your answer with the CRC Handbook formula (Selby, 1973).

(b) What if B^{-1} does not exist? What if E^{-1} does not exist? What if both B^{-1} and E^{-1} do not exist?

Exercise 1.18: The four fundamental subspaces

Find bases for the four fundamental subspaces associated with the following matrices

$$A = \begin{bmatrix} 1 & 2 \\ 3 & 6 \end{bmatrix} \qquad B = \begin{bmatrix} 0 & 0 \\ 0 & 0 \end{bmatrix} \qquad C = \begin{bmatrix} 1 & 1 & 0 & 0 \\ 0 & 1 & 0 & 1 \end{bmatrix}$$

Exercise 1.19: Zero is orthogonal to many vectors

Prove that if

$$x \cdot z = y \cdot z \qquad \text{for all } z \in \mathbb{R}^n$$

then

$$x = y$$

or, equivalently, prove that if

$$u \cdot v = 0 \qquad \text{for all } v \in \mathbb{R}^n$$

then

$$u = 0$$

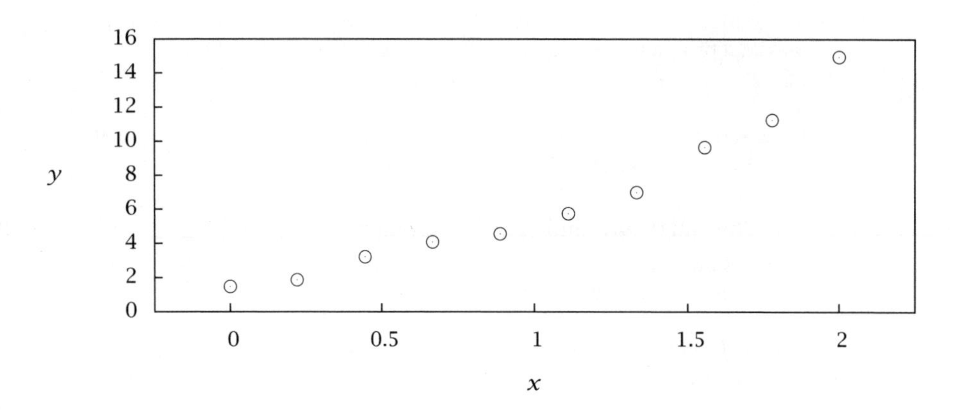

Figure 1.8: Experimental measurements of variable y versus x.

Exercise 1.20: Existence and uniqueness

Find matrices A for which the number of solutions to $Ax = b$ is

(a) 0 or 1, depending on b.

(b) ∞, independent of b.

(c) 0 or ∞, depending on b.

(d) 1, independent of b.

Exercise 1.21: Fitting and overfitting functions with least squares

One of your friends has been spending endless hours in the laboratory collecting data on some obscure process, and now wants to find a function to describe the variable y's dependence on the independent variable, x.

x	0.00	0.22	0.44	0.67	0.89	1.11	1.33	1.56	1.78	2.00
y	2.36	2.49	2.67	3.82	4.87	6.28	8.23	9.47	12.01	15.26

Not having a good theory to determine the form of this expression, your friend has chosen a polynomial to fit the data.

(a) Consider the polynomial model

$$y(x) = a_0 + a_1 x + a_2 x^2 + \ldots + a_n x^n$$

Express the normal equations for finding the coefficients a_i that minimize the sum of squares of errors in y.

(b) Using the x- and y-data shown above and plotted in Figure 1.8, solve the least-squares problem and find the a that minimize

$$\Phi = \sum_{i=1}^{n_d} \left(y_i - \sum_{j=0}^{n} a_j x_i^j \right)^2$$

in which n_d is the number of measurements and n is the order of the polynomial. Do this calculation for all polynomials of order $0 \le n \le 9$.

(c) For each n, also calculate the least-squares objective Φ_n, and plot Φ_n versus n.

(d) Plot the data along with your fitted polynomial curves for each value of n. In particular, plot the data and fits for $n = 2$ and $n = 9$ on one plot. Use the range $-0.25 \le x \le 2.25$ to get an idea about how well the models extrapolate.

(e) Based on the values of Φ_n and the appearance of your plots, what degree polynomial would you choose to fit these data? Why not choose $n = 9$ so that the polynomial can pass through every point and $\Phi = 0$?

Exercise 1.22: Least-squares estimation of activation energy

Assume you have measured a rate constant, k, at several different temperatures, T, and wish to find the activation energy (divided by the gas constant), E/R, and the preexponential factor, k_0, in the Arrhenius model

$$k = k_0 e^{-E/RT} \tag{1.28}$$

The data are shown in Figure 1.9 and listed here.

T(K)	300	325	350	375	400	425	450	475	500
k	1.82	1.89	2.02	2.14	2.12	2.17	2.15	2.21	2.26

(a) Take logarithms of (1.28) and write a model that is linear in the parameters $\ln(k_0)$ and E/R. Summarize the data and model with the linear algebra problem

$$Ax = b$$

in which x contains the parameters of the least-squares problem

$$x = \begin{bmatrix} \ln(k_0) \\ E/R \end{bmatrix}$$

What are A and b for this problem?

(b) Find the least-squares fit to the data. What are your least-squares estimates of $\ln(k_0)$ and E/R?

(c) Is your answer unique? How do you know?

(d) Plot the data and least-squares fit in the original variables k versus T. Do you have a good fit to the data?

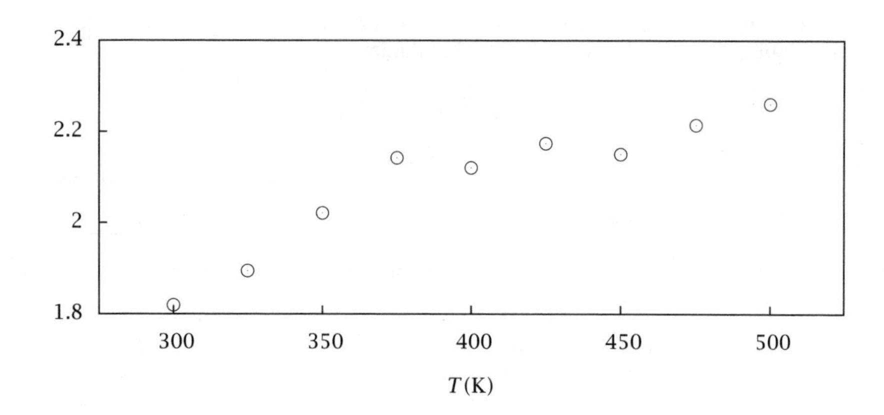

Figure 1.9: Measured rate constant at several temperatures.

Exercise 1.23: Existence and uniqueness of linear equations

Consider the following partitioned A matrix, $A \in \mathbb{R}^{m \times n}$

$$A = \begin{bmatrix} A_1 & 0 \\ 0 & 0 \end{bmatrix}$$

in which $A_1 \in \mathbb{R}^{p \times p}$ is of rank p and $p < \min(m, n)$.

(a) What are the dimensions of the three zero matrices?

(b) What is the rank of A?

(c) What is the dimension of the null space of A? Compute a basis for the null space of A.

(d) Repeat for A^T.

(e) For what b can you solve $Ax = b$?

(f) Is the solution for these b unique? If not, given one solution x_1, such that $Ax_1 = b$, specify all solutions.

Exercise 1.24: Reaction rates from production rates

Consider the following set of reactions.

$$CO + \frac{1}{2}O_2 \rightleftharpoons CO_2$$

$$H_2 + \frac{1}{2}O_2 \rightleftharpoons H_2O$$

$$CH_4 + 2O_2 \rightleftharpoons CO_2 + 2H_2O$$

$$CH_4 + \frac{3}{2}O_2 \rightleftharpoons CO + 2H_2O$$

(a) Given the species list, $A = \begin{bmatrix} CO & O_2 & CO_2 & H_2 & H_2O & CH_4 \end{bmatrix}^T$ write out the stoichiometric matrix, ν, for the reactions relating the four reaction rates to the six production rates

$$R = \nu^T r \tag{1.29}$$

(b) How many of the reactions are linearly independent.

(c) In a laboratory experiment, you measured the production rates for all the species and found

$$R_{\text{meas}} = \begin{bmatrix} -2 & -2 & 3 & 2 & 0 & -1 \end{bmatrix}^T$$

Is there a set of reaction rates r_{ex} that satisfies (1.29) exactly? If not, how do you know? If so, find an r_{ex} that satisfies $R_{\text{meas}} = \nu^T r_{\text{ex}}$.

(d) If there is an r_{ex}, is it unique? If so, how do you know? If not, characterize all solutions.

Exercise 1.25: Least-squares estimation

A colleague has modeled the same system as only the following three reactions

$$CO + \frac{1}{2}O_2 \rightleftharpoons CO_2$$

$$H_2 + \frac{1}{2}O_2 \rightleftharpoons H_2O$$

$$CH_4 + 2O_2 \rightleftharpoons CO_2 + 2H_2O$$

(a) How many of these reactions are linearly independent?

(b) In another laboratory experiment, you measured the production rates for all the species and found

$$R_{\text{meas}} = \begin{bmatrix} 1 & -4.5 & 1 & -2 & 6 & -2.5 \end{bmatrix}^T$$

Is there a set of reaction rates r_{ex} in this second model that satisfies (1.29) exactly? If so, find an r_{ex} that satisfies $R_{\text{meas}} = \nu^T r_{\text{ex}}$. If not, how do you know?

(c) If there is not an exact solution, find the least-squares solution, r_{est}. What is the least-squares objective value?

(d) Is this solution unique? If so, how do you know? If not, characterize all solutions that achieve this value of the objective function.

Exercise 1.26: Controllability

Consider a linear **discrete time** system governed by the difference equation

$$x(k+1) = Ax(k) + Bu(k) \tag{1.30}$$

in which $x(k)$, an n-vector, is the state of the system, and $u(k)$, an m-vector, is the manipulatable input at time k. The goal of the controller is to choose a sequence of inputs that force the state to follow some desirable trajectory.

(a) What are the dimensions of the A and B matrices?

(b) Not all systems can be controlled. Consider the case in which B is the zero matrix. Then the input has no effect on the state and we cannot control it. We should redesign the system before trying to design a controller for it. This is an example of an uncontrollable system.

A system is said to be **controllable** if n input values exist

$$u(0), u(1), u(2), \ldots, u(n-1)$$

that can move the system from any initial condition, x_0, to any final state $x(n)$. By using (1.30), show that $x(n)$ can be expressed as

$$x(n) = A^n x_0 + A^{n-1}Bu(0) + A^{n-2}Bu(1) + \cdots + ABu(n-2) + Bu(n-1)$$

Stack all of the $u(k)$ on top of each other and rewrite this expression in partitioned-matrix form,

$$x(n) = A^n x_0 + C \begin{bmatrix} u(0) \\ u(1) \\ \vdots \\ u(n-1) \end{bmatrix} \tag{1.31}$$

What is the C matrix and what are its dimensions?

(c) What must be true of the rank of C for a system to be controllable, i.e., for there to be a solution to (1.31) for every $x(0)$ and $x(n)$?

(d) Consider the following two systems with 2 states ($n = 2$) and 1 input ($m = 1$)

$$x(k+1) = \begin{bmatrix} 1 & 0 \\ 2 & 1 \end{bmatrix} x(k) + \begin{bmatrix} 0 \\ 1 \end{bmatrix} u(k)$$

$$x(k+1) = \begin{bmatrix} 1 & 0 \\ 2 & 1 \end{bmatrix} x(k) + \begin{bmatrix} 1 \\ 0 \end{bmatrix} u(k)$$

Notice that the input only directly affects one of the states in both of these systems. Are either of these two systems controllable? If not, show which $x(n)$ **cannot** be reached with n input moves starting from $x(0) = 0$.

Exercise 1.27: A vector/matrix derivative

Consider the following derivative for $A, C \in \mathbb{R}^{m \times n}, x, b \in \mathbb{R}^n$

$$C = \frac{d}{dx}(Axx^T b)$$

Or expressed in component form

$$C_{ij} = \frac{d}{dx_j}(Axx^T b)_i \quad i = 1, \ldots, m, \quad j = 1, \ldots, n$$

Find an expression for this derivative (C) in terms of A, x, b.

Exercise 1.28: Rank equality with matrix products

Given arbitrary $B \in \mathbb{R}^{m \times n}$, and full rank $A \in \mathbb{R}^{m \times m}$ and $C \in \mathbb{R}^{n \times n}$, establish the following two facts

$$\text{rank}(AB) = \text{rank}(B) \qquad \text{rank}(BC) = \text{rank}(B)$$

Use these to show

$$\text{rank}(ABC) = \text{rank}(B)$$

Exercise 1.29: More matrix products

Find examples of 2 by 2 matrices such that

(a) $LU \neq UL$,

(b) $A^2 = -I$, with A a real matrix,

(c) $B^2 = 0$, with no zeros in B,

(d) $CD = -DC$, not allowing $CD = 0$.

Exercise 1.30: Programming LU decomposition

Write a program to solve $Ax = b$ using LU decomposition. It should be able to handle matrices up to $n = 10$, read in A and b from data files, and write the solution x to a file. Using this program, solve the problem where

$$A = \begin{bmatrix} 1 & 1 & 1 & 1 \\ 2 & -2 & -1 & 2 \\ 3 & -1 & 2 & 2 \\ 1 & -1 & -1 & 1 \end{bmatrix} \qquad b = \begin{bmatrix} 3 \\ 0 \\ 2 \\ 0 \end{bmatrix}$$

Exercise 1.31: Normal equations

Write the linear system of equations whose solution $x = (x_1, x_2)^T$ minimizes

$$P(x) = \frac{1}{2}(x_1^2 + 2x_1x_2 + 2x_2^2) - x_1 + x_2$$

Find the solution x and the corresponding value of $P(x)$.

Exercise 1.32: Cholesky decomposition

A symmetric matrix A can be factorized into LDL^T where L is lower triangular and D is *diagonal*, i.e., only its diagonal elements are nonzero.

(a) Perform this factorization for the matrix

$$\begin{bmatrix} 2 & -1 & 0 \\ -1 & 2 & -1 \\ 0 & -1 & 2 \end{bmatrix}$$

(b) If all the diagonal elements of D are positive, the matrix can be further factorized into $\hat{L}\hat{L}^T$—this is called the CHOLESKY DECOMPOSITION of A. Find $\hat{L}$ for the matrix of part (a).

Exercise 1.33: A singular matrix

For the system

$$A = \begin{bmatrix} 3 & 6 \\ 6 & q \end{bmatrix} \qquad b = \begin{bmatrix} 1 \\ 4 \end{bmatrix}$$

(a) Find the value of q for which elimination fails (i.e., no solution to $Ax = b$ exists). If you are thoughtful, you won't need to perform the elimination to find out.

(b) For this value of q what happens to the first geometrical interpretation of the problem (intersecting lines)?

(c) What happens to the second (superpositions of column vectors)?

(d) What value should replace 4 in b to make the problem solvable for this q?

Exercise 1.34: LU factorization of nonsquare matrices

(a) Find the LU factorization of

$$A = \begin{bmatrix} 2 & 1 \\ 1 & 1 \\ 3 & 2 \end{bmatrix}$$

(b) If $b = (1, p, q)^T$, find a necessary and sufficient condition on p and q so that $Ax = b$ has a solution.

(c) Given values of p and q for which a solution exists, will the algorithm from Section 1.3.2 solve it? If not, pinpoint the difficulty.

(d) Find the LU factorization of A^T.

(e) Use this factorization to find two LI solutions of $A^T x = b$, where $b = (2, 5)^T$. Since there are fewer equations than unknowns in this case, there are infinitely many solutions, forming a line in $\mathbb{R}^2$. Are there any values of b for which this problem has no solution?

Exercise 1.35: An inverse

Under what conditions on u and v does $(I - auv^T) = (I + auv^T)^{-1}$? Here a is an arbitrary nonzero scalar.

Exercise 1.36: LU decomposition

Write the first step of the LU decomposition process of a matrix A as $A' = (I - auv^T)A$. In other words, what are a, u, and v so that $A'_{21} = 0$?

Exercise 1.37: Newton-Raphson

Write a program that uses the Newton-Raphson method to solve this pair of equations

$$y - (x-1)^2 = 0 \qquad (y+4)^2 - \tan x = 0$$

Do not reduce the pair of equations to a single equation. With this program, find at least one solution.

Exercise 1.38: The QR decomposition

In this exercise, we construct the QR decomposition introduced in Section 1.2.4. Consider an $m \times n$ matrix A with columns a_i. Observe that if $A = BC$, with B an $m \times n$ matrix and C and $n \times n$ matrix, where b_i are the columns of B, then we can write each column of A as a linear combination of the columns of B, as follows

$$\underset{A}{\begin{bmatrix} & a_i & \end{bmatrix}} = \underset{B}{\begin{bmatrix} b_1 & b_2 & \cdots & b_n \end{bmatrix}} \underset{C}{\begin{bmatrix} & c_{1i} & \\ & c_{2i} & \\ & \vdots & \\ & c_{ni} & \end{bmatrix}}$$

The ith column of A is a linear combination of all the columns of B, and the coefficients in the linear combination are the elements of the ith column of matrix C. This result will be helpful in solving the following problem. Let A be an $m \times n$ matrix whose columns a_i are linearly independent (thus $m \geq n$). We know that using the Gram-Schmidt procedure allows us to construct an ON set of vectors from the a_i. Define a matrix Q whose columns are these basis vectors, q_i, where $q_i^T q_j = \delta_{ij}$.

(a) Express each a_i in the basis formed by the q_i. Hint: because the set of q_i are constructed from the set of a_i by Gram-Schmidt, a_1 has a component only in the q_1 direction, a_2 has components only in the q_1 and q_2 directions, etc.

(b) Use the above result to write $A = QR$, i.e., find a square matrix R such that each column of A is written in terms of the columns of Q. You should find that R is upper triangular.

Exercise 1.39: Orthogonal subspace decomposition

Let S be an $r \leq n$ dimensional subspace of $\mathbb{R}^n$ with a basis $\{a_1, a_2, \dots, a_r\}$. Consider the subspace $S^\perp$, the orthogonal complement to S.

(a) Prove that $S^\perp$ has dimension $n - r$. Do not use the fundamental theorem of linear algebra in this proof because this result is used to prove the fundamental theorem.

(b) Show that any vector $x \in \mathbb{R}^n$ can be uniquely expressed as $x = a + b$ in which $a \in S$ and $b \in S^\perp$.

Exercise 1.40: The QR and thin QR decompositions

For $A \in \mathbb{R}^{m \times n}$ with independent columns we have used in the text what is sometimes called the "thin" QR with $Q_1 \in \mathbb{R}^{m \times n}$ and $R_1 \in \mathbb{R}^{n \times n}$ satisfying

$$A = Q_1 R_1$$

It is possible to "fill out" Q_1 by adding the remaining $m - n$ columns that span $\mathbb{R}^m$. In this case $A = QR$ and $Q \in \mathbb{R}^{m \times m}$ is orthonormal, and $R \in \mathbb{R}^{m \times n}$. In the "thin" QR, Q_1 is the shape of A and R_1 is square (of the smaller dimension n), and in the full QR, Q is square (of the larger dimension m) and R is the shape of A.

(a) Is the "thin" QR unique?

(b) Show how to construct the QR from the thin QR. Is the full QR unique?

Exercise 1.41: Uniqueness of solutions to least-squares problems

Prove the following proposition

Proposition 1.19 (Full rank of $A^T A$). *Given matrix $A \in \mathbb{R}^{m\times n}$, the $n \times n$ matrix $A^T A$ has full rank if and only if A has linearly independent columns.*

Note that this proof requires our first use of the fundamental theorem of linear algebra. Since most undergraduate engineers have limited experience doing proofs, we provide a few hints.

1. The "if and only if" statement requires proof of *two* statements: (i) $A^T A$ having full rank implies A has linearly independent columns *and* (ii) A having linearly independent columns implies $A^T A$ has full rank.
2. The statement that S implies T is logically equivalent to the statement that not T implies not S. So one could prove this proposition by showing (ii) and then showing: (i') A *not* having linearly independent columns implies that $A^T A$ is *not* full rank.
3. The fundamental theorem of linear algebra is the starting point. It tells us (among other things) that *square* matrix B has full rank if and only if B has linearly independent rows and columns. Think about what that tells you about the null space of B and B^T. See also Figure 1.1.

Exercise 1.42: A useful decomposition

Let $A \in \mathbb{C}^{n\times n}$, $B \in \mathbb{C}^{p\times p}$, and $X \in \mathbb{C}^{n\times p}$ satisfy

$$AX = XB \qquad \text{rank}(X) = p$$

Show that A can be decomposed as

$$Q^*AQ = T = \begin{array}{c} \\ p \\ n-p \end{array} \begin{array}{c} \begin{array}{cc} p & n-p \end{array} \\ \begin{bmatrix} T_{11} & T_{12} \\ 0 & T_{22} \end{bmatrix} \end{array}$$

in which $\text{eig}(T_{11}) = \text{eig}(B)$, and $\text{eig}(T_{22}) = \text{eig}(A) \setminus \text{eig}(B)$, i.e., the eigenvalues of T_{22} are the eigenvalues of A that are not eigenvalues of B. Also show that $\text{eig}(B) \subseteq \text{eig}(A)$.

Hint: use the QR decomposition of X.

Exercise 1.43: The Schur decomposition

Prove that the Schur decomposition has the properties stated in Theorem 1.15.

Hint: the result is obviously true for $n = 1$. Use induction and the result of Exercise 1.42.

Exercise 1.44: Norm and matrix rotation

Given the following A matrix

$$A = \begin{bmatrix} 0.46287 & 0.11526 \\ 0.53244 & 0.34359 \end{bmatrix}$$

invoking `[u,s,v]=svd(A)` in MATLAB or Octave produces

```
u =                          s =                        v =
  -0.59540  -0.80343          0.78328   0.00000           -0.89798  -0.44004
  -0.80343   0.59540          0.00000   0.12469           -0.44004   0.89798
```

(a) What vector x of unit norm *maximizes* $\|Ax\|$? How large is $\|Ax\|$ for this x?

(b) What vector x of unit norm *minimizes* $\|Ax\|$? How large is $\|Ax\|$ for this x?

(c) What is the definition of $\|A\|$? What is the value of $\|A\|$ for this A?

(d) Denote the columns of v by v_1 and v_2. Draw a sketch of the unit circle traced by x as it travels from $x = v_1$ to $x = v_2$ and the corresponding curve traced by Ax.

(e) Let's find an A, if one exists, that rotates all $x \in \mathbb{R}^2$ counterclockwise by θ radians. What do you choose for the singular values σ_1 and σ_2? Choose $v_1 = e_1$ and $v_2 = e_2$ for the V matrix in which e_i, $i = 1, 2$ is the ith unit vector. What do you want u_1 and u_2 to be for this rotation by θ radians? Form the product USV^T and determine the A matrix that performs this rotation.

Exercise 1.45: Linear difference equation model

Consider the following discrete-time model

$$x(k+1) = Ax(k)$$

in which

$$A = \begin{bmatrix} 0.798 & 0.051 \\ -0.715 & 1.088 \end{bmatrix} \qquad x_0 = \begin{bmatrix} 1 \\ 0 \end{bmatrix}$$

(a) Compute the eigenvalues and singular values of A. See the Octave or MATLAB commands `eig` and `svd`. Are the magnitudes of the eigenvalues of A less than one? Are the singular values less than one?

(b) What is the steady state of this system? Is the steady state asymptotically stable?

(c) Make a two-dimensional plot of the two components of $x(k)$ (phase portrait) as you increase k from $k = 0$ to $k = 200$, starting from the $x(0)$ given above. Is $x(1)$ bigger than $x(0)$? Why or why not?

(d) When the largest eigenvalue of A is less than one but the largest singular value of A is greater than one, what happens to the evolution of $x(k)$?

(e) Now plot the values of x for 50 points uniformly distributed on a unit circle and the corresponding Ax for these points. For the SVD corresponding to Octave and MATLAB convention

$$A = USV^*$$

mark u_1, u_2, v_1, v_2, s_1, and s_2 on your plot. Figure 1.10 gives you an idea of the appearance of the set of points for x and Ax to make sure you are on track.

Exercise 1.46: Is the SVD too good to be true?

Given $A \in \mathbb{R}^{m\times n}$ with rank$(A) = r$ and the SVD of $A = U\Sigma V^*$, if we partition the first r columns of U and V and call them U_1 and V_1 we have

$$A = \begin{bmatrix} U_1 & U_2 \end{bmatrix} \begin{bmatrix} \Sigma_r & 0 \\ 0 & 0 \end{bmatrix} \begin{bmatrix} V_1^* \\ V_2^* \end{bmatrix}$$

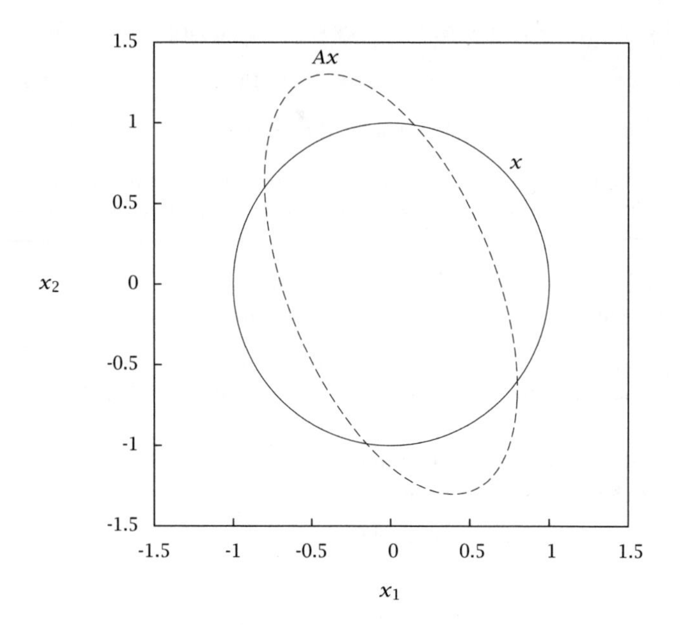

Figure 1.10: Plot of Ax as x moves around a unit circle.

and $A = U_1 \Sigma_r V_1^*$.

Then to solve (possibly in the least-squares sense) $Ax = b$ we have

$$U_1 \Sigma_r V_1^* x = b$$

which motivates the pseudoinverse formula

$$A^+ = V_1 \Sigma_r^{-1} U_1^*$$

and the "solution"

$$x = A^+ b$$

If we form the residual for this "solution"

$$\begin{aligned}
r &= Ax - b \\
&= AA^+ b - b \\
&= \underbrace{U_1 \overbrace{\Sigma_r \underbrace{V_1^* V_1}_{I_r} \Sigma_r^{-1}}^{I_r} U_1^*}_{I_m} \; b - b \\
&= I_m b - b \\
r &= 0
\end{aligned}$$

which seems to show that $r = 0$. We know that we cannot solve $Ax = b$ for every b and every A matrix, so something must have gone wrong. What is wrong with this argument leading to $r = 0$?

Exercise 1.47: SVD and worst-case analysis

Consider the process depicted in Figure 1.11 in which u is a manipulatable input and d is a disturbance. At steady state, the effects of these two variables combine at the measurement y in a linear relationship

$$y = Gu + Dd$$

The steady-state goal of the control system is to minimize the effect of d at the measurement y by adjusting u.

For this problem we have 3 inputs, $u \in \mathbb{R}^3$, 2 disturbances, $d \in \mathbb{R}^2$, and 2 measurements, $y \in \mathbb{R}^2$, and G and D are matrices of appropriate dimensions. We have the following two singular value decompositions available

$$G = \begin{bmatrix} U \end{bmatrix} \begin{bmatrix} S & 0 \end{bmatrix} \begin{bmatrix} V_1^T \\ V_2^T \end{bmatrix} \qquad D = \begin{bmatrix} X \end{bmatrix} \begin{bmatrix} \Sigma \end{bmatrix} \begin{bmatrix} Z^T \end{bmatrix}$$

$$U = \begin{bmatrix} -0.75 & -0.66 \\ -0.66 & 0.75 \end{bmatrix} \qquad S = \begin{bmatrix} 1.57 & 0.00 \\ 0.00 & 0.21 \end{bmatrix} \qquad V_1 = \begin{bmatrix} -0.89 & 0.37 \\ -0.45 & -0.81 \\ -0.085 & 0.46 \end{bmatrix}$$

$$X = \begin{bmatrix} -0.98 & -0.19 \\ -0.19 & 0.98 \end{bmatrix} \qquad \Sigma = \begin{bmatrix} 0.71 & 0.00 \\ 0.00 & 0.13 \end{bmatrix} \qquad Z = \begin{bmatrix} -0.94 & -0.33 \\ -0.33 & 0.94 \end{bmatrix}$$

(a) Can you exactly cancel the effect of d on y using u for all d? Why or why not?

(b) In terms of U, S, V_1, X, Σ, Z, what input u minimizes the effect of d on y? In other words, if you decide the answer is linear

$$u = Kd$$

What is K in terms of U, S, V_1, X, Σ, Z? Give the symbolic and numerical results.

(c) What is the worst d of unit norm, i.e., what d requires the largest response in u? What is the response u to this worst d?

Exercise 1.48: Worst-case disturbance

Consider the system depicted in Figure 1.11 in which we can manipulate an input $u \in \mathbb{R}^2$ to cancel the effect of a disturbance $d \in \mathbb{R}^2$ on an output $y \in \mathbb{R}^2$ of interest. The steady-state relationship between the variables is modeled as a linear relationship

$$y = Gu + Dd$$

and y, u, d are in deviation variables from the steady state at which the system was linearized. Experimental tests on the system have produced the following model parameters

$$G = \begin{bmatrix} 1 & 1 \\ 0 & 1 \end{bmatrix} \qquad D = \begin{bmatrix} 2.857 & 3.125 \\ 0.991 & 2.134 \end{bmatrix}$$

If we have measurements of the disturbance d available, we would like to find the input u that exactly cancels d's effect on y, and we would like to know ahead of time what is the worst-case disturbance that can hit the system.

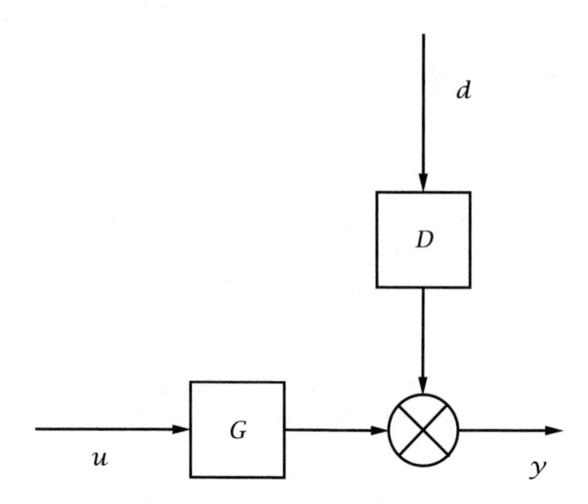

Figure 1.11: Manipulated input u and disturbance d combine to affect output y.

(a) Find the u that cancels d's effect on y.

(b) For d on the unit circle, plot the corresponding value of u.

(c) What d of norm one requires the largest control action u? What d of norm one requires the smallest control action u? Give the exact values of $d_{\max}$ and $d_{\min}$, and the corresponding $u_{\max}$ and $u_{\min}$.

(d) Assume the input is constrained to be in the box

$$\begin{bmatrix} -1 \\ -1 \end{bmatrix} \leq u \leq \begin{bmatrix} 1 \\ 1 \end{bmatrix} \tag{1.32}$$

What is the size of the disturbance so that all disturbances less than this size can be rejected by the input without violating these constraints? In other words find the largest scalar α such that

$$\text{if } \|d\| \leq \alpha \text{ then } u \text{ satisfies (1.32)}$$

Use your plot from the previous part to estimate α.

Exercise 1.49: Determinant, trace, and eigenvalues

Use the Schur decomposition of matrix $A \in \mathbb{C}^{n \times n}$ to prove the following facts

$$\det A = \prod_{i=1}^{n} \lambda_i \tag{1.33}$$

$$\text{tr} A = \sum_{i=1}^{n} \lambda_i \tag{1.34}$$

in which $\lambda_i \in \text{eig}(A), i = 1, 2, \ldots, n$.

Exercise 1.50: Repeated eigenvalues

The self-adjoint matrix

$$A = \begin{bmatrix} 2 & 0 & 0 \\ 0 & 1 & 1 \\ 0 & 1 & 1 \end{bmatrix}$$

has a repeated eigenvalue. Find the eigenvalues of the system and show that despite the repeated eigenvalue the system has a complete orthogonal set of eigenvectors.

Exercise 1.51: More repeated eigenvalues

The non-self-adjoint matrix

$$A = \begin{bmatrix} 1 & 1 & 2 & 0 \\ 0 & 1 & 3 & 0 \\ 0 & 0 & 2 & 2 \\ 0 & 0 & 0 & 1 \end{bmatrix}$$

also has repeated eigenvalues.

(a) Find the eigenvalues and eigenvectors (there are only two) of A.

(b) Denote the eigenvector corresponding to the repeated eigenvalue as v_1 and the other eigenvector as v_4. The GENERALIZED EIGENVECTORS v_2 and v_3 can be found by solving

$$(A - \lambda_1 I)v_2 = v_1$$
$$(A - \lambda_1 I)v_3 = v_2$$

where λ_1 is the repeated eigenvalue. Show that $\{v_1, \ldots, v_4\}$ is necessarily an LI set.

(c) Determine the set, construct the transformation matrix M, and show that $J = M^{-1}AM$ is indeed in Jordan form.

Exercise 1.52: Solution to a singular linear system

Consider a square matrix A that has a complete set of LI eigenvectors and a single zero eigenvalue.

(a) Write the solution to $Ax = \mathbf{0}$ in terms of the eigenvectors of A.

(b) In the problem $Ax = b$, use the eigenvectors to determine necessary and sufficient conditions on b for existence of a solution.

Exercise 1.53: Example of a singular problem

Consider the problem $Ax = b$, where

$$A = \begin{bmatrix} 1 & 2 & 3 \\ 1 & 2 & 3 \\ 1 & 2 & 3 \end{bmatrix}$$

(a) Perform LU decomposition on this matrix. Give L and U.

(b) Find two linearly independent vectors in the nullspace of A.

(c) Use the LU decomposition to find a solution when

$$b = \begin{bmatrix} 4 \\ 4 \\ 4 \end{bmatrix}$$

(d) This solution is not unique. Find another.

(e) Find the eigenvalues and eigenvectors of A. How are these related to your answers to parts b and c above?

Exercise 1.54: Linearly independent eigenvectors

Show that if A has n distinct eigenvalues, its eigenvectors are linearly independent. This result is required to ensure the existence of Q^{-1} in $A = Q\Lambda Q^{-1}$ in (1.11).
Hint: set $\sum_{i=1}^{n} \alpha_i q_i = 0$ and multiply by $(A - \lambda_1 I)(A - \lambda_2 I) \cdots (A - \lambda_{n-1} I)$ to establish that $\alpha_n = 0$. With $\alpha_n = 0$ what can you do next to show that $\alpha_{n-1} = 0$? Continue this process.

Exercise 1.55: General results for eigenvalue problems

Prove the following statements:

(a) If A is nonsingular and has eigenvalues λ_i, the eigenvalues of A^{-1} are $1/\lambda_i$.

(b) Let S be a matrix whose columns form a set of linearly independent but nonorthogonal basis vectors: the mth column is the vector u_m. Find a matrix S' whose columns u_n' satisfy $u_m^T u_n' = \delta_{mn}$. A pair of basis sets whose vectors satisfy this condition are said to be BIORTHOGONAL.

(c) Assume that A has a complete set of eigenvectors. Show that the eigenvectors of A and A^T are biorthogonal.

(d) Show that if the eigenvectors of A are orthogonal, then $AA^T = A^T A$. Such matrices are called NORMAL. (The converse is also true (Horn and Johnson, 1985).)

(e) Show that the eigenvalues of $A = -A^T$ are imaginary and that its eigenvectors are orthogonal.

Exercise 1.56: Eigenvalues of a dyad

Let u and v be unit vectors in $\mathbb{R}^n$, with $u^T v \neq 0$. What are the eigenvalues and eigenvectors of uv^T?

Exercise 1.57: The power method for finding largest eigenvalues

Consider the matrix

$$A = \begin{bmatrix} 0 & 0 & 1 \\ 0 & 0 & 1 \\ 1 & 1 & 1 \end{bmatrix}$$

(a) Let $x_0 = (1, 0, 0)^T$ and consider the iteration procedure $x_{i+1} = Ax_i$. Perform several steps of this procedure by hand and observe the result.

(b) Can you understand what is happening here by writing x in the eigenvector basis? In particular, show that for a self-adjoint matrix with distinct eigenvalues, this iteration procedure yields the eigenvalue of largest absolute value and the corresponding eigenvector.

(c) Write an Octave or MATLAB function to perform this process on a real symmetric matrix, outputting the largest eigenvalue (to within a specified tolerance) of A and the corresponding eigenvector, scaled so that its largest component is 1. Present results for a test case. This is the POWER METHOD. It is much faster than finding all of the eigenvalues and can be generalized to other types of matrices. Google's "PageRank" algorithm is built around this method.

Exercise 1.58: Markov chain models

Imagine that there are three kinds of weather: sunny, rainy, and snowy. Thus a vector $w_0 \in \mathbb{R}^3$ defines today's weather: $w_0 = [1,0,0]^T$ is sunny, $w_0 = [0,1,0]^T$ is rainy, and $w_0 = [0,0,1]$ is snowy. Imagine that tomorrow's weather w_1 is determined only by today's and more generally, the weather on day $n+1$ is determined by the weather on day n. A probabilistic model for the weather then takes the form

$$w_{n+1} = Tw_n$$

where T is called a *transition matrix* and the elements of w_n are the probabilities of having a certain type of weather on that day. For example, if $w_5 = [0.2, 0.1, 0.7]^T$, then the probability of snow five days from now is 70%. The sequence of probability vectors on subsequent days, $\{w_0, w_1, w_2, \ldots\}$ is called a MARKOV CHAIN. Because w is a vector of probabilities, its elements must sum to one, i.e., $\sum_{i=1,3} w_{n,i} = 1$ for all n.

(a) Given that $\sum_{i=1,3} w_{n,i} = 1$, what condition must the elements of T satisfy such that $\sum_{i=1,3} w_{n+1,i}$ is also 1?

(b) Assume that T is a constant matrix, i.e., it is independent of n. What conditions of the eigenvalues of T must hold so that the Markov chain will reach a constant state w_∞ as $n \to \infty$? How is w_∞ related to the eigenvectors of T?

Exercise 1.59: Real Jordan form for a real matrix with complex conjugate eigenvalues

For a 2×2 real matrix A with a complex conjugate pair of eigenvalues $\lambda_1 = \sigma + i\omega$, $\lambda_2 = \bar{\lambda}_1$ with eigenvectors v_1 and v_2:

(a) Derive the result that $v_1 = \bar{v}_2$.

(b) Write the general solution to $\dot{x} = Ax$ in terms of the real and imaginary parts of v_1 and sines and cosines, so that the only complex numbers in the solution are the arbitrary constants.

(c) For the specific matrix

$$A = \begin{bmatrix} -2 & -2 \\ 2 & 1 \end{bmatrix}$$

show that the similarity transformation $S^{-1}AS$, where the columns of S are the real and imaginary parts of v_1, has the form

$$S^{-1}AS = \begin{bmatrix} \sigma & -\omega \\ \omega & \sigma \end{bmatrix}$$

This result can be generalized, showing how a real matrix with complex eigenvalues can be put into Jordan form without introducing any imaginary numbers.

Exercise 1.60: Solving a boundary-value problem by eigenvalue decomposition

Consider the reaction

$$\mathrm{A} \underset{k_{-1}}{\overset{k_1}{\rightleftharpoons}} \mathrm{B} \underset{k_{-2}}{\overset{k_2}{\rightleftharpoons}} \mathrm{C}$$

occurring in a membrane. At steady state the appropriate reaction-diffusion equations for this system are

$$D_A \frac{d^2 c_A}{dx^2} - k_1 c_A + k_{-1} c_B = 0$$

$$D_B \frac{d^2 c_B}{dx^2} + k_1 c_A - k_{-1} c_B - k_2 c_B + k_{-2} c_C = 0$$

$$D_C \frac{d^2 c_C}{dx^2} + k_2 c_B - k_{-2} c_C = 0$$

where the $k_i, i = \pm(1,2)$ are rate constants and the $D_j, j = (\mathrm{A},\mathrm{B},\mathrm{C})$ are the species diffusivities. The boundary conditions are

$$c_A = 1 \quad c_B = c_C = 0 \quad \text{at } x = 1$$

$$\frac{dc_A}{dx} = \frac{dc_B}{dx} = \frac{dc_C}{dx} = 0 \quad \text{at } x = 0$$

Convert this set of second-order equations into a set of first-order differential equations. Write a MATLAB or Octave code to find the solution to this problem in terms of eigenvalues and eigenvectors of the relevant matrix for a given set of parameters. Have the program plot the concentrations as functions of position. Show results for parameter values $D_A = D_B = D_C = 20$, $k_1 = k_2 = 10$, $k_{-1} = k_{-2} = 0.1$, and also for the same rate constants but with the diffusivities set to 0.05.

Exercise 1.61: Nullspaces of nonsquare matrices

Consider a *nonsquare* $m \times n$ matrix A. Show that $A^T A$ is symmetric positive semidefinite. If A were square we could determine its nullspace from the eigenvectors corresponding to zero eigenvalues. How can we determine the nullspace of a nonsquare matrix A? What about the nullspace of A^T?

Exercise 1.62: Stability of an iteration

Consider the iteration procedure $x(i+1) = Ax(i)$, where A is diagonalizable.

(a) What conditions must the eigenvalues of A satisfy so that $x(i) \to 0$ as $i \to \infty$?

(b) What conditions must the eigenvalues satisfy for this iteration to converge to a steady state, i.e., so that $x(i) \to x(i+1)$ as $i \to \infty$?

Exercise 1.63: Cayley-Hamilton theorem

Suppose that A is an $n \times n$ diagonalizable matrix with characteristic equation

$$\det(A - \lambda I) = \lambda^n + a_{n-1}\lambda^{n-1} + \ldots + a_0 = 0$$

(a) Show that

$$A^n + a_{n-1}A^{n-1} + \ldots + a_0 I = 0$$

This result shows that A satisfies its own characteristic equation; it is known as the Cayley-Hamilton theorem.

(b) Let

$$A = \begin{bmatrix} 1 & 2 \\ 2 & 1 \end{bmatrix}$$

Use the theorem to express A^2, A^3, and A^{-1} as linear combinations of A and I.

Exercise 1.64: Solving the nonunique least-squares problem

We have established that the least-squares solution to $Ax = b$ is unique if and only if A has linearly independent columns. Let's treat the case in which the columns are not linearly independent and the least-squares solution is not unique. Consider again the SVD for real-valued A

$$A = \begin{bmatrix} U_1 & U_2 \end{bmatrix} \begin{bmatrix} \Sigma & 0 \\ 0 & 0 \end{bmatrix} \begin{bmatrix} V_1^T \\ V_2^T \end{bmatrix}$$

(a) Show that all solutions to the least-squares problem are given by

$$x_{\text{ls}} = V_1 \Sigma^{-1} U_1^T b + V_2 \alpha \qquad \alpha \in \mathbb{R}^{n-r}$$

in which α is an arbitrary vector.

(b) Show that the unique, minimum-norm solution to the least-squares problem is given by

$$x_{\text{ls}}^0 = V_1 \Sigma^{-1} U_1^T b$$

This minimum-norm solution is the one returned by many standard linear algebra packages. For example, this is the solution returned by Octave and MATLAB when invoking the shorthand command `x = A \ b`.

Exercise 1.65: Propagating zeros in triangular matrices

When multiplying two partitioned (upper) triangular matrices, if the first one has k leading columns of zeros, and the second one has a $0_{p \times p}$ matrix on the second element of the diagonal, show that the product is a triangular matrix with $k + p$ leading columns of zeros. In pictures

$$\begin{array}{c} \\ k \\ p \\ r \end{array} \begin{array}{c} \begin{matrix} k & p & r \end{matrix} \\ \begin{bmatrix} 0 & * & * \\ 0 & T_1 & * \\ 0 & 0 & T_2 \end{bmatrix} \end{array} \begin{array}{c} \begin{matrix} k & p & r \end{matrix} \\ \begin{bmatrix} T_3 & * & * \\ 0 & 0 & * \\ 0 & 0 & T_4 \end{bmatrix} \end{array} = \begin{array}{c} \begin{matrix} k & p & r \end{matrix} \\ \begin{bmatrix} 0 & 0 & * \\ 0 & 0 & * \\ 0 & 0 & T_5 \end{bmatrix} \end{array} \begin{array}{c} \\ k \\ p \\ r \end{array}$$

in which $T_i, i = 1, \ldots, 4$ are arbitrary triangular matrices, T_5 is triangular, and $*$ represents arbitrary (full) matrices. This result is useful in proving the Cayley-Hamilton theorem in the next exercise.

Exercise 1.66: Cayley-Hamilton theorem holds for all matrices

Revisit Exercise 1.63 and establish that *all* matrices $A \in \mathbb{C}^{n\times n}$ satisfy their characteristic equation. We are removing the assumption that A is diagonalizable in the Cayley-Hamilton theorem so that it holds also for defective matrices.

Hint: use the Schur form to represent A and the result of Exercise 1.65.

Exercise 1.67: Small matrix approximation

For x a scalar, consider the Taylor series for $1/(1+x)$

$$\frac{1}{1+x} = 1 - x + x^2 - x^3 + \cdots$$

which converges for $|x| < 1$.

(a) Using this scalar Taylor series, establish the analogous series for matrix $X \in \mathbb{R}^{n\times n}$

$$(I+X)^{-1} = I - X + X^2 - X^3 + \cdots$$

You may assume the eigenvalues of X are unique. For what matrix X does this series converge?

(b) What is the corresponding series for

$$(R+X)^{-1}$$

in which $R \in \mathbb{R}^{n\times n}$ is a full-rank matrix. What conditions on X and R are required for the series to converge?

Exercise 1.68: Matrix exponential, determinant and trace

Use the Schur decomposition of matrix $A \in \mathbb{C}^{n\times n}$ to prove the following fact

$$\det e^A = e^{\text{tr}(A)}$$

Exercise 1.69: Logarithm of a matrix

If $A \in \mathbb{C}^{n\times n}$ is nonsingular, there exists a $B \in \mathbb{C}^{n\times n}$ such that

$$A = e^B$$

and B is known as the logarithm of A

$$B = \ln A$$

If A is positive definite, B can be uniquely defined (the principal branch of the logarithm).

Given this definition of the logarithm, if $A \in \mathbb{C}^{n\times n}$ is nonsingular, show that

$$\det A = e^{\text{tr}(\ln A)} \tag{1.35}$$

Exercise 1.70: Some differential equations, sines, cosines, and exponentials

(a) Solve the following vector, second-order ordinary differential equation with the given initial conditions for $y \in \mathbb{R}^2$

$$\frac{d^2y}{dt^2} + Ay = 0$$

$$A = \begin{bmatrix} 3 & 1 \\ 1 & 3 \end{bmatrix} \qquad y(0) = \begin{bmatrix} 1 \\ -\frac{1}{2} \end{bmatrix} \qquad \frac{dy}{dt}(0) = \begin{bmatrix} 0 \\ 0 \end{bmatrix}$$

Use the solution of the scalar version of this differential equation as your guide.

(b) We can always reduce a high-order differential equation to a set of first-order differential equations. Define $x = dy/dt$ and let

$$z = \begin{bmatrix} x \\ y \end{bmatrix}$$

and show that the above equation can be written as a single first-order differential equation

$$\frac{dz}{dt} + Bz = 0$$

with $z \in \mathbb{R}^4$. What are B and the appropriate initial conditions $z(0)$? What is the solution to this problem?

(c) Plot, on a single graph, the trajectories of the two y components versus time for the given initial conditions.

(d) Show that the result of (a) is the same as the result of (b), even though the functions exp and cos are different.

Exercise 1.71: Bounding the matrix exponential

Given the bound for $\left\|e^{At}\right\|$ in (1.17), establish the validity of the bound in (1.18).

Hints: first, for any $k > 0$ and $\epsilon > 0$, show that there exists a $c > 0$ such that for all $t \geq 0$

$$ce^{\epsilon t} \geq t^k$$

Use this result to show that for any $\epsilon > 0$, $N \in \mathbb{C}^{n\times n}$, there exists $c > 0$ such that for all $t \geq 0$

$$ce^{\epsilon t} \geq \sum_{k=0}^{n-1} \frac{\|Nt\|^k}{k!}$$

Exercise 1.72: Strictly convex quadratic function and positive curvature

Consider the quadratic function

$$f(x) = (1/2)x^T Ax + b^T x + c$$

(a) Show that $f(\cdot)$ is strictly convex if and only if $A > 0$.

(b) For the quadratic function, show that if a minimizer of $f(\cdot)$ exists, it is unique if *and only if* $A > 0$. The text shows the "if" part for any strictly convex function. So you are required to show the "only if" part with the additional restriction that $f(\cdot)$ is quadratic.

(c) Show that $f(\cdot)$ is convex if and only if $A \geq 0$.

Exercise 1.73: Concave functions and maximization

A function $f(\cdot)$ is defined to be (STRICTLY) CONCAVE (concave downward) if $-f(\cdot)$ is (strictly) convex (Rockafellar and Wets, 1998, p. 39). Show that a solution to $\max_x f(x)$ is unique if $f(\cdot)$ is strictly concave.

Exercise 1.74: Solutions to minmax and maxmin problems

Consider again the quadratic function $f(x) = (1/2)x^T Ax$ and the two games given in (1.19). Confirm that Figure 1.6 (c) corresponds to the A matrix

$$A = \sqrt{2}\begin{bmatrix} -1 & 1 \\ 1 & 1 \end{bmatrix}$$

(a) Show that $x_1 = x_2 = 0$ is the *unique* solution to both games in (1.19). Hint: with the outer variable fixed, solve the inner optimization problem and note that its solution exists and is unique. Then substitute the solution for the inner problem, solve the outer optimization problem, and note that its solution also exists and is unique.

(b) Show that neither of the following problems has a solution

$$\max_{x_2}\min_{x_1} f(x) \qquad \min_{x_1}\max_{x_2} f(x)$$

in which we have interchanged the goals of the two players. So obviously the goals of the players matter a great deal in the existence of solutions to the game.

Exercise 1.75: Games with nonunique solutions and different solution sets

Sketch the contours for $f(x) = (1/2)x^T Ax$ with the following A matrix

$$A = \begin{bmatrix} 0 & 1 \\ 1 & 0 \end{bmatrix}$$

What are the eigenvalues of A?

Show that $x_1 = x_2 = 0$ is still a solution to both games in (1.19), but that it is not unique. Find the complete solution *sets* for both games in (1.19). Establish that the solution sets are *not* the same for the two games.

Exercise 1.76: Who plays first?

When the solutions to all optimizations exist, show that

$$\max_{x_2}\min_{x_1} f(x_1, x_2) \leq \min_{x_1}\max_{x_2} f(x_1, x_2)$$

This inequality verifies that the player who goes first, i.e., the inner optimizer, has the advantage in this noncooperative game. Note that the function $f(\cdot)$ is arbitrary, so long as the indicated optimizations all have solutions.

Exercise 1.77: Solving linear matrix equations

Consider the linear matrix equation

$$AXB = C \tag{1.36}$$

in which $A \in \mathbb{R}^{m\times n}$, $X \in \mathbb{R}^{n\times p}$, $B \in \mathbb{R}^{p\times q}$, and $C \in \mathbb{R}^{m\times q}$. We consider A, B, C fixed matrices and X is the unknown matrix. The number of equations is the number of elements in C. The number of unknowns is the number of elements of X. Taking the vec of both sides gives

$$(B' \otimes A)\text{vec}X = \text{vec}C \tag{1.37}$$

We wish to explore how to solve this equation for $\text{vec}X$.

(a) For the solution to exist for all vec$C$, and be unique, we require that $(B' \otimes A)$ has linearly independent rows and columns, i.e., it is square and full rank. Using the rank result (1.27) show that this is equivalent to A and B being square and full rank.

(b) For this case show that the solution

$$\text{vec}X = (B^T \otimes A)^{-1}\text{vec}C$$

is equivalent to that obtained by multiplying (1.36) by A^{-1} on the left and B^{-1} on the right,

$$X = A^{-1}CB^{-1}$$

(c) If we have more equations than unknowns, we can solve (1.37) for vec$X$ as a least-squares problem. The least-squares solution is unique if and only if $B^T \otimes A$ has linearly independent columns. Again, use the rank result to show that this is equivalent to: (i) A has linearly independent columns, and (ii) B has linearly independent rows.

(d) We know that A has linearly independent columns if and only if $A^T A$ has full rank, and B has linearly independent rows if and only if BB^T has full rank (see Proposition 1.19 in Exercise 1.41). In this case, show that the least-squares solution of (1.37)

$$\text{vec}X_{\text{ls}} = (B^T \otimes A)^{\dagger}\text{vec}C$$

is equivalent to that obtained by multiplying (1.36) by $A^{\dagger}$ on the left and $B^{\dagger}$ on the right,

$$X_{\text{ls}} = A^{\dagger}CB^{\dagger}$$

Note that the superscript † denotes the Moore-Penrose pseudoinverse discussed in Section 1.3.7.

Exercise 1.78: Solving the matrix Lyapunov equation

Write a function `S = yourlyap(A,Q)` using the Kronecker product to solve the matrix Lyapunov equation

$$A^T S + SA = -Q$$

Test your function with some A with negative eigenvalues and positive definite Q by comparing to the function `lyap` in Octave or MATLAB.

Bibliography

R. B. Bird, W. E. Stewart, and E. N. Lightfoot. *Transport Phenomena.* John Wiley & Sons, New York, second edition, 2002.

G. H. Golub and C. F. Van Loan. *Matrix Computations.* The Johns Hopkins University Press, Baltimore, Maryland, third edition, 1996.

N. J. Higham. *Functions of Matrices: Theory and Computation.* SIAM, Philadelphia, 2008.

R. A. Horn and C. R. Johnson. *Matrix Analysis.* Cambridge University Press, 1985.

C. C. Lin and L. A. Segel. *Mathematics Applied to Deterministic Problems in the Natural Sciences.* Macmillan, New York, 1974.

J. R. Magnus and H. Neudecker. *Matrix Differential Calculus with Applications in Statistics and Econometrics.* John Wiley, New York, 1999.

J. Nash. Noncooperative games. *Ann. Math.*, 54:286–295, 1951.

W. H. Press, S. A. Teukolsky, W. T. Vetterling, and B. T. Flannery. *Numerical Recipes in C: The Art of Scientific Computing.* Cambridge University Press, Cambridge, 1992.

R. T. Rockafellar and R. J.-B. Wets. *Variational Analysis.* Springer-Verlag, 1998.

S. M. Selby. *CRC Standard Mathematical Tables.* CRC Press, twenty-first edition, 1973.

G. Strang. *Linear Algebra and its Applications.* Academic Press, New York, second edition, 1980.

L. N. Trefethen and D. Bau III. *Numerical Linear Algebra.* Society for Industrial and Applied Mathematics, 1997.

C. F. Van Loan. The sensitivity of the matrix exponential. *SIAM J. Numer. Anal.*, 14:971–981, 1977.

J. von Neumann and O. Morgenstern. *Theory of Games and Economic Behavior.* Princeton University Press, Princeton and Oxford, 1944.

2

Ordinary Differential Equations

2.1 Introduction

Differential equations arise in all areas of chemical engineering. In this chapter we consider ORDINARY differential equations (ODEs), that is, equations that have only one independent variable. For example, for reactions in a stirred-tank reactor the independent variable is time, while in a simple steady-state model of a plug-flow reactor, the independent variable is position along the reactor. Typically, ODEs appear in one of two forms

$$\frac{dx}{dt} = f(x,t), \quad x \in \mathbb{R}^n \tag{2.1}$$

or

$$a_n(x)\frac{d^n y}{dx^n} + a_{n-1}(x)\frac{d^{n-1}y}{dx^{n-1}} + \cdots + a_1(x)\frac{dy}{dx} + a_0(x)y = g(x), \quad y \in \mathbb{R} \tag{2.2}$$

We have intentionally written the two forms in different notation, as the first form typically (but not always) appears when the independent variable is time, and the second form often appears when the independent variable is spatial position. These two forms usually have different boundary conditions. When t is the independent variable, we normally know the conditions at $t = 0$ (e.g., initial reactant concentration) and must solve for the behavior for all $t > 0$. This is called an INITIAL-VALUE PROBLEM (IVP). In a transport problem, on the other hand, we know the temperature, for example, at the boundaries and must find it in the interior. This is a BOUNDARY-VALUE PROBLEM (BVP).

2.2 First-Order Linear Systems

2.2.1 Superposition Principle for Linear Differential Equations

An arbitrary linear differential equation can be written

$$Lu = g$$

where L is a linear differential operator (e.g., $L = d/dt - A$, where A is a matrix), u is the solution to be determined, and g is a known function. Section 1.2 introduced the following general properties of linear operators, which we now write in terms of L

$$L(u + v) = Lu + Lv$$
$$L(\alpha u) = \alpha(Lu)$$

Leaving aside for the moment the issue of boundary conditions, the following two SUPERPOSITION properties follow directly from linearity

1. Homogeneous problem. Let $g = 0$. If u_1 and u_2 are both solutions to $Lu = 0$, then $\alpha u_1 + \beta u_2$ is also a solution, for any scalars α and β.

2. Inhomogeneous problem. Let u_1 be a solution to $Lu = g_1$ and u_2 be a solution to $Lu = g_2$. Then $\alpha u_1 + \beta u_2$ is a solution to $Lu = \alpha g_1 + \beta g_2$.

With regard to boundary conditions, linearity also implies the following.

3. Let u_1 be a solution to $Lu = g_1$ with boundary condition $Bu = h_1$ on a particular boundary, where B is an appropriate operator, e.g., a constant for a DIRICHLET boundary condition, a first derivative d/dx for a NEUMANN boundary condition, or a combination $B = \gamma + \delta\, d/dx$ for a ROBIN boundary condition. Let u_2 solve $Lu = g_2$ with boundary condition $Bu = h_2$. Then $\alpha u_1 + \beta u_2$ satisfies $Lu = \alpha g_1 + \beta g_2$ with boundary condition $Bu = \alpha h_1 + \beta h_2$.

These simple results are very powerful and will be implicitly and explicitly used throughout the book, as they allow complex solutions to be constructed as sums (or integrals) of simple ones.

2.2.2 Homogeneous Linear Systems with Constant Coefficients—General Results for the Initial-Value Problem

Consider (2.1), where t denotes time. The function f is often called a VECTOR FIELD; for each point x in the PHASE SPACE or STATE SPACE of the system, $f(x)$ defines a vector giving the rate of change of x at that point. The system is called AUTONOMOUS if f is not an explicit function of t. The trajectory $x(t)$ traces out a curve in the state space, starting from the initial condition $x(0) = x_0$.

The most general linear first-order system can be written

$$\frac{dx}{dt} = A(t)x + g(t), x \in \mathbb{R}^n \tag{2.3}$$

In the present section we further narrow the focus and consider only the linear, autonomous, homogeneous system

$$\dot{x} = Ax \qquad x \in \mathbb{R}^n \quad A \in R^{n\times n} \tag{2.4}$$

where A is a constant matrix. Note that many dynamics problems are posed as second-order problems: if x is a position variable then Newton's second law takes the form $\ddot{x} = F(x)$. Letting $u_1 = x, u_2 = \dot{x}$, we recover a first-order system

$$\begin{aligned} \dot{u}_1 &= u_2 \\ \dot{u}_2 &= F(u_1) \end{aligned}$$

More generally, a single high-order differential equation can always be written as a system of first-order equations.

Unless A is diagonal, all of the individual scalar equations in the system (2.4) are coupled. The only practical way to find a solution to the system is to try to decouple it. But we already know how to do this—we use the eigenvector decomposition $A = MJM^{-1}$, where J is the Jordan form for A (Section 1.4). Letting $y = M^{-1}x$ be the solution vector in the eigenvector coordinate system, we write

$$\dot{y} = Jy$$

If A can be completely diagonalized, then $J = \Lambda = \text{diag}(\lambda_1, \lambda_2, \ldots, \lambda_n)$ and the equations in the y coordinates are completely decoupled. The solution is

$$y_i(t) = e^{\lambda_i t} c_i$$

or

$$y = e^{\Lambda t} c$$

where c is a vector of arbitrary constants. For an initial-value problem where $x(0)$ is a known vector x_0, $c = y(0) = M^{-1}x_0$. Recall from Section 1.5 that the matrix $e^{\Lambda t}$ is called the MATRIX EXPONENTIAL of Λ. It is defined for a general matrix A as

$$e^{At} = I + At + \frac{1}{2!}A^2t + \frac{1}{3!}A^3t + \cdots$$

For the diagonal matrix Λ, this is simply a diagonal matrix with entries $e^{\lambda_i t}$. Since $y_i(t) = e^{\lambda_i t}c_i$, we see that the eigenvalues of A determine the growth or decay rates λ_i and the eigenvectors (columns of M) determine the directions v_i along which this growth or decay occurs. Converting back to the original coordinates, we have the general solution

$$x(t) = Se^{\Lambda t}c = \sum_i c_i e^{\lambda_i t} v_i$$

where v_i is the eigenvector corresponding to λ_i. This expression shows explicitly that the solution when A has a complete LI set of eigenvectors is a simple combination of exponential growth and decay in the directions defined by the eigenvectors.

An important general consequence of this result is that an initial condition x_0 that lies on the line defined by the kth eigenvector leads to $c_i = \alpha\delta_{ik}$ and thus to a solution $x(t) = \alpha e^{\lambda_k t}v_k$. This solution will never leave the line defined by the eigenvector v_k. This line is thus an INVARIANT SUBSPACE for the dynamics: an initial condition that starts in an invariant subspace never leaves it. Similarly, each pair of eigenvectors defines a plane that is invariant, each triple defines a three-dimensional space that is invariant and so on.

A particularly relevant special case of an invariant plane arises when A has a complex conjugate pair of eigenvalues $\sigma \pm i\omega$ with corresponding eigenvectors v and $\bar{v}$; see Exercise 1.59. A solution with initial conditions in this subspace has the form

$$x(t) = c_1 e^{\sigma t}e^{i\omega t}v + c_2 e^{\sigma t}e^{-i\omega t}\bar{v}$$

If the initial conditions are real, then $c_2 = \bar{c}_1$ (to cancel out the imaginary parts of the two terms in this equation). Equivalently, we can write that

$$x(t) = \operatorname{Re}\left(c_1 e^{\sigma t}e^{i\omega t}v\right)$$

where Re denotes the real part of an expression. Now writing $c_1 = c_r + ic_i$, $v = v_r + iv_i$ and $e^{i\omega t} = \cos\omega t + i\sin\omega t$, this can be written in real form as

$$x(t) = e^{\sigma t}(c_r \cos \omega t + c_i \sin \omega t)v_r + e^{\sigma t}(-c_i \cos \omega t + c_r \sin \omega t)v_i$$

Thus for real initial conditions, the invariant subspace corresponding to a pair of complex conjugate eigenvalues is the plane spanned by v_r and v_i.

If A cannot be diagonalized the situation is not as simple, but is still not really very complicated. We still have that $\dot{y} = Jy$, but J is triangular rather than diagonal. Triangular systems have one-way coupling, so we can solve from the bottom up, back substituting as we go. To illustrate, we consider the case

$$J = \begin{bmatrix} \lambda & 1 \\ 0 & \lambda \end{bmatrix}$$

We can solve the equation $\dot{y}_2 = \lambda y$ first and then back substitute, getting an inhomogeneous problem for y_1. The inhomogeneous term prevents the behavior from being purely exponential, and the general solution becomes (after converting back to the original coordinates)

$$x(t) = c_1 e^{\lambda t} v_1 + c_2 e^{\lambda t}(v_2 + t v_1) \tag{2.5}$$

where v_1 is the eigenvector corresponding to λ and v_2 is the generalized eigenvector; compare with Example 1.13. The line defined by the eigenvector v_1 is an invariant subspace, as is the plane defined by v_1 and v_2. However, the line defined by the generalized eigenvector v_2 is not invariant.

Note the $te^{\lambda t}$ term that appears in (2.5). In initial-value problems, this term allows solutions to grow initially even when all of the eigenvalues have negative real parts. As $t \to \infty$, though, the exponential factor dominates. Thus even when A is defective, its eigenvalues determine the long-time dynamics and, in particular, the stability. The issue of stability is addressed at length in Section 2.5; for the present we note that the steady state $x = 0$ of (2.4) is ASYMPTOTICALLY STABLE—initial conditions approach it as $t \to \infty$ if and only if all the eigenvalues of A have negative real parts.

To summarize, the above results show that every homogeneous constant-coefficient problem $\dot{x} = Ax$ can be rewritten as $\dot{y} = Jy$, where J has a block diagonal structure exemplified by the following template

$$J = \begin{bmatrix} \lambda_1 & 0 & \dots & & & & & \\ 0 & \lambda_2 & 0 & \dots & & & & \\ \dots & 0 & \sigma & -\omega & 0 & \dots & & \\ \dots & 0 & \sigma & \omega & 0 & \dots & & \\ & & \dots & 0 & \lambda_5 & 1 & 0 & \dots \\ & & & \dots & 0 & \lambda_5 & 0 & \dots \\ & & & & \dots & 0 & \lambda_7 & 0 \\ & & & & & \dots & 0 & \lambda_8 \end{bmatrix}$$

The dynamics corresponding to each block are decoupled from those of all the others and the associated eigenvectors define invariant subspaces; the dynamics in each invariant subspace are decoupled from those in all the others.

2.2.3 Qualitative Dynamics of Planar Systems

In a general n-dimensional system, there is a large range of possible combinations of eigenvalues, real and complex, with positive or negative real parts. For $n = 2$, a simple and general classification of the possible dynamics is possible. Such systems are called PLANAR, because all of the dynamics occur on a simple plane (sometimes called the PHASE PLANE) defined by two eigenvectors (or an eigenvector and generalized eigenvector, if A is defective). Writing

$$\dot{x} = Ax = \begin{bmatrix} a & b \\ c & d \end{bmatrix} x$$

the characteristic equation for A is

$$\lambda^2 - (a + d)\lambda + (ad - bc) = 0$$

Notice that $a + d = \text{tr}A$ and $ad - bc = \det A$, which we call T and D, respectively. Recall that $T = \lambda_1 + \lambda_2$ and $D = \lambda_1\lambda_2$. In two dimensions, the eigenvalues are determined only by the trace and determinant of the matrix. When $\text{Re}(\lambda_1) < 0$ and $\text{Re}(\lambda_2) < 0$, any initial condition decays exponentially to the origin—the origin is ASYMPTOTICALLY STABLE. These conditions are equivalent to $T < 0, D > 0$.

Figure 2.1 shows the dynamical regimes that are possible for the planar system as characterized by T and D; asymptotically stable steady-state solutions occupy the second quadrant, excluding the axes. Each regime on Figure 2.1 shows a small plot of the dynamics on the phase

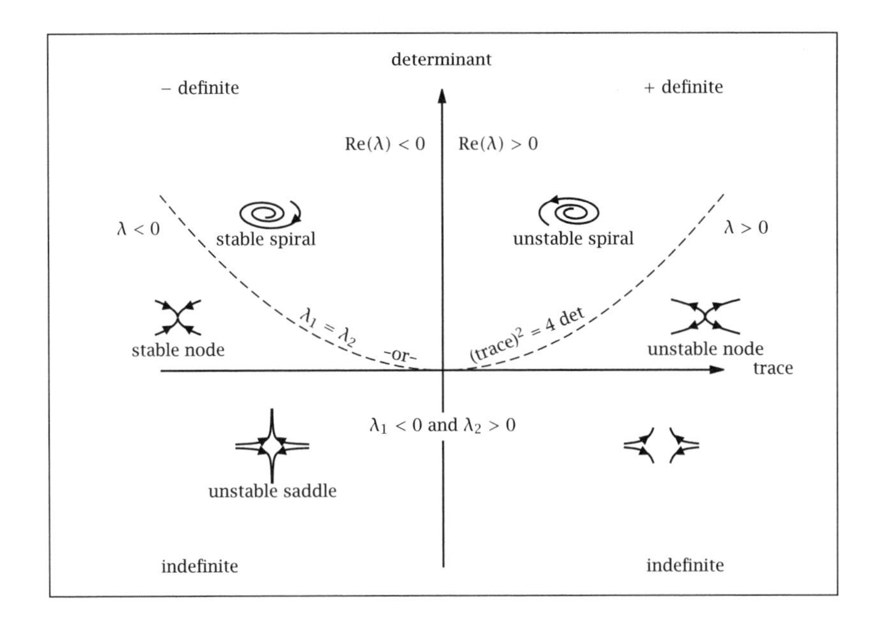

Figure 2.1: Dynamical regimes for the planar system $dx/dt = Ax$, $A \in \mathbb{R}^{2\times2}$ parametrized in the determinant and trace of A; see also Strang (1986, Fig. 6.7).

plane in that regime; the axes correspond to the eigenvectors (or real and imaginary parts of the eigenvectors in the case of complex conjugates) and trajectories $x(t)$ on this plane are shown with time as the parameter. The arrows on the trajectories indicate the direction of time. An important curve on this diagram is $T^2 - 4D = 0$, where the two eigenvalues are equal. This parabola is also the boundary between oscillatory solutions (SPIRALS on the phase plane) and exponential ones (NODES); a spiral arises from a complex conjugate pair of eigenvalues while a node arises from the case of two real eigenvalues with the same sign. In the lower half of the figure, $D < 0$, the eigenvalues are real and with opposite signs. The steady states in this regime are called SADDLE POINTS, because they have one stable direction and one unstable. Figure 2.2 shows the dynamic behavior that occurs on the boundaries between the different regions.

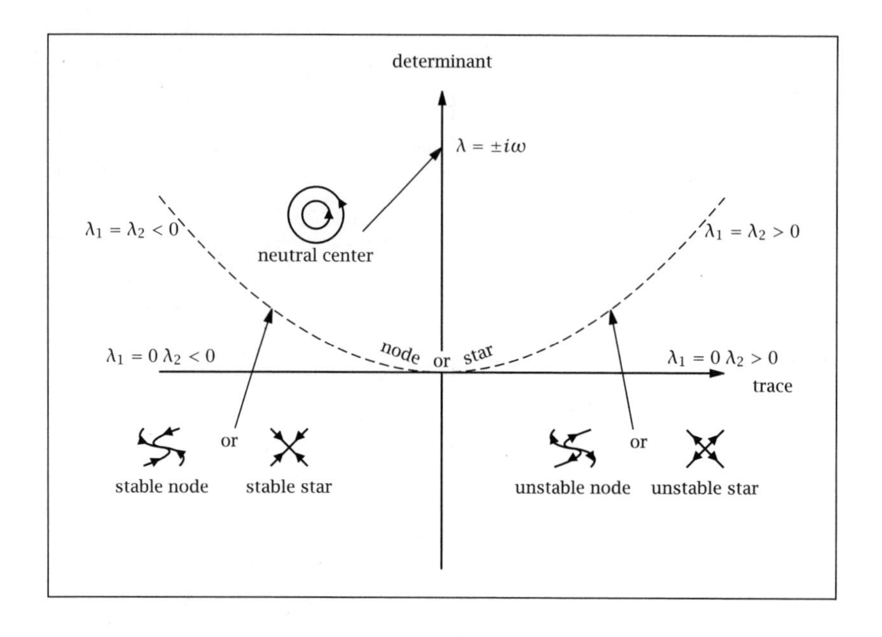

Figure 2.2: Dynamical behavior on the region boundaries for the planar system $dx/dt = Ax$, $A \in \mathbb{R}^{2\times 2}$; see also Strang (1986, Fig. 6.10).

2.2.4 Laplace Transform Methods for Solving the Inhomogeneous Constant-Coefficient Problem

Inhomogeneous constant-coefficient systems also can be decoupled by transformation into Jordan form: $\dot{x} = Ax + g(t)$ becomes $\dot{y} = Jy + h(t)$, where $h(t) = M^{-1}g(t)$. Accordingly, once we understand how to solve the scalar inhomogeneous problem, we will have learned what we need to know to address the vector case. A powerful approach to solving inhomogeneous problems relies on the LAPLACE TRANSFORM.

Definition

Consider functions of time $f(t)$ that vanish for $t < 0$. If there exists a real constant $c > 0$ such that $f(t)e^{-ct} \to 0$ sufficiently fast as $t \to \infty$, we can define the Laplace transform of $f(t)$, denoted $\overline{f}(s)$, for all complex-valued s such that $\mathrm{Re}(s) \geq c$

$$\mathcal{L}(f(t)) = \int_0^\infty e^{-st} f(t)dt \qquad \mathrm{Re}(s) \geq c \tag{2.6}$$
$$= \overline{f}(s)$$

The inverse formula is given by

$$\mathcal{L}^{-1}(\overline{f}(s)) = \frac{1}{2\pi i} \int_{c-i\infty}^{c+i\infty} e^{st}\, \overline{f}(s)ds \qquad t \geq 0 \tag{2.7}$$
$$= f(t)$$

Properties

1. The Laplace transform operator is linear. For every scalar α, β and functions $f(t), g(t)$, the following holds

$$\mathcal{L}\left\{\alpha f(t) + \beta g(t)\right\} = \alpha\overline{f}(s) + \beta\overline{g}(s)$$

   The inverse transform is also linear

$$\mathcal{L}^{-1}\left\{\alpha\overline{f}(s) + \beta\overline{g}(s)\right\} = \alpha f(t) + \beta g(t)$$

2. Transform of derivatives

$$\mathcal{L}\left(\frac{df(t)}{dt}\right) = s\overline{f}(s) - f(0)$$
$$\mathcal{L}\left(\frac{d^2f(t)}{dt^2}\right) = s^2\overline{f}(s) - sf(0) - f'(0)$$
$$\mathcal{L}\left(\frac{d^nf(t)}{dt^n}\right) = s^n\overline{f}(s) - s^{n-1}f(0) - s^{n-2}f'(0) - \cdots$$
$$- sf^{(n-2)}(0) - f^{(n-1)}(0)$$

3. Transform of integral

$$\mathcal{L}\left(\int_0^t f(t')dt'\right) = \frac{1}{s}\overline{f}(s)$$

4. Derivative of transform with respect to s

$$\mathcal{L}(t^n f(t)) = (-1)^n \frac{d^n\overline{f}(s)}{ds^n}$$

5. Time delay and s delay

$$\mathcal{L}(f(t-a)H(t-a)) = e^{-as}\overline{f}(s)$$
$$\mathcal{L}(e^{at}f(t)) = \overline{f}(s-a)$$

where the Heaviside or unit step function is defined as

$$H(t) = \begin{cases} 0 & t < 0 \\ 1 & 0 < t \end{cases}$$

6. Laplace convolution theorem

$$\mathcal{L}\left(\int_0^t f(t')g(t-t')dt'\right) = \overline{f}(s)\overline{g}(s)$$
$$\mathcal{L}\left(\int_0^t f(t-t')g(t')dt'\right) = \overline{f}(s)\overline{g}(s)$$

7. Final value theorem

$$\lim_{s\to 0} s\overline{f}(s) = \lim_{t\to\infty} f(t)$$

if and only if $s\overline{f}(s)$ is bounded for all $\mathrm{Re}(s) \geq 0$

8. Initial-value theorem

$$\lim_{s\to\infty} s\overline{f}(s) = \lim_{t\to 0^+} f(t)$$

We can readily compute the Laplace transform of many simple $f(t)$ by using the definition and performing the integral. In this fashion we can construct Table 2.1 of Laplace transform pairs. Such tables prove useful in solving differential equations. We next solve a few examples using the Laplace transform.

Example 2.1: Particle motion

Consider the motion of a particle of mass m connected to a spring with spring constant K and experiencing an applied force $F(t)$ as depicted in Figure 2.3.

Let y denote the displacement from the origin and model the spring as applying force $F_s = -Ky$. Newton's equation of motion for this system is then

$$m\frac{d^2y}{dt^2} = F - Ky$$

$f(t)$	$\overline{f}(s)$
$\delta(t)$	1
1	$\dfrac{1}{s}$
t	$\dfrac{1}{s^2}$
$t^n \quad -1 < n$	$\dfrac{\Gamma(n+1)}{s^{n+1}}$
$\cos \omega t$	$\dfrac{s}{s^2 + \omega^2}$
$\sin \omega t$	$\dfrac{\omega}{s^2 + \omega^2}$
$\sinh \omega t$	$\dfrac{\omega}{s^2 - \omega^2}$
$\cosh \omega t$	$\dfrac{s}{s^2 - \omega^2}$
e^{at}	$\dfrac{1}{s - a}$
te^{at}	$\dfrac{1}{(s-a)^2}$
$e^{at} \cos \omega t$	$\dfrac{s - a}{(s-a)^2 + b^2}$
$e^{at} \sin \omega t$	$\dfrac{b}{(s-a)^2 + b^2}$

Table 2.1: Small table of Laplace transform pairs. A more extensive table is found in Appendix A.

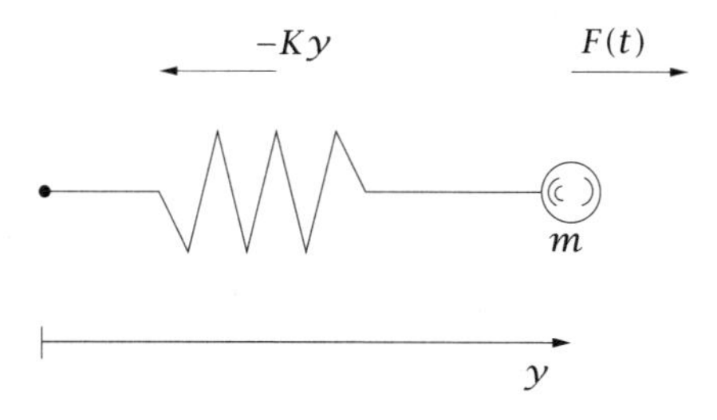

Figure 2.3: Particle of mass m at position y experiences spring force $-Ky$ and applied force $F(t)$.

We require two boundary conditions for this second-order differential equation. If we assume the particle is initially at rest at the origin, then the boundary conditions are both specified at $t = 0$ and these initial conditions are

$$y(0) = 0 \qquad \frac{dy}{dt}(0) = 0$$

If we divide by the mass of the particle we can express the model as

$$\frac{d^2y}{dt^2} + k^2 y = f$$

$$y(t) = 0 \qquad t = 0$$

$$\frac{dy(t)}{dt} = 0 \qquad t = 0$$

in which $k^2 = K/m$ and $f = F/m$. Take the Laplace transform of the model and find the position of the particle versus time $y(t)$, for arbitrary applied force $f(t)$.

Solution

Taking the Laplace transform of the equation of motion and substituting in the two initial conditions gives

$$s^2\overline{y}(s) - sy(0) - y'(0) + k^2\overline{y}(s) = \overline{f}(s)$$

$$s^2\overline{y}(s) + k^2\overline{y}(s) = \overline{f}(s)$$

Solving this equation for $\overline{y}(s)$ gives

$$\overline{y}(s) = \frac{\overline{f}(s)}{s^2 + k^2}$$

We see the transform is the product of two functions of s. The inverse of each of these is available

$$\mathcal{L}^{-1}(\overline{f}(s)) = f(t) \qquad \mathcal{L}^{-1}\left(\frac{1}{s^2 + k^2}\right) = \frac{1}{k}\sin kt$$

The first follows by the definition of $\overline{f}(s)$ and the second follows from Table 2.1. Using the convolution theorem then gives

$$y(t) = \frac{1}{k}\int_0^t f(t')\sin k(t - t')dt'$$

and we have the complete solution. We see that the particle position is a convolution integral of the applied force with the sine function. The reader may wish to check that this solution indeed satisfies the differential equation and both initial conditions as claimed. □

Example 2.2: A forced first-order differential equation

Consider the first-order differential equation with forcing term

$$\frac{dx}{dt} = ax + bu(t)$$
$$x(0) = x_0$$

Use the Laplace transform to find $x(t)$ for any forcing $u(t)$.

Solution

Taking the Laplace transform, substituting the initial condition, and solving for $\overline{x}(s)$, give

$$s\overline{x}(s) - x_0 = a\overline{x}(s) + b\overline{u}(s)$$
$$\overline{x}(s) = \frac{x_0}{s+a} + b\frac{\overline{u}(s)}{s-a}$$

We can invert the first term directly using Table 2.1, and the second term using the table and the convolution theorem giving

$$x(t) = x_0 e^{at} + b\int_0^t e^{a(t-t')}u(t')dt'$$

We see the effect of the initial condition x_0 and the forcing term $u(t)$. If $a < 0$ so the system is asymptotically stable, the effect of the initial condition decays exponentially with time. The forcing term affects the solution through the convolution of u with the time-shifted exponential. □

Example 2.3: Sets of coupled first-order differential equations

Consider next the inhomogeneous constant coefficient system (2.3), with $g(t) = Bu(t)$

$$\frac{dx}{dt} = Ax + Bu(t)$$
$$x(0) = x_0$$

in which $x \in \mathbb{R}^n$, $u \in \mathbb{R}^m$, $A \in \mathbb{R}^{n\times n}$, $B \in \mathbb{R}^{n\times m}$. In systems and control applications, x is known as the state vector and u is the manipulated variable vector. Use Laplace transforms to find $x(t)$ for this problem.

Solution

Again taking the Laplace transform, substituting the initial condition, and solving for $\overline{x}(s)$ gives

$$\begin{aligned} s\overline{x}(s) - x_0 &= A\overline{x}(s) + B\overline{u}(s) \\ (sI - A)\overline{x}(s) &= x_0 + B\overline{u}(s) \\ \overline{x}(s) &= (sI - A)^{-1}x_0 + (sI - A)^{-1}B\overline{u}(s) \end{aligned}$$

We next require the matrix version of the Laplace transform pair

$f(t)$		$\overline{f}(s)$
e^{at}	$a \in \mathbb{R}$	$\dfrac{1}{s-a}$
e^{At}	$A \in \mathbb{R}^{n\times n}$	$(sI - A)^{-1}$

which can be checked by applying the definition of the Laplace transform. Using this result and the convolution theorem gives

$$x(t) = e^{At}x_0 + \int_0^t e^{A(t-t')}Bu(t')dt'$$

Notice we cannot move the constant matrix B outside the integral as we did in the scalar case because the indices in the matrix multiplications must conform as shown below

$$\underbrace{x(t)}_{n\times 1} = \underbrace{e^{At}}_{n\times n}\underbrace{x_0}_{n\times 1} + \int_0^t \underbrace{e^{A(t-t')}}_{n\times n}\underbrace{B}_{n\times m}\underbrace{u(t')}_{m\times 1}dt'$$

□

2.2.5 Delta Function

The DELTA FUNCTION, also known as the Dirac delta function (Dirac, 1958, pp. 58-61) or the unit impulse, is an idealization of a narrow and tall "spike." Two examples of such functions are

$$g_\alpha(x) = \frac{1}{\sqrt{4\pi\alpha}} e^{-x^2/4\alpha} \tag{2.8}$$

$$g_\alpha(x) = \frac{\alpha}{\pi\,(\alpha^2 + x^2)} \tag{2.9}$$

where $\alpha > 0$. Setting $x = 0$ and then taking the limit $\alpha \to 0$ shows that $\lim_{\alpha\to 0} g_\alpha(0) \to \infty$, while setting $x = x_0 \neq 0$ and taking the same limit shows that for any nonzero x_0, $\lim_{\alpha\to 0} g_\alpha(x_0) = 0$. These functions become infinitely high and infinitely narrow. Furthermore, they both have unit area

$$\int_{-\infty}^{\infty} g_\alpha(x)\, dx = 1$$

A set of functions depending on a parameter and obeying the above properties is called a DELTA FAMILY. The delta function $\delta(x)$ is the limiting case of a delta family as $\alpha \to 0$. It has infinite height, zero width, and unit area. It is most properly thought of as a GENERALIZED FUNCTION OR DISTRIBUTION; the mathematical theory of these objects is described in Stakgold (1998).

Operationally, the key feature of the delta function is that when integrated against a "normal" function $f(x)$ the delta function extracts the value of f at the x value where the delta function has its singularity

$$\int_{-\infty}^{\infty} f(x)\delta(x)\, dx = \lim_{\alpha\to 0} \int_{-\infty}^{\infty} f(x) g_\alpha(x)\, dx = f(0) \tag{2.10}$$

The delta function also can be viewed as the generalized derivative of the discontinuous unit step or Heaviside function $H(x)$

$$\delta(x) = \frac{dH(x)}{dx}$$

Also note that the interval of integration in (2.10) does not have to be $(-\infty, \infty)$. The integral over *any* interval containing the point of singularity for the delta function produces the value of $f(x)$ at the point of singularity. For example

$$\int_{a-\epsilon}^{a+\epsilon} f(x)\delta(x-a)dx = f(a) \qquad \text{for all } \epsilon > 0 \quad \text{for all } a \in \mathbb{R}$$

Finally, by changing the variable of integration we can show that the delta function is an even function

$$\delta(-x) = \delta(x)$$

Derivatives of the Delta Function

Doublet. An interesting property of the delta function is that it is also *differentiable*. The first derivative is termed the doublet or dipole, usually denoted $\delta'(x)$

$$\delta'(x) = \frac{d\delta(x)}{dx}$$

Sometimes we see the dot notation $\dot{\delta}(x)$ to denote the doublet instead of $\delta'(x)$. If we perform integration by parts on the integral

$$\int_{-\infty}^{\infty} f(x)\delta'(x)dx$$

we find that the doublet selects the negative of the first derivative of f evaluated at the location of the doublet's singularity

$$\int_{-\infty}^{\infty} f(x)\delta'(x)dx = -f'(0) \tag{2.11}$$

Note the sign in this equation. We also find by changing the variable of integration that, unlike the delta function, or singlet, which is an even function, the doublet is odd

$$\delta'(x) = -\delta'(-x)$$

Higher-order derivatives. Repeated integration by parts produces the following higher-order formulas for triplets, quadruplets, etc.

$$\int_{-\infty}^{\infty} f(x)\delta^{(n)}(x)dx = (-1)^n f^{(n)}(0) \qquad n \geq 0$$

As with the singlet and doublet, we can change the variable of integration and shift the location of the singularity to obtain the general formula

$$\int_{-\infty}^{\infty} f(x)\delta^{(n)}(x-a)dx = (-1)^n f^{(n)}(a) \qquad n \geq 0 \quad a \in \mathbb{R}$$

Finally we can use the definition of the Laplace transform to take the transform of the delta function and its derivatives to obtain the transform pairs listed in Table 2.2.

2.3 Linear Equations with Variable Coefficients

2.3.1 Introduction

In many chemical engineering applications, equations like this one are encountered

$$x^2\frac{d^2y}{dx^2} + x\frac{dy}{dx} + (x^2 - \nu^2)y = 0 \tag{2.12}$$

$f(t)$	$\overline{f}(s)$
$H(t)$	$\frac{1}{s}$
$\delta(t)$	1
$\delta'(t)$	s
$\delta''(t)$	s^2
$\delta^{(n)}(t)$	s^n

Table 2.2: Laplace transform pairs involving δ and its derivatives.

This is called BESSEL'S EQUATION OF ORDER ν, and arises in the study of diffusion and wave propagation via the Laplacian operator in cylindrical coordinates. Since the coefficients in front of the derivative terms are not constant, the exponential functions that solved constant-coefficient problems does not work here. Typically, variable-coefficient problems must be solved by power series methods or by numerical methods, as they have no simple closed-form solution. We focus here on second-order equations, as they arise most commonly in applications.

2.3.2 The Cauchy-Euler Equation

The CAUCHY-EULER equation, also called the EQUIDIMENSIONAL equation, has a simple exact solution that illustrates many important features of variable-coefficient problems and arises during the solution of many problems. The second-order Cauchy-Euler equation has the form

$$a_0 x^2 y'' + a_1 x y' + a_2 y = 0 \tag{2.13}$$

where $y' = dy/dx$. Its defining feature is that the term containing the nth derivative is multiplied by the nth power of x. Because of this, guessing that the form of the solution is $y = x^\alpha$ yields the quadratic equation $a_0\alpha(\alpha - 1) + a_1\alpha + a_2 = 0$. If this equation has distinct roots α_1 and α_2, then each root leads to a solution and thus the general solution is found

$$y = c_1 x^{\alpha_1} + c_2 x^{\alpha_2} \tag{2.14}$$

For example, let $a_0 = 1, a_1 = 1, a_2 = -9$, yielding the equation $\alpha^2 - 9 = 0$, which has solutions $\alpha = \pm 3$. Thus the equation has two solutions of the form $y = x^\alpha$: the general solution is $y = c_1 x^3 + c_2 x^{-3}$. Notice that this solution can blow up at $x = 0$; this singular behavior does not arise in constant-coefficient (linear) problems, but is frequently found in variable-coefficient and nonlinear problems.

In the case of a repeated root, the general solution does not have the form given above. Instead, the technique of REDUCTION OF ORDER, also called VARIATION OF PARAMETERS, can be used. In this technique, given one solution to a second-order linear problem $y_1(x)$, the second can be found in the form $y_2(x) = A(x)y_1(x)$. For example, let $a_0 = 1, a_1 = -2, a_2 = 1$, yielding the repeated root $\alpha = 1$. Thus $y_1 = x$, and $y_2 = A(x)x$, which, upon substitution into the differential equation, yields

$$A''x^3 + 2A'x^2 - A'x^2 - Ax + Ax = 0$$

which simplifies to

$$A''x + A' = 0$$

Letting $A' = w$ leads to a simple first-order equation for w

$$w'x + w = 0$$

so that $w = c/x$ and thus $A = c \ln x + d$, where c and d are arbitrary constants. Thus the general solution for this problem can be written

$$y(x) = c_1 x + c_2 x \ln x = x(c_1 + c_2 \ln x)$$

It can be shown in general that (second-order) Cauchy-Euler equations with repeated roots have the general solution

$$y(x) = x^\alpha (c_1 + c_2 \ln x) \tag{2.15}$$

2.3.3 Series Solutions and the Method of Frobenius

A general linear second-order problem can be written

$$p(x)y'' + q(x)y' + r(x)y = 0 \tag{2.16}$$

or

$$y'' + \frac{q(x)}{p(x)}y' + \frac{r(x)}{p(x)}y = 0 \tag{2.17}$$

If $\frac{q(x)}{p(x)}$ and $\frac{r(x)}{p(x)}$ are ANALYTIC, i.e., they have a convergent Taylor series expansion, at some point $x = a$, then a is an ORDINARY POINT. Otherwise, $x = a$ is a SINGULAR POINT.

If $x = a$ is an ordinary point, there exist solutions in the form of power series

$$y(x) = \sum_{n=0}^{\infty} c_n (x-a)^n \tag{2.18}$$

Two such solutions can be found, thus yielding the general solution. Letting ρ be the distance between a and the nearest singular point of the differential equation, which might be at a complex rather than a real value of x, the series converge[1] for $|x - a| < \rho$. Accordingly ρ is called the RADIUS OF CONVERGENCE of the series. The exception to this is when a series solution truncates after a finite number of terms, i.e., $c_M = 0$ for $M > M_0$; in this case the sum is always finite for finite x.

Example 2.4: Power series solution for a constant-coefficient equation

Let $p(x) = 1$, $q(x) = 0$ and $r(x) = k^2$, resulting in the equation

$$y'' + k^2 y = 0$$

Solve this by power series expansion.

Solution

We seek a solution by expanding around the ordinary point $a = 0$. For this simple example, every point is an ordinary point. Inserting the solution form, (2.18), into this equation yields

$$\sum_{n=2}^{\infty} n(n-1)c_n x^{n-2} + k^2 \sum_{n=0}^{\infty} c_n x^n = 0$$

The two sums can be combined if we can make their lower limits the same. Thus we set $n = m + 2$ in the first series and $n = m$ in the second, obtaining

$$\sum_{m=0}^{\infty} \left[(m+2)(m+1)c_{m+2} + k^2 c_m\right] x^m = 0$$

This can only hold if the term inside the square brackets is zero for all m, requiring that

$$c_{n+2} = \frac{-c_n k^2}{(n+2)(n+1)}$$

[1]A full understanding of convergence of power series requires knowledge of functions of complex variables, see, e.g., Ablowitz and Fokas (2003).

(where we have now reverted to using n as the index). Leaving c_0 and c_1 arbitrary, we find that

$$\begin{aligned} c_2 &= \frac{-c_0 k^2}{2!} \\ c_3 &= \frac{-c_1 k^2}{3!} \\ c_4 &= \frac{-c_2 k^2}{4 \cdot 3} = \frac{c_0 k^4}{4!} \\ c_5 &= \frac{-c_3 k^2}{5 \cdot 4} = \frac{c_1 k^4}{5!} \\ &\vdots \end{aligned}$$

Absorbing a factor of $1/k$ into c_1 (recall that it is arbitrary), the series solution becomes

$$y(x) = c_0 \left(1 - \frac{k^2 x^2}{2!} + \frac{k^4 x^4}{4!} - \ldots\right) + c_1 \left(kx - \frac{k^3 x^3}{3!} + \frac{k^5 x^5}{5!} - \ldots\right)$$

Note that this has two arbitrary constants c_0 and c_1, so it is the general solution. The two infinite series can be recognized as the Taylor expansions of two familiar functions, and we can thus rewrite the general solution as

$$y(x) = c_0 \cos kx + c_1 \sin kx$$

□

If $p(x) \to 0$ at some point $x = a$, the situation is more complex. We set $a = 0$ from now on for convenience. Now $q(x)/p(x)$ and $r(x)/p(x)$ are not analytic and $x = 0$ is called a SINGULAR POINT. If $x\ (q(x)/p(x))$ and $x^2\ (r(x)/p(x))$ *are* analytic, i.e., the singularity in $p(x)$ is not very strong, then the point is a REGULAR SINGULAR POINT. Observe that $x = 0$ is a regular singular point for the Cauchy-Euler equation. In fact, by multiplying (2.17) by x^2 and Taylor-expanding the coefficients, one can see that when the conditions for a regular singular point are satisfied, this general case reduces precisely to a Cauchy-Euler equation as $x \to 0$. This observation motivates the METHOD OF FROBENIUS, which seeks solutions of the form

$$y(x) = x^{\alpha} \sum_{n=0}^{\infty} c_n x^n \tag{2.19}$$

The power series has the same convergence properties as described above for ordinary points.

Example 2.5: Frobenius solution for Bessel's equation of order zero

Bessel's equation (2.12) with $\nu = 0$ is

$$xy'' + y' + xy = 0 \tag{2.20}$$

Here $x = 0$ is a regular singular point. Solve by the method of Frobenius.

Solution

Observe that this equation can be written $x^2y'' + xy' + (0 + x^2)y = 0$ so the corresponding Cauchy-Euler equation is thus $x^2y'' + xy' + 0y = 0$. Seeking a solution $y = x^\alpha$ yields the repeated root $\alpha = 0$ and thus a general solution $y(x) = c_1 + c_2 \ln x$. As we will see, this structure is reflected in the form of the solution to Bessel's equation.

Inserting the Frobenius solution form, (2.19) into (2.20) yields that

$$\sum_{n=0}^{\infty}(n+\alpha)(n+\alpha-1)c_n x^{n+\alpha-1} + \sum_{n=0}^{\infty}(n+\alpha)c_n x^{n+\alpha-1} + \sum_{n=0}^{\infty} c_n x^{n+\alpha+1} = 0$$

To simplify this series, set $n = m + 2$ in the first two sums and $m = n$ in the third. Then set all the ms back to n. This yields a summation starting at $n = -2$, which is fine as long as we make $c_{-2} = c_{-1} = 0$. The formula becomes

$$\sum_{n=-2}^{\infty}\left[(n+\alpha+2)^2 c_{n+2} + c_n\right]x^{n+\alpha+1} = 0$$

Since x can vary, the equality can only hold if the terms in the brackets are all zero. This is the recursion formula for the coefficients c_n. The first term ($n = -2$) picks out the Cauchy-Euler behavior and is called the INDICIAL EQUATION. Since $c_{-2} = 0$, it reduces to $(n + \alpha + 2)^2 = \alpha^2 = 0$. As we anticipated above with the corresponding Cauchy-Euler equation, this has the repeated root $\alpha = 0$. The general recursion relation for the coefficients reads

$$c_{n+2} = -\frac{c_n}{(n+2)^2}$$

Since $c_{-1} = 0$, all the coefficients with n odd are zero. Therefore, only one of the two solutions to the problem has the form of (2.19), again, in parallel with the Cauchy-Euler analysis. With some rearrangements, this solution becomes

$$y_1(x) = \sum_{n=0}^{\infty}\frac{(-1)^n}{(n!)^2}\left(\frac{x}{2}\right)^{2n}$$

This function has the special symbol $J_0(x)$ and is called the "Bessel function of the first kind and order zero." For general ν, the solutions are denoted $J_\nu(x)$. A second solution can be found for this problem by variation of parameters; see Exercise 2.31. It is not of Frobenius form, having a logarithmic singularity as $x \to 0$ (again as anticipated from the solution to the corresponding Cauchy-Euler equation). It is called $Y_0(x)$ and is the "Bessel function of the second kind and order zero." Singular solutions for general ν are denoted $Y_\nu(x)$. The general solution is

$$y(x) = c_1 J_0(x) + c_2 Y_0(x) \tag{2.21}$$

□

See Table 2.3 for a graph of functions J_0 and Y_0. Note that for comparison purposes, the table also shows the solution for the radial part of $\nabla^2 y \pm y = 0$ in rectangular, cylindrical, and spherical coordinates.

In the previous example, the indicial equation yielded a single repeated root for α and one solution of Frobenius form. Other cases are possible. Here are the possibilities and their consequences.

1. If the indicial roots are equal, only one Frobenius solution is obtained. This is what occurred in the above example.

2. If the roots differ by a noninteger constant, then each root leads to a solution and the general solution is obtained.

3. If the roots differ by an integer then the (algebraically) larger root leads to a Frobenius solution and either

   (a) the smaller root also leads to a Frobenius solution and the general solution is obtained, or

   (b) the smaller root does not lead to a second solution of Frobenius form. A second solution can be found by reduction of order and have a logarithmic singularity just as in the Cauchy-Euler case.

2.4 Function Spaces and Differential Operators

2.4.1 Functions as Vectors

One of the main tasks of mathematical modeling is the exact or approximate representation of functions. Here we extend the ideas of vectors and bases into the regime where each vector is a function, so the space the vectors live in is a FUNCTION SPACE.

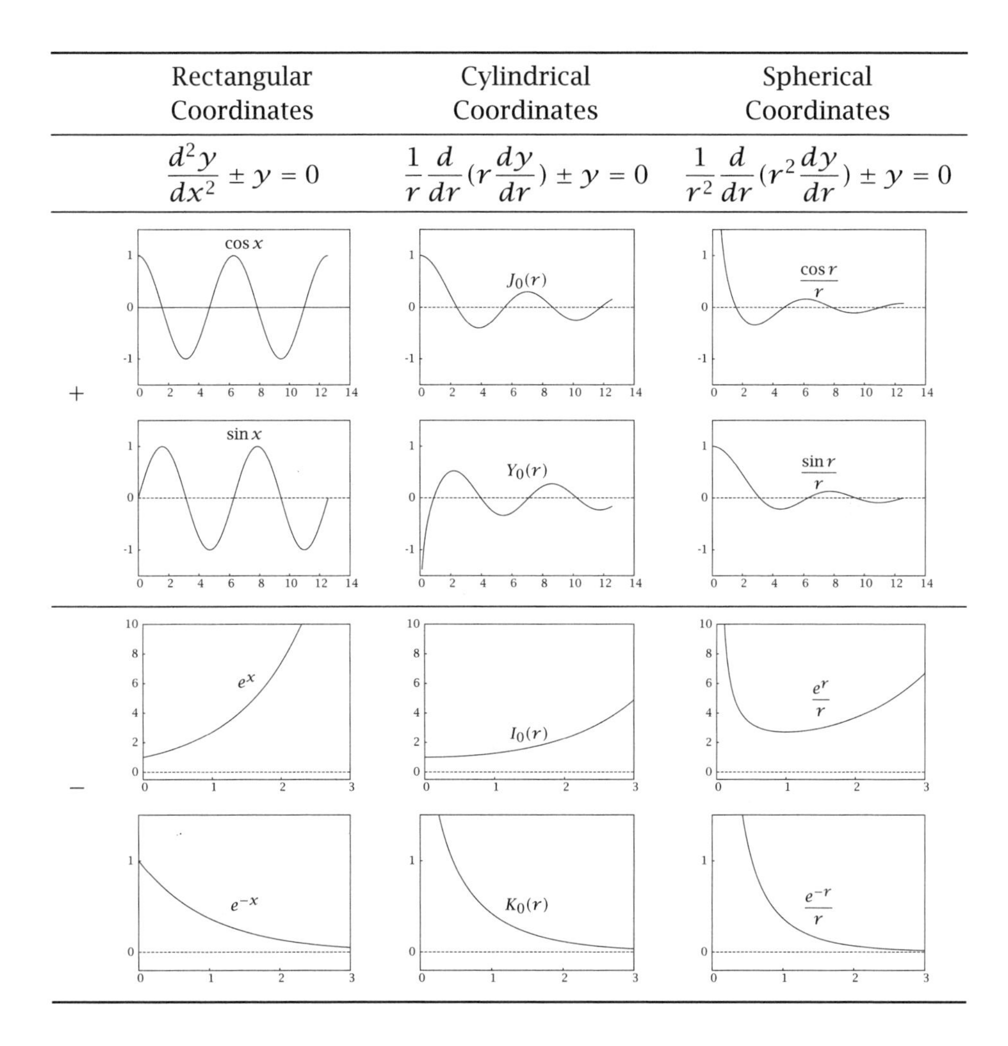

Table 2.3: The linear differential equations arising from the radial part of $\nabla^2 y \pm y = 0$ in rectangular, cylindrical, and spherical coordinates. Bessel functions (J_0, Y_0) and modified Bessel functions (I_0, K_0) are two linearly independent solutions in cylindrical coordinates for the plus and minus signs, respectively. The solutions in spherical coordinates are called spherical Bessel functions.

In the finite-dimensional space $\mathbb{C}^n$, the usual inner product of vectors u and v is simply the n-dimensional version of the dot product

$$(u, v) = \sum_{i=1}^{n} u_i \overline{v}_i$$

For functions $u(x)$ and $v(x)$ in a domain $a \le x \le b$, a natural analog to this relation is

$$(u(x), v(x)) = \int_a^b u(x)\overline{v}(x)\, dx$$

This is the usual inner product for functions defined on the interval $[a, b]$. From this inner product, we can obtain a norm

$$\|u(x)\| = \sqrt{(u, u)} = \left[\int_a^b u(x)\overline{u}(x)\, dx\right]^{1/2}$$

Another inner product, which plays an important role shortly, is given by the formula

$$(u(x), v(x))_w = \int_a^b u(x)\overline{v}(x)w(x)\, dx$$

where $w(x)$ is a so-called weight function and must be positive in (a, b). Finally, with these definitions, a bounded function is one that satisfies

$$\int_a^b |u(x)|^2\, w(x)\, dx = \|u\|_w^2 < \infty$$

With these definitions in hand, we can define an important function space. The set of functions $u(x)$ that satisfy $(u, u) = \|u\| < \infty$ with the usual inner product ($w = 1$) is the LEBESGUE SPACE $L_2(a, b)$. If we had used a nonunit weight function $w(x)$ in the inner product, we would have $L_{2,w}(a, b)$. Lebesgue spaces are examples of HILBERT SPACES. A Hilbert space is essentially identical to a space of vectors with infinitely many components, so that *all of our intuition about directions, lengths and angles carries over from two dimensions into an infinite number of dimensions!*

Basis Sets and Fourier Series

In a finite-dimensional space, any vector can be represented in an orthogonal basis $\{e_1, e_2, \dots\}$ as

$$u = \sum_{i=1}^{n} \frac{(u, e_i)}{(e_i, e_i)} e_i$$

The same is true in a Hilbert space, except that each basis vector is now a function $\phi_i(x)$ and the sum is infinite[2], e.g.,

$$u(x) = \sum_{i=1}^{\infty} \frac{(u(x), \phi_i(x))}{(\phi_i(x), \phi_i(x))} \phi_i(x)$$

Two of the most important basis sets for L_2 are the trigonometric functions and the Legendre polynomials.

Consider the space $L_2(-\pi, \pi)$, i.e., the Lebesgue space defined as above, except on the interval[3] from $-\pi$ to π. The functions

$$e^{ikx} = \cos kx + i \sin kx, k = -\infty, \ldots, -2, -1, 0, 1, 2, \ldots, \infty$$

are in this space. In addition, they satisfy

$$(e^{ikx}, e^{ilx}) = 2\pi\delta_{kl}$$

That is, they are orthogonal. A natural question, then, is whether this set can be used as a basis for $L_2(-\pi, \pi)$. Specifically, we examine the proposition that every function in $L_2(-\pi, \pi)$ can be represented as

$$f(x) = \sum_{k=-\infty}^{\infty} c_k e^{ikx} \tag{2.22}$$

This is the trigonometric FOURIER SERIES representation of $f(x)$. The c_k are the FOURIER COEFFICIENTS and are given by the standard formula for expansion of a vector in an orthogonal basis

$$c_k = \frac{(f, e_k)}{(e_k, e_k)} = \frac{1}{2\pi} \int_{-\pi}^{\pi} f(x) e^{-ikx} dx \tag{2.23}$$

The equality (2.22) cannot possibly hold at every point x for every function $f(x) \in L_2(-\pi, \pi)$, simply because trigonometric functions are continuous and smooth, and functions in $L_2(-\pi, \pi)$ are allowed to have discontinuities. Distance in $L_2(-\pi, \pi)$ is not measured pointwise, however, but rather via the L_2 norm. To address the issue of the distance between a function and its Fourier series representation, consider the finite trigonometric series expansion

$$p_K(x) = \sum_{k=-K}^{K} g_k e^{ikx}$$

[2]Depending on the specific situation, the sum's lower limit might be $0, 1$, or $-\infty$.
[3]The interval $(0, 2\pi)$ might also be used.

and recall that in L_2, the distance between f and p_k is given by

$$\|f - p_K\|^2 = \int_{-\pi}^{\pi} \left| f(x) - \sum_{k=-K}^{K} g_k e^{ikx} \right|^2 dx$$

We can now ask the question: given integer K, what coefficients g_k minimize the L_2 distance between f and p_K? It can be shown that the solution to this minimization problem is

$$g_k = c_k$$

for $k = 1, 2, \ldots, K$, with the c_k given by (2.23) (Gasquet and Witomski, 1999). Because the c_k do not depend on the number of terms, K, if we decide to increase the order of the approximation, we do not need to recalculate the lower-order coefficients. We can now consider the truncated Fourier series

$$f_K(x) = \sum_{k=-K}^{K} c_k e^{ikx}$$

The question of convergence of this series to the function f is nontrivial; we state without proof that for functions in $L_2(-\pi, \pi)$

$$\|f(x) - f_K(x)\|^2 \to 0 \qquad \text{as } K \to \infty$$

The rate of convergence of f_K to f depends on the behavior of the Fourier coefficients c_k as $|k| \to \infty$. Returning to (2.23) and integrating by parts

$$2\pi c_k = (f, e^{ikx}) \tag{2.24}$$

$$= \int_{-\pi}^{\pi} f(x) e^{-ikx}\, dx \tag{2.25}$$

$$= \frac{-1}{ik}\left(f(\pi) - f(-\pi)\right)(-1)^k + \frac{1}{ik}\int_{-\pi}^{\pi} f'(x) e^{-ikx}\, dx \tag{2.26}$$

Therefore $|c_k|$ decays at least as fast as k^{-1} as $k \to \infty$. This is often written as $c_k = O(k^{-1})$: "c_k is order k^{-1}." If, additionally, $f(\pi) = f(-\pi)$ and $f'(x)$ is differentiable, then the first term in (2.26) vanishes and we can repeat the integration by parts procedure on the remaining integral to conclude that $c_k = O(k^{-2})$. Iterating this argument, we can conclude that if $f(x)$ is m-times continuously differentiable in $(-\pi, \pi)$, i.e., the mth derivative $f^{(m)}$ is continuous, and that $f^{(j)}$ is periodic for all $j \leq m - 2$, then

$$c_k = O(k^{-m})$$

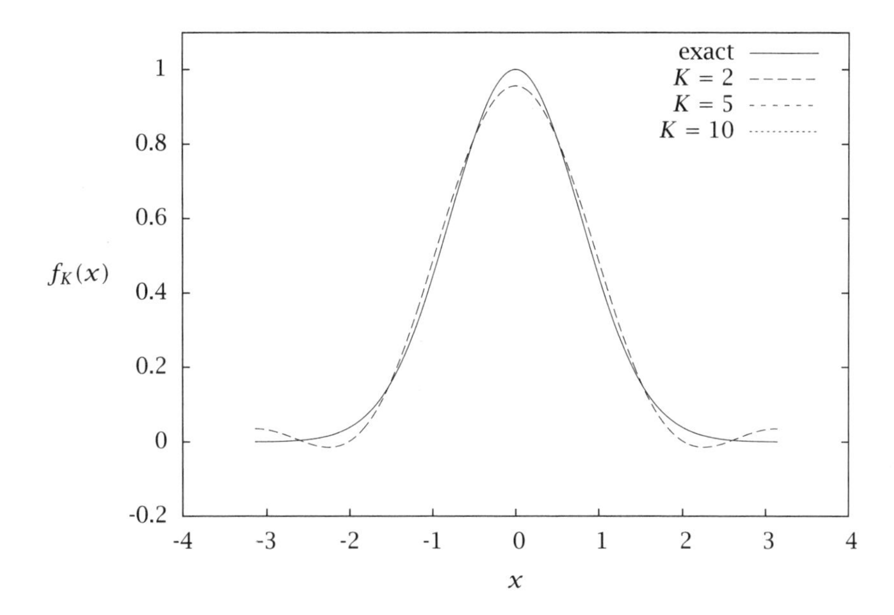

Figure 2.4: Function $f(x) = \exp\left(-8\left(\frac{x}{\pi}\right)^2\right)$ and truncated trigonometric Fourier series approximations with $K = 2, 5, 10$. The approximations with $K = 5$ and $K = 10$ are visually indistinguishable from the exact function.

The case just discussed, in which $c_k = O(k^{-2})$, corresponds to $j = 0, m = 2$. For infinitely smooth periodic functions, this argument implies that the Fourier coefficients decay faster than any finite negative power of k. This is called exponential or SPECTRAL convergence. Figure 2.4 shows truncated Fourier series approximations to the function $f(x) = \exp\left(-8\left(\frac{x}{\pi}\right)^2\right)$ with several values of K. Although this function is not exactly periodic, its function values and derivatives at $x = \pm\pi$ are extremely small, so convergence is rapid.

If $f(x)$ is discontinuous or $f(-\pi) \neq f(\pi)$, then $c_k = O(k^{-1})$—convergence is very slow. The most obvious characteristic of Fourier series representations of discontinuous functions is the GIBBS PHENOMENON, the rapid oscillation of the truncated series f_n in the vicinity of the discontinuity. For further discussion of the convergence of Fourier series see Gasquet and Witomski (1999); and Canuto, Hussaini, Quarteroni, and Zang (2006).

Example 2.6: Fourier series of a nonperiodic function

What is the Fourier series expansion of $f(x) = x$?

Solution

Application of (2.26) immediately yields that

$$c_k = \frac{i(-1)^k}{k} - \frac{i}{k}\delta_{k0}$$

Observe that $c_{-k} = \bar{c}_k$ (see Exercise 2.5), so we can write the Fourier series as

$$f_K(x) = c_0 + 2\sum_{k=1}^{K} (\mathrm{Re}(c_k)\cos kx - \mathrm{Im}(c_k)\sin kx)$$

which in the present case reduces to

$$f_K(x) = \sum_{k=1}^{K} \frac{-2(-1)^k}{k} \sin kx$$

This series contains only sines, not cosines, reflecting the fact that the function $f(x) = x$ is odd. Figure 2.5 shows the approximation for $K = 5, 10$, and 50, which exhibits Gibbs phenomenon as expected for a nonperiodic function.

The plot remains essentially the same if the discontinuity is in the interior rather than on the boundary. For example, the function

$$f(x) = \begin{cases} x + \pi & -\pi \le x < 0 \\ x - \pi & 0 \le x < \pi \end{cases}$$

is periodic (along with all its derivatives) but has a discontinuity at the origin. The Fourier series of this function is the same as that for the previous, except shifted by π

$$f_K(x) = \sum_{k=1}^{K} \frac{-2(-1)^k}{k} \sin k(x+\pi) = \sum_{k=1}^{K} \frac{-2}{k} \sin kx$$

For trigonometric Fourier series, Gibbs phenomenon occurs whether the discontinuity occurs on the boundary or in the interior of the domain.

□

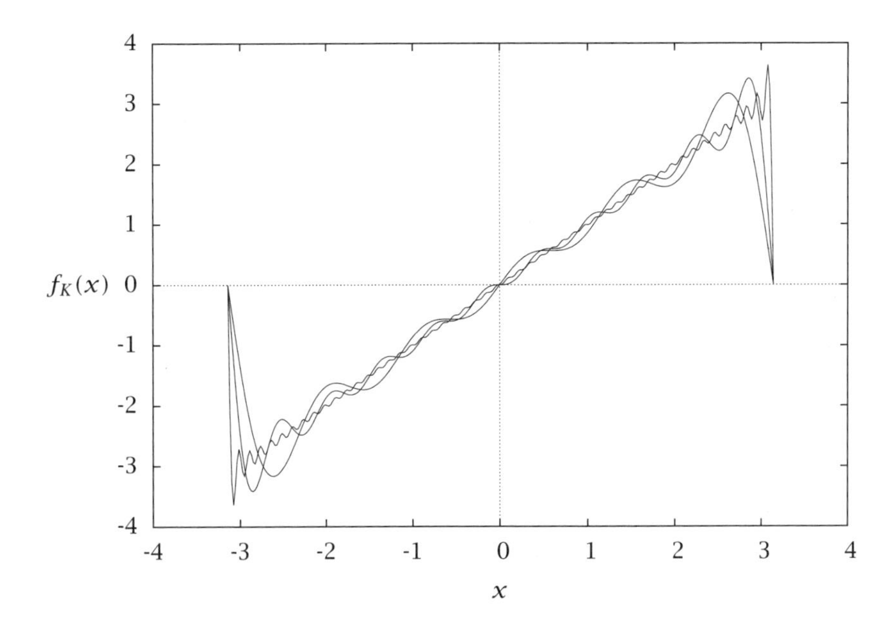

Figure 2.5: Truncated trigonometric Fourier series approximation to $f(x) = x$, using $K = 5, 10, 50$. The wiggles get finer as K increases.

Implicitly, the trigonometric basis assumes that the function is periodic, with the period being the length of the interval. This is why the Gibbs phenomenon occurs if the boundary values of the function are not the same. Another basis that does not make this implicit assumption is given by the so-called LEGENDRE POLYNOMIALS. This basis can be constructed by performing Gram-Schmidt orthogonalization on the set $\{1, x, x^2, x^3, \ldots\}$. The first several of these polynomials, now in the space $L_2(-1, 1)$, the usual setting for polynomial basis functions, are

$$P_0(x) = 1 \tag{2.27}$$

$$P_1(x) = x \tag{2.28}$$

$$P_2(x) = (3x^2 - 1)/2 \tag{2.29}$$

$$P_{j+1}(x) = \frac{2j+1}{j+1} x P_j(x) - \frac{j}{j+1} P_{j-1}(x) \tag{2.30}$$

and the Legendre-Fourier series representation of a function is

$$f(x) = \sum_{i=0}^{\infty} \frac{(f(x), P_i(x))}{(P_i(x), P_i(x))} P_i(x)$$

Note that the sum starts with the index $i = 0$, which is conventional for polynomial bases.

As written, this basis is not orthonormal; instead each polynomial has been scaled so that its value is 1 at $x = 1$. The function $f(x) = x$ can be represented exactly, since $P_1(x) = x$. Convergence for Fourier series based on Legendre polynomials is analogous to that for trigonometric functions; in particular, spectral convergence is found for functions that have infinitely many derivatives, whether they are periodic or not. We refer the interested reader to Canuto et al. (2006) for detailed analysis.

Figure 2.6 shows Legendre-Fourier series approximations to the function $f(x) = \exp(-8x^2)$ truncated at $n + 1$ terms, i.e., including polynomials up to degree n. As with the trigonometric Fourier series approximation of this function, convergence is rapid. Figure 2.7 shows Legendre-Fourier Series approximations to the unit step function $f(x) = H(x)$; because this function is discontinuous, the Legendre-Fourier series also displays Gibbs phenomenon.

The trigonometric and Legendre basis sets are very important, but there are many others that also are important and widely seen in applications. The following section introduces an entire class of equations, each of whose members generates a basis set.

2.4.2 Self-Adjoint Differential Operators and Sturm-Liouville Equations

When we studied linear algebra, we learned that self-adjoint matrix operators in $\mathbb{R}^n$ have special properties, namely that their eigenvalues are real and their eigenvectors form an orthogonal basis for $\mathbb{R}^n$. Self-adjoint *differential* operators also generate basis vectors (functions).

Recall the definition of the adjoint L^* of an operator L

$$(Lu, v) = (u, L^*v)$$

Let us apply this definition to the operator $L = d/dx$ in the interval

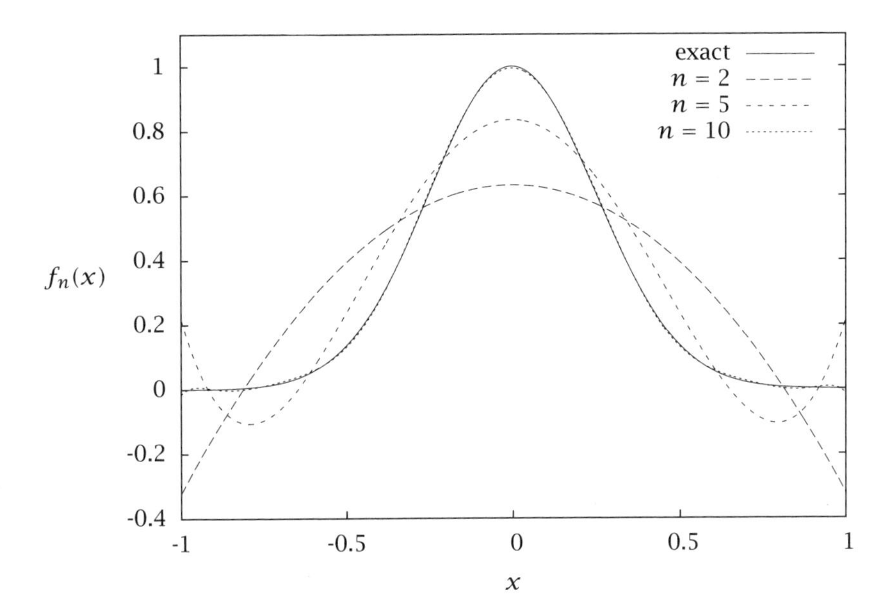

Figure 2.6: Function $f(x) = \exp(-8x^2)$ and truncated Legendre-Fourier series approximations with $n = 2, 5, 10$.

$[0, 1]$ and the usual, i.e., uniformly weighted, inner product

$$\begin{aligned}(Lu, v) &= \int_0^1 u'(x)v(x)dx \\ &= u(1)v(1) - u(0)v(0) - \int_0^1 u(x)v'(x)dx\end{aligned}$$

Since L is here a first derivative, any differential equation involving it requires specification of one boundary condition. As an example, we require that $u(0) = 0$. Now the boundary term at $x = 0$ vanishes. Now observe that if we require that $v(1) = 0$, the boundary term at $x = 1$ also vanishes, leaving the result

$$\begin{aligned}(Lu, v) &= -\int_0^1 u(x)v'(x)dx \\ &= (u, L^*v)\end{aligned}$$

where $L^* = -d/dx$. Therefore, if L is d/dx, operating on functions that vanish at $x = 0$, then from the above equation, $L^* = -d/dx$, operating

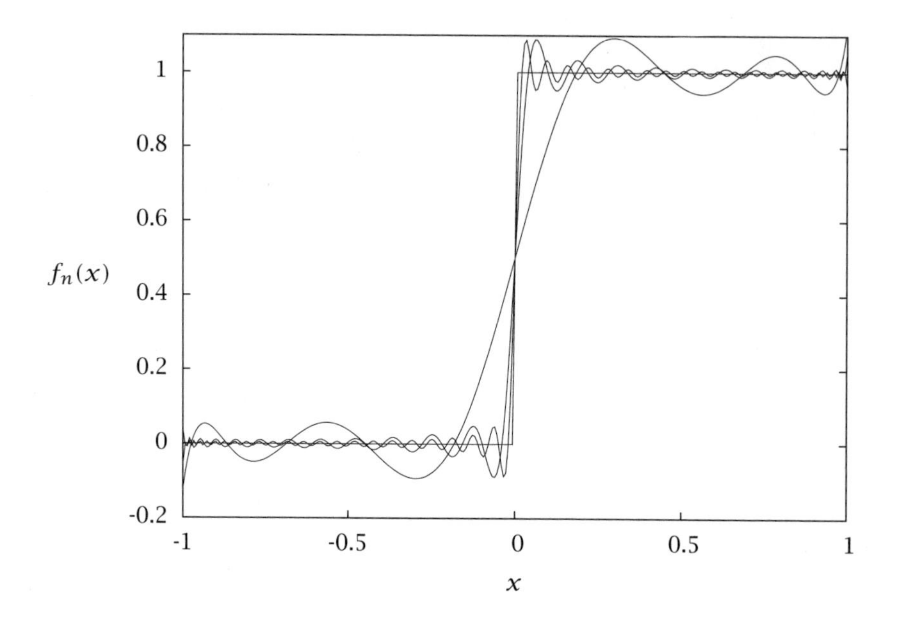

Figure 2.7: Function $f(x) = H(x)$ and truncated Legendre-Fourier series approximations with $n = 10, 50, 100$.

on functions that vanish at $x = 1$. The first derivative operator is *not* self-adjoint.

If, however, we let $L = d^2/dx^2$ and require that $u(0) = u(1) = 0$, then the same procedure (but using integration by parts twice) shows that L^* is also d^2/dx^2 operating on the same domain. The second-derivative operator, therefore, with appropriate boundary conditions, *is* self-adjoint. More generally, consider a class of second-order differential operators called STURM-LIOUVILLE operators. These operators have the general form

$$Lu = \frac{1}{w(x)}\left(\frac{d}{dx}\left[p(x)\frac{du}{dx}\right] + r(x)u\right) \tag{2.31}$$

in the domain $a < x < b$, with homogeneous boundary conditions

$$\alpha u(a) + \beta u'(a) = 0, \qquad \gamma u(b) + \delta u'(b) = 0$$

To avoid the possibility of singular points, $p(x)$ must be positive in the domain. Furthermore, take the inner product to be

$$(u, v)_w = \int_a^b u(x)v(x)w(x)\,dx$$

The function $w(x)$ here is the same as in (2.31). For this integral to be a proper inner product, we must require that $w(x) > 0$ in the domain.

We now show that Sturm-Liouville operators are self-adjoint. Repeated integration by parts yields

$$(Lu, v)_w = \int_a^b \frac{1}{w(x)}\left(\frac{d}{dx}\left[p(x)\frac{du}{dx}\right] + r(x)u\right) v\, w\, dx \qquad (2.32)$$

$$= p(b)\left(u'(b)v(b) - u(b)v'(b)\right)$$

$$- p(a)\left(u'(a)v(a) - u(a)v'(a)\right)$$

$$+ \int_a^b u\frac{1}{w(x)}\left(\frac{d}{dx}\left[p(x)\frac{dv}{dx}\right] + r(x)v\right) w\, dx \qquad (2.33)$$

If the boundary terms vanish, then this expression satisfies the self-adjointness condition $(Lu, v) = (u, Lv)$. This is the case if the above boundary conditions apply on both u and v. The restriction on the boundary conditions can be relaxed if $p(x)$ vanishes at one or both boundaries, in which case only boundedness of the function and its derivative is required at that boundary. The latter case is called a *singular* Sturm-Liouville operator, because it has a singular point at the boundary or boundaries where p vanishes. Finally, the boundary terms also vanish if $p(a) = p(b)$ and PERIODIC BOUNDARY CONDITIONS are imposed: $u(a) = u(b), u'(a) = u'(b)$ and likewise for v.

Next consider the eigenvalue problem associated with the Sturm-Liouville operator[4]

$$Lu + \lambda u = 0$$

As with all self-adjoint operators, the eigenvalues λ are real and the eigenvectors—now called eigenfunctions because they are elements of a function space—are orthogonal with respect to the inner product weighted by $w(x)$. Furthermore, and very importantly, there are an infinite number of eigenfunctions and they form a complete basis for $L_{2,w}(a, b)$. We next consider three Sturm-Liouville operators that produce some famous eigenfunctions that are popular choices for use as basis functions.

[4]This is the conventional form for writing differential eigenvalue problems. Unfortunately, it is different from the convention for algebraic problems.

Example 2.7: Generating trigonometric basis functions

Consider the operator $L = d^2/dx^2$, with boundary conditions $u(0) = u(l) = 0$. The eigenvalue problem for this operator is

$$u'' + \lambda u = 0 \tag{2.34}$$

What are the eigenvalues and eigenfunctions?

Solution

This equation has the general solution

$$u(x) = c_1 \sin\sqrt{\lambda}x + c_2 \cos\sqrt{\lambda}x$$

We have thus taken $\lambda \geq 0$: a negative value of λ would lead to a general solution consisting of growing and decaying exponentials, which cannot satisfy homogeneous boundary conditions on both boundaries, as can be easily checked. The boundary condition $u(0) = 0$ requires that $c_2 = 0$. Setting $c_1 = 0$ leaves only the trivial solution $u = 0$, so to satisfy the remaining boundary condition, we require that

$$\sin\sqrt{\lambda}l - 0$$

This is the characteristic equation for this eigenvalue problem; it has infinitely many roots $\lambda = n^2\pi^2/l^2$ for $n = 1, 2, 3, \ldots, \infty$. The case $n = 0$ does not result in an eigenvalue since $\sin 0 = 0$. Thus the eigenfunctions are

$$u_n(x) = \sin\frac{n\pi x}{l}$$

with $(u_m, u_n) = \frac{l}{2}\delta_{mn}$. The result that Sturm-Liouville eigenfunctions form a basis for functions in $L_2(0, l)$ implies that we can write any function in that space as a Fourier series

$$f(x) = \sum_{n=1}^{\infty} c_n(x) \sin\frac{n\pi x}{l}$$

where

$$c_n = \frac{\left(f(x), \sin\frac{n\pi x}{l}\right)}{\left(\sin\frac{n\pi x}{l}, \sin\frac{n\pi x}{l}\right)} = \frac{2}{l}\left(f(x), \sin\frac{n\pi x}{l}\right)$$

This is the FOURIER SINE SERIES of $f(x)$.

Now consider the same operator but with periodic boundary conditions $u(0) = u(l), u'(0) = u'(l)$. The boundary terms in (2.33) also vanish in this case, because here $p(a) = p(b) = 1$. Now the solution to (2.34) is

$$u = \exp i\sqrt{\lambda}x$$

which satisfies the periodicity requirement if $\lambda = \left(\frac{2n\pi}{l}\right)^2$ for any integer n. Thus the eigenfunctions of d^2/dx^2 with periodic boundary conditions in $(0, l)$ are

$$u_n = \exp i\frac{2n\pi x}{l}$$

Taking $l = 2\pi$, we recover the first set of basis functions we considered in Section 2.4.1. □

Example 2.8: Bessel's equation revisited

The operator

$$Lu = \frac{1}{x}\frac{d}{dx}\left(x\frac{d}{dx}\right)$$

arises in many differential equations originating in problems in polar coordinates, e.g., diffusion in a cylinder. It has Sturm-Liouville form with $w = p = x$, $r = 0$. The eigenvalue problem for this operator can be written

$$u'' + \frac{1}{x}u' + \lambda u = 0$$

or, multiplying through by x^2, as

$$x^2u'' + xu' + \lambda x^2 u = 0$$

What are its eigenfunctions and eigenvalues?

Solution

This is a variable-coefficient problem with a regular singular point at $x = 0$, so we can seek solutions by the method of Frobenius. Alternately, in the present case we can make the substitution $z = x\sqrt{\lambda}$, thus rewriting the equation as

$$z^2\frac{d^2u}{dz^2} + z\frac{du}{dz} + z^2u = 0$$

which is in fact Bessel's equation of order zero. We already found that this equation has the general solution $u(z) = c_1J_0(z) + c_2Y_0(z)$, or, reverting to the original independent variable,

$$u(x) = c_1J_0(\sqrt{\lambda}x) + c_2Y_0(\sqrt{\lambda}x)$$

To complete the specification of the eigenvalue problem requires choosing the domain and imposing specific boundary conditions. Consider the domain $0 < x < l$. We require that $u(0)$ be bounded; this is all that is required, since $p(0) = 0$ and $u(l) = 0$. Boundedness requires that $c_2 = 0$, because Y_0 diverges logarithmically at the origin. Satisfaction of $u(l) = 0$ requires that

$$J_0(\sqrt{\lambda} l) = 0$$

The top center plot of Table 2.3 shows $J_0(x)$; the positions of its zeros determine the eigenvalues λ. The first several of these are at approximately $x = 2.4, 5.5, 8.7, 11.8, \ldots$ and are tabulated in many places, including Abramowitz and Stegun (1970). Thus $\lambda_1 \approx (2.4/l)^2$, etc. The functions

$$u_n(x) = J_0(\sqrt{\lambda_n} x)$$

form an orthogonal basis for $L_{2,w}(0, l)$. Referring again to Table 2.3, u_1 is the function J_0 scaled so that its first zero is at $x = l$, u_2 is the same function, but scaled so that its *second* zero is at $x = l$, etc.

Other boundary conditions could be chosen. For example, one could require $u(a) = 0, u(b) = 0$. In this case the eigenfunctions involve both J_0 and Y_0, and the eigenfunctions and eigenvalues are determined by the solution to the coupled nonlinear equations

$$J_0(\sqrt{\lambda} a) + c_2 Y_0(\sqrt{\lambda} a) = 0$$
$$J_0(\sqrt{\lambda} b) + c_2 Y_0(\sqrt{\lambda} b) = 0$$

Since c_1 is arbitrary, it has been set to unity for convenience. Here c_2 and λ are the unknowns. Solution of these highly nonlinear equations is nontrivial. □

Example 2.9: Legendre's differential equation and Legendre polynomials

Consider the Sturm-Liouville eigenvalue problem with $p(x) = 1 - x^2$, $w(x) = 1$, $r(x) = 0$ in the domain $-1 < x < 1$

$$\left(1 - x^2\right) u'' - 2xu' + \lambda u = 0$$

It has regular singular points at $x = \pm 1$ while the origin is an ordinary point. Because $p(x) = 0$ at $x = \pm 1$, only boundedness at these points is required of the eigenfunctions. What are the eigenvalues and eigenfunctions?

Solution

Seeking a series solution around $x = 0$ reveals that, if $\lambda = l(l+1)$ with $l \geq 0$ an integer, then one of the solutions is a Legendre polynomial of degree l (Exercise 2.35) and using the method of Frobenius one can learn that the other has logarithmic singularities at $x = \pm 1$. Otherwise, because the radius of convergence of a power series solution is given by the distance to the nearest singular point(Ablowitz and Fokas, 2003), there is no solution that is bounded at both $x = 1$ and $x = -1$. Therefore, the eigenvalues of (2.9) are $\lambda = l(l+1)$ with $l = 0, 1, 2, \ldots$ and the corresponding eigenfunctions are the Legendre polynomials $P_l(x)$. Legendre polynomials are the simplest of a broad class of ORTHOGONAL POLYNOMIALS that come from Sturm-Liouville eigenvalue problems and are orthogonal with respect to various weighted inner products. Some examples are given in the exercises. □

2.4.3 Existence and Uniqueness of Solutions

Homogeneous Boundary Conditions

Consider the nonhomogeneous second-order differential equation with the homogeneous boundary conditions

$$\begin{gathered} Lu = f \\ B_1 u = 0 \qquad B_2 u = 0 \end{gathered} \tag{2.35}$$

Define the null space of the operator

$$N(L) = \{u \mid Lu = 0, \quad B_1 u = 0, \quad B_2 u = 0\}$$

and the null space of the adjoint operator

$$N(L^*) = \{v \mid L^* v = 0, \quad B_1^* v = 0, \quad B_2^* v = 0\}$$

then the following theorem characterizes existence and uniqueness of solutions to (2.35) (Stakgold, 1998, p. 210–211).

Theorem 2.10 (Alternative theorem). *For the boundary-value problem in* (2.35), *we have the following two alternatives.*

(a) Either—
$N(L)$ contains only the zero function in which case $N(L^*)$ contains only the zero function and (2.35) *has exactly one solution for every f.*

(b) Or—
$N(L)$ contains n linearly independent functions, in which case $N(L^*)$ contains n linearly independent functions

$$N(L) = \{u_1, u_2, \ldots, u_n\} \qquad N(L^*) = \{v_1, v_2, \ldots, v_n\}$$

and (2.35) *has a solution if and only if*

$$(f, v_k) = 0, \quad k = 1, 2, \ldots, n$$

and the general solution is

$$u(x) = u_p(x) + \sum_{k=1}^{n} \alpha_k u_k(x)$$

in which $u_p(x)$ is any particular solution and α_k are arbitrary scalars.

Next we present two heat-conduction problems that display the two alternatives.

Example 2.11: Steady-state temperature profile with fixed end temperatures

Apply the alternative theorem to the steady-state heat-conduction problem with heat generation $\hat{f}(x)$ and specified end-temperature boundary conditions

$$-k\frac{d^2T(x)}{dx^2} = \hat{f}(x)$$
$$T(x) = T_0 \quad x = 0$$
$$T(x) = T_1 \quad x = 1$$

What can you conclude about existence and uniqueness of the steady-state temperature profile?

Solution

First it is convenient to make the boundary conditions homogeneous by defining

$$u(x) = T(x) - T_0(1 - x) - T_1 x$$

and dividing by the thermal conductivity to give

$$Lu = f$$
$$B_1 u = 0 \qquad B_2 u = 0$$

in which $f = -\hat{f}/k$ and

$$L = \frac{d^2}{dx^2} \qquad B_1 u = u(0) \qquad B_2 u = u(1)$$

Next we compute $N(L)$. Setting $Lu = 0$ gives

$$u(x) = ax + b$$

Applying the boundary conditions gives

$$B_1 u = u(0) = b = 0 \qquad B_2 u = u(1) = a = 0$$

and we see that $u = 0$ is the only element of $N(L)$. We can therefore conclude that $N(L^*)$ also contains only the zero element, and the steady-state temperature profile exists and is unique for any heat-removal rate f. □

Example 2.11 illustrates the first alternative in Theorem 2.10. The following example illustrates the second alternative.

Example 2.12: Steady-state temperature profile with insulated ends

Replace the fixed-temperature boundary conditions in Example 2.11 with insulated-end boundary conditions. What can you conclude about existence and uniqueness of the steady-state temperature profile for these boundary conditions? What is the physical interpretation of the existence condition. Why is the solution not unique?

Solution

The boundary conditions for insulated ends are

$$T_x(x) = 0 \quad x = 0$$
$$T_x(x) = 0 \quad x = 1$$

and since the boundary conditions already are homogeneous, we have

$$LT = f$$
$$B_1 T = 0 \qquad B_2 T = 0$$

in which $f = -\hat{f}/k$ and

$$L = \frac{d^2}{dx^2} \qquad B_1 T = T_x(0) \qquad B_2 T = T_x(1)$$

Next we compute $N(L)$. Setting $LT = 0$ gives

$$T(x) = ax + b$$

as before. Applying the boundary conditions gives

$$B_1 T = T_x(0) = a = 0 \qquad B_2 T = T_x(1) = a = 0$$

and now we have that $T(x) = b$ is in $N(L)$. With these boundary conditions L has a one-dimensional null space consisting of the constant function. Normalizing this element gives $\{1\}$ as the basis function for $N(L)$ and the one-dimensional nullspace

$$N(L) = \alpha \cdot 1 \qquad \alpha \in \mathbb{R}$$

Since the problem is self-adjoint, $N(L^*)$ is identical to $N(L)$. Applying the alternative theorem, we conclude that a steady-state temperature exists only if

$$(f, 1) = \int_0^1 f(x)dx = 0$$

and the general solution is

$$T(x) = T_p(x) + \alpha$$

where T_p is any particular solution. Since f corresponds to a rate of heat removal (or addition when $f < 0$) to the domain, the restriction on f provides the physically intuitive fact that if the ends are insulated, just as much heat must be removed from the domain as is added for a steady-state temperature to exist. For f satisfying this restriction, the general solution indicates that a constant can be added to any steady-state solution to provide another steady-state solution. □

Nonhomogeneous Boundary Conditions

Next consider the nonhomogeneous second-order problem for $u(x)$ on $x \in [a, b]$ with the nonhomogeneous boundary conditions

$$Lu = f$$
$$B_1 u = \gamma_1 \qquad B_2 u = \gamma_2 \tag{2.36}$$

The null spaces of the operator and the adjoint are defined as in the case with homogeneous boundary conditions

$$N(L) = \{u \mid Lu = 0, \quad B_1 u = 0, \quad B_2 u = 0\}$$

$$N(L^*) = \{v \mid L^*v = 0, \quad B_1^*v = 0, \quad B_2^*v = 0\}$$

When we define the adjoint operator, we perform integration by parts

$$(Lu, v) - (u, L^*v) = J(u, v)\big|_a^b$$

From the integration by parts, we have that $J(u,v)$ is linear in both $u(x)$ and $v(x)$ and involves lower-order derivatives of u, v evaluated at the two ends of the interval. Setting $J(u,v)\big|_a^b$ to zero is what determines the adjoint boundary functionals

$$J(u,v)\big|_a^b = 0 \qquad \begin{array}{l} \forall u \text{ such that } B_1u = 0, B_2u = 0 \\ \forall v \text{ such that } B_1^*v = 0, B_2^*v = 0 \end{array}$$

To find the solvability condition for the *nonhomogeneous* boundary conditions, we take the difference

$$(Lu, v_k) - (u, L^*v_k) = J(u, v_k)\big|_a^b$$

in which v_k is any element of the null space of the adjoint and u is the solution to (2.36). Then, because $Lu = f$ and $L^*v_k = 0$, we have

$$(f, v_k) = J(u, v_k)\big|_a^b \tag{2.37}$$

Evaluating $J(u, v_k)$ for u satisfying $B_1u = y_1$ and $B_2u = y_2$, and v_k satisfying $B_1^*v_k = 0$, $B_2^*v_k = 0$, gives the solvability conditions for the nonhomogeneous problem. The next example and Exercise 2.40 derive the solvability conditions for problems with nonhomogeneous boundary conditions.

Example 2.13: Steady-state temperature profile with fixed flux

Consider again Example 2.12, but replace the insulated ends with fixed, nonzero fluxes at the ends

$$T_x(x) = y_1 \quad x = 0$$
$$T_x(x) = y_2 \quad x = 1$$

For what f does the solution exist?

Solution

This fully nonhomogeneous problem can be written as

$$LT = f$$
$$B_1T = y_1 \qquad B_2T = y_2$$

in which $f = -\hat{f}/k$ and

$$L = \frac{d^2}{dx^2} \qquad B_1 T = T_x(0) \qquad B_2 T = T_x(1)$$

The null space $N(L)$ is unchanged, so the constant function $\{1\}$ is a basis function and

$$N(L) = \alpha \cdot 1 \qquad \alpha \in \mathbb{R}$$

The problem was shown to be self-adjoint so $N(L^*)$ is one dimensional and $v_k(x) = 1$. Next we compute $J(u, v)$ for this problem. Integration by parts gives

$$\begin{aligned}(Lu, v) - (u, L^* v) &= J(u, v)\,\big|_0^1 \\ &= v(1)u_x(1) - v(0)u_x(0) - v_x(1)u(1) + v_x(0)u(0)\end{aligned}$$

For T satisfying the boundary conditions and v_k in $N(L^*)$, we have

$$B_1 T = T_x(0) = \gamma_1 \qquad B_2 T = T_x(1) = \gamma_2$$
$$B_1^* v_1 = \frac{dv_1}{dx}(0) = 0 \qquad B_2^* v_1 = \frac{dv_1}{dx}(1) = 0$$

Substituting these into J gives

$$\begin{aligned}J(T, v_1)\,\big|_0^1 &= \underbrace{v_1(1)}_{1}\,\underbrace{T_x(1)}_{\gamma_2} - \underbrace{v_1(0)}_{1}\,\underbrace{T_x(0)}_{\gamma_1} - \underbrace{\frac{dv_1}{dx}(1)}_{0}\,T(1) + \underbrace{\frac{dv_1}{dx}(0)}_{0}\,T(0) \\ &= \gamma_2 - \gamma_1\end{aligned}$$

Substituting this into the solvability condition, (2.37), gives

$$(f, 1) = \int_0^1 f(x)dx = \gamma_2 - \gamma_1$$

and the general solution remains

$$T(x) = T_p(x) + \alpha$$

The restriction on f now stipulates that the net heat generation must exactly balance the heat removed through the two ends. Again, for f satisfying this restriction, a constant can be added to any steady-state solution to provide another steady-state solution. □

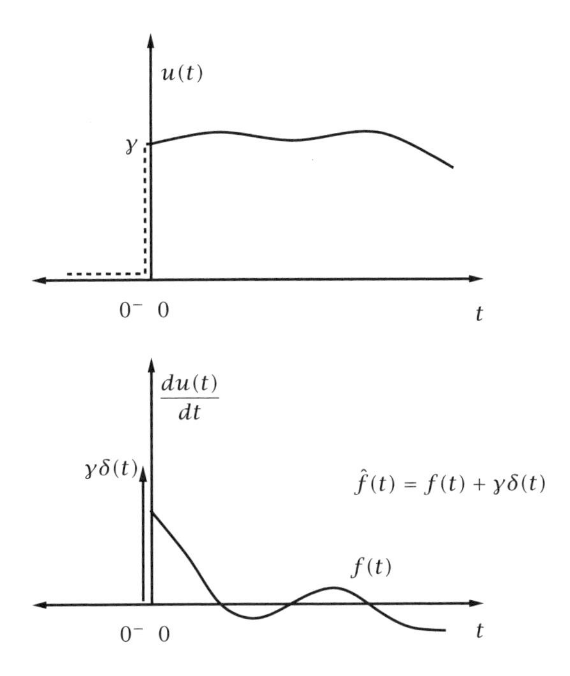

Figure 2.8: Solution to the initial-value problem with nonhomogeneous boundary conditions; top figure shows $u(t)$ with step introduced at $t = 0$, and bottom figure shows resulting du/dt with impulse at $t = 0$.

Nonhomogeneous Boundary Conditions Revisited

We can use the delta function and its derivatives introduced in Section 2.2.5 to streamline the treatment of the nonhomogeneous case. Basically we replace the nonhomogeneous boundary conditions with homogeneous ones, but then compensate for this change by adding appropriate impulsive terms to the forcing term of the differential equation. In this way, we have to recall only how to solve problems with homogeneous boundary conditions, and we can use Theorem 2.10 to analyze existence and uniqueness even when a problem has nonhomogeneous boundary conditions.

It is perhaps easiest to introduce the approach with an example. Let's say we are interested in solving the first-order nonhomogeneous differential equation, with forcing term $f(t)$, and nonhomogeneous boundary (initial) condition

$$\frac{du}{dt} = f(t)$$
$$u(0) = \gamma \qquad \gamma \neq 0$$

The solution is sketched in Figure 2.8. Imagine instead that we solve the problem with the *homogeneous* boundary condition $u(0^-) = 0$, and we push the boundary at $t = 0$ slightly to the left of zero. Now we wish to make the solution jump to value $u(0) = \gamma$ just after time 0^- so that it agrees with the solution to the problem with nonhomogeneous boundary condition at $t = 0$. This idea is also sketched in Figure 2.8. To make $u(t)$ jump discontinuously by amount γ at $t = 0$, we require du/dt to have an impulse of strength γ at $t = 0$, which is $\gamma\delta(t)$. Since $du/dt = f(t)$, we introduce a modified forcing term $\hat{f}$ and choose it to be

$$\hat{f}(t) = f(t) + \gamma\delta(t)$$

We conjecture that solving the problem with this modified forcing term $\hat{f}$ and homogeneous boundary condition should give us the solution to the problem with the original f and nonhomogeneous boundary condition. Let's check this conjecture. By inspection, the solution to the differential equation is obtained by integration

$$\begin{aligned} \frac{du}{dt} &= \hat{f}(t) \\ du &= \hat{f}(t)dt \\ u(t)\big|_{0^-}^{t} &= \int_{0^-}^{t} \hat{f}(\tau)d\tau \\ u(t) - u(0^-) &= \int_{0^-}^{t} \hat{f}(\tau)d\tau \\ u(t) &= \int_{0^-}^{t} \hat{f}(\tau)d\tau \end{aligned}$$

Note that this solution satisfies the homogeneous boundary condition $u(0^-) = 0$ as desired. Now we substitute the definition of $\hat{f}$ to obtain the solution of the original problem

$$\begin{aligned}
u(t) &= \int_{0^-}^{t} (f(\tau) + \gamma\delta(\tau))d\tau \\
&= \int_{0^-}^{t} f(\tau)d\tau + \int_{0^-}^{t} \gamma\delta(\tau)d\tau \\
&= \gamma\int_{0^-}^{t} \delta(\tau)d\tau + \int_{0}^{t} f(\tau)d\tau \\
u(t) &= \gamma + \int_{0}^{t} f(\tau)d\tau \qquad t \geq 0
\end{aligned}$$

By inspection, the last equation is indeed the solution to the original problem with forcing term f and *nonhomogeneous* boundary condition $u(0) = \gamma$.

We can generalize this approach to cover any nonhomogeneity in the boundary conditions by adding appropriate impulsive forcing terms to the original problem's differential equation. We revisit Example 2.13 to illustrate this technique.

Example 2.14: Fixed flux revisited

Rederive the existence and uniqueness conditions for Example 2.13 using the alternative theorem, which applies only to homogeneous problems.

Solution

We replace the nonhomogeneous boundary conditions of Example 2.13 with the homogeneous version

$$\begin{aligned}
B_1T &= T_x(0^-) = 0 \\
B_2T &= T_x(1^+) = 0
\end{aligned}$$

In this example we require that T_x jump from zero to value γ_1 at the left boundary, $x = 0$. That requires an impulse to be added to f so that T_{xx} sees an impulse and T_x sees a jump at $x = 0$. We also require for T_x to jump *from* value γ_2 *to* zero as x passes through $x = 1$ at the right boundary. We add $-\gamma_2\delta(x-1)$ to f to cause T_x to jump by this amount. The modified $\hat{f}$ is therefore[5]

$$\hat{f}(x) = f(x) + \gamma_1\delta(x) - \gamma_2\delta(x-1)$$

[5]Note that if we had nonhomogeneous boundary conditions on T rather than T_x, we would required T_x to have an impulse and T_{xx} to have a *doublet*, and we would add $\gamma_1\delta'(x) - \gamma_2\delta'(x-1)$ to f.

Next we apply the alternative theorem. We have already computed the null space of L for this problem. It is $N(L) = 1$. The problem is self-adjoint so this is also $N(L^*)$. The solvability condition applied to $\hat{f}$ gives

$$\begin{aligned}
0 &= (\hat{f}, 1) \\
&= \int_{0^-}^{1^+} \hat{f}(x)dx \\
&= \int_{0^-}^{1^+} (f(x) + y_1\delta(x) - y_2\delta(x-1))dx \\
0 &= \int_0^1 f(x)dx + y_1 - y_2
\end{aligned}$$

The last equation implies the solution exists for f satisfying

$$\int_0^1 f(x)dx = y_2 - y_1$$

and the general solution remains

$$T(x) = T_p(x) + \alpha$$

□

We see that we have reached the same solvability condition found in Example 2.13. By introducing $\hat{f}$ and using homogeneous boundary conditions, we avoid the additional complication of introducing and evaluating $J(u, v)$ as explained in Section 2.4.3. Evaluating $J(u, v)$ is about the same work as determining the appropriate $\hat{f}$. But using delta functions expands the applicability of Theorem 2.10, and allows this one theorem to cover both homogeneous and nonhomogeneous boundary condition cases, which is not an insignificant benefit.

Example 2.15: Nonhomogeneous boundary-value problem and the Green's function

The following second-order nonhomogeneous boundary-value problem arises in solving the transient wave equation for propagation of sound. We wish to solve the following BVP for $u(x)$, $x \in [0, 1]$

$$Lu = f$$

$$B_1 u = 0 \qquad B_2 u = 0$$

in which the second-order differential operator is $Lu = d^2u/dx^2 - k^2u$, and the two boundary functionals are $B_1u = u(0)$, $B_2u = u(1)$. The constant k is real and the function $f(x)$ is an arbitrary forcing function.

(a) Take the Laplace transform of the BVP with the x variable playing the role of time. Note that the value of $u(0)$ and $u_x(0)$ shows up in the transform. Evaluate $u(0)$ and leave $u_x(0)$ as an unknown constant.

(b) Invert the transform to obtain $u(x)$.

(c) Solve for $u_x(0)$ using the solution in the previous part and the other boundary condition. Plug the expression for $u_x(0)$ back into your solution to obtain the complete solution to the problem.

(d) Next express the solution as

$$u(x) = \int_0^1 G(x,\xi) f(\xi) d\xi$$

The function $G(x,\xi)$ is known as the Green's function for the nonhomogeneous problem.[6] Write out the Green's function $G(x,\xi)$ for this problem.

(e) Establish that the Green's function $G(x,\xi)$ is *symmetric* for this boundary-value problem, i.e., $G(x,\xi) = G(\xi,x)$.
Hint: you may find the hyperbolic difference formula useful: $\sinh(a-b) = \sinh a \cosh b - \cosh a \sinh b$.

Solution

(a) Taking the Laplace transform of the differential equation gives

$$\begin{aligned} s^2\overline{u}(s) - su(0) - u_x(0) - k^2\overline{u}(s) &= \overline{f} \\ (s^2 - k^2)\overline{u}(s) &= \overline{f} + u_x(0) \\ \overline{u}(s) &= \frac{\overline{f}}{s^2 - k^2} + \frac{u_x(0)}{s^2 - k^2} \end{aligned}$$

[6]The Green's function concept is explored in greater detail in Chapter 3, Section 3.3.5.

(b) Using the transform pair

$$\mathcal{L}^{-1}\left(\frac{1}{s^2 - k^2}\right) = \frac{1}{k}\sinh kx$$

and the convolution theorem gives

$$u(x) = \frac{1}{k}\int_0^x \sinh(k(x-\xi))f(\xi)d\xi + \frac{u_x(0)}{k}\sinh kx$$

(c) Evaluating the solution at $x = 1$ and solving for the unknown $u_x(0)$ gives

$$\begin{aligned} 0 &= u(1) \\ &= \frac{1}{k}\int_0^1 \sinh(k(1-\xi))f(\xi)d\xi + \frac{u_x(0)}{k}\sinh k \\ u_x(0) &= \frac{-1}{\sinh k}\int_0^1 \sinh(k(1-\xi))f(\xi)d\xi \end{aligned}$$

Substituting $u_x(0)$ into the previous solution gives

$$u(x) = \frac{1}{k}\int_0^x \sinh(k(x-\xi))f(\xi)d\xi - \frac{\sinh kx}{k\sinh k}\int_0^1 \sinh(k(1-\xi))f(\xi)d\xi$$

(d) Combining these two integrals into one gives

$$u(x) = \int_0^1 G(x,\xi)f(\xi)d\xi$$

with

$$G(x,\xi) = \begin{cases} \dfrac{1}{k}\sinh(k(x-\xi)) - \dfrac{\sinh kx}{k\sinh k}\sinh(k(1-\xi)) & \xi < x \\ -\dfrac{\sinh kx \sinh k(1-\xi)}{k\sinh k} & \xi > x \end{cases}$$

(e) We work on the first part of $G(x,\xi)$ using the sinh difference formula

$$\sinh(a-b) = \sinh a\cosh b - \cosh a\sinh b$$

We have for $\xi < x$ that

$$\begin{aligned} G(x,\xi) &= \frac{1}{k}\sinh(k(x-\xi)) - \frac{\sinh kx}{k\sinh k}\sinh(k(1-\xi)) \\ &= \frac{1}{k\sinh k}(\sinh k\sinh(k(x-\xi)) - \sinh kx\sinh(k(1-\xi))) \end{aligned}$$

Using the sinh difference formula on the term in parentheses gives

$$\sinh k \sinh(k(x-\xi)) - \sinh kx \sinh(k(1-\xi)) = \\ \sinh k(\sinh kx \cosh k\xi - \cosh kx \sinh k\xi) - \\ \sinh kx(\sinh k \cosh k\xi - \cosh k \sinh k\xi)$$

Canceling the $\cosh k\xi$ terms gives

$$\sinh k \sinh(k(x-\xi)) - \sinh kx \sinh(k(1-\xi)) = \\ -\sinh k \cosh kx \sinh k\xi + \sinh kx \cosh k \sinh k\xi$$

Factoring out the $\sinh k\xi$ term and using the difference formula again gives

$$\begin{aligned} &\sinh k \sinh(k(x-\xi)) - \sinh kx \sinh(k(1-\xi)) \\ &\quad = \sinh k\xi(\sinh kx \cosh k - \cosh k \sinh k) \\ &\quad = \sinh k\xi \sinh(kx-k) \\ &\quad = -\sinh k\xi \sinh(k(1-x)) \end{aligned}$$

Substituting this result into the equation for $G(x,\xi)$ gives

$$G(x,\xi) = \begin{cases} -\dfrac{\sinh k\xi \sinh(k(1-x))}{k \sinh k} & \xi < x \\ -\dfrac{\sinh kx \sinh k(1-\xi)}{k \sinh k} & \xi > x \end{cases}$$

and we have established that $G(x,\xi) = G(\xi,x)$; the Green's function for this operator is symmetric, a consequence of the self-adjointness of L in this case. □

2.5 Lyapunov Functions and Stability

2.5.1 Types of Stability

Consider a system model of interest to be an autonomous initial-value problem

$$\frac{dx}{dt} = f(x) \qquad x(0) = x_0 \tag{2.38}$$

We are interested in the behavior of solutions to this system. Since the solution depends on the initial condition, we denote by $\phi(t;x)$ the

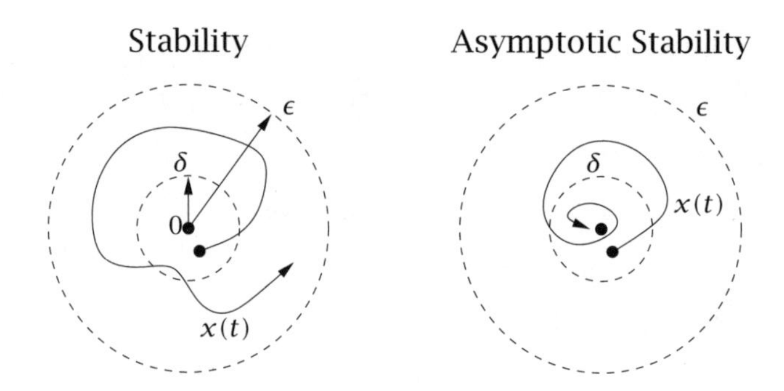

Figure 2.9: Solution behavior; stability (left) and asymptotic stability (right).

solution to the initial-value problem at time $t \geq 0$, which has value x at time $t = 0$. So the solution to the initial-value problem above is given by $\phi(t; x_0), t \geq 0$. But we are also interested in the solution as we vary the initial value x. Steady-state solutions to the model, if any exist, satisfy

$$f(x_s) = 0$$

We can always shift a steady state of interest to the origin by defining a new coordinate, $\widetilde{x} = x - x_s$, and $\widetilde{f}(\widetilde{x}) = f(\widetilde{x} + x_s)$ so that

$$\frac{d\widetilde{x}}{dt} = \frac{dx}{dt} = f(x) = f(\widetilde{x} + x_s)$$
$$\frac{d\widetilde{x}}{dt} = \widetilde{f}(\widetilde{x}) \qquad \widetilde{f}(0) = 0$$

So we assume without loss of generality that $x_s = 0$, i.e., *the origin* is the steady state of interest. Unlike a linear system, when dealing with a nonlinear system, stability depends on the solution of interest, and we may have some solutions that are stable, while others are unstable. For a given linear system, the stability of all solutions are identical, and to reflect this special situation, we often refer to stability of the *system*, rather than stability of a *solution*.

There are several aspects to stability, and we define these next. The first most basic characteristic of interest is whether a small perturbation to x away from the steady-state solution results in a small subsequent deviation for all future times. The general term *stability* is commonly reserved for this most basic notion; we use the more precise term *Lyapunov stability* or *stable in the sense of Lyapunov* if we need to ensure that there is no confusion. The definition is as follows.

Definition 2.16 ((Lyapunov) Stability). The origin is (Lyapunov) stable if for every $\epsilon > 0$, there exists $\delta > 0$ such that $\|x\| \leq \delta$ implies $\|\phi(t; x)\| \leq \epsilon$ for all $t \geq 0$.

The stability concept is illustrated on the left side of Figure 2.9. A solution that is not stable is termed UNSTABLE. The next characteristic of interest is whether small perturbations to the initial state die away as time increases. The idea here is whether the origin *attracts* solutions starting nearby.

Definition 2.17 (Attractivity). The origin is attractive if there exists $\delta > 0$ such that $\|x\| \leq \delta$ implies that

$$\lim_{t \to \infty} \|\phi(t; x)\| = 0$$

Asymptotic stability is then the combination of these two properties.

Definition 2.18 (Asymptotic stability). The origin is asymptotically stable if it is (i) stable and (ii) attractive.

The right side of Figure 2.9 shows a representative solution trajectory when the origin is asymptotically stable.[7] One might wonder why Lyapunov stability is a requirement of asymptotic stability, or even whether the origin can be attractive, and not Lyapunov stable. The answer is yes, the origin in a nonlinear system may be *globally* attractive and still not Lyapunov stable. The problem with these systems is that there exist starting points, arbitrarily close to the origin, for which the resulting trajectories become *large* before they asymptotically approach zero as time tends to infinity. Because we cannot bound how large the solution transient becomes by constraining the size of its initial value, we classify the origin as *unstable*.[8] Note that the system must

[7]Asymptotic stability is probably the most common notion of stability that people have in mind, and sometimes it is referred to simply as *stability*. Of course, this usage may cause confusion because now the term stability is being used in two ways: as Lyapunov stability and as asymptotic stability; and one is a subset of the other.

[8]One is obviously free to define words as one pleases, but defining asymptotic stability in this way precludes a possible solution behavior that is not expected of "nice" or "stable" solutions. Regardless of terminology, the important point is to *be aware* that solutions can be globally attractive and *not* Lyapunov stable.

be nonlinear for a solution to be attractive and unstable. For linear systems, attractivity and asymptotic stability are identical; see Exercise 2.60.

A stronger form of asymptotic stability known as exponential stability is often useful, especially when dealing with linear dynamics. It is defined as follows.

Definition 2.19 (Exponential stability)**.** The origin is exponentially stable if there exists $\delta > 0$ such that $\|x\| \leq \delta$ implies that there exist $c, \lambda > 0$ for which

$$\|\phi(t;x)\| \leq c \, \|x\| \, e^{-\lambda t} \quad \text{for all } t \geq 0$$

We leave it as an exercise for the reader to show that the definition of exponential stability implies also Lyapunov stability.

2.5.2 Lyapunov Functions

Now we consider a scalar function of x, denoted $V(x)$, whose characteristics are going to enable us to analyze the stability of the origin without requiring us to first solve completely the model $\dot{x} = f(x)$. The motivation for this class of functions is the role that mechanical energy plays in a mechanical system. Consider mechanical energy to be the sum of kinetic and potential energies, T and K, and let total energy be the sum of mechanical energy and internal energy

$$E = U + E_M \qquad E_M = T + K$$

If we start an isolated mechanical system, such as the particle on a track depicted in Figure 2.10, at some system temperature with some initial kinetic and potential energies, and monitor the mechanical energy with time, we observe that although the *total* energy E is conserved, the mechanical energy E_M steadily drops as some of that form of energy is converted into heat by friction.[9] The temperature of the system slowly increases due to the conversion of energy into heat, and the internal energy U of the system increases to maintain the total energy constant. If we define the height of the track at its lowest point as $h = 0$, we then have $E_M = (1/2)mv^2 + mgh$, and since $h \geq 0$, $m > 0$, and $v^2 \geq 0$, we have that $E_M \geq 0$. The mechanical energy is therefore a scalar function satisfying

$$E_M \geq 0 \qquad \dot{E}_M \leq 0$$

[9]This conversion of mechanical energy into heat is what causes the system's entropy to increase.

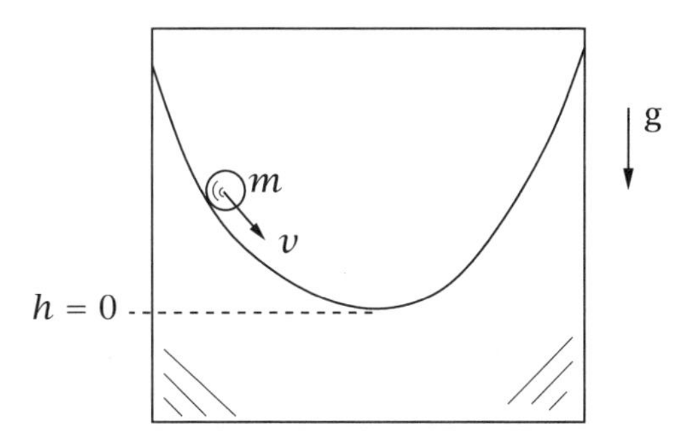

Figure 2.10: A simple mechanical system with total energy E, internal energy U, kinetic energy $T = (1/2)mv^2$, and potential energy $K = mgh$. The mechanical energy is $E_M = T + K$, and the total energy is $E = E_M + U$.

Because E_M decreases with time and is bounded below by zero, we expect that its only possible steady state is $E_M = 0$, and $E_M = 0$ implies both $v = 0$ and $h = 0$. So by analyzing the energy function E_M in this fashion, we conclude that the marble at rest at the bottom of the track is an asymptotically stable steady state, and we do not have to solve the complicated equations of motion of the system to deduce this fact.

We wish to generalize this concept, and the key idea is to define $V(x)$ to be a nonnegative scalar function $V : \mathbb{R}^n \to \mathbb{R}_{\geq 0}$, with a negative time derivative $\dot{V}(x(t)) \leq 0$. To compute the time derivative of $V(x(t))$, we apply the chain rule giving[10]

$$\dot{V}(x) = \left(\frac{\partial V}{\partial x}\right)^T \frac{dx}{dt}$$

$$\dot{V}(x) = \left(\frac{\partial V}{\partial x}\right)^T f(x) \tag{2.39}$$

This generalization of mechanical energy is the concept of a Lyapunov function for the system $\dot{x} = f(x)$. A precise definition is as follows.

Definition 2.20 (Lyapunov function)**.** Consider a compact (closed and bounded) set $D \subset \mathbb{R}^n$ containing the origin in its interior and let func-

[10]See Appendix A for various notations for derivatives with respect to vectors. Some readers may be more familiar with this equation in the form $\dot{V}(x) = \nabla V \cdot \dot{x}$ or $\dot{V}(x) = (\nabla V)^T \dot{x}$ or $\dot{V}(x) = \frac{\partial V}{\partial x_i} \dot{x}_i$.

tion $V : \mathbb{R}^n \to \mathbb{R}_{\geq 0}$ be continuously differentiable and satisfy[11]

$$V(0) = 0 \text{ and } V(x) > 0 \text{ for } x \in D \setminus 0 \tag{2.40}$$
$$\dot{V}(x) \leq 0 \text{ for } x \in D \tag{2.41}$$

Then $V(\cdot)$ is a Lyapunov function for the system $\dot{x} = f(x)$.

The big payoff for having a Lyapunov function for a system is the immediate stability analysis that it provides. We present next a few representative theorems stating these results. We mainly follow Khalil (2002) in the following presentation, and the interested reader may wish to consult that reference for further results on Lyapunov functions and stability theory. We require two fundamental results from real analysis to prove the Lyapunov stability theorems. The first concerns a nonincreasing function of time that is bounded below, which is a property we shall establish for $V(x(t))$ considered as a function of time. One of the fundamental results from real analysis is that such a function *converges* as time tends to infinity (Bartle and Sherbert, 2000, Theorems 3.3.2 and 4.3.11). The second result is that a continuous function defined on a compact (closed and bounded) set achieves its minimum and maximum values on the set. For *scalar* functions, i.e., $f : \mathbb{R} \to \mathbb{R}_{\geq 0}$, this "extreme-value" or "maximum-minimum" theorem is a fundamental result in real analysis (Bartle and Sherbert, 2000, p. 130), and is often associated with Weierstrass or Bolzano. The result also holds for multivariate functions like the Lyapunov function $V : \mathbb{R}^n \to \mathbb{R}_{\geq 0}$, which we require here, and is a highly useful tool in optimization theory (Mangasarian, 1994, p. 198) (Polak, 1997, Corollary 5.1.25) (Rockafellar and Wets, 1998, p. 11) (Rawlings and Mayne, 2009, Proposition A.7).

Theorem 2.21 (Lyapunov stability). *Let $V(\cdot)$ be a Lyapunov function for the system $\dot{x} = f(x)$. Then the origin is (Lyapunov) stable.*

Proof. Given $\epsilon > 0$ choose $r \in (0, \epsilon]$ such that

$$B_r = \{x \in \mathbb{R}^n \mid \|x\| \leq r\} \subseteq D$$

The symbol B_r denotes a BALL of radius r. Such an $r > 0$ exists since D contains the origin in its *interior*. The sets D and B_r are depicted in Figure 2.11. Define α by

$$\alpha = \min_{x \in D, \|x\| \geq r} V(x)$$

[11]For two sets A and B, the notation $A \setminus B$ is defined to be the elements of A that are not elements of B, or, equivalently, the elements of A remaining after removing the elements of B.

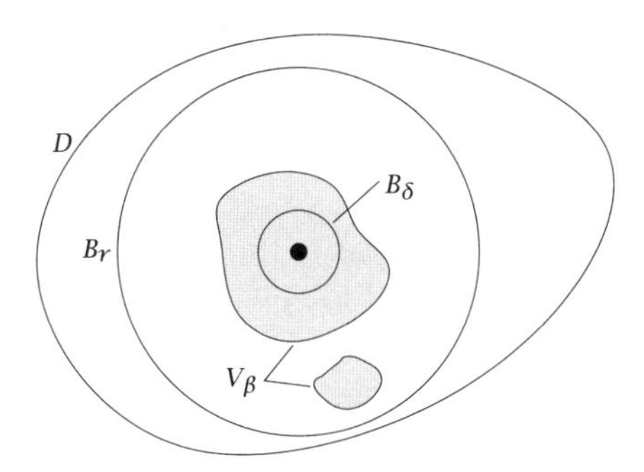

Figure 2.11: The origin and sets D, B_r, V_β (shaded), and B_δ.

Note that α is well defined because it is the minimization of a continuous function on a compact set, and $\alpha > 0$ because of (2.40). Choose $\beta \in (0, \alpha)$ and consider the sublevel set

$$V_\beta = \{x \mid V(x) \leq \beta\}$$

Note that, as shown in Figure 2.11, sublevel sets do not need to be *connected*. Regardless, we can readily establish that V_β is contained in the interior of B_r as follows. A point p not in the interior of B_r has $\|p\| \geq r$ and therefore satisfies $V(p) \geq \alpha$ due to α's definition, and is therefore not in the set V_β since $\beta < \alpha$. Notice also that any solution starting in V_β remains in V_β for all $t \geq 0$, which follows from (2.41) since $\dot{V}(x(t)) \leq 0$ implies that $V(x(t)) \leq V(x(0)) \leq \beta$ for all $t \geq 0$. A set with this property is called an INVARIANT SET, or sometimes a POSITIVE INVARIANT SET, to indicate that the set is invariant for time running in the *positive* direction. Next notice that V_β contains the origin in its interior since $\beta > 0$. Therefore we can choose $\delta > 0$ such that the ball B_δ is contained in V_β. Therefore, if we choose initial $x \in B_\delta$, we have for all $t \geq 0$

$$\|x\| \leq \delta \implies x \in V_\beta \implies$$
$$\phi(t;x) \in V_\beta \implies \phi(t;x) \in B_r \implies \|\phi(t;x)\| \leq \epsilon$$

and Lyapunov stability is established. ■

Theorem 2.22 (Asymptotic stability). *Let $V(\cdot)$ be a Lyapunov function for the system $\dot{x} = f(x)$. Moreover, let $V(\cdot)$ satisfy*

$$\dot{V}(x) < 0 \quad \text{for } x \in D \setminus 0 \tag{2.42}$$

Then the origin is asymptotically stable.

Proof. We conclude that the origin is stable from the previous theorem. So to prove asymptotic stability we need to show only that the origin is attractive. Since $V(\cdot)$ is continuous and vanishes only at zero, it is sufficient to establish that $V(\phi(t;x))$ goes to zero as $t \to \infty$ for all x satisfying $\|x\| \leq \eta$. We choose η as in the proof of Lyapunov stability so $B_\eta \subseteq D$. Since $\dot{V}(x(t)) \leq 0$ for all $x(t)$, $V(x(t))$ is a nonincreasing function of time, and it is bounded below by zero. Therefore it converges. We need to show that it converges to zero. Assume the contrary, that $V(x(t))$ converges to some $c > 0$, and we establish a contradiction. Consider the level set $V_c = \{x \mid V(x) = c\}$. This level set does not contain the origin, so we can choose $d > 0$ such that $\max_{\|x\| \leq d} V(x) < c$. Since $V(x(t))$ is nonincreasing and approaches c as $t \to \infty$, we have that $x(t)$ is outside B_d for all $t \geq 0$. Next define γ as

$$\gamma = -\max_{d \leq \|x\| \leq \eta} \dot{V}(x)$$

Note that γ is well defined because $\dot{V}(x)$ is continuous due to (2.39) and the fact that $\partial V(x)/\partial x$ and $f(x)$ are continuous. We know $\gamma > 0$ due to (2.42). Therefore

$$V(x(t)) = V(x(0)) + \int_0^t \dot{V}(x(\tau))d\tau \leq V(x(0)) - \gamma t$$

The right-hand side becomes negative for finite t for any $x(0) \in B_\eta$, which contradicts nonnegativity of $V(\cdot)$, and we conclude $c = 0$ and $V(x(t)) \to 0$, and hence $x(t) \to 0$, as $t \to \infty$. ■

Under the stronger assumption of Theorem 2.22, i.e., (2.42), establishing *continuity* of the solution $\phi(t;x)$ in t for all $t \geq 0$ and all x in a level set V_β contained in B_η also implies that the level set V_β *is connected.* This follows because every point $x \in V_\beta$ is then connected to the origin by a *continuous* curve $\phi(t;x)$ that remains in the positive invariant set V_β for all $t \geq 0$.

Next we consider a further strengthening of the properties of the Lyapunov function to ensure exponential stability. We have the following result.

Theorem 2.23 (Exponential stability). *Let $V(\cdot)$ be a Lyapunov function for the system $\dot{x} = f(x)$. Moreover, let $V(\cdot)$ satisfy for all $x \in D$*

$$a\|x\|^\sigma \leq V(x) \leq b\|x\|^\sigma \tag{2.43}$$

$$\dot{V}(x) \leq -c\|x\|^\sigma \tag{2.44}$$

for some $a, b, c, \sigma > 0$. Then the origin is exponentially stable.

Proof. Consider an arbitrary $r > 0$ and define function $\beta(\cdot)$ by $\beta(r) = \max_{\|x\|\leq r} V(x)$. We have that $\beta(\cdot)$ is positive definite and $\beta(0) = 0$. Now choose $r > 0$ small enough so that $V_{\beta(r)} \subseteq D$. Such an r exists since $V(\cdot)$ is continuous and $V(0) = 0$. We know that trajectories starting in V_β remain in V_β and hence D, so the inequalities stated in the theorem hold for solutions $\phi(t;x)$ for all $t \geq 0$ and $x \in V_\beta$. The upper-bounding inequality on $V(\cdot)$ implies that $\|x\|^\sigma \geq V(x)/b$, which combined with the bound on the time derivative of $V(x(t))$ gives

$$\dot{V} \leq -\frac{c}{b} V$$

Notice that the scalar time function $v(t) = V(x(t))$ satisfies the ODE $\dot{v} \leq -(c/b)v$ and therefore $v(t) \leq v(0)e^{-(c/b)t}$. Translating this statement back to $V(\cdot)$ gives $V(\phi(t;x)) \leq V(x)e^{-(c/b)t}$ for all $t \geq 0$ and $x \in V_\beta$. Using the lower-bounding inequality for $V(\cdot)$ gives $\|\phi(t;x)\|^\sigma \leq V(x)/a \; e^{-(c/b)t}$. Using the upper-bounding inequality again gives for all $x \in V_\beta$ and all $t \geq 0$

$$\|\phi(t;x)\| \leq \left(\frac{b}{a}\right)^{1/\sigma} \|x\| \, e^{-(c/(b\sigma))t}$$

We can choose $\delta > 0$ such that the ball B_δ is contained in V_β as shown in Figure 2.11. We then have that for all $\|x\| \leq \delta$

$$\|\phi(t;x)\| \leq c \, \|x\| \, e^{-\lambda t} \qquad \text{for all } t \geq 0$$

in which $c = (b/a)^{1/\sigma} > 0$ and $\lambda = c/(b\sigma) > 0$, and exponential stability of the origin is established. ■

2.5.3 Application to Linear Systems

Lyapunov function analysis of stability can of course be applied to linear systems, but this is mainly for illustrative purposes. We have many ways to analyze stability of linear systems because we have the analytical solution available. The true value of Lyapunov functions lies in analysis of nonlinear systems, for which we have few general purpose alternatives. To build up some expertise in using Lyapunov functions, we consider again the linear continuous time differential equation

$$\frac{dx}{dt} = Ax \qquad x(0) = x_0 \tag{2.45}$$

in which $x \in \mathbb{R}^n$ and $A \in \mathbb{R}^{n\times n}$. We have already discussed in Section 2.2.2 the stability of this system and shown that $x(t) = 0$ is an asymptotically stable steady state if and only if $\text{Re}(\text{eig}(A)) < 0$, i.e, all eigenvalues of A have strictly negative real parts. Let's see how we construct a Lyapunov function for this system. Consider as a candidate Lyapunov function

$$V(x) = x^T S x$$

in which $S \in \mathbb{R}^{n \times n}$ is positive definite, denoted $S > 0$. With this choice we have that $V : \mathbb{R}^n \to \mathbb{R}_{\geq 0}$, which is the first requirement, i.e., $V(0) = 0$ and $V(x) > 0$ for $x \neq 0$. We wish to evaluate the evolution of $V(x(t))$ with time as x evolves according to (2.45). Taking the time derivative of V gives

$$\begin{aligned}
\frac{d}{dt} V(x(t)) &= \frac{d}{dt} x^T S x \\
&= \frac{dx^T}{dt} S x + x^T S \frac{dx}{dt} \\
&= x^T A^T S x + x^T S A x \\
\frac{d}{dt} V(x(t)) &= x^T (A^T S + SA) x
\end{aligned}$$

and the initial condition is $V(0) = x_0^T S x_0$. One means to ensure that $V(x(t))$ is decreasing with time when $x \neq 0$ is to enforce that the matrix $A^T S + SA$ is negative definite. We choose some positive definite matrix $Q > 0$ and attempt to find a positive definite S that satisfies

$$A^T S + SA = -Q \tag{2.46}$$

so that

$$\frac{d}{dt} V = -x^T Q x$$

Equation (2.46) is known as the matrix Lyapunov equation. It says that given a $Q > 0$, if we can find a positive definite solution $S > 0$ of (2.46), then $V(x) = x^T S x$ is a Lyapunov function for linear system (2.45), and the steady-state solution $x = 0$ is asymptotically (in fact, exponentially) stable. This requirement can be shown to be also *necessary* for the system to be asymptotically (exponentially) stable, which we verify shortly. We seem to have exactly characterized the stability of the linear system (2.45) without any reference to the eigenvalues of matrix A. Of course, since the condition on the eigenvalues as well as the condition on the matrix Lyapunov equation are both necessary and sufficient conditions for asymptotic stability, they must be equivalent. Indeed, we have the following result stating this equivalence.

Theorem 2.24 (Lyapunov function for linear systems). *The following statements are equivalent (Sontag, 1998, p. 231).*

(a) *A is asymptotically stable, i.e.,* $\text{Re}(\text{eig}(A)) < 0$.

(b) *For each $Q \in \mathbb{R}^{n\times n}$, there is a unique solution S of the matrix Lyapunov equation*

$$A^T S + SA = -Q$$

and if $Q > 0$ then $S > 0$.

(c) *There is some $S > 0$ such that $A^T S + SA < 0$.*

(d) *There is some $S > 0$ such that $V(x) = x^T Sx$ is a Lyapunov function for the system $\dot{x} = Ax$.*

Exercise 2.62 asks you to establish the equivalence of (a) and (b).

2.5.4 Discrete Time Systems

Next we consider discrete time systems modeled by

$$x(k+1) = f(x(k)) \qquad x(0) = x_0$$

in which the sample time k is an *integer* $k = 0, 1, 2, \ldots$. To streamline the presentation we assume throughout that $f(\cdot)$ is continuous on its domain of definition. Steady states are now given by solutions to the equation $x_s = f(x_s)$, and we again assume without loss of generality that $f(0) = 0$ so that the origin is a steady state of the discrete time model. Discrete time models arise when *time* is discretized, as in digital control systems for chemical plants. But discrete time models also arise when representing the behavior of an iterative *algorithm*, such as the Newton-Raphson method for solving nonlinear algebraic equations discussed in Chapter 1. In these cases, the integer k represents the algorithm iteration number rather than time. We compress notation by defining the superscript $^+$ operator to denote the variable at the next sample time (or iteration), giving

$$x^+ = f(x) \qquad x(0) = x_0 \tag{2.47}$$

Notice that this notation also emphasizes the similarity with the continuous time model $\dot{x} = f(x)$ in (2.38). We again denote solutions to (2.47) by $\phi(k; x)$ with $k \geq 0$ that start at state x at $k = 0$. The discrete time definitions of stability, attractivity, and asymptotic stability of the origin are then *identical* to their continuous time counterparts given in

Definitions 2.16, 2.17, and 2.18, respectively, with integer $k \geq 0$ replacing real-valued time $t \geq 0$. In discrete time, the definition of exponential stability is modified slightly to the following.

Definition 2.25 (Exponential stability (discrete time)). The origin is exponentially stable if there exists $\delta > 0$ such that $\|x\| \leq \delta$ implies that there exist $c > 0$, $\lambda \in (0,1)$ for which

$$||\phi(k;x)|| \leq c \, \|x\| \, \lambda^k \quad \text{for all } k \geq 0$$

We see that λ^k with $\lambda < 1$ is the characteristic rate of solution decay for exponentially stable discrete time systems.

Lyapunov functions. The main difference in constructing Lyapunov functions for discrete time systems compared to those for continuous time systems is that we compare the value of V at two successive sample times, i.e., $V(x(k+1)) - V(x(k))$. If this change is negative, then we have the analogous behavior in discrete time that we have when $\dot{V}$ is negative in continuous time, i.e., $V(x(k))$ is decreasing when evaluated along the solution $x(k)$. We define the ΔV notation

$$\Delta V(x) = V(f(x)) - V(x) = V(x^+) - V(x)$$

to denote the change in V starting at state x and proceeding to successor state $x^+ = f(x)$. Another significant change is that we do not require differentiability of the Lyapunov function $V(\cdot)$ in discrete time since we do not require the chain rule to compute the time derivative. We do require continuity of $V(\cdot)$ at the origin, however. For consistency with the earlier continuous time results, we assume here that $V(\cdot)$ is continuous everywhere on its domain of definition.[12] The definition of the (continuous) Lyapunov function for discrete time is as follows.

Definition 2.26 (Lyapunov function (discrete time)). Consider a compact (closed and bounded) set $D \subset \mathbb{R}^n$ containing the origin in its interior and let $V : \mathbb{R}^n \to \mathbb{R}_{\geq 0}$ be continuous on D and satisfy

$$V(0) = 0 \text{ and } V(x) > 0 \text{ for } x \in D \setminus 0 \tag{2.48}$$

$$\Delta V(x) \leq 0 \text{ for } x \in D \tag{2.49}$$

Then $V(\cdot)$ is a Lyapunov function for the system $x^+ = f(x)$.

[12]For those needing discontinuous $V(\cdot)$ for discrete time systems, see Rawlings and Mayne (2009, Appendix B) for the required extension.

Notice that $\Delta V(x)$ also is continuous on its domain of definition since both $V(\cdot)$ and $f(\cdot)$ are assumed continuous.

Theorem 2.27 (Lyapunov stability (discrete time)). *Let $V(\cdot)$ be a Lyapunov function for the system $x^+ = f(x)$. Then the origin is (Lyapunov) stable.*

Theorem 2.28 (Asymptotic stability (discrete time)). *Let $V(\cdot)$ be a Lyapunov function for the system $x^+ = f(x)$. Moreover, let $V(\cdot)$ satisfy*

$$\Delta V(x) < 0 \quad \text{for } x \in D \setminus 0 \tag{2.50}$$

Then the origin is asymptotically stable.

Theorem 2.29 (Exponential stability (discrete time)). *Let $V(\cdot)$ be a Lyapunov function for the system $x^+ = f(x)$. Moreover, let $V(\cdot)$ satisfy for all $x \in D$*

$$a \|x\|^\sigma \leq V(x) \leq b \|x\|^\sigma \tag{2.51}$$

$$\Delta V(x) \leq -c \|x\|^\sigma \tag{2.52}$$

for some $a, b, c, \sigma > 0$. Then the origin is exponentially stable.

The proofs of Theorems 2.27, 2.28, and 2.29 are essentially identical to their continuous time counterparts, Theorems 2.21, 2.22, and 2.23, respectively, with integer k replacing real t and the difference ΔV replacing the derivative $\dot{V}$. An essential difference between continuous and discrete time cases is that the solution of the continuous time model $\phi(t; x)$ is *continuous* in t, and the solution of the discrete time model $\phi(k; x)$ has no continuity with index k since k takes on discrete values. Notice that in the proofs of the continuous time results, we did not follow the common practice of appealing to continuity of $\phi(t; x)$ in t, so the supplied arguments are valid for both continuous and discrete cases.

Linear systems. The time-invariant discrete time linear model is

$$x^+ = Ax \qquad x(0) = x_0$$

and in analogy with the continuous time development, we try to find a Lyapunov function of the form $V(x) = x^T S x$ for some positive definite matrix $S > 0$. Computing the change in the Lyapunov function at state x gives

$$\begin{aligned} \Delta V(x) &= V(x^+) - V(x) = (Ax)^T S (Ax) - x^T S x \\ &= x^T (A^T S A - S) x \end{aligned}$$

Choosing a positive definite $Q > 0$, if we can find $S > 0$ that satisfies

$$A^T SA - S = -Q \tag{2.53}$$

then we have succeeded in finding a $V(\cdot)$ with the desired properties: $V(x) = x^T Sx \geq 0$ and $\Delta V(x) = -x^T Qx \leq 0$ for all x. Equation (2.53) is known as the *discrete* matrix Lyapunov equation. Exercise 2.63 asks you to state the discrete time version of Theorem 2.24, listing the connections between the solution of the discrete Lyapunov equation and the eigenvalues of A. These connections often come in handy when analyzing the stability of discrete linear systems.

2.6 Asymptotic Analysis and Perturbation Methods

2.6.1 Introduction

Typical mathematical models have a number of explicit parameters. Often we are interested in how the solution to a problem depends on a certain parameter. Asymptotic analysis is the branch of applied mathematics that deals with the construction of precise approximate solutions to problems in asymptotic cases, i.e., when a parameter of the problem is large or small. In chemical engineering problems, small parameters often arise as ratios of time or length scales. Important limiting cases arise for example in the limits of large or small Reynolds, Péclet, or Damköhler numbers. In many cases, an analytical solution can be found, even if the problem is nonlinear. In others, the scaling behavior of the solution (e.g., the correct exponent for the power-law dependence of one quantity on another) can be found without even solving the problem. In still others, the asymptotic analysis yields an equation that must be solved numerically, but is much less complicated than the original model. The goal here is to provide a background on the basic concepts and techniques of asymptotic analysis, beginning with some notation and basic ideas about series approximations.

2.6.2 Series Approximations: Convergence, Asymptoticness, Uniformity

As this section deals extensively with how one function approximates another, we begin by introducing symbols that describe degrees of identification between different functions.

$a = b$	a is equal to b
$a \sim b$	a is asymptotically equal to b (in some given/implied limit)
$a \approx b$	a is approximately equal to b (in any useful sense)
$a \propto b$	a is proportional to b

It is important to note that $\sim$ implies a limit process, while $\approx$ does not. In this section we will be careful to use the symbol "$\sim$" in the precise manner defined here, though one must be aware that it often means different things in different contexts (and different parts of this book). Closely related to these symbols are ORDER SYMBOLS, which provide a qualitative description of the relationships between functions in limiting cases. Consider a function $f(\epsilon)$ whose behavior we wish to describe relative to another function (a GAUGE FUNCTION) $\delta(\epsilon)$. The order symbols "O", "o" and "ord" describe the relationships

$$
\begin{aligned}
f(\epsilon) &= O(\delta(\epsilon)) \text{ as } \epsilon \to 0 && \text{if } \lim_{\epsilon\to 0} \frac{f(\epsilon)}{\delta(\epsilon)} < \infty \\
f(\epsilon) &= o(\delta(\epsilon)) \text{ as } \epsilon \to 0 && \text{if } \lim_{\epsilon\to 0} \frac{f(\epsilon)}{\delta(\epsilon)} = 0 \\
f(\epsilon) &= \operatorname{ord}(\delta(\epsilon)) \text{ as } \epsilon \to 0 && \text{if } f(\epsilon) = O(\delta(\epsilon)) \text{ but not } o(\delta(\epsilon))
\end{aligned}
$$

In the latter case, f is said to be *strictly* order δ. Often, authors write "$f(\epsilon) \sim \delta(\epsilon)$" to mean "$f(\epsilon) = O(\delta(\epsilon))$", though the latter only implies equality to within a multiplicative constant as $\epsilon \to 0$, while as defined here the former implies equality.

Asymptotic approximations take the form of series, the most familiar of which is the truncated Taylor series approximation that forms the basis of many engineering approximations. An infinite series

$$f(x) = \sum_{n=0}^{\infty} f_n(x)$$

CONVERGES at a particular value of x if and only if, for every $\epsilon > 0$, there exists N_0 such that

$$\left| \sum_{n=M}^{N} f_n(x) \right| < \epsilon \text{ for all } M, N > N_0$$

In contrast, an ASYMPTOTIC SERIES

$$f(\epsilon) \approx \sum_{n=0}^{N} f_n(\epsilon)$$

satisfies

$$\lim_{\epsilon \to 0} \frac{f(\epsilon) - \sum_{n=0}^{M} f_n(\epsilon)}{f_M(\epsilon)} = 0 \text{ for each } M \leq N$$

In words, the remainder is much smaller than the last term kept. This property is the source of the usefulness of asymptotic series. If this property is satisfied, we write

$$f(\epsilon) \sim \sum_{n=0}^{N} f_n(\epsilon) \text{ as } \epsilon \to 0$$

In general, we *do not care* whether the series converges if we let $N \to \infty$. Often it does not. The important point is that the *finite* sum—often the first term or two—provides a useful approximation to a function for small ϵ. This is in stark contrast to convergent infinite series, which, although they converge, often require a large number of terms to be evaluated to obtain a reasonably accurate approximation.

We typically construct asymptotic series in this form

$$f(\epsilon) \sim \sum_{n} a_n \delta_n(\epsilon) \tag{2.54}$$

where

$$\delta_0(\epsilon) \gg \delta_1(\epsilon) \gg \delta_2(\epsilon) \gg \cdots$$

for small ϵ. We also require that $\delta_{n+1}(\epsilon) = o(\delta_n(\epsilon))$ as $\epsilon \to 0$. In practice, the δs are not generally known *a priori*, but must be determined as part of the solution procedure to satisfy the requirement that the coefficients a_n be ord(1). This procedure is best illustrated by example as we do in several instances below. In principle, we can construct a series approximation with N as large as we like, as long as the a_n remain ord(1) and $\delta_{N-1}(\epsilon) \ll \delta_N(\epsilon)$ at the value of ϵ of interest. However, the most common application of asymptotic analysis is to the construction of a one- or two-term approximation that captures the most important behavior as $\epsilon \to 0$.

As an example of the difference between convergent and asymptotic series, we look at the error function erf (z), written here as

$$\text{erf }(x) = 1 - \frac{2}{\sqrt{\pi}} \int_x^{\infty} e^{-t^2} dt$$

By Taylor expanding the integrand around the origin and integrating term by term, a power series convergent for all x can be constructed for this function

$$\text{erf }(x) = \frac{2}{\sqrt{\pi}} \sum_0^{\infty} \frac{(-1)^n x^{2n+1}}{(2n+1)n!}$$

Although convergent, this expression may require many terms for reasonable accuracy to be obtained, especially when x is large. One could try setting $w = 1/x$ and Taylor expanding e^{-1/w^2} around $w = 0$. This leads to immediate difficulty because

$$\lim_{w \to 0} \frac{d^n}{dw^n} e^{-1/w^2} = 0$$

for all n; his Taylor expansion is identically zero! This difficulty arises because e^{-x^2} decays to zero faster than any negative power of x as $x \to \infty$.

On the other hand, for $x \gg 1$, an asymptotic series for the function may be constructed by repeated integration by parts (a common trick for the asymptotic approximation of integrals). This approximation is

$$\begin{aligned}
\text{erf }(x) &= 1 - \frac{2}{\sqrt{\pi}} \int_x^{\infty} e^{-t^2} dt \\
&= 1 - \frac{2}{\sqrt{\pi}} \left(\int_x^{\infty} \frac{-1}{2t} \, de^{-t^2} \right) \\
&= 1 - \frac{2}{\sqrt{\pi}} \left(\frac{e^{-x^2}}{2x} - \int_x^{\infty} \frac{1}{2t^2} e^{-t^2} \, dt \right) \\
&= 1 - \frac{2}{\sqrt{\pi}} \left(\frac{e^{-x^2}}{2x} - \int_x^{\infty} \frac{-1}{4t^3} \, de^{-t^2} \right) \\
&= 1 - \frac{2}{\sqrt{\pi}} \left(\frac{e^{-x^2}}{2x} - \frac{e^{-x^2}}{4x^3} + \int_x^{\infty} \frac{3}{4t^4} e^{-t^2} \, dt \right) \\
\text{erf }(x) &\sim 1 - \frac{e^{-x^2}}{x\sqrt{\pi}} \left(1 - \frac{1}{2x^2} + \frac{1 \cdot 3}{(2x^2)^2} + O(x^{-6}) \right)
\end{aligned}$$

If continued indefinitely, this series would diverge. The truncated series, however, is useful. In particular, the "leading order" term $1 - \frac{e^{-x^2}}{x\sqrt{\pi}}$, the expression that includes the first correction for finite but large x, precisely indicates the behavior of erf (x) for large values of x. Furthermore, the truncated series can be used to provide accurate numerical values and is the basis of modern algorithms for doing so (Cody, 1969).

Now consider a function f of ϵ and some other parameter or variable, x

$$f(x, \epsilon) \sim \sum a_n \delta_n(\epsilon)$$

If the approximation is asymptotic as $\epsilon \to 0$ for each fixed x, then we say it is POINTWISE ASYMPTOTIC. Now consider the particular case

$$f(x,\epsilon) \sim \delta_1(\epsilon) + \delta_2(\epsilon) = 1 + \epsilon/\sqrt{x}$$

This is pointwise asymptotic, but for fixed ϵ, the second term blows up as $x \to 0$. So obviously, it cannot remain much smaller than the first term, which is our requirement for asymptoticness. The approximation is not UNIFORMLY VALID. Put another way

$$\lim_{\epsilon\to 0}\lim_{x\to 0} \epsilon/\sqrt{x} \neq \lim_{x\to 0}\lim_{\epsilon\to 0} \epsilon/\sqrt{x}$$

To be precise, a function $u(x,\epsilon)$ CONVERGES UNIFORMLY to $u(x,0)$ on the interval $x \in [0,a]$, if, given $E > 0$, there is a $D > 0$ such that

$$|u(x,\epsilon) - u(x,0)| < E, \text{for } \epsilon < D \text{ and all } x \in [0,a]$$

Nonuniformity is a feature of many practical singular perturbation problems. A major challenge of asymptotic analysis is the construction of UNIFORMLY VALID approximations. We shall see a number of techniques for doing this. They all have a general structure that looks like this

$$f(x,\epsilon) \sim \sum a_n(x,\epsilon)\delta_n(\epsilon)$$

2.6.3 Scaling, and Regular and Singular Perturbations

Before proceeding to discuss perturbation methods for differential equations, we introduce some important concepts in the context of algebraic equations. First, consider the quadratic equation

$$x^2 + \epsilon x - 1 = 0, \quad \epsilon \ll 1 \tag{2.55}$$

If $\epsilon = 0, x = \pm 1$. We would like to characterize how these solutions are perturbed when $0 < \epsilon \ll 1$. To to so, we posit a solution of the form (2.54)

$$x(\epsilon) = \delta_0 x_0 + \delta_1(\epsilon)x_1 + \delta_2(\epsilon)x_2 + o(\delta_2) \tag{2.56}$$

where $x_i = \mathrm{ord}(1)$ (independent of ϵ) and the functional forms of $\delta_1(\epsilon)$ and $\delta_2(\epsilon)$ remain to be determined. Substituting into the quadratic and neglecting the small $o(\delta_2)$ terms yields

$$(\delta_0 x_0+\delta_1(\epsilon)x_1+\delta_2(\epsilon)x_2)^2+\epsilon(\delta_0 x_0+\delta_1(\epsilon)x_1+\delta_2(\epsilon)x_2)-1 = 0 \tag{2.57}$$

At $\epsilon = 0$, the solution is $x = x_0 = \pm 1$. So we let $\delta_0 = 1$ and for the moment consider the root $x_0 = 1$. Now (2.57) becomes

$$\epsilon + 2\delta_1 x_1 + \delta_1 \epsilon x_1 + \delta_1^2 x_1^2 + 2\delta_2 x_2 + \delta_2 \epsilon x_2 + 2\delta_1 \delta_2 x_1 x_2 + \delta_2^2 x_2^2 = 0 \quad (2.58)$$

Observe that all but the first two terms are $o(\epsilon)$ or $o(\delta_1)$. Neglecting these, we would find that

$$\epsilon + 2\delta_1 x_1 = 0 \quad (2.59)$$

Since x_1 is independent of ϵ, we set $\delta_1 = \epsilon$, in which case $x_1 = -\frac{1}{2}$. Now (2.58) becomes

$$-\frac{\epsilon^2}{4} + 2\delta_2 x_2 + \delta_2^2 x_2^2 = 0 \quad (2.60)$$

Now, since $\delta_2^2 = o(\delta_2)$, we neglect the term containing it to get

$$-\frac{\epsilon^2}{4} + 2\delta_2 x_2 = 0 \quad (2.61)$$

for which there is a solution if $\delta_2(\epsilon) = \epsilon^2$ and $x_2 = \frac{1}{8}$. Thus we have constructed an asymptotic approximation

$$x = 1 - \frac{1}{2}\epsilon + \frac{1}{8}\epsilon^2 + o(\epsilon^2) \quad (2.62)$$

Observe that to determine $\delta_1(\epsilon)$ and $\delta_2(\epsilon)$, we have found a DOMINANT BALANCE: a self-consistent choice of $\delta_k(\epsilon)$, where it is comparable in size to the largest term not containing a δ_k and where all the terms containing δ_ks and ϵs are smaller as $\epsilon \to 0$.

To find how the second root $x_0 = -1$ depends on ϵ, we use the lessons learned in the previous paragraph to streamline the solution process. That analysis suggests that $\delta_k(\epsilon) = \epsilon^k$ so we seek a solution

$$x = -1 + \epsilon x_1 + \epsilon^2 x_2 + O(\epsilon^3)$$

which upon substitution into (2.55) yields

$$-\epsilon - 2\epsilon x_1 + \epsilon^2 x_1 + \epsilon^2 x_1^2 - 2\epsilon^2 x_2 + O(\epsilon^3) = 0$$

Since by assumption the x_k are independent of ϵ, this expression can only hold in general if it holds power by power in ϵ. We have already zeroed out the ϵ^0 term by setting $x_0 = -1$. The ϵ^1 and ϵ^2 terms yield

$$\epsilon^1: \quad -2x_1 - 1 = 0$$
$$\epsilon^2: \quad -2x_2 + x_1^2 + x_1 = 0$$

There is one-way coupling between these equations: the equation for x_k depends on x_l with $l < k$. The solutions to these are $x_1 = -\frac{1}{2}$ and $x_2 = -\frac{1}{8}$, so the second root is

$$x = -1 - \frac{1}{2}\epsilon - \frac{1}{8}\epsilon^2 + O(\epsilon^3) \tag{2.63}$$

In the limit $\epsilon \to 0$ both solutions (2.62) and (2.63) reduce to the solutions when $\epsilon = 0$. Cases such as this are called REGULAR PERTURBATION problems.

The situation is much more interesting when $\epsilon = 0$ is qualitatively different from $\epsilon \ll 1$. Cases like this are called SINGULAR PERTURBATION problems. Consider the equation

$$\epsilon x^2 + x - 1 = 0 \tag{2.64}$$

When $\epsilon = 0$, this has the unique exact solution $x = 1$, while for any $\epsilon \neq 0$, it has two solutions. This problem is singular because the small parameter multiplies the highest power in the equation - when the parameter is zero the polynomial becomes lower degree so it has one fewer root. To analyze this problem, we define a scaled variable $X = x/\delta$ where $\delta = \delta_0$ and

$$X \sim x_0 + \frac{\delta_1}{\delta_0}x_1 + \frac{\delta_2}{\delta_0}x_2 = \text{ord}(1)$$

Thus δ measures the size of x as $\epsilon \to 0$. Substitution into (2.64) yields

$$\epsilon\delta^2 X^2 + \delta X - 1 = 0$$

Now we examine the possibility of finding a dominant balance between different terms with various guesses for δ. If we let $\delta = 1$, then the second and third of these terms balance as $\epsilon \to 0$ while the first is small. This scaling gives the root $x = 1 + O(\epsilon)$. If we let $\delta = o(1)$ then the first and second terms are small, while the third term is still $\text{ord}(1)$. There is no balance of terms for this scaling. On the other hand if we let $\delta = \epsilon^{-1}$ then we can get the first and second terms to balance. Applying this scaling yields

$$X^2 + X + \epsilon = 0$$

which clearly has the solution $X = 1 + O(\epsilon)$ or $x = \frac{1}{\epsilon} + O(1)$. As $\epsilon \to 0$ the root goes off to infinity. Although the first term in (2.64) contains the small parameter, it can multiply a large number so that the term overall is not small. This characteristic is typical of singular perturbation problems.

A more subtle singular perturbation problem is

$$(1 - \epsilon)x^2 - 2x + 1 = 0 \tag{2.65}$$

When $\epsilon = 0$ this has double root $x = 1$. When $\epsilon < 0$ there are no real solution whereas when $\epsilon > 0$ there are two. Clearly $\delta_0 = 0$, $x_0 = 1$ so we seek a solution

$$x = 1 + \delta_1 x_1 + \delta_2 x_2 + o(\delta_2)$$

Substitution into (2.65) gives

$$\delta_1^2 x_1^2 + 2\delta_1\delta_2 x_1 x_2 + \delta_2^2 x_2^2 - \epsilon - 2\epsilon\delta_1 x_1 - \epsilon\delta_1^2 x_1^2 - 2\epsilon\delta_2 x_2 + \ldots = 0$$

Since $1 \gg \delta_1 \gg \delta_2$, we can conclude that $\delta_1^2 x_1^2$ and ϵ are the largest (dominant) terms. These balance if $\delta_1^2 = O(\epsilon)$. Thus we set $\delta_1 = \epsilon^{1/2}$, which implies that $x_1 = \pm 1$. So the solutions are

$$x = 1 \pm \epsilon^{1/2} + O(\delta_2) \tag{2.66}$$

As an exercise, find that $\delta_2 = \epsilon$ and that the solutions to (2.65) can be written as an asymptotic series in powers of $\epsilon^{1/2}$.

2.6.4 Regular Perturbation Analysis of an ODE

One attractive feature of perturbation methods is their capacity to provide analytical, albeit approximate solutions to complex problems. For regular perturbation problems, the approach is rather straightforward. As an illustration we consider the problem of second-order reaction occurring in a spherical catalyst pellet, which we can model at steady state by

$$0 = \frac{1}{r^2}\frac{d}{dr}r^2\frac{dc}{dr} - \mathrm{Da}\, c^2 \tag{2.67}$$

with $c = 1$ at $r = 1$ and c bounded at the origin. If D, R, k, and c_B are the diffusivity, particle radius, rate constant, and dimensional surface concentration respectively, then $\mathrm{Da} = kc_B R^2/D$ is the DAMKÖHLER NUMBER. The problem is nonlinear, so a simple analytical solution is unavailable. An approximate solution for $\mathrm{Da} \ll 1$ can be constructed, however, using a regular perturbation approach.

Let $\epsilon = \text{Da}$. We seek a solution of the form $c(r) = c_0 + \epsilon c_1 + \epsilon^2 c_2 + O(\epsilon^3)$. Substituting into (2.67) and equating like powers yields

$$\begin{aligned}
\epsilon^0 : 0 &= \frac{1}{r^2}\frac{d}{dr}r^2\frac{dc_0}{dr}, \quad c_0(1) = 1 \\
\epsilon^1 : 0 &= \frac{1}{r^2}\frac{d}{dr}r^2\frac{dc_1}{dr} - c_0^2, \quad c_1(1) = 0 \\
\epsilon^2 : 0 &= \frac{1}{r^2}\frac{d}{dr}r^2\frac{dc_2}{dr} - 2c_1c_0, \quad c_2(1) = 0
\end{aligned}$$

Observe that the solution at each order has the same operator but different "forcing" from the solution at lower order. This structure is typical of regular perturbation problems. The solution at ϵ^0 is trivial: $c_0 = 1$ for all r. At ϵ^1, we have

$$0 = \frac{1}{r^2}\frac{d}{dr}r^2\frac{dc_1}{dr} - 1, \quad c_1(1) = 0$$

The solution to this equation is $c_1 = (r^2 - 1)/6$. The solution to the ϵ^2 problem is left to Exercise 2.66. Although this problem is nonlinear, the regular perturbation method provides a simple approximate closed-form solution.

2.6.5 Matched Asymptotic Expansions

The regular perturbation approach above provided an approximate solution for $\text{Da} \ll 1$. We can also pursue a perturbation solution in the opposite limit, $\text{Da} \gg 1$. Now letting $\epsilon = \text{Da}^{-1}$ we have

$$0 = \epsilon\frac{1}{r^2}\frac{d}{dr}r^2\frac{dc}{dr} - c^2 \tag{2.68}$$

If we naively seek a regular perturbation solution $c = c_0 + \epsilon c_1 + O(\epsilon^2)$, the leading-order equation will be

$$0 = c_0^2$$

This has solution $c_0 = 0$, which satisfies the boundedness condition at $r = 0$ and makes physical sense for the interior of the domain because when $\text{Da} \gg 1$, reaction is fast compared to diffusion so we expect the concentration in the particle to be very small. On the other hand, this solution cannot be complete, as it cannot satisfy the boundary condition $c = 1$ at $r = 1$. The inability of the solution to satisfy the boundary condition arises from the fact that the small parameter

ϵ multiplies the highest derivative in the equation. It is thus absent from the leading-order problem, so the arbitrary constants required to satisfy the boundary conditions are not available.

The resolution to this issue lies in proper scaling. Although ϵ is small, it multiplies a second derivative. If the gradient of the solution is large in some region, then the product of the small parameter and large gradient may result in a term that is not small. In the present case, we can use physical intuition to guess where the gradients are large. At high Da the reaction occurs rapidly, so we expect the concentration to be small in most of the catalyst particle. Near $r = 1$, however, reactant is diffusing in from the surroundings and indeed right at $r = 1$ the concentration must be unity. Thus we define a new spatial variable $\eta = (1 - r)/\zeta(\epsilon)$ where ζ is a length scale that is yet to be determined. Applying this change of variable to (2.68) yields

$$0 = \frac{\epsilon}{\zeta^2} \frac{2}{(1 - \zeta\eta)^2} \frac{d}{d\eta} (1 - \zeta\eta)^2 \frac{dc}{d\eta} - c^2 \tag{2.69}$$

The first term contains $\epsilon\zeta^{-2}$. If we take $\zeta = \epsilon^{1/2}$ then this term is ord(1) as $\epsilon \to 0$ and can balance the term c^2 to yield a nontrivial solution. This scaling implies that near $r = 1$ the steepness of the concentration gradient scales as $\zeta^{-1} = \epsilon^{-1/2}$. Proceeding with this scaling, (2.69) becomes

$$0 = \frac{1}{(1 - \epsilon^{1/2}\eta)^2} \frac{d}{d\eta} (1 - \epsilon^{1/2}\eta)^2 \frac{dc}{d\eta} - c^2 \tag{2.70}$$

Now we seek a perturbation solution of this rescaled problem: $c(\eta) = c_0 + \epsilon^{1/2} c_1 + O(\epsilon)$. The choice of $\epsilon^{1/2}$ comes from the observation that the Taylor expansion of $(1 - \epsilon^{1/2}\eta)^{\pm 2} = 1 \mp 2\epsilon^{1/2}\eta + O(\epsilon)$. This gives the leading-order problem

$$0 = \frac{d^2 c_0}{d\eta^2} - c_0^2 \tag{2.71}$$

Although this equation is nonlinear, it has a special form that facilitates solution.[13] Let $w = c'$ where $'$ denotes $d/d\eta$. Now we can write

$$w' = c_0^2 = -\frac{\partial H}{\partial c_0} \qquad c_0' = w = \frac{\partial H}{\partial w}$$

with

$$H = \frac{1}{2} w^2 - \frac{1}{3} c_0^3$$

[13]If we had considered first-order kinetics instead, the solution would be simple.

As constructed, this system has the special property that

$$\frac{dH}{d\eta} = \frac{\partial H}{\partial w}w' + \frac{\partial H}{\partial c_0}c_0' = \frac{\partial H}{\partial w}\left(-\frac{\partial H}{\partial c_0}\right) + \frac{\partial H}{\partial c_0}\frac{\partial H}{\partial w} = 0$$

Therefore, curves of $H = \frac{1}{2}c'^2 - \frac{1}{3}c_0^3 = K$, where K is a constant, are solutions. As η becomes large, i.e., at positions much larger than a distance $\epsilon^{1/2}$ from the interface, we expect the concentration and its gradient to go to zero, we take $K = 0$, so

$$c_0' = \pm\sqrt{\frac{2}{3}}c_0^{3/2}$$

The negative sign must be chosen so that c_0 decays with increasing η. This equation can be integrated and the boundary condition $c_0(\eta = 0) = 1$ applied to yield

$$c_0(\eta) = \left(1 + \sqrt{\frac{1}{6}}\eta\right)^{-2}$$

In terms of the original variables this becomes

$$c_0(r) = \left(1 + \sqrt{\frac{1}{6\epsilon}}(1 - r)\right)^{-2} \tag{2.72}$$

This decays to zero once $1 - r$ is larger than $O(\epsilon^{1/2})$. Thus the concentration changes rapidly in a BOUNDARY LAYER with thickness of $O(\epsilon^{1/2})$ that is located near the catalyst particle surface. Outside this thin boundary layer, in the interior of the particle, the concentration is very small, going to zero as $\epsilon \to 0$. One can carry this analysis to higher order terms. For example, the first effects of the particle shape on the result appear at $O(\epsilon^{1/2})$ but it should be clear that the primary structure of the solution behavior has been captured by this leading-order solution.

This example is a simple instance of a singular perturbation method. The solution $c_0 = 0$ that we obtained before rescaling is called the OUTER SOLUTION. It is valid away from the boundary $r = 1$. The solution (2.72) that we obtained after rescaling is called the INNER SOLUTION. In this simple example the inner solution decays to zero, automatically matching the outer solution. In general, the outer solution is not simply a constant, and a MATCHING CONDITION must be imposed to properly connect the two solutions to one another. This process is the origin of the term MATCHED ASYMPTOTIC EXPANSIONS.

Matching can be accomplished with a number of different procedures. We describe here a simple approach that works for many problems. More sophisticated and general approaches are described in Hinch (1991). In the simple approach, we denote the outer solution with terms up to δ_P as $u_P(x)$ and the inner solution as $U_P(\xi)$ where x and $\xi = x/\epsilon$ are the outer and inner variables, respectively. The simple matching procedure just requires that at each order $n = 0, 1, \ldots, N$

$$\lim_{x \to 0} u_n(x) = \lim_{\xi \to \infty} U_n(\xi) \tag{2.73}$$

In words, the inner limit of the outer solution equals the outer limit of the inner solution. In the above case, taking the outer variable x as $1-r$ and the inner variable ξ as η, this expression is satisfied trivially. In general, neither the inner nor the outer solution is valid throughout the entire domain, but the matching procedure provides a means to construct a uniformly valid solution. This so-called COMPOSITE SOLUTION is given by

$$u_{nc}(x) = u_n(x) + U_n(\xi) - \lim_{\xi \to \infty} U_n(\xi) \tag{2.74}$$

The last term avoids double counting of the overlapping parts of the two solutions. These ideas are illustrated in the following example.

Example 2.30: Matched asymptotic expansion analysis of the reaction equilibrium assumption

Consider the following reactions

$$\mathrm{A} \underset{k_{-1}}{\overset{k_1}{\rightleftharpoons}} \mathrm{B}, \qquad \mathrm{B} \xrightarrow{k_2} \mathrm{C}$$

in which rate constants k_1, k_{-1} are much larger than the rate constant k_2, so the first reaction equilibrates quickly. In a batch system where c_A, c_B, and c_C are the concentrations, the governing equations are

$$\frac{dc_A}{dt} = -k_1 c_A + k_{-1} c_B$$
$$\frac{dc_B}{dt} = k_1 c_A - k_{-1} c_B - k_2 c_B$$
$$\frac{dc_C}{dt} = k_2 c_B$$

The reaction equilibrium assumption takes c_A and c_B to be in equilibrium so that $c_B = K c_A$ where $K = k_1/k_{-1}$. Further assume that k_{-1} is the largest rate constant. Initial concentrations in the reactor are $c_A(0) = c_{A0}, c_B(0) = c_C(0) = 0$.

(a) Find a proper nondimensionalization so that a systematic perturbation expansion can be performed.

(b) Use matched asymptotic expansions to show that the reaction equilibrium approximation corresponds to the leading-order outer solution of the kinetic equations. Also find the equations for the $O(\epsilon^1)$ terms in the outer solution.

(c) Find the leading-order inner solution for the dynamics on the fast time scale $1/k_{-1}$, match the inner and outer solutions, and find a composite solution that is uniformly valid for all time.

Solution

(a) Let $u = c_A c_A(0), v = c_B/c_A(0), w = c_C/c_A(0)$, so $u(0) = 1, v(0) = w(0) = 0$. Define a scaled "slow" time variable $t_s = k_2 t$ so that an $O(1)$ change in t_1 corresponds to a time interval of $O(1/k_2)$, and define the small parameter $\epsilon = k_2/k_{-1}$. In these variables, the rate equations are

$$\frac{du}{dt_s} = -\frac{K}{\epsilon}u + \frac{1}{\epsilon}v$$

$$\frac{dv}{dt_s} = \frac{K}{\epsilon}u - \frac{1}{\epsilon}v - w$$

$$\frac{dw}{dt_s} = v$$

Since c_C is determined completely by c_B we do not include its evolution in the following development.

(b) Multiplying the dimensionless equations by ϵ yields

$$\epsilon\frac{du}{dt_s} = -Ku + v \qquad \epsilon\frac{dv}{dt_s} = Ku - v - \epsilon v$$

Assuming a power series form, the outer solution is obtained by letting

$$u(t_s) = u_0(t_s) + \epsilon u_1(t_s) + O(\epsilon^2)$$

$$v(t_s) = v_0(t_s) + \epsilon v_1(t_s) + O(\epsilon^2)$$

Substituting and considering only the terms of $O(\epsilon^0)$ yields

$$Ku_0 = v_0$$

for both of these equations. This is the reaction equilibrium assumption in dimensionless form. Although physically reasonable, observe that this assumption is not consistent with the initial conditions $u(0) = 1, v(0) = 0$. Similarly, because the time derivatives are multiplied by ϵ, they do not appear in the leading-order outer problem, so we do not have differential equations whose solutions include the arbitrary constants that are determined by the initial conditions. Keeping this issue in mind, we collect $O(\epsilon^1)$ terms to yield

$$\frac{du_0}{dt_s} = -Ku_1 + v_1 \qquad \frac{dv_0}{dt_s} = Ku_1 - v_1 - v_0$$

Although this equation is valid, it is not yet useful because we do not know the values of of u_0 and v_0. To obtain these we consider the inner solution.

(c) The problem with the outer solution can be traced to the loss of the time-derivative terms. Recognizing that the derivatives can be large at short times because k_1 and k_{-1} and much larger than k_2, we define a new fast time scale $t_f = t/k_{-1} = t_s/\epsilon$. Now t_f changes an $O(1)$ amount in a dimensional time of about $1/k_{-1}$. Rewriting the equations with this new time scaling yields

$$\frac{du}{dt_f} = -Ku + v \qquad \frac{dv}{dt_f} = Ku - v - \epsilon v$$

Now we seek an inner solution

$$u(t_f) = U_0(t_f) + \epsilon U_1(t_f) + O(\epsilon^2)$$
$$v(t_f) = V_0(t_f) + \epsilon V_1(t_f) + O(\epsilon^2)$$

Substituting and extracting the $O(\epsilon^0)$ terms yields

$$\frac{dU_0}{dt_f} = -KU_0 + V_0 \qquad \frac{dV}{dt_f} = KU_0 - V_0$$

with initial condition $U_0(0) = 1, V_0(0) = 0$. This coupled pair of equations could be solved, for example, by Laplace transforms or by rewriting as a system $dx/dt_f = Ax$ and diagonalizing A, but in this case we can use the observation that $dU_0/dt_f + dV_0/dt_f = 0$ so $U_0 + V_0 = 1$ to just solve for U_0

$$\frac{dU_0}{dt_f} = -KU_0 + 1 - U_0$$

which has solution

$$U_0 = e^{-(1+K)t_f} + \frac{1}{1+K}\left(1 - e^{-(1+K)t_f}\right)$$

Using this, we obtain

$$V_0 = 1 - U_0 = \frac{K}{1+K}\left(1 - e^{-(1+K)t_f}\right)$$

By analogy with the reaction-diffusion example above, this inner solution corresponds to a boundary layer in time, rather than space.

With inner and outer solutions in hand, we can use (2.73) to match them. The "outer limit" of the inner solution is

$$\lim_{t_f \to \infty} U_0 = \frac{1}{1+K} \qquad \lim_{t_f \to \infty} V_0 = \frac{K}{1+K}$$

which satisfies the equilibrium assumption $Ku = v$. The inner limit of the outer solution is simply $u_0(0)$, $v_0(0)$ and using the previous result yields

$$u_0(0) = \frac{1}{1+K} \qquad v_0(0) = \frac{K}{1+K}$$

Now we have initial conditions for the outer solution. Adding the two differential equations at $O(\epsilon^1)$ and differentiating the algebraic equation (reaction equilibrium result) at $O(\epsilon^0)$ give

$$\frac{du_0}{dt_s} + \frac{dv_0}{dt_s} = -v_0 \qquad -K\frac{du_0}{dt_s} + \frac{dv_0}{dt_s} = 0$$

Solving these two equations for the two time derivatives gives

$$\frac{du_0}{dt_s} = -\frac{K}{1+K}u_0 \qquad \frac{dv_0}{dt_s} = -\frac{K}{1+K}v_0$$

Solving these with their respective initial (matching) conditions $u_0(0) = 1, v_0(0) = 0$ gives the full leading-order outer solution

$$u_0 = \frac{1}{1+K}e^{-\frac{K}{1+K}t_s} \qquad v_0 = \frac{K}{1+K}e^{-\frac{K}{1+K}t_s}$$

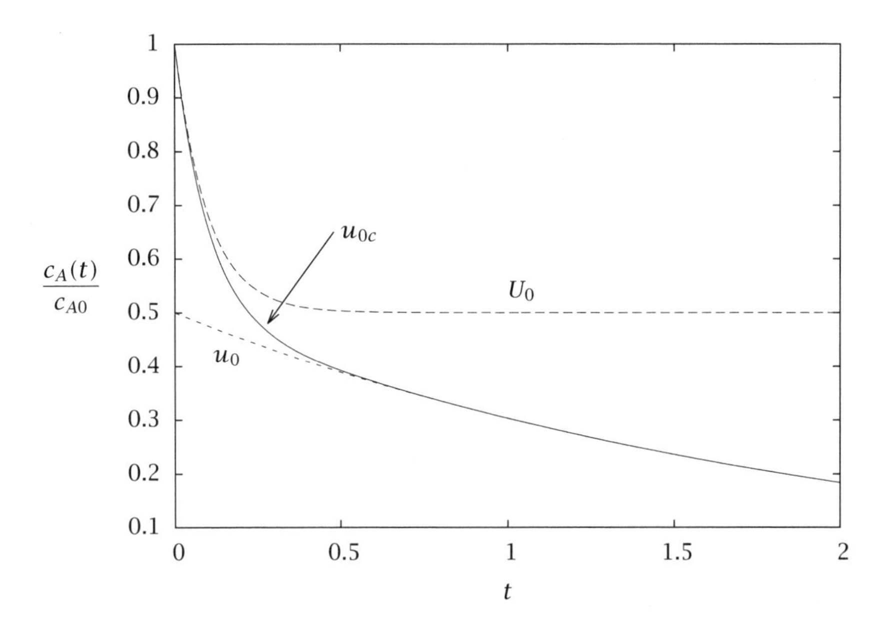

Figure 2.12: Leading-order inner U_0, outer u_0, and composite solutions u_{0c}, for Example 2.30 with $\epsilon = 0.2, K = 1$, and $k_2 = 1$.

or, reverting to dimensional form

$$c_A(t) = \frac{1}{1+K} e^{-\frac{K}{1+K} k_2 t} \qquad c_B(t) = \frac{K}{1+K} e^{-\frac{K}{1+K} k_2 t}$$

This is precisely the solution that would be obtained via uncritical application of the reaction equilibrium approximation. Now we see this approximation in more precise terms.

Finally, we construct a uniformly valid composite solution via (2.74). To leading order in dimensional variables

$$\frac{c_A(t)}{c_{A0}} = \frac{1}{1+K} e^{-\frac{K}{1+K} k_2 t} + \left(1 - \frac{1}{1+K}\right) e^{-(1+K)k_2 t/\epsilon}$$
$$\frac{c_B(t)}{c_{A0}} = \frac{K}{1+K} \left(e^{-\frac{K}{1+K} k_2 t} - e^{-(1+K)k_2 t/\epsilon}\right)$$

Figure 2.12 shows the leading-order inner, outer and composite solutions for $u(t) = c_A(t)/c_{A0}$ for $\epsilon = 0.2, K = 1$ and $k_2 = 1$. □

2.6.6 Method of Multiple Scales

The method of matched asymptotic expansions deals with problems in which different time or length scales dominate in distinct regions of the solution domain. In many problems, however, processes occur *concurrently* on disparate scales, a situation that requires a different approach, the METHOD OF MULTIPLE SCALES. Problems amenable to this approach include dynamical systems with multiple natural frequencies or decay times[14], nonlinear systems with widely separated timescales and problems of propagation (wavelike or diffusive) in inhomogeneous media.

As an introduction to this approach, we consider a weakly damped oscillator modeled by the linear equation

$$\ddot{x} + \epsilon \dot{x} + \omega^2 x = 0, \quad x(0) = 1, \dot{x}(0) = 0$$

On physical grounds, we expect two time scales to act simultaneously in this problem: harmonic oscillation, with natural period $2\pi/\omega$ (assumed to be $\mathrm{ord}(1)$), and the exponential decay, with time scale of $\mathrm{ord}(\epsilon)$. If we proceed naively, looking for a regular perturbation solution $x(t) = x_0(t) + \epsilon x_1(t) + O(\epsilon^2)$, we find that

$$\begin{aligned} \ddot{x}_0 + \omega^2 x_0 &= 0, \qquad & x_0(0) = 1, \dot{x}_0(0) = 0 \\ \ddot{x}_1 + \omega^2 x_1 &= -\dot{x}_0, \qquad & x_1(0) = \dot{x}_1(0) = 0 \end{aligned}$$

with solution

$$x(t) \sim \cos \omega t + \epsilon \left(\frac{1}{2\omega} \sin \omega t - \frac{1}{2} t \cos \omega t \right)$$

The equation at $O(\epsilon)$ has the same differential operator as does the zeroth order problem, but has a resonant forcing term $\dot{x}_0 = \cos \omega t$ that leads to the $t \cos \omega t$ SECULAR term in the solution. When $t = \mathrm{ord}(1/\epsilon)$, this term destroys the asymptoticness of the expansion; the approximation is not uniformly valid, failing at large times. The method of multiple scales avoids this nonuniformity by explicitly recognizing the existence of two time scales in the problem, by letting $t_0 = t, t_1 = \epsilon t$ and looking for a solution of the form $x(t_0, t_1; \epsilon)$. Now

$$\frac{dx}{dt} = \frac{\partial x}{\partial t_0} + \epsilon \frac{\partial x}{\partial t_1}$$

[14] For extensive application of the method in this context, see: Nayfeh and Mook (1979).

and we look for a solution of the form

$$x(t_0, t_1; \epsilon) \sim x_0(t_0, t_1) + \epsilon x_1(t_0, t_1)$$

Defining $D_0 = \partial/\partial t_0$ and $D_1 = \partial/\partial t_1$, the leading-order equation becomes a partial differential equation

$$D_0^2 x_0 + \omega^2 x_0 = 0, x_0(0) = 1, D_0 x_0(0) = 0$$

This has the solution $x_0 = A(t_1) \cos \omega t_0$, where $A(0) = 1$, but is as yet otherwise undetermined. At the next order, we have

$$D_0^2 x_1 + \omega^2 x_1 = 2D_1 A(t_1) \omega \sin \omega t_0 + A(t_1) \omega \sin \omega t_0$$
$$x_1(0) = 0 \qquad D_0 x_1(0) = 0$$

Again, a resonant forcing term is present on the right-hand side. Unless this is zero, a secular term again shows up in the equation and the approximation will not be asymptotic. However, we now have the possibility of eliminating this term. Notice that if

$$\frac{dA}{dt_1} = -\frac{1}{2} A$$

the resonant term vanishes. This equation is called the SOLVABILITY CONDITION or secularity condition or integrability condition. It is an amplitude equation, determining the evolution of the amplitude of the solution over the slow time scale t_1. From the leading-order result, we have that $A(0) = 1$, so $A = \exp(-\frac{1}{2} t_1)$. At leading order, the solution is therefore

$$x_0 = e^{-\frac{1}{2}\epsilon t} \cos \omega t$$

This is the type of solution we expect intuitively: a very slowly decaying harmonic oscillation. A couple final comments on this example: the solution x_1 is identically zero, but another resonance term shows up in the equation for x_2. This nonuniformity does not show up until $t = \text{ord}(1/\epsilon^2)$, by which time the amplitude has nearly decayed to zero, but if desired, it could be eliminated by including a "superslow" scale $t_2 = \epsilon^2 t$. This time scale arises because the damping causes a very small ($O(\epsilon^2)$) change in the frequency of oscillation.

This simple example illustrates the procedure and resulting structure. The recurring theme is the existence of a secularity condition, whose satisfaction requires the solution of an amplitude equation. This amplitude equation determines the evolution of the system at its largest scale. If the underlying problem is linear, so is the amplitude equation; a nonlinear equation leads to a nonlinear amplitude equation. The following example illustrates this.

Example 2.31: Oscillatory dynamics of a nonlinear system

From Section 2.2.2, we have a complete understanding of the linear system $\dot{x} = Ax$. When A has complex conjugate eigenvalues $\lambda = \sigma \pm i\omega$, the origin is a stable or unstable spiral depending on the sign of σ. When $|\sigma| \ll |\omega|$, the growth or decay of solutions occurs on a time scale much longer than the period of oscillation. In this situation, the method of multiple scales can be used to show very generally the dynamics of the *nonlinear* system $\dot{x} = Ax + N(x)$, where $N(x)$ contains no linear part. In this example we apply the method of multiple scales to the system of equations

$$\frac{dx}{dt} = \begin{bmatrix} \epsilon\mu & -\omega \\ \omega & \epsilon\mu \end{bmatrix} x + \begin{bmatrix} x_1^2 - x_1 x_2 + x_1^3 \\ x_1^2 x_2 \end{bmatrix}$$

where $\sigma = \epsilon\mu$, with $\epsilon \ll 1$, while μ and ω are $\text{ord}(1)$. The steady state $x = 0$ of this system is very weakly stable or unstable, depending on the sign of μ. Since the problem is nonlinear, finding the proper scaling of x is an important part of the solution procedure. The oscillatory nature of the linearized equation suggests that a solution can be found in terms of amplitude $\|x\|$ and phase ϕ.

Solution

Although we consider here a specific form for the nonlinearity, the multiple-scales solution will lead to equations for r and ϕ whose general structure is both extremely simple and extremely general. The time scaling of this problem is similar to that of the linear example above, so we consider a multiple-scales expansion with $t_0 = t, t_1 = \epsilon t$. The scale t_0 reflects the time scale of the oscillation, while the scale t_1 reflects the scale for growth or decay of the amplitude of the solution. To determine the proper scaling of the solution amplitude, we let $x = \delta X$, where $X = \text{ord}(1)$ as $\epsilon \to 0$. Now the equation becomes

$$\delta D_0 X + \delta\epsilon D_1 X = \delta \begin{bmatrix} 0 & -\omega \\ \omega & 0 \end{bmatrix} X + \delta\epsilon \begin{bmatrix} \mu & 0 \\ 0 & \mu \end{bmatrix} X + \delta^2 N_2(X, X) + \delta^3 N_3(X, X, X)$$

where $N_2(X, X)$ and $N_3(X, X, X)$ are the quadratic and cubic terms written in a form convenient for perturbation expansions. For general vectors $u = [u_1, u_2]^T, v = [v_1, v_2]^T, w = [w_1, w_2]^T$, the nonlinear terms for this problem are

$$N_2(u, v) = \begin{bmatrix} u_1 u_1 - u_1 v_2 \\ 0 \end{bmatrix} \qquad N_3(u, v, w) = \begin{bmatrix} u_1 v_1 w_1 \\ u_1 v_1 w_2 \end{bmatrix}$$

Any polynomial nonlinearity can be written as a sum of terms with this structure.

If we tentatively let $\delta = \epsilon$ and $X = X_0 + \epsilon X_1 + O(\epsilon^2)$ then the problems at $O(\epsilon^0)$ and $O(\epsilon^1)$ become, respectively,

$$D_0 X_0 - \begin{bmatrix} 0 & -\omega \\ \omega & 0 \end{bmatrix} X_0 = 0$$

and

$$D_0 X_1 - \begin{bmatrix} 0 & -\omega \\ \omega & 0 \end{bmatrix} X_1 = \begin{bmatrix} \mu & 0 \\ 0 & \mu \end{bmatrix} X_0 - D_1 X_0 + N_2(X_0)$$

The solution at $O(\epsilon^0)$ is

$$X_0(t_0, t_1) = r(t_1) \begin{bmatrix} \cos(\omega t_0 - \phi(t_1)) \\ \sin(\omega t_0 - \phi(t_1)) \end{bmatrix}$$

Turning to the $O(\epsilon)$ equation, the term $N_2(X_0, X_0)$ does not lead to resonance because it is quadratic and thus contains no terms with frequency ω. The solvability condition for this choice of scaling is thus

$$D_1 X_0 = \begin{bmatrix} \mu & 0 \\ 0 & \mu \end{bmatrix} X_0$$

This equation is linear, leading to exponential growth on the time scale t_1 when $\mu > 0$ and eventually violating the scaling assumption $x = O(\epsilon)$. Thus the choice $\delta = \epsilon$ is not self consistent.

We need a different guess for δ. The term N_2 does not lead to resonance at leading order, but the term N_3 can. For example $\sin^3 \omega t_0 = (3 \sin \omega t_0 - \sin 3\omega t_0)/4$. A balance between the linear term, which is $O(\epsilon\delta)$ and this cubic term, which is $O(\delta^3)$ would imply that $\delta = \epsilon^{1/2}$. Thus we seek a solution of the form $x = \epsilon^{1/2}(X_0 + \epsilon^{1/2} X_1 + \epsilon X_2 + O(\epsilon^{3/2}))$. With this scaling the $O(\epsilon^0)$ problem and its solution remain the same as above, while the $O(\epsilon^{1/2})$ equation becomes

$$L X_1 = N_2(X_0, X_0)$$

where

$$L = D_0 - \begin{bmatrix} 0 & -\omega \\ \omega & 0 \end{bmatrix}$$

As noted above, $N_2(X_0, X_0)$ contains no resonant terms. A particular solution can be found

$$X_1(t_0, t_1) = \frac{r(t_1)^2}{\omega} \begin{bmatrix} \frac{1}{3}(\cos 2\theta + \sin 2\theta) \\ \frac{1}{2}\left(1 - \frac{1}{3}(\cos 2\theta + \sin 2\theta)\right) \end{bmatrix}$$

where $\theta = \omega t_0 - \phi(t_1)$. Observe that X_1 has frequency 2ω.

The leading-order amplitude and phase, $r(t_1)$ and $\phi(t_1)$, have not yet been determined, so we turn to the equation at $O(\epsilon)$, which determines X_2

$$LX_2 = N_2(X_0, X_1) + N_2(X_1, X_0) + N_3(X_0, X_0, X_0) + \begin{bmatrix} \mu & 0 \\ 0 & \mu \end{bmatrix} X_0 - D_1 X_0$$

For brevity, we denote the right-hand side as R. Resonance will occur if it has terms of the form $[\sin \omega t_0, -\cos \omega t_0]^T$ or $[\cos \omega t_0, \sin \omega t_0]^T$, which in general it does. To obtain the solvability conditions, we thus require that R be orthogonal to these terms

$$\int_0^{2\pi/\omega} R^T \begin{bmatrix} \sin \omega t_0 \\ -\cos \omega t_0 \end{bmatrix} dt_0 = \int_0^{2\pi/\omega} R^T \begin{bmatrix} \cos \omega t_0 \\ \sin \omega t_0 \end{bmatrix} dt_0 = 0$$

Omitting the detailed calculation, which involves elementary but extensive trigonometric manipulations, we find that

$$\frac{dr}{dt_1} = \mu r + a r^3$$

$$\frac{d\phi}{dt_1} = b r^2$$

where for the nonlinearity given here

$$a = \frac{4\omega - 1}{8\omega} \qquad b = \frac{5}{24\omega}$$

This is a remarkably general result. These simple differential equations govern the leading-order behavior for small ϵ and their form is completely insensitive to the nature of the nonlinearity—the entire structure of N_2 and N_3 is distilled into the constants a and b. Furthermore, even for a more general nonlinearity containing higher powers, only the quadratic and cubic terms contribute. For example, a quartic term does not appear until $O(\epsilon^2)$ and thus does not contribute to R. Because of its generality, it is known as the normal or canonical form for this class of nonlinear problems[15].

The equation for the oscillation amplitude r is the most important. It has steady-state solutions $r = 0$ and $r = \sqrt{-\mu/a}$. Including the scaling $\delta = \epsilon^{1/2}$, the latter solution becomes $\|x\| = \sqrt{-\mu\epsilon/a}$. Therefore real nontrivial solutions exist if $\mu\epsilon/a < 0$. We return to this example and related ones in a more general context in Section 2.7.5.

□

2.7 Qualitative Dynamics of Nonlinear Initial-Value Problems

2.7.1 Introduction

The dynamics of nonlinear differential equations can be extremely complex. In this section we introduce a number of the issues that arise in these systems. Questions that we address include

- How do nonlinear systems differ from linear ones?
- What general qualitative (geometrical) structure can be found in nonlinear systems?
- What kinds of steady-state and time-dependent behaviors are typical?
- How do solutions change as parameters change?

2.7.2 Invariant Subspaces and Manifolds

We begin with an introduction to the geometry of differential equations, by describing *invariant manifolds*, regions of phase space in which solutions to an equation remain for all time. We shall see that these regions organize the dynamics of initial-value problems. For linear constant-coefficient systems, thinking of the solution to $\dot{x} = Ax$ in terms of the eigenvectors leads toward a geometric view of solutions to differential equations. An important point to notice is this: a point lying on a line

[15]Guckenheimer and Holmes (1983) give a general formula for construction of the normal form, including explicit formulas for a and b, derived using a rigorous and elegant method of nonlinear coordinate transformations.

defined by one of the eigenvectors v_i of A never leaves this line and never has left. These lines are *invariant* under the solution operator e^{At}. Recall from Section 2.2.2 that if $x(0) = cv_i$, then

$$x(t) = e^{At}cv_i = ce^{\lambda_i t}v_i \quad \forall t$$

Thus we call the line defined by v_i an invariant subspace of the phase space. The most important invariant subspaces of a linear system are defined as follows. Let $u_1, \ldots, u_{n_s}$ be the (possibly generalized) eigenvectors whose eigenvalues have negative real parts; $v_1, \ldots, v_{n_u}$ be those whose eigenvalues have positive real parts and $w_1, \ldots, w_{n_c}$ those whose eigenvalues have zero real parts. Now we can define three invariant subspaces

$$\begin{aligned} E^s &= \text{span}\{u_1, \ldots u_{n_s}\} && \text{stable subspace} \\ E^u &= \text{span}\{v_1, \ldots v_{n_u}\} && \text{unstable subspace} \\ E^c &= \text{span}\{w_1, \ldots w_{n_c}\} && \text{center subspace} \end{aligned}$$

An initial condition in E^s will remain in E^s and eventually decay to zero, one in E^u will remain in E^u and grow exponentially with time, and one in E^c will remain in E^c, staying the same magnitude or growing with at most a polynomial time dependence. Figure 2.13 shows some examples of these invariant subspaces in linear systems. In general, a system with eigenvalues with zero real parts is not robust: a small change in the system, e.g., the parameters, moves the eigenvalues off the imaginary axis and the invariant subspace E^c vanishes. A system like this, for which an arbitrarily small change in the system changes the qualitative behavior, is said to be STRUCTURALLY UNSTABLE. In contrast, if all the eigenvalues have nonzero real parts, no qualitative change occurs if the system is changed slightly. Such a system is said to be STRUCTURALLY STABLE. Similarly, if a system linearized around a steady state has no eigenvalues with zero real parts, the steady state is said to be HYPERBOLIC. Otherwise it is nonhyperbolic.

So what happens when we allow nonlinearity to creep in? Consider a nonlinear system in the vicinity of a steady state x_s. Letting $z = x - x_s$, we can write the system $\dot{x} = f(x)$ as

$$\dot{z}_i = \left.\frac{\partial f_i}{\partial z_j}\right|_{z=0} z_j + \frac{1}{2}\left.\frac{\partial^2 f_i}{\partial z_j \partial z_k}\right|_{z=0} z_j z_k + O(\|z\|^3)$$

or

$$\dot{z} = Jz + N_2(z, z) + O(\|z\|^3)$$

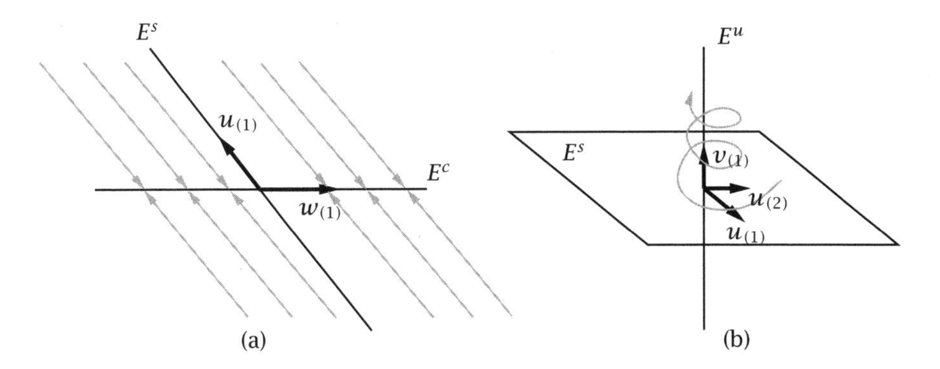

Figure 2.13: Examples of invariant subspaces for linear systems: (a) $\lambda_1 = -1, \lambda_2 = 0$; (b) $\lambda_{1,2} = -1 \pm i, \lambda_3 = 1$.

where $J_{ij} = \partial f_i / \partial x_j \big|_{z=0}$ is the Jacobian of $f(x_s)$ and $N_2(z,z)$ contains all terms that are quadratic in z. Since $Jz = O(z)$ and $N_2(z,z) = O(z^2)$, the leading-order behavior for small z is determined by the linearized system, as long as all the eigenvalues of J have nonzero real parts, i.e., the steady state x_s is hyperbolic. The rigorous and general statement of this fact is called the HARTMAN-GROBMAN THEOREM (Guckenheimer and Holmes, 1983). If there is an eigenvalue with zero real part, then the linearized problem gives that

$$\frac{d}{dt} \|z\|^2 = 0$$

for some values of z, in which case the quadratic term $N_2(z,z)$ appears at leading order.

Restricting ourselves to the usual situation, when x_s is hyperbolic, we now generalize the ideas of the stable and unstable subspaces to the nonlinear case. We define the STABLE AND UNSTABLE MANIFOLDS[16] of x_s as follows:

- The stable manifold $W^s(x_s)$ is the set of points that tend to x_s as $t \to +\infty$.
- The unstable manifold $W^u(x_s)$ is the set of points that tend to x_s as $t \to -\infty$.

[16] A manifold for our purposes is simply a curve or surface. We use the term because W^s and W^u are not generally linear subspaces of $\mathbb{R}^n$, while E^s and E^u are.

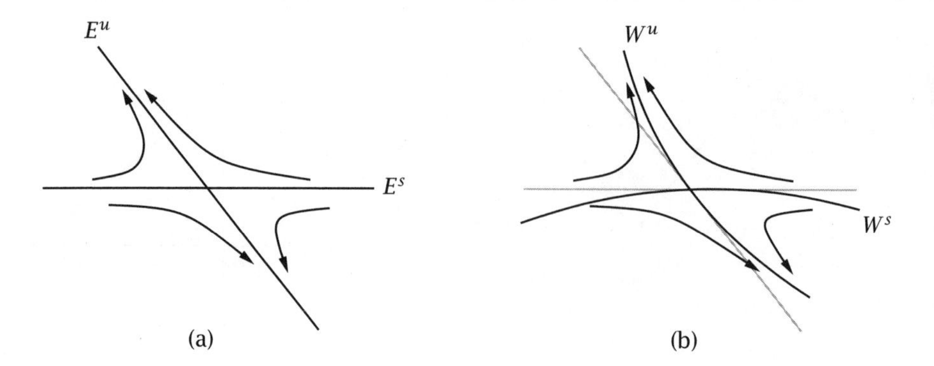

Figure 2.14: Invariant subspaces of the linearized system (a) and invariant manifolds of the nonlinear system (b).

These have the same dimensions n_s and n_u as the subspaces E^s and E^u of the linearized system, and are tangent to them at $x = x_s$. The relationship between them is shown in Figure 2.14. Convince yourself that the definitions of W^s and W^u are equivalent to those given above for E^s and E^u for a hyperbolic linear system.

For many interesting situations, a steady state of interest is stable; there is no unstable manifold. Recall, however, that in the linear case each individual eigendirection defines an invariant manifold so E^s contains within it further invariant subspaces. This fact gives us a tool for understanding the approach to a steady state and possibly for constructing simplified models of the dynamics near a steady state. As an example, consider the following pair of differential equations, with $\epsilon \ll 1$

$$\epsilon \dot{x}_1 = f_1(x_1, x_2)$$
$$\dot{x}_2 = f_2(x_1, x_2)$$

Let $[x_1, x_2]^T = (0, 0)$ be a stable steady state. Furthermore, assume that we have written the equation in coordinates where

$$L = \begin{bmatrix} -\epsilon^{-1} & 0 \\ 0 & -1 \end{bmatrix}$$

Thus the eigenvalues are $-\epsilon^{-1}$ and -1, with corresponding eigenvectors $u_f = [1, 0]^T$ and $u_s = [0, 1]^T$. The "s" and "f" stand for "slow" and "fast" respectively, because the dynamics in the u_s direction occur on an $O(1)$ time scale, while those in the u_f direction occur on an $O(\epsilon)$ time scale. We can thus define a "slow" subspace E^{ss} and a fast subspace E^{sf}, with nonlinear extensions W^{ss} and W^{sf}. Initial conditions (sufficiently close to the origin) approach W^{ss} in a time of $O(\epsilon)$ so after this transient all the dynamics are along the slow manifold W^{ss}. To leading order in ϵ, W^{ss} is defined by the equation $f_1(x_1, x_2) = 0$. This is the result we would get by just setting ϵ to zero above and can be found as an outer solution in a matched asymptotic expansions analysis.

Close to the origin, this equation can be rewritten $x_1 = h(x_2)$, so we can reduce the pair of equations to a single equation

$$\dot{x}_2 = f_2(h(x_2), x_2)$$

What we have just done is a form of the quasi-steady-state approximation used in all areas of chemical engineering analysis. It illustrates a very important and general property of initial-value problems: beyond initial transients, solutions often evolve on a subspace or manifold that has many fewer dimensions than the entire phase space. This fact is both conceptually and computationally important. It means that the behavior of large systems can often be understood by only considering a few dimensions, and also that computations might be performed with many fewer degrees of freedom than formally required.

2.7.3 Some Special Nonlinear Systems

Gradient Systems

Imagine a small particle suspended in a viscous fluid, and moving under the influence of a force that can be written as the gradient of a scalar potential function $U(x)$. This situation is described by

$$\dot{x} = -\nabla U \tag{2.75}$$

In general, systems of this form are called *gradient systems.* Recall that the vector ∇U is always normal to surfaces of constant U so trajectories of this type of system are always moving "downhill" on the "energy landscape" defined by U. In other words, the potential U is a Lyapunov function for (2.75). The only steady states of gradient systems are sources, saddle points, and sinks (can you show this?). A two-dimensional example is shown in Figure 2.15. Some more insight into the behavior of a gradient system is gained by asking how the "potential energy" of a trajectory evolves with time. The rate of change of

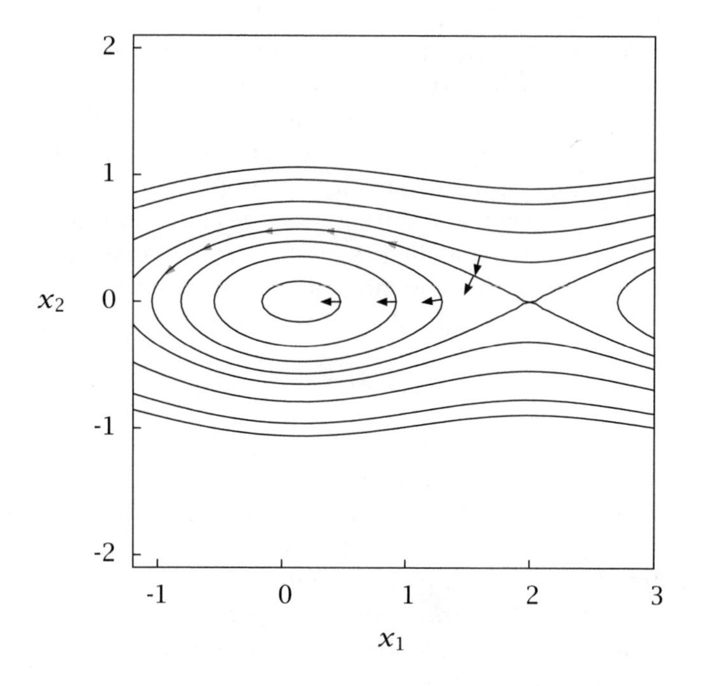

Figure 2.15: Contours of an energy function $V(x_1, x_2)$ or $H(x_1, x_2)$. Black arrows denote directions of motion on the energy surface for a gradient system, while gray ones denote motion for a Hamiltonian system; $V(x_1, x_2) = 0.2x_1^2 - 2\cos(\pi/2)x + 10x_2^2$.

U on a trajectory is

$$\frac{dU(x(t))}{dt} = \frac{\partial U}{\partial x_i}\frac{dx_i}{dt} = -\frac{\partial U}{\partial x_i}\frac{\partial U}{\partial x_i} = -\left\|\nabla U\right\|^2 \leq 0$$

with equality only at steady states, where $\nabla U = 0$. So whatever the trajectory of the vector equation for $\dot{x}$, it satisfies this scalar equation showing that the rate of change of potential energy is the square of the gradient of the potential. Trajectories roll downhill until they reach a minimum in U. In a high-dimensional problem, the potential surface can be very complex, with many minima, and saddle points where there are many "downhill" directions for the system to choose from.

Hamiltonian Systems

Consider again the landscape of Figure 2.15. Call the energy function H rather than V. Now imagine a dynamical system where trajectories are not normal to H, but are *tangent* to them. So we modify the gradient system by rotating ∇H by $\pi/2$ to get

$$\dot{x} = \begin{bmatrix} 0 & -1 \\ 1 & 0 \end{bmatrix} \nabla H = \begin{bmatrix} -\dfrac{\partial H}{\partial x_2} \\ \dfrac{\partial H}{\partial x_1} \end{bmatrix} \tag{2.76}$$

By the same exercise we performed above for U along trajectories, we can show that for (2.76), trajectories satisfy

$$\frac{dH(x,t)}{dt} = 0$$

so the energy function H is conserved on trajectories that follow (2.76).

The above situation is a special case of a very general and important class of equations. Consider a system of particles, e.g., molecules. For each particle, there is a number of coordinates (typically three) that describes the position of the particle, and for each coordinate there is an associated momentum. We denote the full set of coordinates as q and the momenta as p. The total (kinetic plus potential) energy H of the system is a function of the positions and momenta and is called the HAMILTONIAN. In the absence of friction (always true at the atomic level), the sum of kinetic and potential energy is conserved, so

$$\frac{dH}{dt} = \frac{\partial H}{\partial p_i}\frac{dp_i}{dt} + \frac{\partial H}{\partial q_i}\frac{dq_i}{dt} = 0$$

In general, this holds only if

$$\dot{q}_i = \frac{\partial H}{\partial p_i}$$
$$\dot{p}_i = -\frac{\partial H}{\partial q_i}$$

These equations are called HAMILTON'S EQUATIONS. A system whose dynamics are described by a model of this form is said to be *Hamiltonian* (Goldstein, 1980).

In addition to the property that H is constant along trajectories, Hamiltonian systems have another important attribute: phase space volume is conserved along trajectories. In other words, a "blob" of initial conditions may deform and rotate with time, but it cannot shrink. We can see this by looking at the divergence[17] in phase space of the vector field for a Hamiltonian system

$$\nabla \cdot f = \begin{bmatrix} \frac{\partial}{\partial q_i} & \frac{\partial}{\partial p_i} \end{bmatrix} \begin{bmatrix} \frac{\partial H}{\partial p_i} \\ -\frac{\partial H}{\partial q_i} \end{bmatrix} = 0$$

This result is known as LIOUVILLE'S THEOREM. In general vector fields with $\nabla \cdot f = 0$ are said to be *conservative*; if $\nabla \cdot f < 0$ the system is *dissipative*.[18] What is $\nabla \cdot f$ for a gradient system?

An important class of conservative vector fields is the velocity fields of incompressible flows. In two dimensions, equations for motion of a fluid element are Hamiltonian, with the Hamiltonian function being simply the stream function. A three-dimensional incompressible flow field, although conservative, cannot generally be Hamiltonian. Why not?

Single Degree-of-Freedom Hamiltonian Systems

A mechanical system with only one degree of freedom, such as a particle moving along a line or a pendulum restricted to swing in a single plane, illustrates some of the important features of nonlinear differential equations. In this case p and q are scalars. Often the Hamiltonian can be written in this simple form

$$H(p, q) = \frac{1}{2}p^2 + V(q)$$

Along any trajectory, H is constant, so we can solve for the momentum in terms of the position

$$p = \pm\sqrt{2(H - V(q))}$$

Trajectories in phase space are thus symmetric across $p = 0$. Furthermore, this formula can be used to construct the energy landscape, the curves of H = constant on the (q, p) plane. From Hamilton's equations

[17]See Section 3.2.

[18]Grmela and Öttinger (1997) have developed a formalism for continuum models of materials, in which the vector field is simply a sum of a Hamiltonian part and a gradient part.

and the expression for p, we can see that steady states occur when $H = V(q)$ and $\partial V/\partial q = 0$. For the pendulum, $V(q) = -\kappa \cos q$, where κ is a constant; the energy landscape and phase-plane trajectories are shown in Figure 2.16 for $\kappa = 2$. Note in particular the trajectories that round the "hill" or "valleys," connecting two saddle points. These special trajectories are called HETEROCLINIC ORBITS. Denoting the two steady states involved as P and Q, the heteroclinic orbit is part of both the unstable manifold of P and the stable manifold of Q. If the potential energy is changed to $V(q) = -\frac{1}{2}q^2 + \frac{1}{4}q^4$, the landscape and trajectories are as shown in Figure 2.17. Now we have two trajectories connecting a saddle point to itself, called HOMOCLINIC ORBITS. The homoclinic orbit is part of *both* the unstable and stable manifold of the steady state. Homoclinic and heteroclinic orbits are examples of *global* features of a dynamical system, because their existence cannot be deduced by only looking at behavior in a small neighborhood of a particular point. Hamiltonian systems are not structurally stable; physically we can understand this by noting that any dissipation of energy leads to "downhill" motion on the energy landscape and the special properties that $H =$ constant on trajectories and $\nabla \cdot f = 0$ are lost. Similarly, homoclinic and heteroclinic orbits are not structurally stable features, but they remain important for general systems because they can arise at particular points in parameter space, called GLOBAL BIFURCATIONS (Guckenheimer and Holmes, 1983).

2.7.4 Long-Time Behavior and Attractors

A question of significant practical interest when studying a mathematical model of a process is: what happens to the dynamics after a long time, i.e., as $t \to \infty$? In one- or two-dimensional phase spaces, the possibilities are quite limited and we describe them essentially completely. In three or more dimensions, very complex behavior is possible and we shall only touch on the topic.

One Dimension

If x is a scalar, then the autonomous equation

$$\dot{x} = f(x)$$

can always be written in gradient system form

$$\dot{x} = -\frac{dV}{dx}$$

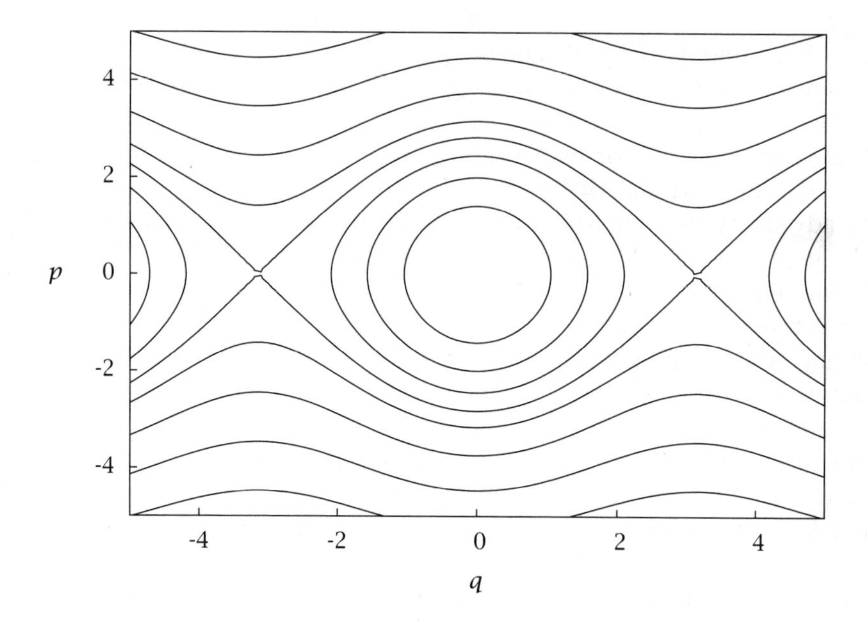

Figure 2.16: Energy landscape for a pendulum; $H = \frac{1}{2}p^2 - \kappa \cos q$; $\kappa = 2$.

where $V(x) = \int f(x')\,dx'$. All initial conditions must end up at a steady state, or roll downhill forever toward $\pm\infty$.

Two Dimensions—Planar Systems

Not every two-dimensional vector field can be written as the gradient of a potential, so two-dimensional (or PLANAR[19]) systems are not quite as restricted as one-dimensional ones. Nevertheless, they are still fairly constrained by the topology of the plane. Let us write a two-dimensional system as

$$\dot{x}_1 = f_1(x_1, x_2) \qquad \dot{x}_2 = f_2(x_1, x_2)$$

[19]Not all two-dimensional systems are planar. For example, consider a system whose trajectories are restricted to the surface of a torus, i.e., a doughnut with a hole. This surface cannot be mapped onto an unbounded plane. We discuss this case when considering three-dimensional systems. On the other hand, it turns out that the surface of a sphere can be mapped onto a plane, but the mapping is singular. Another nontrivial two-dimensional surface is a Möbius strip.

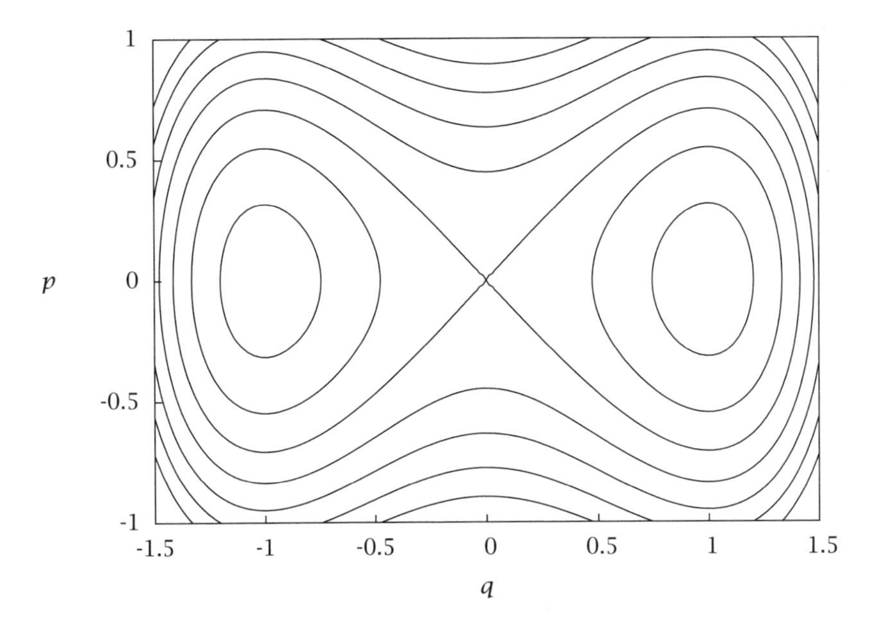

Figure 2.17: Landscape for $H = \frac{1}{2}p^2 + \frac{1}{4}q^4 - \frac{1}{2}q^2$.

where $(x_1, x_2)^T \in \mathbb{R}^2$. The steady states of this system are simply the intersections of the curves $f_1(x_1, x_2) = 0$ and $f_2(x_1, x_2) = 0$. Near these steady states, the behavior is described by the linearizations, if the eigenvalues have nonzero real parts. In addition to steady states, we know that closed trajectories (oscillations) can arise, as we saw in the Hamiltonian examples described previously. Can anything else happen as $t \to \infty$? Can, for example, a periodic orbit have a figure-eight shape? The answer to this is no; for trajectories in phase space to cross would require two values of the vector field $(f_1, f_2)^T$ for the same point $(x_1, x_2)^T$, which cannot occur. This prohibition on trajectories crossing applies in any number of dimensions, but in two dimensions it severely constrains the possible behavior. One very important consequence of the constraint is the POINCARÉ-BENDIXSON THEOREM

Theorem 2.32 (Poincaré-Bendixson). *If D is a closed region of $\mathbb{R}^2$ and a solution $(x_1(t), x_2(t))^T \in D$ for all $t > t_0$, then the solution either is a closed path, approaches a closed path as $t \to \infty$, or approaches a fixed point (steady state).*

As an application of this theorem, consider the system

$$\dot{x} = x - y - x(x^2 + 2y^2)$$
$$\dot{y} = x + y - y(x^2 + y^2)$$

Transforming to polar coordinates gives

$$\dot{\theta} = 1 + \frac{1}{2}r^2 \sin^2\theta \sin 2\theta$$
$$\dot{r} = r(1 - r^2(1 + \frac{1}{4}\sin^2 2\theta))$$

From this form of the equations we see that the origin is the only steady state and that it is unstable. So where do the trajectories go? Note that $\dot{r} < 0$ for all $r > 1$ so the trajectories must be bounded. Furthermore, $\dot{r} \geq 0$ for all θ on the circle $r = r_1 = 2/\sqrt{5}$ and $\dot{r} \leq 0$ on the circle $r = r_2 = 1$. So all trajectories entering this annulus (let us call it D) never leave it. This region is the area between the two gray circles on Figure 2.18. Since $\dot{\theta} > 0$ throughout D, there can be no steady states in this region. The Poincaré-Bendixson theorem thus requires that there be at least one closed path (periodic orbit) in this region. Numerical integration reveals that for this problem there is one asymptotically stable periodic orbit, which is also known as a LIMIT CYCLE. Part of a trajectory that starts near the origin, as well as the limit cycle it approaches, are shown in Figure 2.18.

At this point we have seen two types of behavior that trajectories may tend to as $t \to \infty$: a steady state and a limit cycle. These are simple examples of ATTRACTORS. A good working definition of an attractor is the following.

> An attractor A of a dynamical system is a set of points that is invariant under time evolution of the system, and that is the ultimate destination as $t \to \infty$ of all initial conditions that begin sufficiently near it, i.e., in a neighborhood U.[20]

For planar systems, the Poincaré-Bendixson theorem dictates that the only attractors are steady states and limit cycles. Note that the two-dimensional Hamiltonian systems discussed above also have periodic orbits; these are not attractors because an initial condition close to one such orbit does not approach it as $t \to \infty$. The fact that trajectories of Hamiltonian systems lie on constant energy surfaces precludes them from having attractors.

[20] See, for example, Guckenheimer and Holmes (1983) for a discussion of various definitions of attractors and the difficulties in developing a satisfactory general definition.

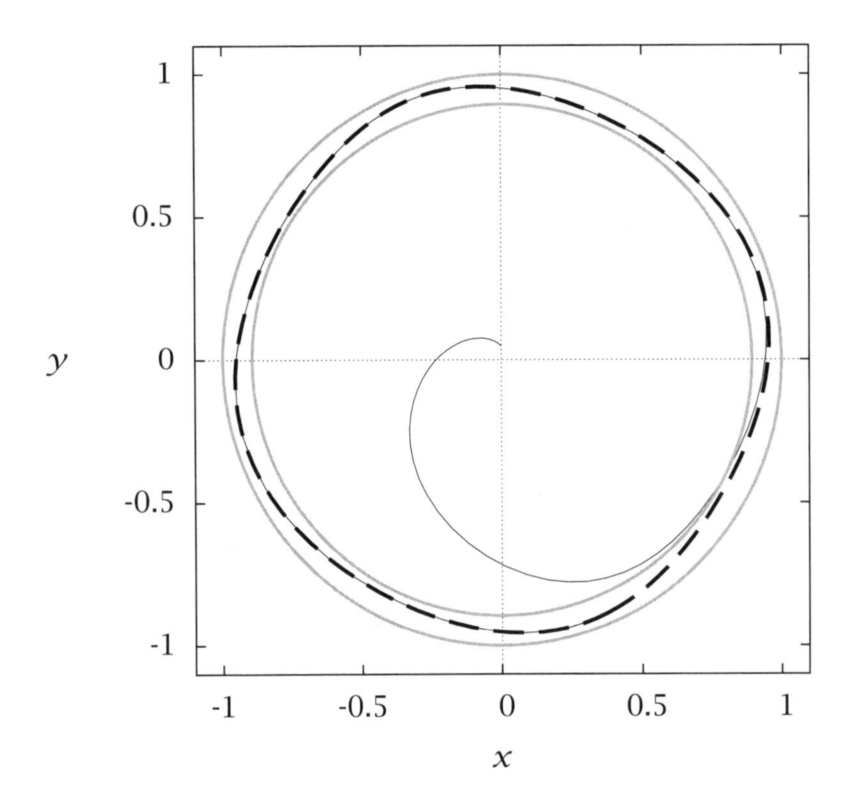

Figure 2.18: A limit cycle (thick dashed curve) and a trajectory (thin solid curve) approaching it. The region D is bounded by the two gray curves.

Three Dimensions

Trajectories in the three-dimensional phase space $\mathbb{R}^3$ are much less topologically constrained than are those in one or two dimensions. There is no three-dimensional analog of the Poincaré-Bendixson theorem and thus no restriction that all attractors be either steady states or periodic orbits. We look first at a simple, geometrically defined example. Consider a torus (a donut-shaped surface) floating in three dimensions and assume that all trajectories asymptotically approach the surface of the torus, so that we only need consider what happens on the torus itself. Further assume that there are no steady states on the

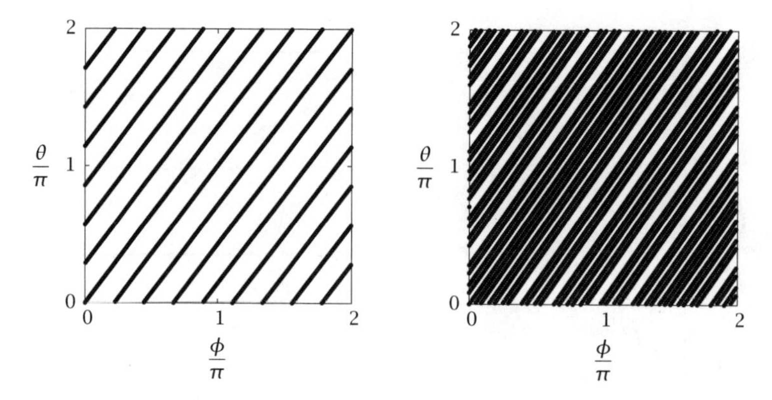

Figure 2.19: Periodic (left) and quasiperiodic (right) orbits on the surface of a torus. The orbit on the right eventually passes through every point in the domain.

torus. Now any point on the torus can be represented by two angular positions, $\theta \in [0, 2\pi)$ and $\phi \in [0, 2\pi)$, so we can represent the phase space by a square—any trajectory that leaves one side of the square reenters on the opposite side. Consider a very simple evolution of these variables

$$\dot{\theta} = p \qquad \dot{\phi} = q$$

where p and q are constants. Eliminating time and integrating we find an explicit solution for the trajectories: $\theta = \frac{p}{q}(\phi - \phi_0) + \theta_0$, where (θ_0, ϕ_0) is the value of (θ, ϕ) at a chosen value of t. Now since θ and ϕ are in $[0, 2\pi)$, the trajectory will return to (θ_0, ϕ_0) if $\theta - \theta_0 = 2\pi m$ when $\phi - \phi_0 = 2\pi n$, where m and n are (as yet unspecified) integers. This requires that $qm = pn$, which can only hold if $p/q = m/n$ for some pair of integers m and n. That is, p/q must be a rational number; this situation is a form of resonance. Otherwise, the trajectory will never repeat and will eventually pass through *every* point on the torus! Such an orbit is called QUASIPERIODIC, because $\theta(t)$ and $\phi(t)$ are individually time periodic, but the pair $(\theta(t), \phi(t))$ is not. Figure 2.19 shows trajectories for the cases $p/q = 9/7$ (left) and $p/q = \sqrt{2}$ (right). The qualitative distinction should be clear. From this example we see a new type of dynamical behavior, the quasiperiodic orbit.

Finally, we present one example of an even more complex type of attractor that can occur in phase spaces of dimension 3 or higher. Consider the system

$$\dot{x} = -y - z$$
$$\dot{y} = x + ay$$
$$\dot{z} = b + z(x - c)$$

known as the RÖSSLER system. If $a = b = 0.2$, $c = 1$, the system displays a limit cycle, as shown in Figure 2.20. If $c = 5.7$, however, the system has the attractor shown in Figure 2.21. This attractor is neither periodic nor quasiperiodic; in fact, nearby initial conditions will not follow similar paths but will eventually diverge from one another. This property is known as SENSITIVITY TO INITIAL CONDITIONS and is characteristic of CHAOTIC dynamics. Loosely speaking, an attractor on which the dynamics are chaotic is called a STRANGE ATTRACTOR (Guckenheimer and Holmes, 1983; Strogatz, 1994).

2.7.5 The Fundamental Local Bifurcations of Steady States

We now have seen a variety of possible behaviors for nonlinear dynamical systems: steady states, periodic orbits, quasiperiodic orbits, strange attractors, heteroclinic orbits, homoclinic orbits.... Our focus now shifts to understanding the ways in which the qualitative behavior of a system changes as we change parameters. This branch of the theory of differential equations is called BIFURCATION THEORY(Iooss and Joseph, 1990).

We begin the discussion just by thinking generally about the steady states of

$$\dot{x} = f(x; \mu)$$

where we now explicitly indicated the dependent of the vector field f on the parameter μ. For definiteness, assume that derivatives of f of all orders exist. Let $x_s(\mu)$ be a steady state, i.e., $f(x_s(\mu); \mu) = 0$. We can determine from the linearization of f at x_s whether this steady state is hyperbolic. If it is, then we know, from the Hartman-Grobman theorem, that a small change in μ does not change the qualitative behavior near x_s. Thus our attention focuses on behavior near values of μ where x_s is not hyperbolic—where the linearization has eigenvalues with zero real part. This is where qualitative changes in the local behavior near x_s can occur.[21] We denote a value μ_0 where x_s is not hyperbolic as a BIFURCATION POINT.

[21] If the system has a special structure, like a Hamiltonian, then additional conditions must be satisfied for bifurcation to occur.

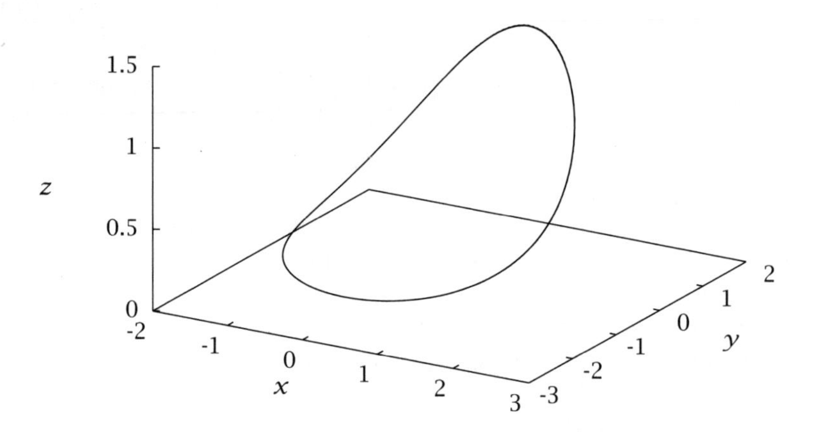

Figure 2.20: A limit cycle for the Rössler system, $a = b = 0.2$, $c = 1$.

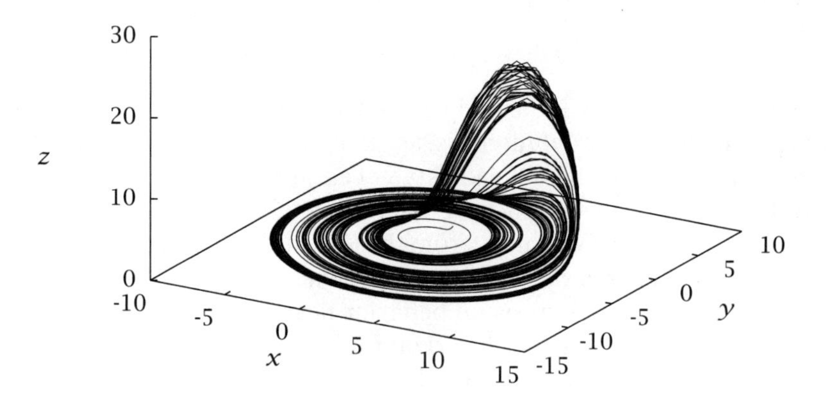

Figure 2.21: A strange attractor for the Rössler system, $a = b = 0.2$, $c = 5.7$.

It will be productive to begin our examination of bifurcations with one-dimensional systems: $x \in \mathbb{R}^1, \mu \in \mathbb{R}^1$. We shall see later how this discussion generalizes to higher-dimensional systems. Without loss of generality, we can specify that the bifurcation point is at $\mu_0 = 0$ and define a new dependent variable $y = x - x_s(\mu_0)$. Taylor expanding about $y = 0, \mu = 0$ and using the facts that $f = f_x = 0$ give

$$\dot{y} = f_\mu \mu + \frac{1}{2}\left(f_{xx}y^2 + 2f_{x\mu}\mu y + f_{\mu\mu}\mu^2\right) + \frac{1}{6}\left(f_{xxx}y^3 + 3f_{xx\mu}y^2\mu + 3f_{x\mu\mu}y\mu^2 + f_{\mu\mu\mu}\mu^3\right) + \cdots \tag{2.77}$$

Here the subscript denotes partial derivative, $f_\mu = (\partial f/\partial \mu)\big|_{\mu=0,y=0}$, etc. We now examine the structure of solutions to this in the most important cases.

Saddle-Node Bifurcation

We begin with the most general case: the partial derivatives of f (other than f_x) involved in the leading-order behavior are nonzero at the bifurcation point. This gives the GENERIC bifurcation behavior; the behavior that arises in the absence of special conditions on f. For small μ and y, the dominant balance in (2.77) is

$$\dot{y} = f_\mu \mu + \frac{1}{2} f_{xx} y^2 \tag{2.78}$$

This has steady states

$$y = \pm\sqrt{\frac{-2f_\mu \mu}{f_{xx}}}$$

(To see this, check that when $y = O(\sqrt{\mu})$ the terms in (2.77) that we neglected to get (2.78) are small compared to the ones that we kept.) Therefore, depending on the sign of f_μ/f_{xx}, there are two real solutions for $\mu > 0$ and none for $\mu < 0$ or vice versa. The point $\mu = 0$ is thus quite special in that on one side of it there are *no* steady states near $y = 0$ and on the other there are two. This type of bifurcation point is called variously a LIMIT POINT, TURNING POINT, or SADDLE-NODE bifurcation point. It arises when the conditions $f_x = 0, f_\mu \neq 0, f_{xx} \neq 0$ are satisfied. By rescaling, we can write the NORMAL FORM for this bifurcation as

$$\dot{y} = \mu - y^2 \tag{2.79}$$

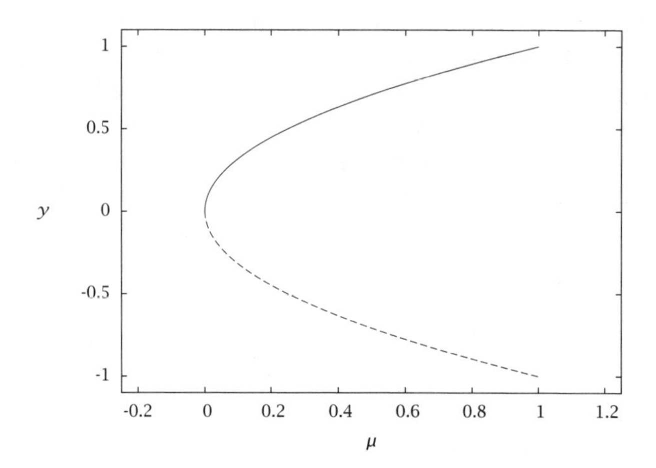

Figure 2.22: Bifurcation diagram for the saddle-node bifurcation. Every bifurcation of this type looks like this modulo a vertical and/or horizontal reflection across $y = 0, \mu = 0$. The branch of stable solutions is the solid curve, the unstable branch is dashed.

Now the steady states are simply $y = \pm\sqrt{\mu}$; the positive root is stable and the negative unstable. When $\mu = 0$, there is a single (repeated) root, which is stable from the right but not the left, and when $\mu < 0$ there is no steady state, although trajectories that pass close to $y = 0$ move very slowly through that region. The time spent in the interval $[-1, 1]$ is $\frac{\pi}{\sqrt{-\mu}}$. The BIFURCATION DIAGRAM associated with the saddle-node bifurcation is shown in Figure 2.22. It summarizes the position and stability of the steady states as μ changes.

Transcritical Bifurcation

In the above scenario, steady states exist only on one side or the other of the bifurcation point. What type of bifurcation do we expect to see if we know, on physical grounds, for example, that solutions exist on both sides of the bifurcation point? To capture this situation, we impose the additional condition that $f_\mu = 0$ at the bifurcation point. Now we find that $y = O(\mu)$ and the leading-order equation for the dynamics becomes

$$\dot{y} = \frac{1}{2}\left(f_{xx}y^2 + 2f_{x\mu}\mu y + f_{\mu\mu}\mu^2\right)$$

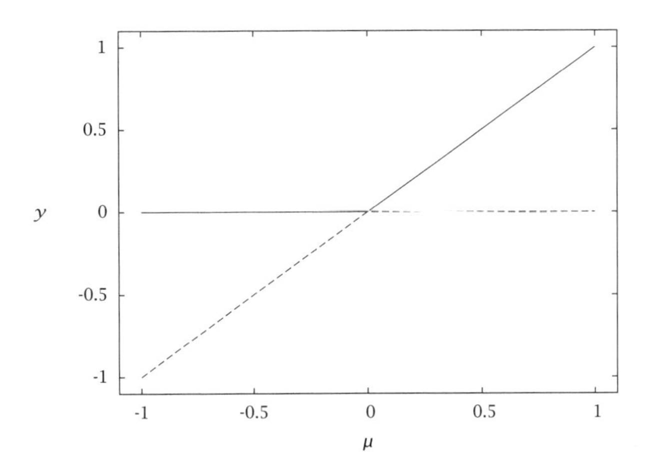

Figure 2.23: Bifurcation diagram for the transcritical bifurcation. Every bifurcation of this type looks like this modulo a vertical and/or horizontal reflection across $y = 0, \mu = 0$. The stable branch is solid, and the unstable branch is dashed.

This has steady states

$$y = \frac{1}{f_{xx}}\left(-f_{x\mu} \pm \sqrt{f_{x\mu}^2 - f_{xx}f_{\mu\mu}}\right)\mu$$

So the steady states are (locally) lines in the (y, μ) space, which cross at $(y, \mu) = (0, 0)$. Since steady states persist on both sides of the bifurcation point, this scenario is called TRANSCRITICAL bifurcation. It arises when the conditions $f_x = 0, f_\mu = 0, f_{x\mu} \neq 0, f_{xx} \neq 0$ are satisfied. We can make the presentation simpler without loss of generality by setting $f_{\mu\mu} = 0$ and rescaling, which gives us the normal form for the transcritical bifurcation

$$\dot{y} = y(\mu + ay), \quad a = \pm 1 \tag{2.80}$$

We can show that the steady state $y = 0$ is stable when $\mu < 0$ and unstable when $\mu > 0$, and the nontrivial steady state $y = -\mu/a$ has the opposite stability characteristics. The solutions are sometimes said to "exchange stability" at the bifurcation point. The bifurcation diagram for the transcritical bifurcation is shown for $a < 0$ in Figure 2.23.

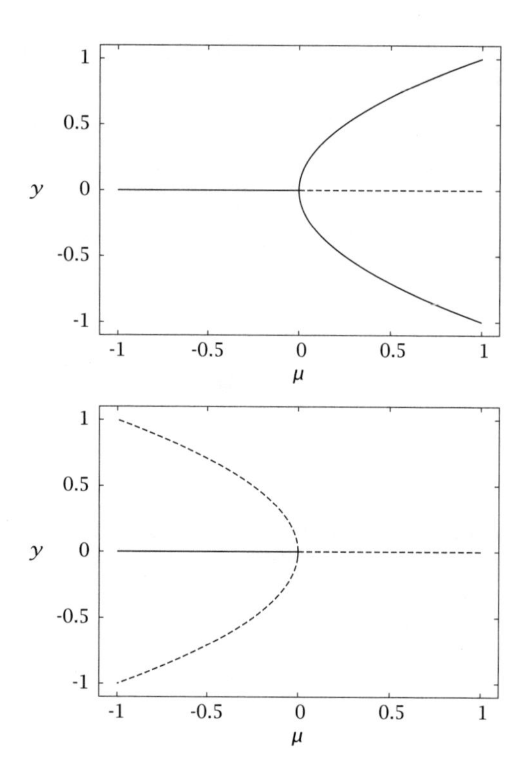

Figure 2.24: Bifurcation diagrams for the pitchfork bifurcation. Top: supercritical bifurcation, $a = -1$. Bottom: subcritical bifurcation, $a = 1$. The stable branches are solid, the unstable are dashed.

Pitchfork Bifurcation

Many physical problems have some symmetry that constrains the type of bifurcation that can occur. For example, for problems with a reflection symmetry, a one-dimensional model may satisfy the condition

$$f(x - x_s; \mu) = -f(-(x - x_s); \mu)$$

for all values of μ. With $y = x - x_s$ we have that $f(y;\mu) = -f(-y;\mu)$ so $y = 0$ is always a solution and f is odd with respect to $y = 0$. Therefore, at a bifurcation point $y = 0, \mu = 0$ we have that $0 = f = f_{xx} = f_{xxxx} = \ldots$ and $0 = f_\mu = f_{\mu\mu} = f_{\mu\mu\mu} = \ldots$. Our Taylor expansion thus becomes

$$\dot{y} = f_{x\mu}\mu y + \frac{1}{6} f_{xxx} y^3$$

with $y = O(\mu^{1/2})$. Rescaling, we find the normal form

$$\dot{y} = y(\mu + ay^2), \quad a = \pm 1 \tag{2.81}$$

This has steady states $y = 0, y = \pm\sqrt{-\mu/a}$. The steady states and stability for this bifurcation are shown in Figure 2.24. For obvious reasons, this scenario is called PITCHFORK BIFURCATION. It arises when the conditions $f_x = 0, f(y;\mu) = -f(-y;\mu), f_{x\mu} \neq 0, f_{xx} = 0, f_{xxx} \neq 0$ are satisfied. If $a = -1$, then the nontrivial steady-state branch exists only for $\mu > 0$ and is stable; this case is said to be SUPERCRITICAL. If $a = +1$, the nontrivial branch exists for $\mu < 0$ and is unstable; this is the SUBCRITICAL case. Note that in the latter case, the linearly stable trivial branch will not be approached by initial conditions with magnitude $|y(0)| > \sqrt{-\mu/a}$; so although small perturbations from the steady state $y = 0$ decay, larger ones grow.

Hopf Bifurcation

In all of the above scenarios, solutions either monotonically approach a steady state or go off to ∞ (or more precisely, to where higher-order terms in the Taylor expansion are important). We now consider the case where we expect oscillatory behavior, i.e., where the linearized version of the problem has complex conjugate eigenvalues $\lambda = \sigma \pm i\omega$. Obviously, we must move from one- to two-dimensional systems for this behavior to occur. As above, we expect a bifurcation when the steady state is nonhyperbolic, so $\sigma = 0$ and the eigenvalues of J are purely imaginary. In this instance, the steady-state solution persists on both sides of the bifurcation point, as long as $\omega \neq 0$ when σ is small. We let $\sigma = \mu\epsilon$ with $\mu = O(1)$ and write the model as

$$\dot{x} = \begin{bmatrix} \mu\epsilon & -\omega \\ \omega & \mu\epsilon \end{bmatrix} x + N_2(x,x) + N_3(x,x,x) + O(|x|^4)$$

The behavior of the linearized system is characterized by oscillation on a time scale of ω^{-1} (which we assume remains finite as $\epsilon \to 0$), and slow growth or decay on an $O(\epsilon^{-1})$ scale. In Example 2.31, we used the method of multiple scales to show that for small ϵ, balancing the linear growth terms with the nonlinearity requires that $x = O(\epsilon^{1/2})$, as in the pitchfork and saddle-node cases above, and that the solution has the form

$$x(t) = \epsilon^{1/2} r(t) \begin{bmatrix} \cos(\omega t - \phi(t)) \\ \sin(\omega t - \phi(t)) \end{bmatrix}$$

where the amplitude r and phase ϕ of the solution are given by

$$\dot{r} = \epsilon\mu r + a\epsilon r^3 \tag{2.82}$$

$$\dot{\phi} = b\epsilon r^2 \tag{2.83}$$

The constants a and b are functions of the nonlinearity and of ω (Guckenheimer and Holmes, 1983; Iooss and Joseph, 1990). These equations comprise the normal form for the so-called HOPF BIFURCATION, the generic bifurcation connecting steady states ($r = 0$) to periodic orbits ($r \neq 0$). Notice that the equation for r is identical in form to that for the pitchfork bifurcation. So if $a < 0$, we have a supercritical Hopf bifurcation, a transition with increasing μ from a stable steady state to a stable limit cycle whose amplitude is $\epsilon^{1/2} r = \sqrt{-\mu\epsilon/a}$. For the subcritical case $a > 0$ there is a periodic solution, but it exists for $\mu < 0$ and is unstable. Turning to the phase equation, we see that on the limit cycle, $\phi = -\mu\epsilon b t/a$, so the frequency of the solution is $\omega + b\mu\epsilon/a$. It changes linearly as μ increases from zero, with a rate determined by b/a.

2.8 Numerical Solutions of Initial-Value Problems

We have seen that for linear constant-coefficient problems, a complete theory exists and the general solution can be found in terms of eigenvalues and eigenvectors. For systems of order greater than four, however, there is no general, exact way to find the eigenvalues. So even in the most well-understood case, numerical approximations must be introduced to find actual solutions. The situation is worse in general, because no simple quantitative theory exists for nonlinear systems. Most of them need to be treated numerically right from the start. Therefore it is important to understand how numerical solutions of ODEs are constructed. Here we consider initial-value problems (IVPs). We focus on the solution of a single first-order equation, because the generalization to a system is usually apparent. The equation

$$\dot{x} = f(x, t)$$

can be formally integrated from a time t to a future time $t + \Delta t$ to read

$$x(t + \Delta t) = x(t) + \int_t^{t+\Delta t} f(x(t'), t')dt' \tag{2.84}$$

The central issue in the numerical solution of IVPs is the approximate evaluation of the integral on the right-hand side of this equation. With a good approximation and a small enough time step Δt, the above formula can be applied repeatedly for as long a time interval as we like, i.e., $x(\Delta t)$ is obtained from $x(0)$, $x(2\Delta t)$ is obtained from $x(\Delta t)$, etc. We use the shorthand notation $x^{(k)} = x(k\Delta t)$.

2.8.1 Euler Methods: Accuracy and Stability

The three key issues in the numerical solution of IVPs are SIMPLICITY, ACCURACY, and STABILITY. We introduce each of these issues in turn, in the context of the so-called Euler methods.

The simplest formula to approximate the integral in (2.84) is the rectangle rule. This can be evaluated at either t or $t + \Delta t$, giving these two approximations for $x(t + \Delta t)$

$$x^{(k+1)} \approx x^{(k)} + \Delta t f(x^{(k)}, t^{(k)}) \tag{2.85}$$

$$x^{(k+1)} \approx x^{(k)} + \Delta t f(x^{(k+1)}, t^{(k+1)}) \tag{2.86}$$

The first of these approximations is the EXPLICIT or FORWARD Euler scheme, and the second is the IMPLICIT or BACKWARD Euler scheme. The explicit Euler scheme is the simplest integration scheme that can be obtained. It simply requires one evaluation of f at each time step. The implicit scheme is not as simple, requiring the solution to an algebraic equation (or system of equations) at each step. Both of these schemes are examples of SINGLE-STEP schemes, as they involve quantities at the beginning and end of only one time step.

To consider the accuracy of the forward Euler method, we rewrite it like this

$$x^{(k+1)} = x^{(k)} + \Delta t f(x^{(k)}, t^{(k)}) + \epsilon \Delta t$$

where ϵ is the LOCAL TRUNCATION ERROR—the error incurred in a single time step. This can be determined by plugging into this expression the Taylor expansion of the exact solution

$$x^{(k+1)} = x^{(k)} + \dot{x}^{(k)} \Delta t + \frac{1}{2} \ddot{x}^{(k)} \Delta t^2 + \frac{1}{3!} \dddot{x}^{(k)} \Delta t^3 + \cdots$$

yielding

$$x^{(k)}+\dot{x}^{(k)}\Delta t+\frac{1}{2}\ddot{x}^{(k)}\Delta t^2+\frac{1}{3!}\dddot{x}^{(k)}\Delta t^3+O(\Delta t^4) = x^{(k)}+\Delta t f(x^{(k)},t^{(k)})+\epsilon\Delta t$$

Using the fact that $\dot{x}^{(k)} = f(x^{(k)}, t^{(k)})$, the first two terms on each side of this equation cancel, and we find that

$$\epsilon = \frac{1}{2}\ddot{x}^{(k)}\Delta t + O(\Delta t^2)$$

Thus ϵ scales as Δt^1 as $\Delta t \to 0$. The implicit Euler method obeys the same scaling. Thus the Euler methods are said to be "first-order accurate." Since the explicit method is simpler, is there any reason to use the implicit method? The answer is yes and arises when we look at the third issue mentioned above, stability.

Consider a single linear equation $\dot{x} = \lambda x$, so $f(x,t) = \lambda x$. If $\mathrm{Re}(\lambda) < 0$, then $x(t) \to 0$ as $t \to \infty$. It is not asking too much that a numerical approximation maintain the same property. The Euler approximations for this special case are

$$x^{(k+1)} = x^{(k)} + \lambda\Delta t x^{(k)}$$
$$x^{(k+1)} = x^{(k)} + \lambda\Delta t x^{(k+1)}$$

For the explicit Euler scheme, the iteration formula can be written in the general form $x^{(k+1)} = Gx^{(k)}$, where in the present case $G = (1 + \lambda\Delta t)$. We call G the GROWTH FACTOR or AMPLIFICATION FACTOR for the approximation. By applying this equation recursively from $k = 0$, we see that $x^{(k)} = G^k x^{(0)}$, so if $|G| > 1$, then $x^{(k)} \to \infty$ as $k \to \infty$. Conversely, if $|G| < 1$, then $x^{(k)} \to 0$ as $k \to 0$. Thus there is a NUMERICAL STABILITY CRITERION: $|G| < 1$. This is equivalent to $G_R^2 + G_I^2 < 1$, where subscripts R and I denote real and imaginary parts, respectively. For explicit Euler, $G_R = 1 + \lambda_R\Delta t, G_I = \lambda_I\Delta t$, yielding stability when

$$(1 + \lambda_R\Delta t)^2 + (\lambda_I\Delta t)^2 < 1$$

On a plane with axes $\lambda_R\Delta t$ and $\lambda_I\Delta t$, this region is the interior of a circle centered at $\lambda_R\Delta t = -1, \lambda_I\Delta t = 0$. If Δt is chosen to be within this circle, the time-integration process is numerically stable; otherwise it is not. If λ is real, instability occurs if $\lambda > 0$; this is as it should be, because the exact solution also blows up. But it also happens if $\lambda < 0$ but $\Delta t < -2/\lambda$, which leads to $G < -1$. This is pathological, because the exact solution decays. This situation is known as NUMERICAL INSTABILITY.

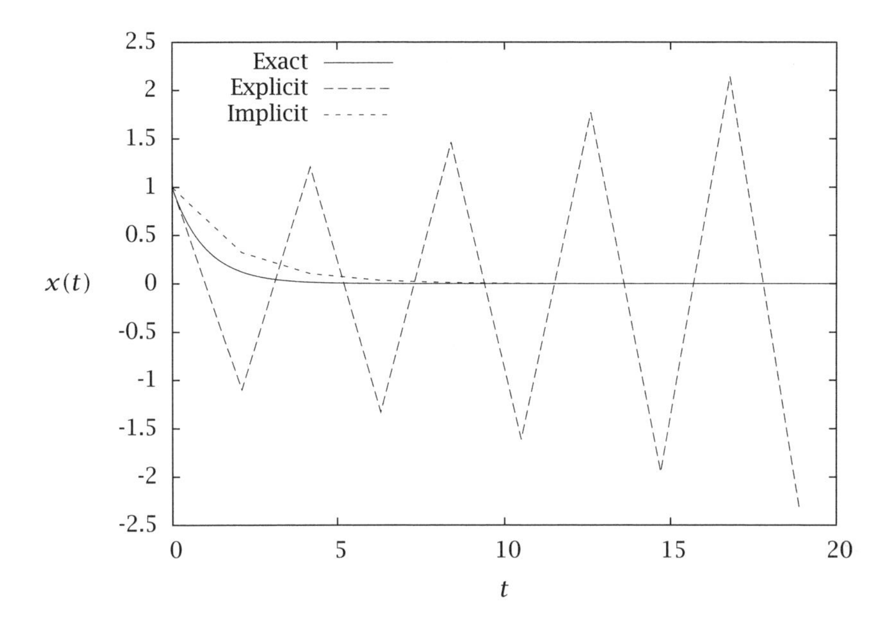

Figure 2.25: Approximate solutions to $\dot{x} = -x$ using explicit and implicit Euler methods with $\Delta t = 2.1$, along with the exact solution $x(t) = e^{-t}$.

A numerically unstable solution is not a faithful approximation of the true behavior of the system.

For a system of equations $\dot{x} = Ax$, numerical stability is obtained only if the time step satisfies the $|G| < 1$ criterion for all of the eigenvalues λ_i. Observe that for systems with purely imaginary eigenvalues, i.e., purely oscillatory solutions, the explicit Euler method is never numerically stable.

Now consider the same analysis for the implicit Euler scheme. We can again write $x^{(k+1)} = Gx^{(k)}$, but now $G = (1 - \lambda\Delta t)^{-1}$. Therefore

$$|G|^2 = G\bar{G} = \left(1 - 2\lambda_R\Delta t + |\lambda|^2 \Delta t^2\right)^{-1} < 1$$

whenever $\lambda_R < 0$. That is, if the exact solution decays, so does the approximation. The stability of this method is independent of Δt, so it is said to be ABSOLUTELY STABLE or A-stable.

Figure 2.25 shows plots of $x(t)$ for the case $\lambda = -1$ starting from initial condition $x_0 = 1$ using explicit and implicit Euler methods with

$\Delta t = 2.1$, along with the exact solution $x(t) = e^{-t}$. The implicit Euler solution is not particularly accurate because the time step is not small relative to the natural time scale of the system ($\sim \min(|1/\lambda_R|, |1/\lambda_I|)$), while the explicit Euler solution displays numerical instability.

2.8.2 Stability, Accuracy, and Stiff Systems

Say we have a differential equation model whose shortest time scale of interest is t_{int}. Obviously, we cannot choose a time step Δt that is greater than t_{int}, or our approximate solution will jump right over the interesting behavior. So accuracy requires that $\Delta t < t_{int}$. But if we use an explicit method, stability requires a time step smaller than t_{min}, which is the smallest time scale of the entire problem. For example, in a kinetics problem, this might be the reaction time for a free-radical intermediate whose kinetics are so fast that its concentration always remains near equilibrium. Problems where $t_{min} \ll t_{int}$ are said to be STIFF. Implicit methods are always used to solve such problems, as explicit methods require unreasonably small time steps.

In general, for the problem

$$\dot{x} = Ax$$

we can write a single-step scheme as

$$x^{(k+1)} = Gx^{(k)}$$

For example, consider the system

$$\dot{x} = Ax = \begin{bmatrix} -3 & 1 \\ 0 & -100 \end{bmatrix} x$$

The matrix A has eigenvalues -3 and -100, so its characteristic time scales are $1/3$ and $1/100$. In fact $x_2(t) \sim e^{-100t}$, so it is negligible after only a very short time. The explicit Euler method must capture this time scale to remain stable. Specifically, $G = I + \Delta t A$, whose eigenvalues are $1 - 3\Delta t$ and $1 - 100\delta t$, giving a stability limit $\Delta t < 2/100$. If implicit Euler is used instead, $G = (I - \Delta t A)^{-1} x^{(k)}$, whose eigenvalues are $1/(1 + 3\Delta t)$ and $1/(1 + 100\Delta t)$, which are both always less than one. Again, the implicit Euler method is always stable.

2.8.3 Higher-Order Methods

The Euler methods are simple to implement and convenient for introducing the concepts of simplicity, accuracy, and stability, but they are not necessarily the most efficient for solving real problems. For example, if an implicit method is required, the ADAMS-MOULTON second-order formula (AM2) is much preferable. This formula uses the *trapezoid rule* rather than the rectangle rule to evaluate the integral and therefore has second-order accuracy. The accuracy of IVP methods is usually given by a number p, the exponent in the expression $\epsilon = O(\Delta t^p)$. Therefore, the Euler methods have $p = 1$ and AM2 has $p = 2$. The formula for this method is

$$x^{(k+1)} = x^{(k)} + \frac{\Delta t}{2}\left(f(x^{(k+1)}, t^{(k+1)}) + f(x^{(k)}, t^{(k)})\right)$$

Like the backward Euler method, this formula requires the solution of an algebraic equation for $x^{(k+1)}$ at each time step. Also like the backward Euler method, it is A-stable. It is preferable to the backward Euler method because it has higher accuracy, the same stability, and requires no more work. AM2 is widely used for stiff problems. Adams-Moulton formulas of arbitrary order are available. The third-order formula, for example, uses $f^{(k+1)}$, $f^{(k)}$, $f^{(k-1)}$, and $f^{(k-2)}$ to estimate the integral in (2.84) by polynomial approximation. These methods are not A-stable, however, and since they are expensive, are rarely used (except in the context described later in this section).

The second-order ADAMS-BASHFORTH (AB2) method is an explicit method that also uses the trapezoid rule, but it *extrapolates* to the point $f(x^{(k+1)}, t^{(k+1)})$, using current and past values of f. Denoting $f(x^{(k)}, t^{(k)})$ by $f^{(k)}$, AB2 approximates $f^{(k+1)}$ by $f^{(k)} + (f^{(k)} - f^{(k-1)})$ (linear extrapolation), so it is a two-step scheme. Using this extrapolation in the trapezoid rule formula above yields

$$x^{(k+1)} = x^{(k)} + \frac{\Delta t}{2}\left(3f^{(k)} - f^{(k-1)}\right)$$

The price that is paid for higher accuracy without more work is a stability limit that is twice as restrictive as the forward Euler limit, e.g., for real λ the stability limit is $\lambda\Delta t < -1$ instead of -2. This stricter limit arises from the extrapolation that Adams-Bashforth uses, as seen in Figure 2.26. Adams-Bashforth formulas of arbitrary order also are available. The third-order formula, for example, uses $f^{(k)}$, $f^{(k-1)}$, and $f^{(k-2)}$.

Stability can be improved by combining an explicit method for "predicting" $x^{(k+1)}$ with an implicit method for "correcting" it. Such approaches are called PREDICTOR-CORRECTOR methods. Often the order of the predictor is chosen to be one less than that of the corrector. We denote APCn as the $n - 1$ order predictor combined with the n-order corrector. For example, APC3 bas the following steps:

1. A predicted value of the solution at the next time step is denoted by $x^{(*)}$ and computed by the AB2 formula

$$x^{(*)} = x^{(k)} + \frac{\Delta t}{2}\left(3f^{(k)} - f^{(k-1)}\right)$$

2. This value is now corrected, using the implicit third-order Adams-Moulton formula

$$x^{(k+1)} = x^{(k)} + \frac{\Delta t}{12}\left(5f^{(*)} + 8f^{(k)} - f^{(k-1)}\right)$$

   where $f^{(*)} = f(x^{(*)}, t^{(k+1)})$

APC3 displays third-order accuracy with only one more function evaluation than explicit Euler and comparable stability. Figure 2.27 shows the stability regions for the APC2, APC3, and APC4 methods. If $\lambda\Delta t$ for each eigenvalue of the Jacobian of $f^{(k)}$ is within the region, the method is stable. If the solutions are expected to be very smooth and function evaluations are expensive, the APC methods are very economical, because of their high-order accuracy with only two function evaluations per time step.

Adams predictor-corrector methods are multistep methods because they use information from prior time steps. RUNGE-KUTTA (RK) methods also have higher-degree accuracy than Euler, but are one-step methods, a useful feature in situations where one may want to change the time-step during the course of the integration. The simplest of these, RK2, uses the trapezoid rule to obtain second-order accuracy, extrapolating to $f^{(k+1)}$ using a simple forward Euler step: $f^{(k+1)} \approx f^{(k)} + \Delta t f^{(k)}$. Letting

$$k_1 = f(x^{(k)}, t^{(k)})$$
$$k_2 = f(x^{(k)} + \Delta t k_1, t + \Delta t)$$

the trapezoid rule formula becomes

$$x^{(k+1)} = x^{(k)} + \frac{\Delta t}{2}(k_1 + k_2)$$

RK2 is in fact identical to APC2 (because a first-order Adams-Bashforth formula is simply an explicit Euler step), but RK4, the fourth-order

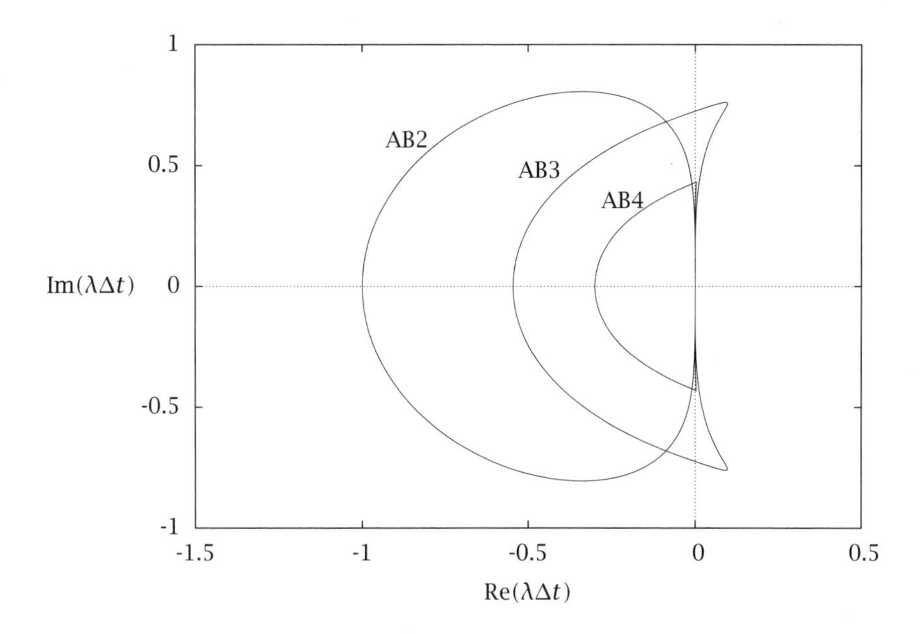

Figure 2.26: Stability regions for Adams-Bashforth methods; $\dot{x} = \lambda x$; see also Canuto et al. (2006, Fig. D.1).

Runge-Kutta formula, has a larger stability limit than the corresponding APC4 method; see Figure 2.28. The RK4 formula is

$$x^{(k+1)} = x^{(k)} + \frac{\Delta t}{6}(k_1 + 2k_2 + 2k_3 + k_4)$$

in which

$$k_1 = f(x^{(k)}, t^{(k)}) \qquad k_2 = f(x^{(k)} + \frac{1}{2}\Delta t k_1, t^{(k)} + \frac{1}{2}\Delta t)$$

$$k_3 = f(x^{(k)} + \frac{1}{2}\Delta t k_2, t^{(k)} + \frac{1}{2}\Delta t) \qquad k_4 = f(x^{(k)} + \Delta t k_3, t^{(k)} + \Delta t)$$

If f were independent of x, this would reduce to the Simpson's rule formula. RK4 requires four function evaluations. Because they have better stability properties than APC formulas, Runge-Kutta methods are generally preferable for nonstiff problems unless evaluation of f is expensive. If f is stiff, AM2 is the method of choice.

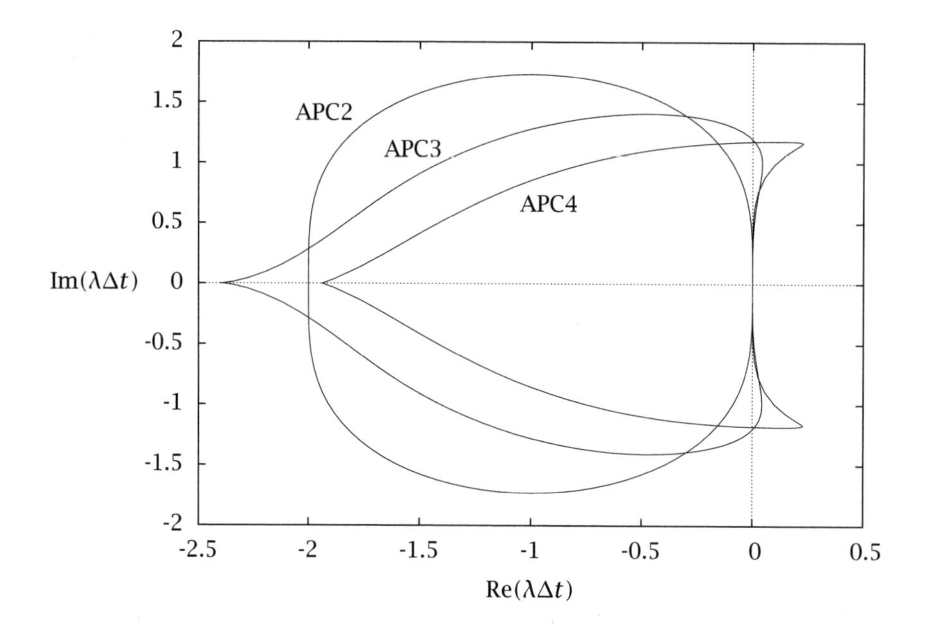

Figure 2.27: Stability regions for Adams predictor-corrector methods; $\dot{x} = \lambda x$; APCn uses $(n-1)$st-order predictor and nth-order corrector; see also Canuto et al. (1988, Fig. 4.7).

2.9 Numerical Solutions of Boundary-Value Problems

2.9.1 The Method of Weighted Residuals

There are basically two ways to make a continuous problem, like an ODE, discrete. One is to choose a finite number of points (values of the independent variable) and find an approximate solution at those points. This is what we did to solve initial-value problems (IVPs). We picked a point a distance Δt from the current time step, and used various approximate integration techniques to find the solution at that point. This is a natural approach for IVPs, because the solution at each time depends only on the solution at the immediately previous time. The situation with boundary-value problems (BVPs) is different. In this case, the solution at any point is coupled to the solution at all other points in the interval because the boundary conditions are imposed at both ends of the interval (think of a diffusion problem). So if the

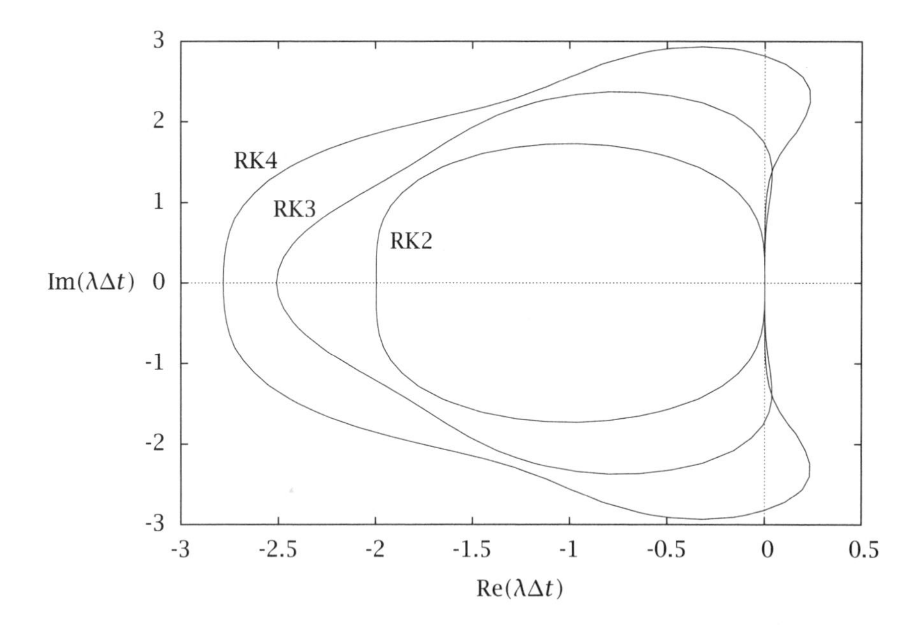

Figure 2.28: Stability regions for Runge-Kutta methods; $\dot{x} = \lambda x$; see also Canuto et al. (2006, Fig. D.2).

solution at a point changes, so does the solution at the neighboring points. A natural way to take this fact into account is to approximate the solution as the sum of a finite number of functions, i.e., to choose a set of functions over the interval and represent the solution as a linear combination of those functions. A general and systematic approach to this approximation process is given by the METHOD OF WEIGHTED RESIDUALS (MWR).

Consider the linear ODE

$$Lu = f(x) \quad x \in [a, b]$$

We choose a set of TRIAL FUNCTIONS $\{\phi_i(x)\}$ in which to represent the solution $u(x)$ and let $u_n(x)$ be the approximate solution

$$u_n(x) = \sum_{j=1}^{n} c_j \phi_j(x)$$

For the moment we require that the solution $u(x)$ and the trial functions satisfy homogeneous boundary conditions, though it is easy to relax this requirement. As $n \to \infty$, we expect u_n to approach the exact solution. For finite n, we expect a finite error, or residual, R, which we define pointwise as

$$R = Lu_n - f$$

Obviously, if $u_n = u$, then $Lu_n = f$, the equation is solved and $R = 0$. In any case, we want R to be as small as possible. In what sense do we want R to be small? We choose a second set of functions, the WEIGHT FUNCTIONS or TEST FUNCTIONS $\{\psi_i(x)\}$, and require that

$$(R, \psi_i) = 0, i = 1, 2, \ldots, n \tag{2.87}$$

This condition is equivalent to requiring that the residual be orthogonal to all of the test functions, with respect to the chosen inner product. We expect that an approximate solution $u_n(x)$ that satisfies these conditions will converge to the exact solution as $n \to \infty$ because a function that is orthogonal to infinitely many basis functions must be zero. Using the expressions for R and u_n, the condition becomes

$$\sum_{j=1}^{n} (L\phi_j(x), \psi_i(x))c_j = (f(x), \psi_i(x)), i = 1, 2, \ldots, n$$

Setting

$$A_{ij} = (L\phi_j(x), \psi_i(x)) \tag{2.88}$$

and

$$b_i = (f(x), \psi_i(x)) \tag{2.89}$$

results in the linear algebraic system $A_{ij}c_j = b_i$. We know, of course, how to solve this. Once we have done so, we have the coefficients c_j in the series for u_n and therefore we have our solution.

As yet, the trial and test functions have been left unspecified. We already have introduced several examples of trial functions and shall shortly see another. As for test functions, there are two common choices, which lead to two types of formulations:

1. **Galerkin:** $\psi_i(x) = \phi_i(x)$. If the trial functions are orthogonal, this approach simply forces the first n terms in the representation of R in the trial function basis to vanish.

2. **Collocation:** $\psi_i(x) = \delta(x - x_i)$, where $\{x_i\}, i = 1, 2, \ldots, n$ is a set of COLLOCATION POINTS. Since $(R, \delta(x - x_i)) = R(x_i)$, the collocation method simply requires the residual to be zero at the chosen set of points.

We introduce a number of specific MWR implementations using the model problem

$$y'' + y = x - 1 \qquad y(0) = -1 \quad y(1) = 1 \tag{2.90}$$

Since the boundary conditions are not homogeneous, let $u = y - (2x - 1)$. Now $u(0) = u(1) = 0$ and the equation becomes

$$u'' + u = -x \tag{2.91}$$

Galerkin Method

Finite element Galerkin method. In this method, the trial functions are low-order piecewise polynomials localized to small subsets of the domain, known as the elements, and are zero elsewhere. Consider the space $L_2(0,1)$ and the set of functions $\phi_i(x)$ where

$$\phi_0(x) = \begin{cases} 1 - \frac{x}{h}, & 0 \le x \le x_1 \\ 0, & \text{otherwise} \end{cases}$$

$$\phi_j(x) = \begin{cases} 1 - j + \frac{x}{h}, & x_{j-1} \le x \le x_j \\ 1 + j - \frac{x}{h}, & x_j \le x \le x_{j+1} \\ 0, & \text{otherwise} \end{cases} \qquad j = 1, N-1$$

$$\phi_N(x) = \begin{cases} 1 - N + \frac{x}{h}, & x_{N-1} \le x \le x_N \\ 0, & \text{otherwise} \end{cases}$$

with $x_j = jh$ and $h = 1/N$. These functions are called "hat" functions and are shown in Fig. 2.29 for $N = 2$. Observe that $\phi_j(x)$ and ϕ_{j+1} are nonzero in overlapping regions—these regions are the "elements" to which the name of the method alludes. These functions are not orthogonal. Attractive features of this set are that the functions are spatially localized (important for multidimensional problems in complicated domains) and simple, and that the coefficients c_j are the actual values of the (approximate) solution at the points x_j: $c_j = u_n(x_j)$.

For (2.91), the boundary conditions $u(0) = u(1) = 0$ obviate the use of ϕ_0 and ϕ_N in the basis, since they do not satisfy the boundary conditions. In the Galerkin approach, $\psi_i(x) = \phi_i(x)$, so the weighted residual conditions become

$$(R, \psi_i) = \sum_{j=1}^{n} \int_0^1 (\phi_j'' + \phi_j)\phi_i dx\, u_j - \int_0^1 -x\phi_i dx = 0, i = 1, 2, \ldots, n$$

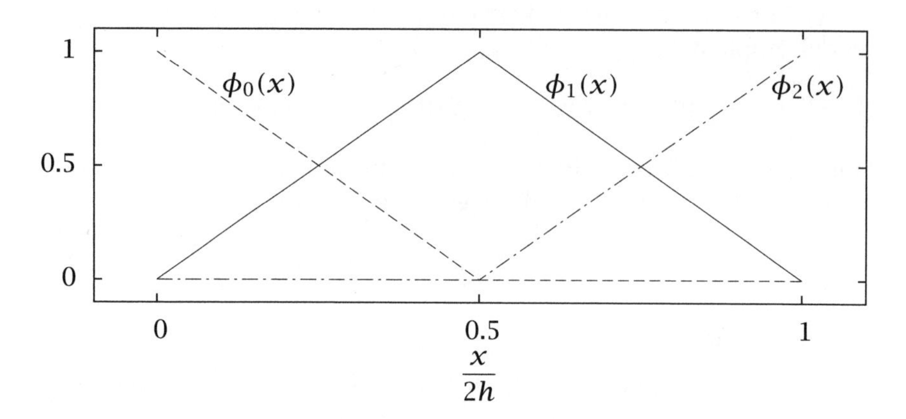

Figure 2.29: Hat functions for $N = 2$.

where $n = N - 1$. Thus

$$\begin{aligned} A_{ij} &= \int_0^1 (\phi_j'' + \phi_j)\phi_i dx \\ &= \phi_j'\phi_i\Big|_0^1 + \int_0^1 (-\phi_j'\phi_i' + \phi_j\phi_i)dx \\ &= \begin{cases} -\frac{2}{h} + \frac{2h}{3}, & i = j \\ \frac{1}{h} + \frac{h}{6}, & j = i \pm 1 \\ 0, & \text{otherwise} \end{cases} \end{aligned}$$

and

$$b_i = \int_0^1 -x\phi_i dx = -ih^2$$

Note that integrating by parts is unnecessary if we are willing to deal with the delta function nature of ϕ'' for the hat functions. Now we have a linear algebra problem $A_{ij}c_j = b_i$, which can be solved by LU decomposition, for example. For this particular choice of basis, A has a special structure: only the diagonal elements and those just above and below the diagonal are nonzero. Such a matrix is called TRIDIAGONAL and can be LU decomposed quickly, i.e., in $O(n)$ operations, since most of its entries are already zero. In general, an $n \times n$ matrix that only has $\sim n$ nonzero elements is said to be SPARSE. Because the trial functions in this case are piecewise linear, the L_2 norm of the error $\|u_n - u\|_2$ decays rather slowly as n increases: $\|u_n - u\|_2 = O(\frac{1}{n^2})$ as $n \to \infty$. The maximum (L_∞) error decays even more slowly: $\|u_n - u\|_\infty = O(\frac{1}{n^1})$

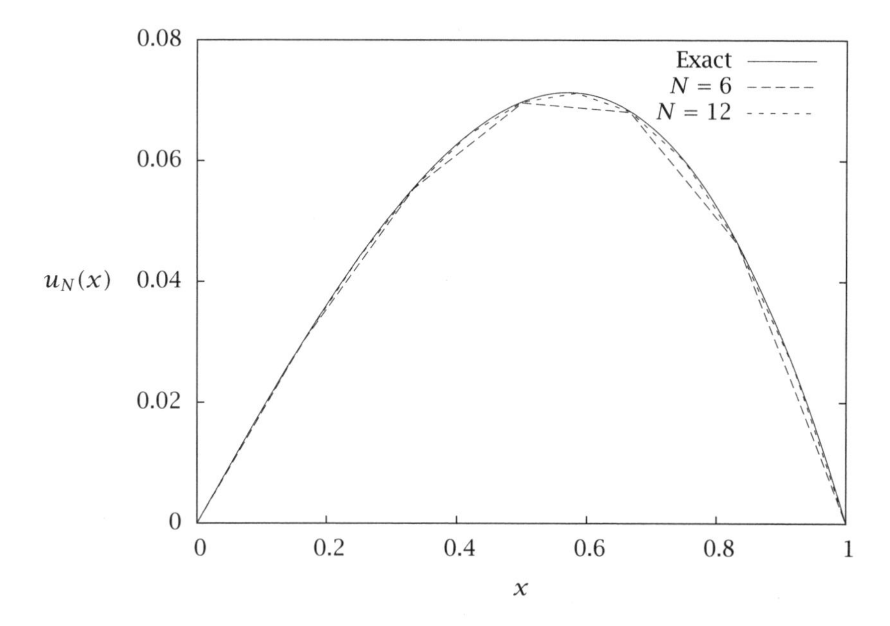

Figure 2.30: Approximate solutions to (2.91) using the finite element method with hat functions for $N = 6$ and $N = 12$. The exact solution also is shown.

as $n \to \infty$ (Hughes, 2000; Strang and Fix, 2008). Figure 2.30 shows finite element solutions for this problem, as well as the exact solution $u(x) = -x + \csc(1)\sin x$.

The finite method bears some similarities to FINITE DIFFERENCE methods, which instead of expanding solutions in basis functions, considers function values at distinct grid points in a domain and replaces derivatives by difference formulas (Press, Teukolsky, Vetterling, and Flannery, 1992). For example, $u'(x)$ can be approximated as

$$u'_f(x_j) = \frac{u(x_{j+1}) - u(x_j)}{h} + O(h) \tag{2.92}$$

or

$$u'_b(x_j) \approx \frac{u(x_j) - u(x_{j-1})}{h} + O(h) \tag{2.93}$$

where x_j and h are defined as above. These two equations are known as FORWARD and BACKWARD difference formulas, respectively. The

CENTRAL difference formula for the first derivative is given by

$$u_c'(x_j) = \frac{u(x_{j+\frac{1}{2}}) - u(x_{j-\frac{1}{2}})}{h} + O(h^2) \tag{2.94}$$

These formulas are easily verified by Taylor expansion. Applying this formula twice gives the central difference formula for the second derivative

$$u_c''(x_j) = \frac{u(x_{j+1}) - 2u(x_j) + u(x_{j-1})}{h^2} + O(h^2) \tag{2.95}$$

Using this formula to approximate the second derivative in (2.91) yields the following set of equations

$$\frac{u(x_{j+1}) - 2u(x_j) + u(x_{j-1})}{h^2} + u(x_j) = -jh, \quad j = 1, 2, \ldots, n$$

with $u(x_0) = u(x_{n+1}) = 0$. For comparison, writing the finite element formulation above in the same format gives

$$\frac{u(x_{j+1}) - 2u(x_j) + u(x_{j-1})}{h^2} + \frac{4u(x_{j-1}) + u(x_j) + 4u(x_{j+1})}{6} = -jh \quad j = 1, 2, \ldots, n$$

Observe that the term corresponding to the second derivative is identical in the two cases, as is the right-hand side. In many situations, finite difference and finite element formulations lead to similar sets of discretized equations. A great advantage of the finite element method, however, is its flexibility in dealing with multidimensional problems in complex geometries, as one does not need to develop multidimensional analogues of the difference formulas.

Fourier-Galerkin method and eigenfunction expansion. Here, instead of the hat functions, we use the sine functions as trial and test functions, i.e., $\phi_j(x) = \sin j\pi x$; we seek a solution in the form of a truncated Fourier sine series. In the present case these trial functions are eigenfunctions of L. Choosing the trial functions to be the eigenfunctions of the linear operator is called EIGENFUNCTION EXPANSION and in this situation the matrix A defined by (2.88) becomes diagonal. For the example

$$A_{ij} = \int_0^1 (\phi_j'' + \phi_j)\phi_i dx = \frac{-j^2\pi^2 + 1}{2}\delta_{ij}$$

$$b_i = \int_0^1 -x\phi_i dx = \frac{(-1)^{i-1}}{i\pi}$$

The diagonal nature of A makes the solution procedure for c simple once the above integrals have been performed

$$c_j = \frac{2(-1)^{j-1}}{(1 - j^2\pi^2)j\pi}$$

Because $c_j \sim j^{-3}$ for large j, the L_2 error scales as $1/n^3$. This error is smaller than for the finite element method, but not as small as it could be, because the solution to the problem is not a smooth periodic function, as the use of the Fourier basis implicitly assumes.

Legendre-Galerkin method. The trigonometric functions used in the previous example were the eigenfunctions of a regular Sturm-Liouville problem. What happens if we instead use the eigenfunctions of a *singular* Sturm-Liouville problem, for example the Legendre polynomials?

Our example problem is set in the domain $(0, 1)$, while the natural domain for Legendre polynomials is $(-1, 1)$, so we change coordinates, letting $z = 2x - 1$, which gives the new equation

$$4\frac{d^2u}{dz^2} + u = -\frac{1}{2}(z + 1), \qquad u(-1) = u(1) = 0$$

We let $\phi_j(z) = P_{j-1}(z)$, so

$$u_n(z) = \sum_{j=0}^{n-1} c_{j+1}P_j(z)$$

The Legendre polynomials do not satisfy the boundary conditions, so we need to use a slightly modified approach, called the GALERKIN TAU method:

1. Impose the weighted residual conditions only for $i = 1, 2, \ldots, n-2$

$$(R, \phi_i) = 0, i = 1, \ldots, n - 2$$

   This gives $n - 2$ equation for the n unknowns c_j.

2. Supplement these equations with the expressions for the boundary conditions on u_n

$$u_n(-1) = 0 \Rightarrow \sum_{j=0}^{n-1} P_j(-1)c_j = 0$$

$$u_n(1) = 0 \Rightarrow \sum_{j=1}^{n} P_j(1)c_j = 0$$

Now the first $n - 2$ rows of A and b contain the weighted residual equations, and the last two rows the equations needed to satisfy the boundary conditions.

To construct the equations resulting from the weighted residual conditions, the following properties of Legendre polynomials are relevant

$$\int_{-1}^{1} P_n(x)P_m(x)\,dx = \frac{2}{2n+1}\delta_{mn}$$

$$P_j'(x) = \sum_{k=0,j-k\text{ odd}}^{j-1} (2k+1)P_k(x)$$

$$P_j''(x) = \sum_{k=0,j-k\text{ even}}^{j-1} \frac{1}{2}(2k+1)(j-k)(j+k+1)P_k(x)$$

These can be derived from (2.27)-(2.30). For the sample problem, these results can be used to yield

$$A_{i+1,j+1} = 4\int_{-1}^{1}\Bigg(\sum_{\substack{k=0\\ j-k\text{ even}}}^{j-1} \frac{1}{2}(2k+1)(j-k)(j+k+1)P_k(z) + P_j(z)\Bigg)P_i(x)\,dz$$

$$= 4\sum_{\substack{k=0\\ j-k\text{ even}}}^{j-1} \frac{1}{2}(2k+1)(j-k)(j+k+1)\frac{2}{2k+1}\delta_{ik} + \frac{2}{2i+1}\delta_{ij}$$

for $i = 0, n-3$, $j = 0, n-1$ and

$$b_{i+1} = \int_{-1}^{1} \frac{-1}{2}(z+1)P_i(z)\,dz$$

$$= \int_{-1}^{1} \frac{-1}{2}(P_0(z) + P_1(z))P_i(z)\,dz$$

$$= -\delta_{i0} - \frac{1}{3}\delta_{i1}$$

for $i = 0, n-3$. The expressions for the boundary conditions lead to

$$A_{n-1,j+1} = (-1)^j \qquad A_{n,j+1} = 1 \qquad b_{n-1} = 0 \qquad b_n = 0$$

We do not plot the comparison between approximate and exact solutions for this case, because even for $n = 5$, the two are visually indistinguishable. Rather, Figure 2.31 shows $|c_j|$ versus j for $n = 10$. For $j \geq 4$, the plot is nearly a straight line on a semilog plot, indicating that c_j decays exponentially with j. This exponential or spectral

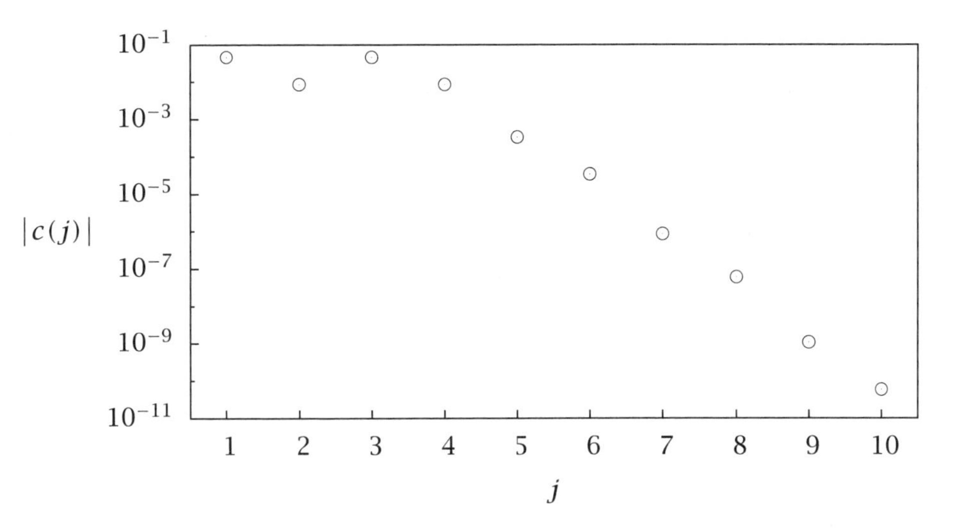

Figure 2.31: Dependence of $|c(j)|$ on j for the Legendre-Galerkin approximation of (2.91) with $n = 10$.

convergence is characteristic of MWR methods that use trial functions chosen to be eigenfunctions of a singular Sturm-Liouville problem (Gottlieb and Orszag, 1977). For this reason these methods are often called SPECTRAL METHODS. The rapid convergence reflects the fact that the Galerkin approximation yields a solution very close to the truncated Fourier series of the solution in the trial function basis. The very high accuracy of spectral methods does come at a cost—the matrix A is not sparse so it cannot generally be factorized in $O(N)$ operations.

Collocation Method

Galerkin methods require evaluation of many integrals of products of trial functions. This fact is particularly cumbersome in nonlinear problems. In the collocation method, the integrals of (2.87) are simplified greatly by the fact that the test functions are delta functions. Another attractive feature of the collocation approach is that the solution can be directly represented by its values at the collocation points, rather than as coefficients in a series. To illustrate the structure of a collocation formulation, consider the trial function set $\{\phi_1(x), \phi_2(x), \phi_3(x)\}$ and three collocation points x_1, x_2, x_3. The approximate solution is thus $u_n(x) = c_1\phi_1(x) + c_2\phi_2(x) + c_3\phi_3(x)$. The coefficients $c_1 - c_3$ are uniquely determined if the values of u_n are known at three points, as we can see by writing in matrix form the equations for the values of u_n at the collocation points

$$\begin{bmatrix} \phi_1(x_1) & \phi_2(x_1) & \phi_3(x_1) \\ \phi_1(x_2) & \phi_2(x_2) & \phi_3(x_2) \\ \phi_1(x_3) & \phi_2(x_3) & \phi_3(x_3) \end{bmatrix} \begin{bmatrix} c_1 \\ c_2 \\ c_3 \end{bmatrix} = \begin{bmatrix} u_n(x_1) \\ u_n(x_2) \\ u_n(x_3) \end{bmatrix}$$

This equation can be written $Sc = U$, where S is the (invertible) transformation that relates the coefficient vector c, with the vector of solution values at the collocation points U

$$c = \begin{bmatrix} c_1 & c_2 & c_3 \end{bmatrix}^T \qquad U = \begin{bmatrix} u_n(x_1) & u_n(x_2) & u_n(x_3) \end{bmatrix}^T$$

We also can write the equations for du_n/dx at the collocation points

$$\begin{bmatrix} \phi_1'(x_1) & \phi_2'(x_1) & \phi_3'(x_1) \\ \phi_1'(x_2) & \phi_2'(x_2) & \phi_3'(x_2) \\ \phi_1'(x_3) & \phi_2'(x_3) & \phi_3'(x_3) \end{bmatrix} \begin{bmatrix} c_1 \\ c_2 \\ c_3 \end{bmatrix} = \begin{bmatrix} u_n'(x_1) \\ u_n'(x_2) \\ u_n'(x_3) \end{bmatrix}$$

or $S_d c = U'$. Using the fact that $c = S^{-1}U$, we can write $U' = S_d S^{-1} U$ or $U' = D_n U$, where $D_n = S_d S^{-1}$ is called the COLLOCATION DIFFERENTIATION MATRIX. With this formula, we can compute the derivative of the function u_n (evaluated at the collocation points) directly from the function values at the collocation points. All of the information about what basis functions have been used is absorbed into the operator D_n. Similarly, the second derivative matrix is simply D_n^2. Note that within the space of functions that are spanned by the set of trial functions, the differentiation is exact. For example, if we use a polynomial basis, the derivative of any quadratic function is evaluated exactly by the collocation differentiation operator constructed above.

The choice of collocation points depends on the basis functions and is based on the following idea. A weighted integral (inner product) of functions

$$(u, v)_w = \int_a^b u(x)\bar{v}(x)w(x)\,dx$$

can be approximated as a sum

$$(u, v)_w \approx \sum_{j=0}^{n} u(x_j)\bar{v}(x_j)w_j \tag{2.96}$$

where $w_j \neq w(x_j)$ in general. It can be shown that for certain choices of $u(x)$ and $v(x)$, the points x_j and weights w_j can be chosen so that (2.96) is *exact*. These points are the ideal choice for collocation points.

For example, let u and v be periodic functions that can be written as truncated Fourier series

$$u(x) = \sum_{k=-n/2}^{n/2-1} \hat{u}_k e^{ikx}, \quad v(x) = \sum_{k=-n/2}^{n/2-1} \hat{v}_k e^{ikx}$$

in the domain $0 < x < 2\pi$. Equation (2.96), modified to exclude the term $j = n$, which is redundant due to periodicity, yields the exact integral if $x_j = 2\pi j/n$ and $w_j = 2\pi/n$. Similarly, if u and v are polynomials of degree $\leq n$, (2.96) can be made exact using the GAUSSIAN INTEGRATION FORMULAS. Canuto et al. (2006) provide a detailed discussion.

Chebyshev collocation. Chebyshev polynomials are a particularly popular choice as trial functions for the collocation method. These functions are the solutions to the Sturm-Liouville equation

$$(1 - x^2)\frac{d^2y}{dx^2} - x\frac{dy}{dx} + \nu^2 y = 0 \tag{2.97}$$

When ν is an integer, this equation always has a polynomial solution called a Chebyshev polynomial (of the first kind) $T_\nu(x)$; see Exercise 2.36. These polynomials form an orthogonal basis in the domain $-1 < x < 1$ with the weight function $w(x) = (1 - x^2)^{-1/2}$ and have the form

$$\begin{aligned} T_0(x) &= 1 \\ T_1(x) &= x \\ T_{\nu+1} &= 2xT_\nu(x) - T_{\nu-1}(x) \end{aligned}$$

As with Legendre polynomials, Chebyshev polynomials also arise from Gram-Schmidt orthogonalization of the set $\{1, x, x^2, \ldots\}$, but now using the weighted inner product. A particularly important property of Chebyshev's equation is that when using the coordinate transformation $x = \cos\theta$, it reduces to

$$\frac{d^2y}{d\theta^2} + \nu^2 y = 0$$

and the Chebyshev polynomials become

$$T_\nu(\theta) = \cos(\nu\theta)$$

in the domain $-\pi < \theta < \pi$. In this domain, the optimal collocation points are uniformly spaced, which in the original domain $-1 < x < 1$ results in the points

$$x_j = \cos\frac{\pi j}{n}, j = 0, \ldots, n$$

These points are very closely spaced near $x = \pm 1$, making Chebyshev collocation an attractive approach for problems in which sharp gradients near boundaries are expected. The differentiation operator is given by

$$D_{n,lj} = \begin{cases} \frac{\bar{c}_l}{\bar{c}_j}\frac{(-1)^{l+j}}{x_l - x_j}, & l \neq j \\ \frac{-x_j}{2(1-x_j^2)}, & 1 \leq l = j \leq n-1 \\ \frac{2n^2+1}{6}, & l = j = 0 \\ -\frac{2n^2+1}{6}, & l = j = n \end{cases}$$

where $\bar{c}_j = 1 + \delta_{j0} + \delta_{jn}$.

As with the Legendre-Galerkin method, the natural setting for Chebyshev collocation is the domain $(-1, 1)$. For our example problem, (2.91) transformed into this domain, the equations of the Chebyshev collocation approximation are

$$4(D_n^2)_{ij}U_j + U_j = -\frac{1}{2}(z_i + 1), i = 1, \ldots, n-1$$

This gives $n-1$ equations; the additional two equations come from the boundary conditions: $U_0 = U_n = 0$. This is a set of $n+1$ algebraic equations in $n+1$ unknowns and can be solved in the usual way. Because it uses orthogonal polynomials as trial functions, the Chebyshev collocation method also achieves the exponential convergence illustrated in the Legendre-Galerkin example.

2.10 Exercises

Exercise 2.1: A linear constant coefficient problem

Find the general solution to

$$\dot{x} = Ax$$

where

$$A = \begin{bmatrix} -1 & -1 & 0 \\ 1 & -1 & -1 \\ 0 & 0 & 2 \end{bmatrix}$$

Express it so that only the arbitrary constants are (possibly) complex. You should be able to solve the problem without explicitly performing any similarity transformations, i.e., you should not need to invert any matrices.

Exercise 2.2: Phase plane dynamics of a linear problem

Find the general solution to

$$\dot{x} = Ax$$

where

$$A = \begin{bmatrix} 19 & -14 \\ 14 & -16 \end{bmatrix}$$

Sketch the dynamics on the phase plane in the original coordinate system, being careful to show the invariant directions and the stability along those directions.

Exercise 2.3: Members of function spaces

Determine which of the following functions are in the linear space spanned by the set $\{1, \cos 2x, \sin 2x\}$.

1. $\cos^2 x$
2. $\cos x(\cos x + \sin x)$
3. $1 + \sin^2 x$
4. $1 + \cos x$

Hint: remember to look at the basic trigonometric identities.

Exercise 2.4: Weighted inner products and approximation of singular functions

Consider the function $f(t) = 1/t$ in the interval $(0, 1]$.

(a) Show that $f(t)$ is not in $L_2(0, 1)$, but that it is in the Hilbert space $L_{2,w}(0, 1)$, where the inner product is given by

$$(x, y)_w = \int_0^1 x(t)\overline{y}(t)w(t)dt$$

and $w(t) = t^2$.

(b) From the set $\{1, t, t^2, t^3, t^4\}$, construct a set of ON basis functions for $L_{2,w}(0, 1)$. These are the first five *Jacobi* polynomials (Abramowitz and Stegun, 1970).

(c) Find a five-term approximation to $1/t$ with this inner product and basis. Plot the exact function and five-term approximation. Compute the error between the exact and approximate solutions using the inner product above to define a norm. This type of inner product is sometimes used in problems where the solution is known to show a singularity. As your analysis will show, polynomials can be used to get a fairly good approximation except very near the singularity.

Hint: this problem is a good excuse to begin using a symbolic manipulation program like *Mathematica*. The calculations are not hard, but they are tedious and that is exactly the kind of problem *Mathematica* is good at.

Exercise 2.5: Fourier series of a real function

For a *real* function $f(x)$ with Fourier series representation $\sum_{k=-\infty}^{\infty} c_k e^{ikx}$, show that the Fourier coefficients satisfy $c_k = \overline{c}_{-k}$.

Exercise 2.6: Fourier series of a sawtooth function

Consider the "sawtooth" function in the domain $0 < x \leq 2\pi$

$$f(x) = \begin{cases} x & \text{if } x < \pi \\ 2\pi - x & \text{if } x \geq \pi \end{cases}$$

Find the Fourier coefficients c_k for this function using the basis functions e^{ikx}. Show that they decay as $1/k^2$ as $|k| \to \infty$.

Exercise 2.7: Fourier series of a square wave

Repeat the above exercise, but for the "square wave" function

$$f(x) = \begin{cases} 1 & \text{if } x < \pi \\ -1 & \text{if } x \geq \pi \end{cases}$$

Avoid redoing all of the integrals by using the fact that this function is simply the derivative of the sawtooth (so its Fourier series is the derivative of that of the sawtooth). Show specifically that the Fourier coefficients decay as $1/k$ as $|k| \to \infty$. Use Octave or MATLAB to plot the 10-term approximation to this function, i.e., $-10 \leq k \leq 10$.

Exercise 2.8: Basis functions of the finite element method

Consider the hat functions described in Section 2.9.1.

(a) For $N = 2$, find the inner products $(\phi_k(x), \phi_l(x)), k = 0, \ldots, N, l = 0, \ldots, N$. Is the set orthogonal?

(b) Approximate (in $L_2(0,1)$) the function $f(x) = 1 + x(1-x)$ in terms of the hat functions with $N = 2$. That is, find and solve a linear system for the coefficients c_i in the expression $f(x) \approx \sum_{i=0}^{2} c_i \phi_i(x)$.

Hint: use symmetry to save time evaluating integrals.

Exercise 2.9: Parseval's equality

Consider a function $f(x)$, represented in an orthonormal basis as a generalized Fourier series: $f(x) = \sum_{i=0}^{\infty} c_i \phi_i(x)$, with $c_i = (f(x), \phi_i(x))$.

(a) Show that $||f||_2^2 = \sum_{i=0}^{\infty} |c_i|^2$. This result is called PARSEVAL'S EQUALITY.

(b) Consider a truncated approximation, $f(x) \approx \sum_{i=0}^{N} \alpha_i \phi_i(x)$, where α_i might be different from c_i. Show that in fact, the truncation error $\left\|f - \sum_{i=0}^{N} \alpha_i \phi_i(x)\right\|^2$ is smallest when $\alpha_i = c_i$. This result shows that the generalized Fourier coefficients are the optimal coefficients for the truncated representation of f.

Exercise 2.10: Fourier series of a triangle wave

Consider the Fourier sine series approximation for the triangle wave depicted in Figure 2.32.

$$f_M(x) = \sum_{n=1}^{M} a_n \sin(n\pi x) \qquad x \in [0,1]$$

(a) Find the coefficients a_n, $n = 1, 2, \ldots$. To save time you may find the following integral formulas useful

$$\int (mx + b)\sin(n\pi x)dx = -\frac{mx + b}{n\pi}\cos(n\pi x) + \frac{m}{(n\pi)^2}\sin(n\pi x)$$

$$\int_0^1 \sin(n\pi x)\sin(m\pi x)dx = \frac{1}{2}\delta_{nm}, \qquad n, m = 1, 2, \ldots$$

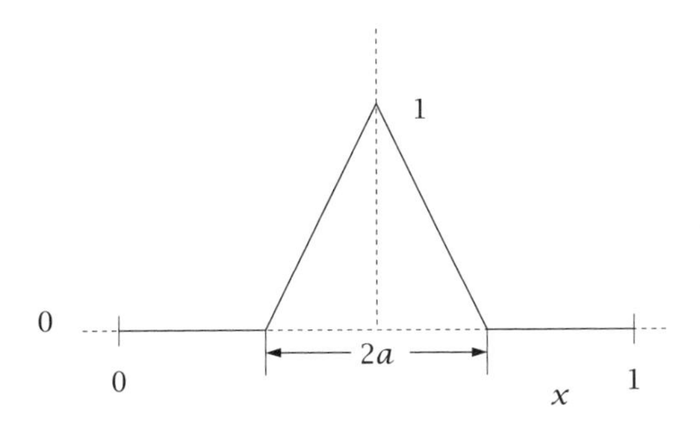

Figure 2.32: Triangle wave on $[0, 1]$.

(b) Plot the function $f_M(x)$ for $M = 5, 10, 50$ with parameter $a = 0.1$ to demonstrate convergence to $f(x)$. How many terms are required to obtain good accuracy?

Exercise 2.11: Differentiating integrals

Use the Leibniz rule for differentiating integrals to solve the following two problems.

(a) Check that the solution to the differential equation

$$\frac{dy}{dt} + p(t)y = q(t)$$

with initial condition $y(0) = y_0$ is

$$y(t) = e^{-\int_0^t p(t')dt'} \left[\int_0^t q(t'')e^{\int_0^{t''} p(\tau)d\tau} dt'' + y_0 \right]$$

Remember to show the solution satisfies both the differential equation and initial condition.

(b) Derive a Leibniz rule for differentiating the double integral

$$f(t) = \int_{a(t)}^{b(t)} \int_{c(t,p)}^{d(t,p)} h(t,p,s)dsdp$$

Your answer should not contain the derivatives of any integrals.

Exercise 2.12: Convolution theorem

(a) Use the definition of the Laplace transform to derive the convolution theorem

$$\mathcal{L}\left\{ \int_0^t f(t')g(t-t')dt' \right\} = \overline{f}(s)\overline{g}(s)$$

(b) Use the definition of the inverse Laplace transform to derive the convolution theorem going in the other direction

$$\mathcal{L}^{-1}\{\overline{f}(s)\overline{g}(s)\} = \int_0^t f(t')g(t-t')dt'$$

Which direction do you prefer and why?

Exercise 2.13: Final and initial-value theorems

Two useful theorems are the final and initial-value theorems

$$\lim_{t \to \infty} f(t) = \lim_{s \to 0} s\overline{f}(s)$$

$$\text{if and only if } s\overline{f}(s) < \infty \text{ for all } s \text{ such that } \mathrm{Re}(s) \geq 0$$

$$\text{otherwise } \lim_{t \to \infty} f(t) \text{ does not exist}$$

and

$$\lim_{t \to 0^+} f(t) = \lim_{s \to \infty} s\overline{f}(s)$$

(a) The conditions on $s\overline{f}(s)$ for the final-value theorem are *crucial.* For the functions below, state which satisfy the conditions and give their final values.

1. $\dfrac{1}{s}$
2. $\dfrac{1}{s^2}$
3. $\dfrac{1}{s(s-a)} \qquad \mathrm{Re}(a) > 0$
4. $\dfrac{1}{s(s+a)} \qquad \mathrm{Re}(a) > 0$

(b) What are the initial values, $f(0^+)$?

(c) Invert each of the transforms to get $f(t)$ and check your results.

Exercise 2.14: Network of four isomerization reactions

Consider the set of reversible, first-order reactions

$$A \underset{k_{-1}}{\overset{k_1}{\rightleftharpoons}} B \underset{k_{-2}}{\overset{k_2}{\rightleftharpoons}} C \underset{k_{-3}}{\overset{k_3}{\rightleftharpoons}} D \underset{k_{-4}}{\overset{k_4}{\rightleftharpoons}} E$$

taking place in a well-mixed, batch reactor. The reactions are all elementary reactions with corresponding first-order rate expressions. Let the concentration of the species be arranged in a column vector

$$c = \begin{bmatrix} c_A & c_B & c_C & c_D & c_E \end{bmatrix}^T$$

(a) Write the mass balance for the well-mixed, batch reactor of constant volume

$$\frac{dc}{dt} = Kc$$

What is K for this problem?

(b) What is the solution of this mass balance for initial condition $c(0) = c_0$? What calculation do you do to find out if this solution is stable?

(c) Determine the rank of matrix K. Hint: focus on the *rows* of K. Justify your answer. From the fundamental theorem of linear algebra, what is the dimension of the null space of K?

(d) What is the condition for a steady-state solution of the model? Is the steady state unique? Why or why not?

Exercise 2.15: Network of first-order chemical reactions

Consider the generalization of Exercise 2.14 to the following set of n reversible, first-order reactions

$$A_1 \underset{k_{-1}}{\overset{k_1}{\rightleftharpoons}} A_2 \underset{k_{-2}}{\overset{k_2}{\rightleftharpoons}} A_3 \underset{k_{-3}}{\overset{k_3}{\rightleftharpoons}} \cdots \underset{k_{-(n-1)}}{\overset{k_{n-1}}{\rightleftharpoons}} A_n$$

taking place in a well-mixed, batch reactor. The reaction rate for the ith reaction is

$$r_i = k_i A_i - k_{-i} A_{i+1}$$

Let the concentration of the species be arranged in a column vector

$$c = \begin{bmatrix} c_{A_1} & c_{A_2} & \cdots & c_{A_n} \end{bmatrix}^T$$

(a) Write the mass balance for the well-mixed, batch reactor of constant volume

$$\frac{dc}{dt} = \cdots$$

(b) What is the solution of this mass balance for initial condition $c(0) = c_0$?

(c) What is the steady-state solution of the model? Is the steady state unique? Why or why not?

(d) What calculation would you do to decide if the steady state is stable?

Exercise 2.16: Using the inverse Laplace transform formula

Establish property 4 of the Laplace transform pair given in Section 2.2.4, which states for $n = 1$

$$\frac{d\overline{f}(s)}{ds} = -\mathcal{L}(tf(t))$$

This formula proves useful in Exercise 3.19.

Exercise 2.17: ODE review

Solve the following ODEs: unless boundary conditions are given, find the general solution:

(a) $y' = e^{x+2y}$ (separable)

(b) $\dot{y} = y^2, y(0) = 1$ (separable)

(c) $(y - 2x)dy - (2y - x)dx = 0$ (exact)

(d) $(x^2 - y^2)dy = 2xydx$ (integrating factor)

(e) $xdy + (y + e^x)dx = 0$ (integrating factor)

Exercise 2.18: General solution to a first-order linear system of ODEs

Find the general solution to $\dot{y} = Jy$, where

$$J = \begin{bmatrix} \lambda & 1 & 0 \\ 0 & \lambda & 1 \\ 0 & 0 & \lambda \end{bmatrix}$$

Exercise 2.19: A linear system—dynamics on the phase plane

Consider the system

$$\dot{x} = \begin{bmatrix} -1 & 1 \\ 1 & -1 \end{bmatrix} x + h(t)$$

(a) Find the general solution to the homogeneous problem, i.e., with $h(t) = 0$. Characterize its stability.

(b) Sketch the qualitative behavior of solutions on the $x_1 - x_2$ plane. Where does this system fit on Figs. 2.1-2.2?

(c) Now solve the inhomogeneous problem with $h(t) = (1, 1)^T$ and characterize its stability.

Exercise 2.20: Dynamics of a freely rotating rigid body

Consider the system of equations

$$I_1 \dot{\omega}_1 = \omega_2 \omega_3 (I_2 - I_3)$$
$$I_2 \dot{\omega}_2 = \omega_3 \omega_1 (I_3 - I_1)$$
$$I_3 \dot{\omega}_3 = \omega_1 \omega_2 (I_1 - I_2)$$

with $I_1 > I_2 > I_3 > 0$. This set of equations describes the motion of a rigid body freely rotating in space. The Is are the moments of inertia of the body relative to each of the principal axes of the body and the ωs are the angular velocities with respect to those axes.

(a) If $\bar{\omega} = (\bar{\omega}_1, \bar{\omega}_2, \bar{\omega}_3)$ is a steady state of this system, find the linearized equation for deviations $\hat{\omega} = (\hat{\omega}_1, \hat{\omega}_2, \hat{\omega}_3)$ from the steady state.

(b) Find three steady states of the system that satisfy $\omega_1^2 + \omega_2^2 + \omega_3^2 = 1$. Which are linearly stable?

(c) Sketch, in the $(\hat{\omega}_1, \hat{\omega}_2, \hat{\omega}_3)$ phase space, the qualitative behavior of trajectories that begin near each of the steady states, using the linearized equations as your guide.

(d) Your results can be tested experimentally. The principal axes of a book are, in order of decreasing moment of inertia, the axis passing through the front and back covers, the right and left sides, and the top and bottom. Experimentally assess the stability of free rotation of a book with respect to these three axes. (You have to do something to keep the covers from flying open while the book spins.) Do the theory and experiment agree?

Exercise 2.21: Duffing's equation

DUFFING'S EQUATION describes the dynamics of an undamped beam

$$\ddot{x} - \beta x + x^3 = 0$$

where x is proportional to the displacement of the middle of the beam. When $\beta > 0$ the beam buckles: the "unbuckled" state $x = \dot{x} = 0$, is unstable.

(a) The two nontrivial steady states are $x = \pm\sqrt{\beta}, \dot{x} = 0$. Find the eigenvalues of the linearizations around those states.

(b) In this model there is no friction so the total energy (kinetic plus elastic) is conserved. The total energy is given by

$$H = \frac{\dot{x}^2}{2} - \beta\frac{x^2}{2} + \frac{x^4}{4}$$

A given initial condition will have a specified value of H, and the resulting trajectory must have the same value of H for all time, so a trajectory in phase space is a curve of constant H. Show that the trajectories near the two nontrivial steady states are closed curves and thus that the linearized equations give the correct qualitative behavior in this case.

Exercise 2.22: Predator-prey model

The following model describes a "predator-prey" system: species 1 eats the grass and species 2 eats species 1

$$\dot{x}_1 = x_1(1 - x_1 - \beta x_2)$$
$$\dot{x}_2 = x_2(\alpha x_1 - 1)$$

In this model, $\beta > 0$ and $\alpha > 1$, and x_1 and x_2 represent the sizes of the prey and predator populations.

(a) There are three steady states to this model. Find them.

(b) Find the linear stability of each of the steady states. Since this is a 2-dimensional system, the trace and determinant criterion can be used.

(c) Draw the phase-plane behavior near each of these steady states.

Exercise 2.23: Cell in shear flow

The following differential equation arises from a model of a cell moving in a shear flow

$$\dot{\theta} = -A + \cos 2\theta$$

where θ is the orientation angle of the cell with respect to the flow direction and A is a parameter that is determined by the geometry and mechanics of the cell (Keller and Skalak, 1982).

(a) For $A = 0$, there are four steady states in the domain $0 < \theta \leq 2\pi$. Find them and determine which ones are linearly stable.

(b) Draw the trajectories in phase space for $A = 0$, along with the steady states. Here phase space is simply the line, and since θ is periodic can alternately be considered to be just a circle with unit radius.

(c) For A larger than a certain value, this equation has no steady-state solutions. What is that value? What do the phase-space dynamics look like, i.e., draw a picture, when A exceeds that critical value?

Exercise 2.24: Steady-state heat conduction in an annulus

Consider the steady-state conduction of heat in a solid annular region shown in Figure 2.33. There is uniform heat generation in the solid. The heat-generation rate is given by

$$\dot{Q} = S_0(1 + \alpha(T - T_0))$$

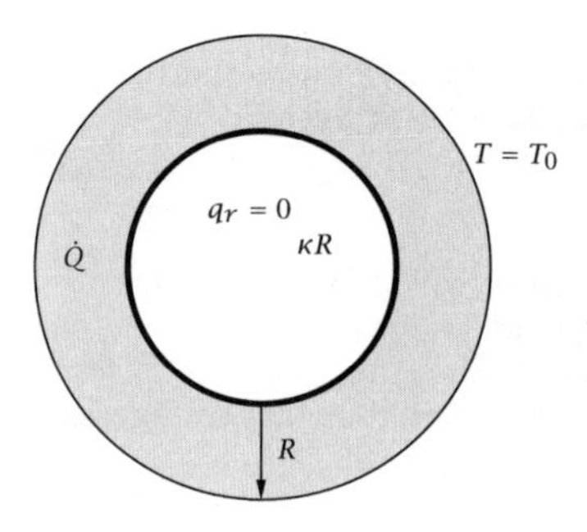

Figure 2.33: Annulus with heat generation in the solid.

in which α is a dimensional constant. The inner wall of the annulus is insulated and the outer wall is at constant temperature T_0. The material has thermal conductivity k.

(a) Write the steady-state heat equation with the source term.

(b) Define dimensionless variables

$$\xi = \frac{r}{R} \qquad \Theta = \frac{k(T - T_0)}{S_0 R^2} \qquad \beta^2 = \frac{\alpha S_0 R^2}{k}$$

Show that the model reduces to

$$\frac{1}{\xi}\frac{d}{d\xi}\left(\xi \frac{d\Theta}{d\xi}\right) + \beta^2 \Theta = -1$$

with boundary conditions

$$\Theta = 0 \qquad \xi = 1$$
$$\frac{d\Theta}{d\xi} = 0 \qquad \xi = \kappa$$

(c) What is the complementary function?

(d) By inspection, what is a particular solution?

(e) Using the two boundary conditions, specify the two unknowns in the complementary function.

(f) Plot $\Theta(\xi)$ for the following values

$$\kappa = 0.8 \qquad \beta = [1, 3, 5, 7, 7.5]$$

Exercise 2.25: Existence of a positive steady-state temperature profile

Consider Exercise 2.24 again.

(a) Plot and compare the solution $\Theta(\xi)$ if you set $\kappa = 0.8$ and $\beta = 7.5, 8.0, 8.5, 10$? What happens as you increase β in this problem?

(b) Look again at how you solve for the constants c_1, c_2. What are you assuming for this solution to exist?

(c) For κ values ranging from 0 to 0.99, find and plot the critical value of β such that the solution for c_1, c_2 does not exist. If you exceed this critical value of β, what do you think happens in the transient heat-conduction problem?

Exercise 2.26: Flow through a porous medium in a tube

Brinkman's modification of Darcy's law for flow in porous media is

$$-\nabla P - \left(\frac{\mu}{\kappa}\right)\mathbf{v} + \mu\nabla^2\mathbf{v} = 0$$

For axial flow in a tube containing a porous medium this becomes

$$\frac{1}{r}\frac{d}{dr}\left(r\frac{dv_z}{dr}\right) - \frac{v_z}{\kappa} = -\frac{\Delta\mathcal{P}}{\mu L}$$

in which

$$\begin{aligned}
\kappa &= \text{permeability of the porous medium} \\
\mu &= \text{viscosity of the fluid} \\
\Delta\mathcal{P} &= \text{pressure difference + gravity driving force} \\
v_z &= \text{z-component of the "superficial velocity" } \mathbf{v} \\
R &= \text{radius of tube} \\
L &= \text{length of tube}
\end{aligned}$$

Reasonable boundary conditions are $v_z(R) = 0$ and $v_z(0) < \infty$.

(a) Introduce a dimensionless velocity and radius

$$\phi = \frac{v_z \mu L}{\Delta\mathcal{P}R^2} \qquad \rho = \frac{r}{R}$$

and rewrite the differential equation and the boundary conditions in terms of the dimensionless variables. How many dimensionless parameters does the new differential equation contain?

(b) Obtain a particular solution of the differential equation obtained in (a) *by inspection.*

(c) Obtain the solution of the homogeneous equation; it should contain two constants. One constant can be immediately evaluated from the boundary condition at $\rho = 0$. Why?

(d) Evaluate the remaining constant using the boundary condition at $\rho = 1$. Write the general solution $\phi(\rho)$ to the differential equation. Plot $\phi(\rho)$ for permeability $\kappa/R^2 = 0.01, 0.1, 0.3, 1.0$. Also include on this plot the velocity profile for Hagen-Poiseuille flow.

(e) Evaluate the average dimensionless velocity $\langle\phi\rangle$ and show that

$$\begin{aligned}
\langle\phi\rangle &= \frac{\int_0^1 \rho\phi(\rho)\,d\rho}{\int_0^1 \rho d\rho} \\
&= \frac{\kappa}{R^2}\left[1 - \frac{2\sqrt{\kappa}}{R}\frac{I_1\left(\frac{R}{\sqrt{\kappa}}\right)}{I_0\left(\frac{R}{\sqrt{\kappa}}\right)}\right]
\end{aligned}$$

Plot $\langle\phi\rangle$ versus κ/R^2 with a log scale for the x-axis for $10^{-4} \leq \kappa/R^2 \leq 10^2$.

(f) Show that in the limit of small permeability, κ, the result in (e) simplifies to $\langle\phi\rangle = \frac{\kappa}{R^2}$ (which is exactly the result from Darcy's law).

(g) Show that in the limit as $\kappa \to \infty$, $\langle\phi\rangle = \frac{1}{8}$ (which is exactly the result for flow in an empty tube).

Exercise 2.27: Laguerre's equation

The ODE

$$xy'' + (1-x)y' + \lambda y = 0$$

where λ is a constant, is called Laguerre's equation. It arises in determining the wave function for the electrons of a hydrogen atom—the orbitals that you learn about in quantum mechanics (and thus the structure of the periodic table) emerge in part from the solutions to this equation.

(a) Show that $x = 0$ is a regular singular point.

(b) Determine the roots of the indicial equation and one solution to this problem (for $x > 0$) using the method of Frobenius.

(c) Show that when λ is a positive integer, this solution reduces to a polynomial. These polynomials are called the Laguerre polynomials.

Exercise 2.28: Hermite's equation

Hermite's differential equation is

$$u'' - 2xu' + 2ku = 0$$

Among other places, it arises in the solution of Schrödinger's equation for a particle in a potential well.

(a) Write Hermite's equation as $Lu + \lambda u = 0$, where $\lambda = 2k$ and L takes the standard form of a Sturm-Liouville operator, $Lu = \frac{1}{w(x)}\left(\frac{d}{dx}\left(p(x)\frac{du}{dx}\right) + r(x)u\right)$, with $w(x) = e^{-x^2}$. What are $p(x)$ and $r(x)$?

(b) Consider the inner product

$$(a,b)_w = \lim_{\ell\to\infty}\int_{-\ell}^{\ell} a(x)b(x)w(x)dx$$

where $w(x)$ is as given above. What boundary conditions must we impose in the limit $\ell \to \infty$ so that L is self-adjoint, i.e., so that $(Lu,v)_w = (u,Lv)_w$?

(c) The point $x = 0$ is an ordinary point for this equation. Find the general solution by series expansion around this point. Show that if k is an integer, one solution to the equation is a polynomial. These polynomials are known as the Hermite polynomials.

Exercise 2.29: Series solution

Find the general solution to the differential equation

$$(x^2 - x)u'' - xu' + u = 0$$

Start by seeking a solution of Frobenius form, expanding around $x = 0$.

Exercise 2.30: Another series solution

Find the general solution to

$$5x^2y'' + xy' + (x^3 - 1)y = 0$$

Expand around $x = 0$ and keep up to quartic terms.

Exercise 2.31: Bessel's equation: singular solution

The Bessel equation of order zero is

$$x^2y'' + xy' + x^2y = 0$$

and the associated Cauchy-Euler equation is

$$x^2y'' + xy' = 0$$

(a) Find the general solution to this Cauchy-Euler equation.

(b) Motivated by this result, seek a second solution to the the Bessel equation, of the form $y_2(x) = J_0(x)\ln x + g(x)$, where $g(x)$ has a power series solution. It will be convenient to note that g is even and write it as $g(x) = \sum_{n=0}^{\infty} c_n \left(\frac{x}{2}\right)^{2n}$. Find the first two terms in the power series for g.

Exercise 2.32: Sturm-Liouville problem with mixed boundary condition

Consider the Sturm-Liouville eigenvalue problem

$$u'' + \lambda u = 0, \qquad u(0) = 0, u(1) + u'(1) = 0$$

Find the eigenfunctions of this problem and the nonlinear algebraic equation that determines the eigenvalues λ. (This equation cannot be solved analytically.) Draw a sketch that indicates that there will be an infinite number of these eigenvalues, and use your sketch to propose an approximation for the eigenvalues that is valid in the situation $\lambda \gg 1$.

Exercise 2.33: A higher-order variable coefficient problem

Find the general solution to the third-order equation

$$x^3y''' + 3x^2y'' - 3xy' = 0$$

Exercise 2.34: A fourth-order variable coefficient ODE

The following differential equation arises in the analysis of time-dependent flow of a polymeric liquid

$$\left(x^2D^2 - x^2 + 2 - 2xD\right)\left(D^2 - 2iD - 3\right)y = 0$$

where $D = \frac{d}{dx}$. This equation has solutions of Frobenius form. Find the roots of the indicial equation.

Exercise 2.35: Legendre's equation

Legendre's equation is

$$(1 - x^2)y'' - 2xy' + l(l + 1)y = 0$$

The point $x = 0$ is an ordinary point for this equation; seek a series solution, expanding around this point. Find the two solutions that make up the general solution. These will have the form

$$y_1(x) = 1 - \frac{l(l+1)}{2!}x^2 + \frac{(l-2)l(l+1)(l+3)}{4!}x^4 - \cdots$$

$$y_2(x) = x - \frac{(l-1)(l+2)}{3!}x^3 + \frac{(l-3)(l-1)(l+2)(l+4)}{5!}x^5 - \cdots$$

By examining the recursion relation, show that for every integer l, one of these series will truncate, becoming a polynomial. These are the Legendre polynomials.

Exercise 2.36: Chebyshev's equation

Chebyshev's equation is

$$(1 - x^2)u'' - xu' + \nu^2 u = 0$$

Its solutions are important in the approximation of functions and in numerical solution methods for boundary-value problems.

(a) Put this in the form of a Sturm-Liouville problem $Lu + \lambda u = 0$ in the domain $[-1, 1]$, with $\lambda = \nu^2, w(x) = (1 - x^2)^{-1/2}$. What boundary conditions must u and u' satisfy at $x = \pm 1$ for self-adjointness to hold?

(b) By expanding in a power series about $x = 0$, obtain two LI solutions of this equation. Show that when ν is a nonnegative integer, one of these is always a polynomial of degree ν. Because these satisfy a Sturm-Liouville problem, these polynomials form an orthogonal basis for $L_{2,w}(-1, 1)$, with $w(x) = (1-x^2)^{-1/2}$.

(c) The points $x = \pm 1$ are regular singular points for this equation. As a first step toward finding the behavior of the solution near these points, find the roots of the indicial equation for a solution in Frobenius form expanded around $x = 1$.

Exercise 2.37: Laplace's equation as second-order, variable coefficient ODEs

Express the radial part of Laplace's equation $\nabla^2 y \pm y = 0$ in the form $a_2(x)y'' + a_1(x)y' + a_0(x)y = 0$.

(a) What are a_0, a_1, a_2 for one-dimensional rectangular coordinates, cylindrical coordinates, and spherical coordinates?

(b) What are two linearly independent solutions for each coordinate system?

Exercise 2.38: How many solutions?

Consider the second-order differential equation

$$\frac{d^2u}{dx^2} = 0 \qquad 0 < x < 1$$

(a) How many linearly independent solutions exist for the single boundary condition

$$u(0) = u(1)$$

(b) How many linearly independent solutions exist for the two boundary conditions

$$u(0) = 0 \qquad u(1) = 0$$

(c) How many linearly independent solutions exist for the two boundary conditions

$$\frac{du}{dx}(0) = 0 \qquad \frac{du}{dx}(1) = 0$$

(d) What can you conclude about the dimension of the null space of this second-order differential operator and the number of boundary conditions?

Exercise 2.39: Heat conduction with equal temperatures at the ends

Consider the differential equation for steady-state heat conduction with heat generation, $f(x)$, in a one-dimensional slab

$$\frac{d^2T}{dx^2} = f, \qquad 0 < x < 1$$

Suppose we set up the problem with a temperature controller that keeps the ends of the body at the same temperature.

(a) Identify the appropriate differential operator, L, and associated boundary functional B_1, so this problem can be written as

$$\begin{aligned} LT &= f \\ B_1 T &= 0 \end{aligned} \tag{2.98}$$

(b) Notice that we do not have enough boundary conditions to expect to be able to solve (2.98) uniquely. Define the adjoint operator and adjoint boundary functionals so that

$$(Lu, v) = (u, L^* v)$$

for every admissible $u(x)$ and $v(x)$. Notice that since you are missing a boundary condition on $u(x)$, you will require three boundary conditions on $v(x)$. What are L^*, B_1^*, B_2^*, B_3^*?

(c) What are the null spaces of L and L^* with their associated boundary conditions? For which f can (2.98) be solved? Is the solution unique? If not, what is the general form of the solution?

(d) Solve (2.98) using any method at your disposal. Laplace transforms would work, for example. Check your solution by substituting into the differential equation and boundary condition. Does your solution agree with the existence and uniqueness result you determined previously?

(e) What is the Green's function for this problem, i.e., identify the function g appearing in the $T(x)$ solution as

$$T(x) = \int_0^1 g(x, \xi) f(\xi) d\xi + \text{terms not involving } f$$

Exercise 2.40: Solvability conditions and general solution for second-order operator

Consider the second-order differential operator and two boundary conditions

$$Lu = -\frac{d^2u}{dx^2} - u \qquad -\pi < x < \pi$$
$$B_1 u = u(\pi) - u(-\pi)$$
$$B_2 u = \frac{du}{dx}(\pi) - \frac{du}{dx}(-\pi)$$

(a) Find the adjoint operator and boundary conditions, L^*, B_1^*, and B_2^*.

(b) Find the null spaces $N(L)$ and $N(L^*)$.

(c) For what f can you solve the nonhomogeneous problem

$$Lu = f(x)$$
$$B_1(u) = \gamma_1 \qquad B_2(u) = \gamma_2$$

answer: $(f, \sin x) = -\gamma_1 \qquad (f, \cos x) = \gamma_2$

(d) For f satisfying this solvability condition, what is the general solution?

answer: $u(x) = -\int_{-\pi}^{x} f(\xi)\sin(x-\xi)d\xi + a\cos x + b\sin x$

Exercise 2.41: Steady-state temperature profile

Solve the steady-state heat-conduction problems in Examples 2.11 and 2.12 using Laplace transforms.

$$T_{xx} = f \qquad\qquad T_{xx} = f$$
$$T(0) = T_0 \qquad\qquad T_x(0) = 0$$
$$T(1) = T_1 \qquad\qquad T_x(1) = 0$$

Exercise 2.42: Heat-transfer boundary conditions

Consider the one-dimensional steady-state heat-conduction problem

$$-k\frac{d^2T(x)}{dx^2} = \hat{f}(x)$$
$$\frac{d^2T}{dx^2} = f \qquad f = -\hat{f}/k$$

Consider Newton's law of cooling boundary conditions

$$h_0(T_{e0} - T(0)) = -kT_x(0)$$
$$h_1(T(1) - T_{e1}) = -kT_x(1)$$

in which h_0, h_1 are the heat-transfer coefficients at the two ends, and T_{e0}, T_{e1} are the temperatures providing the heat-transfer driving forces at the two ends.

(a) Write this problem as

$$DT = f$$
$$B_1 T = \gamma_1 \qquad B_2 T = \gamma_2$$

What are D, B_1 and B_2, and γ_1 and γ_2?

(b) Solve for the steady-state temperature profile.

(c) For what $f(x)$ does the solution exist? For these $f(x)$, is the solution unique?

Exercise 2.43: Orthogonality of Sturm-Liouville eigenfunctions

Show that two eigenfunctions $u1$ and u_2 of a Sturm-Liouville problem $(pu')' + ru + \lambda wu = 0$ are orthogonal if the inner product weighted with w is used. Consider only zero boundary conditions $u(a) = u(b) = 0$. Multiply the equation for u_1 (setting $\lambda = \lambda_1$) by u_2; multiply the equation for u_2 (setting $\lambda = \lambda_2 \neq \lambda_1$) by u_1; subtract and integrate over the interval. Use the boundary conditions and integration by parts to prove orthogonality.

Exercise 2.44: The convection-diffusion operator

For problems with convection and diffusion, an important differential operator is

$$Lu = -\frac{d^2u}{dx^2} + \text{Pe}\frac{du}{dx}$$

with boundary condition $u(0) = u(1) = 0$. Pe is the PECLET number, measuring the relative importance of convection and diffusion.

(a) Find the adjoint of this operator, first with an inner product with a constant weight function $w(x) = 1$, and then with the weight function $w(x) = \exp(-\text{Pe}\ x)$.

(b) Solve the eigenvalue problem $Lu + \lambda u =$ for arbitrary Pe. Hint: since the equation has constant coefficients, express the solution as $y(x) = e^{i\alpha x}$. Plot the eigenfunction corresponding to the first (closest to zero) eigenvalue for Pe = 10.

Exercise 2.45: Testing a CSTR operating condition for stability[22]

The reaction

$$\text{A} \xrightarrow{k} \text{B} \qquad r = kc_A = k_0 e^{-E/T} c_A$$

is carried out in a CSTR. The mass and energy balances are given by

$$\frac{dc_A}{dt} = \frac{c_{Af} - c_A}{\tau} - kc_A$$
$$\frac{dT}{dt} = \frac{U^oA}{V_R\rho\hat{C}_P}(T_a - T) + \frac{T_f - T}{\tau} - \frac{\Delta H_R}{\rho\hat{C}_P}kc_A$$

Find the three steady states corresponding to the conditions in the following table. Determine whether each of these three steady states is stable or unstable.

[22]See also Exercise 6.7 in Rawlings and Ekerdt (2012)

Parameter	Value	Units
E	7550	K
T_f	298	K
c_{Af}	3	$kmol/m^3$
U^o	0	
ΔH_R	-2.09×10^8	J/kmol
k_0	4.48×10^6	1/s
$\hat{C}_P$	4.19×10^3	J/(kg K)
ρ	10^3	kg/m^3
V_R	18×10^{-3}	m^3
Q_f	60×10^{-6}	m^3/s

Exercise 2.46: Choosing an ODE solver

You are given the task of modeling the dynamics of a chemical reactor in which a large number of reactions are occurring. The rate constants for the reactions vary between $1s^{-1}$ and $10^7 s^{-1}$. Will you base your code on a fourth-order Runge-Kutta scheme, an explicit Euler scheme, or an Adams-Moulton scheme? Why?

Exercise 2.47: Numerical stability criterion for RK2

Derive the numerical stability criterion for integrating the single equation $\dot{x} = \lambda x$ with the second-order Runge-Kutta method. Allow λ to be complex. Hint: the general solution to a linear constant-coefficient *difference equation* $a_n x^{(n)} + a_{n-1} x^{(n-1)} + \ldots =$ 0 is of the form $x^{(n)} = \rho^n$.

Exercise 2.48: Dynamics of a nonlinear problem

Consider the pair of ODEs

$$\dot{y}_1 = (1 - y_1) - 10 y_1^2 y_2$$
$$\dot{y}_2 = -0.05 y_1^2 y_2$$

with initial conditions $y_1(0) = 0.2, y_2(0) = 1$.

(a) Find the Jacobian of the RHS at $t = 0$. Show, using the eigenvalues of the Jacobian, that the you expect the problem to be stiff.

(b) Write a computer program to use the Adams-Moulton second-order method to solve the initial-value problem. Integrate the equations out to $t = 20$ and plot the solutions. Can you find any stability limit on the time step?

(c) Write a second-order Runge-Kutta program and attempt to use it for the above problem. What time step do you have to use to get a stable result?

(d) Modify your RK code to use variable time steps. Use the criterion that $\Delta t < t_{min}/5$. Estimate t_{min} from the values of $y/\dot{y}$ at each time step.

Exercise 2.49: Solutions of difference equations

When examining the numerical stability of integration schemes, as well as in many other situations, we run across the linear constant-coefficient *difference* equation

$$a_M y_{n+M} + a_{M-1} y_{n+M-1} + \ldots + a_0 y_n = 0. \tag{2.99}$$

For example, y_n could be the value of y at the nth time step of some process.

(a) Show that this equation can be written in vector form

$$x_{n+1} = Gx_n \tag{2.100}$$

What are x and G in terms of y and a coefficients?

(b) Given the initial condition x_0, find the solution to this equation (i.e., x_n in terms of n and x_0) in the situation where A has distinct eigenvalues λ.

(c) Repeat for the case where

$$G = \begin{bmatrix} \lambda & 1 \\ 0 & \lambda \end{bmatrix}$$

(d) What is the general criterion for asymptotic stability of the steady state $x = 0$?

Exercise 2.50: Numerical integration for undamped oscillations

Second-order initial-value problems $\ddot{u} = f(u)$ are important for many applications. For the specific case $f(u) = -q^2 u$ do the following:

(a) Find the exact general solution.

(b) By letting $\dot{u} = v$ convert the equation to a pair of first-order equations and show that the forward Euler method is always unstable for integrating these.

(c) Consider the following numerical integration formula

$$u_{n+1} - 2u_n + u_{n-1} = (\Delta t)^2 f(u_n)$$

For $f(u) = -q^2 u$ find a quadratic equation for the growth factor G for this method, i.e., look for solutions of the form $u_{n+1} = Gu_n$. Up to what threshold $(q\Delta t)^2$ are the numerical solutions stable?

(d) By expanding all terms in Taylor series around time step n, find the local truncation error p of this formula (the first power of Δt that does not cancel).

Exercise 2.51: The velocity Verlet algorithm of molecular dynamics simulation

The VELOCITY VERLET ALGORITHM is very commonly used to perform numerical time integration for molecular dynamics simulations. Consider the numerical stability problem for a very simple case

$$\dot{x} = v \qquad \dot{v} = ax$$

where $a \in \mathbb{R}$.

(a) What property must a satisfy so that the true solution $x = 0, v = 0$ is stable?

(b) For this problem, the velocity Verlet algorithm becomes

$$x_{n+1} = x_n + v_n \Delta t + \frac{1}{2} a x_n \Delta t^2$$

$$v_{n+1} = v_n + \frac{1}{2}(a x_n + a x_{n+1}) \Delta t$$

Put this expression in the form

$$x_{n+1} = Gx_n$$

where $x = (x, v)^T$.

(c) Find the criteria that $a\Delta t^2$ must satisfy for numerical stability of the algorithm.

Exercise 2.52: Stability of predictor-corrector methods

Denote the general (up to fourth-order) predictor-corrector formulas for the differential equation $\dot{x} = ax$ by

$$x^{(*)} = x^{(n)} + w\left(p_1 x^{(n)} + p_2 x^{(n-1)} + p_3 x^{(n-2)} + p_4 x^{(n-3)}\right)$$

$$x^{(n+1)} = x^{(n)} + w\left(c_1 x^{(*)} + c_2 x^{(n)} + c_3 x^{(n-1)} + c_4 x^{(n-2)}\right)$$

in which $w = a\Delta t$. The coefficient vectors of the first four Adams-Bashforth predictors and Adams-Moulton correctors are as follows

$$\begin{aligned} p\{1\} &= [1,0,0,0] & c\{1\} &= [1,0,0,0] \\ p\{2\} &= (1/2)[3,-1,0,0] & c\{2\} &= (1/2)[1,1,0,0] \\ p\{3\} &= (1/12)[23,-16,10,0] & c\{3\} &= (1/12)[5,8,-1,0] \\ p\{4\} &= (1/24)[55,-59,37,-9] & c\{4\} &= (1/24)[9,19,-5,1] \end{aligned}$$

Show that combining the two steps gives

$$x^{(n+1)} = x^{(n)}(1 + wc_1 + w(c_1 w p_1 + c_2)) + x^{(n-1)}(w(c_1 w p_2 + c_3)) + \\ x^{(n-2)}(w(c_1 w p_3 + c_4)) + x^{(n-3)}(w c_1 w p_4)$$

Let $z^{(n)} = (x^{(n)}, x^{(n-1)}, x^{(n-2)}, x^{(n-3)})$, and find matrix G such that

$$z^{(n+1)} = G(w) z^{(n)}$$

The eigenvalues of $G(w)$ then determine the stability of the method.

Exercise 2.53: Stability boundary of predictor-corrector methods

Given $G(w)$ from the previous exercise, to map out the *boundary* of the stability region, consider $\omega = e^{i\theta}$ for $0 \le \theta \le 2\pi$, so ω has unit magnitude, and solve the single algebraic equation $\det(G(w) - \omega I) = 0$ for the complex value w as a function of parameter θ. The stability boundary of the APC method then comprises these values of w. That is how Figure 2.27 was prepared, for example.

Now consider the class of predictor-corrector methods that use the *same order* in the predictor and corrector. Recall the methods in Figure 2.27 used a predictor with order one less than the corrector. Find the stability boundaries for first-order through fourth-order methods. Compare your calculated results to Figure 2.34. Contrast the stability results displayed in Figures 2.27 and 2.34. From a stability standpoint, which class of methods do you prefer and why?

Hints:

You will need to increase the θ interval to $[0, 4\pi]$ to close the stability boundary. Why do you suppose this increased interval is required? Consider mapping out the square root function on the unit circle using $\theta \in [0, 2\pi]$. Does this boundary close?

You will need to clip off some unstable regions made by loops in the boundary to match Figure 2.34.

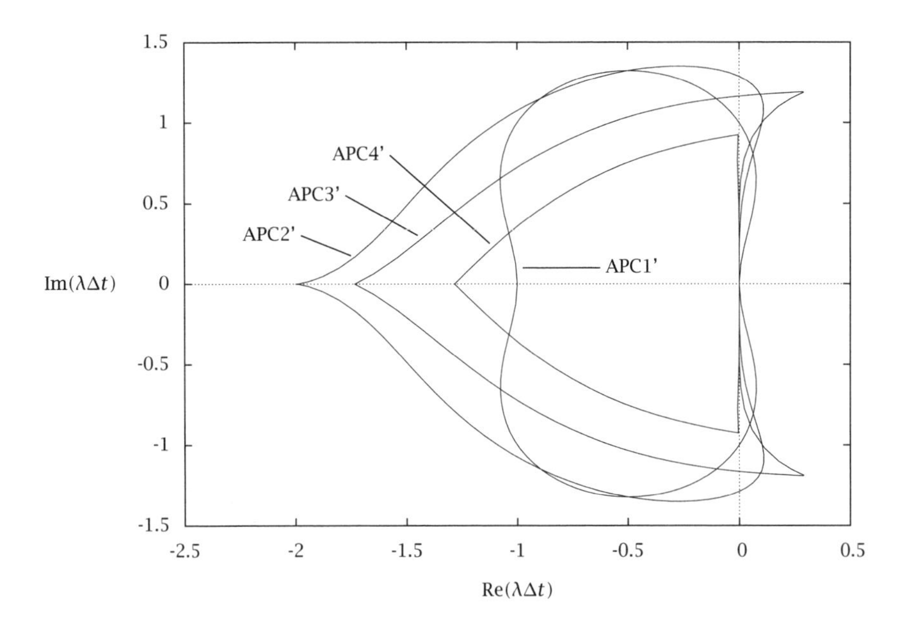

Figure 2.34: Stability regions for Adams predictor-corrector methods; $\dot{x} = \lambda x$; APCn' uses nth-order predictor and nth-order corrector.

Exercise 2.54: Airy's equation

The equation

$$y'' + (x + \lambda)y = 0$$

arises in optics, quantum mechanics, and hydrodynamics, and is known as Airy's equation.

(a) Find an approximate power series solution to this problem (expanding around $x = -\lambda$), keeping terms up to fourth order.

(b) Use this solution to approximate the first two eigenvalues λ of the equation with boundary conditions $y(0) = y(1) = 0$. Use Newton's method if necessary.

(c) Use the finite element method to construct an *algebraic* eigenvalue problem for the Airy equation. Find the approximate eigenvalues and eigenfunctions using six hat functions as the approximate basis and also using 12 functions. Are all of the eigenvalues of the algebraic problem good approximations of the exact eigenvalues? You may use Mathematica or a programming language, whichever you prefer.

Exercise 2.55: Applying Galerkin and collocation methods

Solve the problem

$$x^2y'' + xy' + x^2y = x^2, y'(0) = 0, y(1) = 0$$

using

(a) The Legendre-Galerkin method.

(b) The Chebyshev collocation method. Recall that the Chebyshev collocation points are numbered from right to left, i.e., $x_0 = 1$ and $x_N = -1$.

Exercise 2.56: Modeling a tubular reactor: convection, diffusion, and reaction

The equation

$$2u' = u'' + 1, u(-1) = 0, u'(1) = 0$$

models the temperature profile in a tubular reactor in which an exothermic reaction occurs.

(a) Find the exact solution.

(b) Use the Galerkin tau method to construct an approximate solution. Use the Legendre polynomial basis set: $\phi_0(x) = 1, \phi_1(x) = x, \phi_2(x) = (3x^2 - 1)/2$. Sketch the solution and look at $u'(0)$ and $u(1)$ to compare the approximate and exact solutions.

Exercise 2.57: Converting a differential operator to an algebraic operator

Solve the eigenvalue problem

$$x^2y'' + xy' + x^2\lambda y = 0, y'(0) = y(1) = 0$$

using the Legendre-Galerkin method. You should be able to reduce this problem to a linear algebra problem of the form $Ac + \lambda Bc = 0$. Note that because of the boundary conditions, B will be singular, but A will not. How many basis functions do you need to compute the first three eigenvalues to four-digit accuracy? Plot the first four eigenfunctions. This is the eigenvalue problem for Bessel's equation of order zero. In the chapter, we showed that the eigenvalues of this problem are related to the roots of the Bessel function J_0.

Exercise 2.58: An eigenvalue problem with finite elements

Solve the above problem again, using the finite element method with the "hat functions" described in Section 2.9.1. Study how the approximation converges as the number of node points N increases. Also look at the computation time as a function of N.

Exercise 2.59: Chebyshev collocation for a nonlinear problem

Using the Chebyshev collocation technique, write an Octave or MATLAB program to solve the boundary-value problem (a steady-state reaction-diffusion problem)

$$\epsilon T'' + T - T^3 = 0, T(-1) = T(1) = 0$$

for $\epsilon = 0.05$. Use the initial guess $T = 1$ to find a nontrivial solution. Study how the approximation converges as the number of collocation points $N + 1$ increases. Also look at the computation time as a function of N.

Exercise 2.60: Attractivity and asymptotic stability for linear systems

Show that asymptotic stability and attractivity are identical for linear systems, $\dot{x} = Ax$, and both are equivalent to the condition that $\text{Re}(\text{eig}(A)) < 0$.

Exercise 2.61: Stability and asymptotic stability for linear systems

(a) Consider the linear system

$$\frac{dx}{dt} = \begin{bmatrix} 0 & 1 \\ 0 & 0 \end{bmatrix} x$$

Is this system asymptotically stable? Why or why not?

(b) Is the system (Lyapunov) stable or unstable? Prove it.

(c) Generalize this example and provide a checkable condition to test for (Lyapunov) stability of all linear systems, $\dot{x} = Ax$.

(d) Given this result, characterize the class of linear systems that are stable but not asymptotically stable.

Exercise 2.62: Lyapunov equation and linear systems

Establish the equivalence of (a) and (b) in Theorem 2.24.

Exercise 2.63: Discrete time Lyapunov function for linear systems

State the discrete time version of Theorem 2.24. Show that (a) and (b) are equivalent in the discrete time version.

Exercise 2.64: Nonsymmetric matrices and definition of positive definite

For real, square matrix S, consider redefining $S > 0$ to mean that $x^T S x > 0$ for all $x \in \mathbb{R}^n \neq 0$. We are removing the usual requirement that S is symmetric in the definition of positive definite in Section 1.4.4.

(a) Define the matrix $B = (S + S^T)/2$. Show that B is symmetric and $x^T B x = x^T S x$ for all $x \in \mathbb{R}^n$. Therefore $S > 0$ (new definition) if and only if B is positive definite (standard definition).

(b) What happens to the connection between this new definition of $S > 0$ and the eigenvalues of S? Consider statement 1. from Section 1.4.4

$$S > 0 \text{ if and only if } \lambda > 0, \quad \lambda \in \text{eig}(S)$$

Does this statement remain valid? If so prove it. If not, provide a counterexample.

Exercise 2.65: Stabilities of a linear system

Consider the linear, time-invariant system $\dot{x} = Ax$. Characterize the class of A matrices for which the systems exhibit the following forms of stability.

(a) Stable (in the sense of Lyapunov).

(b) Attractive.

(c) Asymptotically stable.

(d) Exponentially stable.

(e) Which of these forms of stability are *equivalent* for the linear, time-invariant system?

Exercise 2.66: Extending a regular perturbation solution to higher order

For the regular perturbation solution of (2.67) presented in Section 2.6.4, compute the next term in the series, $c_2(r)$.

Exercise 2.67: QSSA as the outer solution in a two-time-scale singular perturbation

Consider the following simple reaction mechanism taking place in a well-mixed, constant-volume, batch reactor

$$\mathrm{A} \xrightarrow{k_1} \mathrm{B} \xrightarrow{k_2} \mathrm{C}$$

and assume $k_2 \gg k_1$ so B is a low-concentration species for which we wish to examine the QSSA.

(a) Solve A's material balance and show

$$c_{As} = c_{A0} e^{-k_1 t}$$

Apply the usual QSSA approach, set $R_B = 0$, and show that

$$c_{Bs} = \frac{k_1}{k_2} c_{As} = c_{A0} \frac{k_1}{k_2} e^{-k_1 t}$$

The concentration of C is always available if desired from the total species balance

$$c_{Cs}(t) = c_A(0) + c_B(0) + c_C(0) - c_{As}(t) - c_{Bs}(t)$$

(b) The B species has two-time-scale behavior. On the fast time scale, it changes rapidly from initial concentration c_{B0} to the quasi-steady-state value for which $R_B \approx 0$. Divide B's material balance by k_2, define the fast time-scale time as $\tau = k_2 t$, and obtain for B's material balance

$$\frac{dc_B}{d\tau} = \epsilon k_1 c_A - c_B \qquad \epsilon = \frac{1}{k_2}$$

We wish to find an asymptotic solution for small ϵ. We try a series expansion in powers of ϵ for the inner solution (fast time scale)

$$c_{Bi} = Y_0 + \epsilon Y_1 + \epsilon^2 Y_2 + \cdots$$

The initial condition, $C_{Bi} = C_{B0}$, must be valid for all ϵ, which gives for the initial conditions of the Y_n

$$Y_0(0) = c_{B0} \qquad Y_n(0) = 0, \quad n = 1, 2, \ldots$$

Substitute the series expansion into B's material balance, collect like powers of ϵ and show the following differential equations govern the Y_n

$$\epsilon^0: \qquad \frac{dY_0}{d\tau} = -Y_0$$

$$\epsilon^1: \qquad \frac{dY_1}{d\tau} = k_1 c_A - Y_1$$

$$\epsilon^n: \qquad \frac{dY_n}{d\tau} = -Y_n \quad n \geq 2$$

(c) Solve these differential equations and show

$$Y_0 = c_{B0}e^{-\tau}$$

$$Y_1 = c_{A0}\frac{k_1}{k_1/k_2 - 1}\left(e^{-\tau} - e^{-k_1\tau/k_2}\right)$$

$$Y_n = 0 \qquad n \geq 2$$

Because Y_n vanishes for $n \geq 2$, show you obtain the exact solution for the B concentration for all ϵ by using the first two terms.

(d) Next we analyze B's large-time-scale behavior, also called the outer solution. Divide B's material balance by k_2 again but do not rescale time and obtain

$$\epsilon\frac{dc_B}{dt} = \epsilon k_1 c_A - c_B$$

Expand c_B again in a power series of ϵ

$$c_{Bo} = B_0 + \epsilon B_1 + \epsilon^2 B_2 + \cdots$$

Substitute the power series into the material balance and collect like powers of ϵ to obtain the following equations

$$\epsilon^0 : \qquad B_0 = 0$$

$$\epsilon^1 : \qquad \frac{dB_0}{dt} = k_1 c_A - B_1$$

$$\epsilon^n : \qquad \frac{dB_n}{dt} = -B_{n+1} \quad n \geq 1$$

Solve these equations and show

$$B_0 = 0$$

$$B_1 = k_1 c_A$$

$$B_n = k_1^n c_A \qquad n \geq 2$$

So we see the zero-order outer solution is $C_{B0} = 0$, which is appropriate for a QSSA species, but a rather rough approximation.

(e) Show that the classic QSSA analysis is the first-order outer solution.

(f) To obtain a uniform solution valid for both short and long times, we add the inner and outer solution and subtract any common terms. Plot the uniform zeroth-order and first-order solutions for the following parameter values

$$c_{A0} = 1 \qquad c_{B0} = 1/2 \qquad k_1 = 1 \qquad k_2 = 10$$

Compare to the exact solution and the first-order outer solution (QSSA solution).

(g) Show that the infinite-order uniform solution is also the exact solution.

Exercise 2.68: QSSA and matching conditions in singular perturbation

Consider again Exercise 2.67 with a slightly more complex reaction mechanism

$$A \underset{k_{-1}}{\overset{k_1}{\rightleftharpoons}} B \overset{k_2}{\longrightarrow} C$$

and assume that either $k_{-1} \gg k_1$ or $k_2 \gg k_1$ (or both) so B is again a low-concentration species for which we wish to examine the QSSA. Notice that either k_{-1} or k_2 may be large with respect to the other without invalidating the QSSA assumption for B. Only if $k_{-1} \gg k_1 \gg k_2$ is the reaction equilibrium assumption also valid for this mechanism.

(a) Apply the QSSA on species B and show

$$c_{As} = \left(c_{A0} + c_{B0}\frac{1}{1+K_2}\right) e^{-\frac{k_1 K_2}{1+K_2} t}$$

$$c_{Bs} = \frac{1}{k_{-1}}\frac{k_1}{1+K_2}\left(c_{A0} + c_{B0}\frac{1}{1+K_2}\right) e^{-\frac{k_1 K_2}{1+K_2} t}$$

in which $K_2 = k_2/k_{-1}$.

(b) With this mechanism, both the A and B species have two-time-scale behavior, so we use a series expansion for both c_A and c_B. Let the inner solution be given by

$$c_{Ai} = X_0 + \epsilon X_1 + \epsilon^2 X_2 + \cdots$$

$$c_{Bi} = Y_0 + \epsilon Y_1 + \epsilon^2 Y_2 + \cdots$$

in which the small parameter ϵ is the inverse of the largest rate constant in the mechanism. In the following we assume k_{-1} is largest and $\epsilon = 1/k_{-1}$. Define $K_2 = k_2/k_{-1}$ and we assume that K_2 is order unity or smaller. If K_2 were large, we should have chosen $\epsilon = 1/k_2$ as the small parameter. Collect terms of like power of ϵ and show

$$\epsilon^0: \quad \frac{dX_0}{d\tau} = Y_0 \qquad \frac{dY_0}{d\tau} = -(1+K_2)Y_0$$

$$\epsilon^1: \quad \frac{dX_1}{d\tau} = -k_1 X_0 + Y_1 \qquad \frac{dY_1}{d\tau} = k_1 X_0 - (1+K_2)Y_1$$

$$\epsilon^n: \quad \frac{dX_n}{d\tau} = -k_1 X_{n-1} + Y_n \qquad \frac{dY_n}{d\tau} = k_1 X_{n-1} - (1+K_2)Y_n \quad n \geq 1$$

What are the initial conditions for the X_n and Y_n variables?

(c) Solve these for the zero-order inner solution and show

$$X_0 = c_{A0} + c_{B0}\frac{1}{1+K_2}\left(1 - e^{-(1+K_2)\tau}\right) \qquad Y_0 = c_{B0}e^{-(1+K_2)\tau}$$

(d) Next we construct the outer solution valid for large times. Postulate a series expansion of the form

$$c_{Ao} = A_0 + \epsilon A_1 + \epsilon^2 A_2 + \cdots$$

$$c_{Bo} = B_0 + \epsilon B_1 + \epsilon^2 B_2 + \cdots$$

Substitute these into the A and B material balances and show

$$\epsilon^0: \quad B_0 = 0 \qquad (1+K_2)B_0 = 0$$

$$\epsilon^1: \quad \frac{dA_0}{dt} = -k_1 A_0 + B_1 \qquad \frac{dB_0}{dt} = k_1 A_0 - (1+K_2)B_1$$

$$\epsilon^n: \quad \frac{dA_{n-1}}{dt} = -k_1 A_n + B_n \qquad \frac{dB_{n-1}}{dt} = k_1 A_{n-1} - (1+K_2)B_n \quad n \geq 1$$

(e) Solve these and show for zero order

$$A_0 = A_0(0)e^{-\frac{k_1 K_2}{1+K_2}t} \qquad B_0 = 0$$

Again we see that to zero order, the B concentration is zero after a short time.

Note also that, unlike in Exercise 2.67, *we require an initial condition for the outer solution A_n differential equations.* We obtain the missing initial condition by matching with the inner solution as follows

$$\lim_{\tau \to \infty} X_0(\tau) = \lim_{t \to 0} A_0(t)$$

In other words, the long-time solution (steady state) on the fast time scale is the short-time solution (initial condition) on the slow time scale. Using this matching condition show

$$A_0(0) = c_{A0} + c_{B0}\frac{1}{1+K_2}$$

(f) Find also the first-order solution, B_1, and show that the QSSA solution corresponds to the zero-order outer solution for c_A and the first-order outer solution for c_B.

Exercise 2.69: Michaelis-Menten kinetics as QSSA

Consider the enzyme kinetics

$$\text{E} + \text{S} \underset{k_{-1}}{\overset{k_1}{\rightleftharpoons}} \text{ES}$$

$$\text{ES} \overset{k_2}{\longrightarrow} \text{P} + \text{E}$$

in which the free enzyme E binds with substrate S to form bound substrate ES in the first reaction, and the bound substrate is converted to product P and releases free enzyme in the second reaction. This mechanism has become known as Michaelis-Menten kinetics (Michaelis and Menten, 1913), but it was proposed earlier by Henri (1901). If the rates of these two reactions are such that either the free or bound enzyme is present in small concentration, the mechanism is a candidate for model reduction with the QSSA.

Assume $k_1 \gg k_{-1}, k_2$ so E is present in small concentration. Apply the QSSA and show that the slow time scale model reduces to a first-order, irreversible decomposition of S to P

$$\text{S} \longrightarrow \text{P} \qquad r$$

(a) For a well-stirred batch reactor, show the total enzyme concentration satisfies

$$c_E(t) + c_{ES}(t) = c_E(0) + c_{ES}(0)$$

(b) Find an expression for the QSS concentration of E. What is the corresponding concentration of ES?

(c) Show the rate expression for the reduced model's single reaction is

$$r = \frac{kc_S}{1+Kc_S} \qquad k = k_2 K E_0 \qquad K = \frac{k_1}{k_{-1}+k_2} \qquad E_0 = c_E(0) + c_{ES}(0) \quad (2.101)$$

which depends solely on the substrate concentration. The inverse of the constant K is known as the Michaelis constant. The production rates of reactant S and product P in the reduced model are then simply

$$R_S = -r \qquad R_P = r$$

Notice we have reduced the number of reactions from two to one; we have reduced the number of rate constants from three (k_1, k_{-1}, k_2) to two (k, K).

(d) Plot the concentrations versus time for the full model and QSSA model for the following values of the rate constants and initial conditions.

$$k_1 = 5 \qquad k_{-1} = 1 \qquad k_2 = 10$$
$$c_E(0) = 1 \qquad c_{ES}(0) = 0 \qquad c_S(0) = 50 \qquad c_P(0) = 0$$

Exercise 2.70: Michaelis-Menten kinetics as reaction equilibrium

Consider again the enzyme kinetics given in Exercise 2.69.

$$\text{E} + \text{S} \underset{k_{-1}}{\overset{k_1}{\rightleftharpoons}} \text{ES}$$
$$\text{ES} \xrightarrow{k_2} \text{P} + \text{E}$$

Now assume the rate constants satisfy $k_1, k_{-1} \gg k_2$ so that the first reaction is at equilibrium on the time scale of the second reaction.

(a) Find the equilibrium concentrations of E and ES

(b) Show the production rate of P is given by

$$R_P = \frac{\tilde{k} c_S}{1 + K_1 c_S} \qquad \tilde{k} = k_2 K_1 E_0 \qquad K_1 = k_1/k_{-1} \tag{2.102}$$

in which K_1 is the equilibrium constant for the first reaction. Notice this form is identical to the production rate of P given in the QSSA approach. For this reason, these two assumptions for reducing enzyme kinetics are often mistakenly labeled as the same approach.

It is interesting to note that in their original work in 1913, Michaelis and Menten proposed the reaction equilibrium approximation to describe enzyme kinetics, in which the second step is slow compared to the first step (Michaelis and Menten, 1913). Michaelis and Menten credit Henri with proposing this mechanism to explain the experimental observations that (i) production rate of P increases linearly with substrate at low substrate concentration and (ii) production rate of P is independent of substrate concentration at high substrate concentration (Henri, 1901).

The QSSA analysis of enzyme kinetics was introduced by Briggs and Haldane in 1925, in which the enzyme concentration is assumed small compared to the substrate (Briggs and Haldane, 1925). Since that time, the QSSA approach has become the more popular explanation of the observed dependence of substrate in the production rate of product R_P in 2.101 and 2.102 (Nelson and Cox, 2000).

The reader should be aware that either approximation may be appropriate depending on the values of the rate constants and initial conditions. Although both reduced models give the same form for the production rate of P, they are often quite different in other respects. Finally, for some values of rate constants, in particular $k_{-1} \gg k_1 \gg k_2$, both the QSS assumption and the reaction equilibrium assumption apply.

(c) Show that the slow-time-scale reduced model for the reaction equilibrium assumption can be summarized by two irreversible reactions

$$\begin{aligned} \mathrm{ES} &\longrightarrow \mathrm{E} + \mathrm{S} \qquad \mathrm{tr}_1 \\ \mathrm{S} &\longrightarrow \mathrm{P} \qquad \mathrm{tr}_2 \end{aligned}$$

with the following rate expressions

$$\mathrm{tr}_1 = \left(\frac{K_1 c_E}{1 + K_1(c_E + c_S)}\right)\left(\frac{\tilde{k} c_S}{1 + K_1 c_S}\right) \qquad \tilde{k} = k_2 K_1 E_0$$

$$\mathrm{tr}_2 = \frac{\tilde{k} c_S}{1 + K_1 c_S} \qquad K_1 = k_1 / k_{-1}$$

Notice here we have not reduced the number of reactions; we still have two reactions, but as before we have reduced the number of rate constants from three (k_1, k_{-1}, k_2) to two $(\tilde{k}, K_1)$. The first rate expression here depends on c_S and c_E rather than only c_S as in the previous QSSA reduction. Therefore the production rates of E, ES, and S depend on c_E as well as c_S. Only the production rate of P ($R_P = \mathrm{tr}_2$) loses the c_E dependence.

(d) Plot the concentrations versus time for the full model and reaction equilibrium model for the following values of the rate constants and initial conditions.

$$k_1 = 0.5 \qquad k_{-1} = 1 \qquad k_2 = 0.5$$
$$c_E(0) = 20 \qquad c_{ES}(0) = 10 \qquad c_S(0) = 50 \qquad c_P(0) = 0$$

Recall that you must modify the initial conditions for the slow-time-scale model by equilibrating the first reaction from these starting values.

Exercise 2.71: Asymptotic expansion of an integral

Find an asymptotic expansion of the integral

$$f(x) = \int_x^\infty t^{-1} e^{x-t} dt$$

for large positive values of x. Use repeated integration by parts. Show that the approximation is asymptotic as $x \to \infty$.

Exercise 2.72: Asymptotic series are not always power series

Find the leading-order approximations to the two solutions of

$$x e^{-x} = \epsilon$$

for $\epsilon \ll 1$. Seek solutions of the form $x = \delta(\epsilon) X$, find two dominant balances: one where $\delta(\epsilon) \ll 1$ and one where $\delta(\epsilon) \gg 1$.

Exercise 2.73: Perturbed eigenvalue problems

Consider the eigenvalue problem

$$Ax + \epsilon B(x) = \lambda x$$

where A is an $n \times n$ matrix and $B(x)$ and x are n-vectors. Assume that the eigenvalues of A are distinct. If the unperturbed problem has an eigenvalue λ_0 with a corresponding eigenvector v_0, find the leading-order corrections to the eigenvalue and eigenvector. Hint: review the existence and uniqueness theory for linear equations.

Exercise 2.74: Multiple-scales analysis of a problem with a pitchfork bifurcation

Consider the system of equations

$$\begin{aligned} \dot{x} &= -x - \epsilon^{1/2} y^2 \\ \dot{y} &= \epsilon \lambda y - \epsilon^{1/2} x y \end{aligned}$$

Assume that x and y are both ord(1). (They have already been scaled by $\epsilon^{1/2}$.) Perform a multiple-scales expansion, letting $t_0 = t, t_1 = \epsilon^{1/2} t, t_2 = \epsilon t$. Show that the solvability conditions require that

$$\begin{aligned} \frac{\partial y_0}{\partial t_0} = \frac{\partial y_0}{\partial t_1} &= 0 \\ \frac{d y_0}{d t_2} &= \lambda y_0 + y_0^3 \end{aligned}$$

when $t_0 \gg 1$. What are the steady-state solutions of the amplitude equation for y_0? Sketch the steady states as λ varies between -1 and $+1$.

Exercise 2.75: Degenerate pitchfork bifurcation

Consider the one-dimensional system

$$\dot{x} = f(x; \mu)$$

where $f(x; \mu) = -f(-x; \mu)$ and $f_{xxx} = 0$ at $x = 0$. Although this equation has the correct symmetry to display a pitchfork bifurcation, (2.81) does not hold because $f_{xxx} = 0$.

(a) Derive the correct normal form in this case and draw the corresponding bifurcation diagram(s).

(b) Now let f_{xxx} be nonzero, but very small. How are the above bifurcation diagrams modified?

Exercise 2.76: Multiple scales to determine stability of a time-periodic solution

Consider the stability of a periodic orbit of a nonlinear system. Let $x_p(t) = x_p(t + T)$ be a time-periodic solution of the differential equation

$$\dot{x} = f(x)$$

Now let $x = x_p(t) + \delta z(t), \delta \ll 1$.

(a) Show that the linearized equation for z takes the form

$$\dot{z} = A(t)z$$

where $A(t) = A(t + T)$ is a matrix operator with time-periodic coefficients.

(b) The DAMPED MATHIEU EQUATION is a particular case of a linear equation with time-periodic coefficients. It is (written as a single second-order equation)

$$\ddot{x} + \epsilon\mu\dot{x} + (\omega^2 + \epsilon\cos 2t)x = 0$$

Letting $\omega = 1, 0 < \epsilon \ll 1, \mu = \text{ord}(1) > 0$, use the multiple-scales method to determine the stability of the point $z = 0$. Show that $z = 0$ is stable when $\mu > 1/2$. (Although this equation can be put in the form $\dot{z} = A(t)z$, it is easier to work with when kept in second-order form.) Use time scales $t_0 = t, t_1 = \epsilon t$ and assume a solution of the form $x(t_0, t_1) = A(t_1)\cos t_0 + B(t_1)\sin t_0 + \epsilon x_1(t_0, t_1) + O(\epsilon^2)$.

Exercise 2.77: Oscillator with slowly varying frequency

Use the multiple-scales approach with $t_1 = t, t_2 = t/\epsilon$ to find the leading-order general solution to the problem of an oscillator with slowly varying frequency

$$\epsilon^2 \frac{d^2y}{dt^2} + (\omega(t))^2 y = 0$$

Assume that $\omega(t) > 0$ in the domain of interest. Show that a leading-order solution of the form $y_0 \sim r(t_1)\exp(\pm i\omega(t_1)t_2)$ will not work, but that a solution of the slightly more general form $y_0 \sim r(t_1)\exp(\pm i\epsilon^{-1}\theta(t_1))$ will. You will see from the multiple-scales result that the quantity $r^2\omega$ is independent of t_1, to leading order: it is a so-called ADIABATIC INVARIANT.

Exercise 2.78: Multiple-scales solution to a nonlinear oscillator problem

Use the method of multiple scales to find a leading-order solution to the nonlinear oscillation problem

$$\ddot{x} + \epsilon(x^2 - 1)\dot{x} + x = 0, \quad x(0) = 1, \dot{x}(0) = 0$$

Use time scales $t_0 = t, t_1 = \epsilon t$.

Exercise 2.79: Synchronization of oscillators

Huygens was the first to observe that two oscillators (mechanical clocks in his case) whose natural frequencies ω_1 and ω_2 are close but not identical can be synchronized ("phase locked") if they are coupled to one another. Such synchronization has since been observed in a diverse range of applications, including coupled chemical reactors. A simple model for a pair of coupled oscillators is

$$\begin{aligned}\dot{\theta}_1 &= \omega_1 + K_1\sin(\theta_2 - \theta_1)\\ \dot{\theta}_2 &= \omega_2 + K_2\sin(\theta_1 - \theta_2)\end{aligned}$$

where θ_1 and θ_2 are the phase variables for the two oscillators. Thus these equations describe trajectories on a torus. Synchronization occurs when the phase difference $\phi = \theta_2 - \theta_1$ attains a stable steady-state value. Analyze the dynamics of ϕ to determine the range of parameters in which the oscillators are synchronized. Draw the bifurcation diagram. Draw what happens on the torus as the system passes from the synchronized to the unsynchronized state.

Bibliography

M. J. Ablowitz and A. S. Fokas. *Complex Variables: Introduction and Applications.* Cambridge University Press, Cambridge, 2003.

M. Abramowitz and I. A. Stegun. *Handbook of Mathematical Functions.* National Bureau of Standards, Washington, D.C., 1970.

R. G. Bartle and D. R. Sherbert. *Introduction to Real Analysis.* John Wiley & Sons, Inc., New York, third edition, 2000.

C. M. Bender and S. A. Orszag. *Advanced Mathematical Methods for Scientists and Engineers. I. Asymptotic Methods and Perturbation Theory.* Springer-Verlag, New York, 1999.

G. E. Briggs and J. B. S. Haldane. A note on the kinetics of enzyme action. *Biochem. J.*, 19:338–339, 1925.

C. Canuto, M. Y. Hussaini, A. Quarteroni, and T. A. Zang. *Spectral Methods in Fluid Dynamics.* Springer-Verlag, Berlin, 1988.

C. Canuto, M. Y. Hussaini, A. Quarteroni, and T. A. Zang. *Spectral Methods: Fundamentals in Single Domains.* Springer-Verlag, Berlin, 2006.

W. J. Cody. Rational Chebyshev approximations for the error function. *Math. Comp.*, 23(107):631–637, 1969.

P. A. M. Dirac. *Principles of quantum mechanics.* Oxford, Clarendon Press, fourth edition, 1958.

M. V. Dyke. *Perturbation Methods in Fluid Mechanics.* Parabolic Press, Stanford, CA, annotated edition, 1975.

C. Gasquet and P. Witomski. *Fourier Analysis and Applications.* Springer-Verlag, New York, 1999.

H. Goldstein. *Classical Mechanics.* Addison–Wesley, Reading, Massachusetts, second edition, 1980.

D. Gottlieb and S. A. Orszag. *Numerical Analysis of Spectral Methods: Theory and Applications.* SIAM, Philadelphia, 1977.

M. D. Greenberg. *Foundations of Applied Mathematics.* Prentice-Hall, New Jersey, 1978.

M. Grmela and H.-C. Öttinger. Dynamics and thermodynamics of complex fluids. I. Development of a general formalism. *Phys. Rev. E*, 56(6):6620–6632, 1997.

J. Guckenheimer and P. Holmes. *Nonlinear Oscillations, Dynamical Systems, and Bifurcations of Vector Fields.* Springer Verlag, New York, New York, 1983.

M. V. Henri. Théorie générale de l'action de quelques diastases. *Comptes Rendu*, 135:916–919, 1901.

E. J. Hinch. *Perturbation Methods.* Cambridge University Press, Cambridge, 1991.

M. W. Hirsch and S. Smale. *Differential Equations, Dynamical Systems and Linear Algebra.* Academic Press, New York, 1974.

T. J. R. Hughes. *The Finite Element Method.* Dover, Mineola, New York, 2000.

E. L. Ince. *Ordinary Differential Equations.* Dover Publications Inc., New York, 1956.

G. Iooss and D. D. Joseph. *Elementary Stability and Bifurcation Theory.* Springer-Verlag, Berlin, second edition, 1990.

S. R. Keller and R. Skalak. Motion of a tank-treading ellipsoidal particle in a shear-flow. *J. Fluid Mech.*, 120:27–47, 1982.

J. Kevorkian and J. D. Cole. *Multiple Scale and Singular Perturbation Methods.* Springer-Verlag, New York, 1996.

H. K. Khalil. *Nonlinear Systems.* Prentice-Hall, Upper Saddle River, NJ, third edition, 2002.

O. Mangasarian. *Nonlinear Programming.* SIAM, Philadelphia, PA, 1994.

L. Michaelis and M. L. Menten. Die Kinetik der Invertinwirkung. *Biochem. Z.*, 49:333–369, 1913.

A. H. Nayfeh and D. T. Mook. *Nonlinear Oscillations.* John Wiley & Sons, New York, 1979.

A. W. Naylor and G. R. Sell. *Linear Operator Theory in Engineering and Science.* Springer-Verlag, New York, 1982.

D. L. Nelson and M. M. Cox. *Lehninger Principles of Biochemistry.* Worth Publishers, New York, third edition, 2000.

E. Polak. *Optimization: Algorithms and Consistent Approximations.* Springer Verlag, New York, 1997.

W. H. Press, S. A. Teukolsky, W. T. Vetterling, and B. T. Flannery. *Numerical Recipes in C: The Art of Scientific Computing.* Cambridge University Press, Cambridge, 1992.

J. B. Rawlings and J. G. Ekerdt. *Chemical Reactor Analysis and Design Fundamentals.* Nob Hill Publishing, Madison, WI, second edition, 2012.

J. B. Rawlings and D. Q. Mayne. *Model Predictive Control: Theory and Design.* Nob Hill Publishing, Madison, WI, 2009.

R. T. Rockafellar and R. J.-B. Wets. *Variational Analysis.* Springer-Verlag, 1998.

E. D. Sontag. *Mathematical Control Theory.* Springer-Verlag, New York, second edition, 1998.

I. Stakgold. *Green's Functions and Boundary Value Problems.* John Wiley & Sons, New York, second edition, 1998.

G. Strang. *Introduction to Applied Mathematics.* Wellesley-Cambridge Press, Wellesley, MA, 1986.

G. Strang and G. J. Fix. *An Analysis of the Finite Element Method.* Wellesley-Cambridge Press, Cambridge, MA, 2008.

S. H. Strogatz. *Nonlinear Dynamics and Chaos: With Applications to Physics, Biology, Chemistry and Engineering.* Westview Press, Cambridge, MA, 1994.

3

Vector Calculus and Partial Differential Equations

3.1 Vector and Tensor Algebra

3.1.1 Introduction

Many of the partial differential equations (PDEs) that we encounter as chemical and biological engineers arise from field equations such as the Navier-Stokes equations of fluid dynamics or the Schrödinger equation of quantum mechanics. These equations govern quantities (velocity, wave function) that vary with position in three-dimensional physical space. In general, such a quantity is known as a FIELD. Therefore, this chapter begins with a discussion of the properties of vectors and related objects (tensors) in physical space. In general, a TENSOR is an object that has an intrinsic geometric definition, independent of coordinate system. It may be a velocity vector, a dot product between two vectors (a scalar) or, as we shall see, even a linear operator.

3.1.2 Vectors in Three Physical Dimensions

In this chapter, we consider only vectors in three-dimensional physical space and following convention in the physics and engineering literature, represent these vectors using bold type. We begin with a brief review of vectors, tensors and their algebra. For now, let us consider only a Cartesian basis for the space, with position independent, orthonormal basis vectors $\boldsymbol{e}_1, \boldsymbol{e}_2, \boldsymbol{e}_3$. Any vector $\boldsymbol{u}$ can be represented as $\boldsymbol{u} = \sum_{i=1}^{3} u_i \boldsymbol{e}_i$, or, using the summation convention, $u_i \boldsymbol{e}_i$. In CARTESIAN TENSOR notation, we streamline the notation even further, denoting the vector as u_i. The unsummed index i on u_i indicates that $\boldsymbol{u}$ is a vector. The length of a vector is $\|\boldsymbol{u}\| = \sqrt{\sum_{i=1}^{3} u_i^2} = \sqrt{u_i u_i}$. The degree of alignment between two vectors is determined by the dot product

$$\boldsymbol{u} \cdot \boldsymbol{v} = \boldsymbol{u}^T \boldsymbol{v} = \sum_{i=1}^{3} \sum_{j=1}^{3} u_i v_j (\boldsymbol{e}_i \cdot \boldsymbol{e}_j) = u_i v_i = \|\boldsymbol{u}\| \, \|\boldsymbol{v}\| \cos\theta$$

Using some elementary geometry, it can be shown that

$$\boldsymbol{u} \cdot \boldsymbol{v} = \frac{1}{2} \left(\|\boldsymbol{u}\|^2 + \|\boldsymbol{v}\|^2 - \|\boldsymbol{v} - \boldsymbol{u}\|^2 \right)$$

This result shows that $\boldsymbol{u} \cdot \boldsymbol{v}$ can be expressed without referring to a coordinate system, but only to the lengths of vectors. Therefore, the dot product is independent of coordinate system; it is a GEOMETRIC INVARIANT. Recall that the inner product of Chapter 1 is the generalization of the dot product.

In Chapter 1 we also introduced the outer product between two vectors, also called the DIRECT PRODUCT or DYADIC PRODUCT. The outer product between vectors $\boldsymbol{u}$ and $\boldsymbol{v}$ is the DYAD[1] $\boldsymbol{uv}$. A dyad is a SECOND-ORDER TENSOR: a quantity that incorporates information regarding two directions. (A vector, which has one magnitude and one direction, is a first-order tensor). A dyad can act as a linear operator

$$(\boldsymbol{uv}) \cdot \boldsymbol{w} = \boldsymbol{u}(\boldsymbol{v} \cdot \boldsymbol{w})$$

Similarly,

$$\boldsymbol{w} \cdot (\boldsymbol{uv}) = (\boldsymbol{w} \cdot \boldsymbol{u})\boldsymbol{v}$$

Note that $\boldsymbol{uv} \neq \boldsymbol{vu}$. Based on this definition, we can write $\boldsymbol{uv}$ out, including basis vectors

$$\boldsymbol{uv} = \sum_{i=1}^{3} \sum_{j=1}^{3} u_i v_j \boldsymbol{e}_i \boldsymbol{e}_j$$

In Cartesian tensor notation, $\boldsymbol{uv}$ is denoted as $u_i v_j$ (the presence of the basis vectors $\boldsymbol{e}_i$ and $\boldsymbol{e}_j$ is implied by the presence of the two subscripts). When a dyad operates on something, the rightmost index (and basis vector) is involved. An example of a useful dyad is the projection operator $\hat{\boldsymbol{u}}\hat{\boldsymbol{u}}$, where $\hat{\boldsymbol{u}}$ is a unit vector. The product $(\hat{\boldsymbol{u}}\hat{\boldsymbol{u}}) \cdot \boldsymbol{v}$ is the component of the vector $\boldsymbol{v}$ in the $\hat{\boldsymbol{u}}$ direction. You can check this by applying the definition of the outer product.

[1]As noted in Chapter 1, sometimes the dyad $\boldsymbol{uv}$ is denoted by $\boldsymbol{uv}^T$ or $\boldsymbol{u} \otimes \boldsymbol{v}$.

A general second-order tensor $\boldsymbol{\tau}$ can be written as a linear combination of the basis dyads $\boldsymbol{e}_i\boldsymbol{e}_j$

$$\boldsymbol{\tau} = \sum_{i=1}^{3}\sum_{j=1}^{3} \tau_{ij}\boldsymbol{e}_i\boldsymbol{e}_j$$

In Cartesian tensor notation, the summations and base vectors are implied and we can denote the tensor by its component matrix τ_{ij}. The dot product $\boldsymbol{u} = \boldsymbol{\tau}\cdot\boldsymbol{v}$ between a second-order tensor and a vector is another vector: $u_i = \tau_{ij}v_j$. Similarly, $\boldsymbol{u} = \boldsymbol{v}\cdot\boldsymbol{\tau}$ is, in Cartesian coordinates: $u_i = v_j\tau_{ji}$. The second-order identity tensor is denoted $\boldsymbol{\delta}$ and satisfies the property $\boldsymbol{\delta}\cdot\boldsymbol{a} = \boldsymbol{a}\cdot\boldsymbol{\delta} = \boldsymbol{a}$ for all $\boldsymbol{a}$. In Cartesian coordinates, the ij component of $\boldsymbol{\delta}$ is simply the Kronecker delta δ_{ij}, or equivalently $\boldsymbol{\delta} = \boldsymbol{e}_1\boldsymbol{e}_1 + \boldsymbol{e}_2\boldsymbol{e}_2 + \boldsymbol{e}_3\boldsymbol{e}_3$.

Also important is the cross product, $\boldsymbol{u}\times\boldsymbol{v}$. Recall that, while the dot product is a scalar, the cross product is a vector, with magnitude $\|\boldsymbol{u}\|\,\|\boldsymbol{v}\|\sin\theta$ and direction orthogonal to both $\boldsymbol{u}$ and $\boldsymbol{v}$ and determined by the "right-hand rule." The cross product is not commutative: $\boldsymbol{u}\times\boldsymbol{v} = -\boldsymbol{v}\times\boldsymbol{u}$. Because of the invocation of the right-hand rule in its definition, the cross product is strictly speaking a PSEUDOVECTOR, because its definition is affected by the handedness of the coordinate system in which it is computed.

It is useful to view the cross product as a matrix-vector multiplication. Using the Cartesian components

$$\boldsymbol{u}\times\boldsymbol{v} = \begin{pmatrix} 0 & -u_3 & u_2 \\ u_3 & 0 & -u_1 \\ -u_2 & u_1 & 0 \end{pmatrix}\begin{pmatrix} v_1 \\ v_2 \\ v_3 \end{pmatrix}$$

We can write the cross product more compactly if we introduce the following operator, called the LEVI-CIVITA SYMBOL

$$\epsilon_{ijk} = \begin{cases} 1, & ijk = 123, 231 \text{ or } 312 \\ -1, & ijk = 132, 321 \text{ or } 213 \\ 0, & i = j, i = k \text{ or } j = k \end{cases}$$

This is the Cartesian coordinate representation of the ALTERNATING UNIT TENSOR or PERMUTATION TENSOR $\boldsymbol{\epsilon}$. As with the cross-product itself, ϵ_{ijk} is not actually a tensor, but rather a pseudotensor, because its definition is based on the use of right-handed Cartesian coordinates. Now the operator $(\boldsymbol{u}\times)$ can be written $\epsilon_{ijk}u_j$. This quantity has two free indices, so it is a second-order pseudotensor. Finally, we can write the cross product as

$$(\boldsymbol{u} \times \boldsymbol{v})_i = \epsilon_{ijk} u_j v_k$$

A useful identity involving ϵ_{ijk} is

$$\epsilon_{ijk}\epsilon_{klm} = \delta_{il}\delta_{jm} - \delta_{im}\delta_{jl}$$

which arises in the computation of double cross products such as $\boldsymbol{a} \times (\boldsymbol{b} \times \boldsymbol{c})$. Since the Kronecker delta is not handedness dependent, the double cross product between three vectors is a true vector.

3.2 Vector Calculus: Differential Operators and Integral Theorems

3.2.1 Divergence, Gradient, and Curl

Consider a vector that is a function of position, $\boldsymbol{v}(\boldsymbol{x})$, a vector FIELD. Physically, this vector field could be a fluid velocity (mass flux) or an electric current (charge flux), for example. An important physical consideration is the total flow into or out of a closed region. We denote this region as V, its boundary surface as S and the outward unit normal vector to S as $\boldsymbol{n}$, as illustrated in Figure 3.1. The volume of V is $\text{Vol}(V) = \int_V dV$. If $\boldsymbol{v}$ is a flux of some quantity, then $\boldsymbol{n} \cdot \boldsymbol{v}\, dS$ is the amount of that quantity crossing the boundary element dS per unit time and thus

$$\frac{1}{\text{Vol}} \int_S \boldsymbol{n} \cdot \boldsymbol{v}\, dS$$

is the amount of that quantity leaving V, per unit volume. Now let the region be centered at a position $\boldsymbol{x}_0$ and let V shrink to zero around that point. The DIVERGENCE of $\boldsymbol{v}$ at point $\boldsymbol{x}_0$ is defined by

$$\text{div}\, \boldsymbol{v} = \lim_{\text{Vol}\to 0} \frac{1}{\text{Vol}} \int_S \boldsymbol{n} \cdot \boldsymbol{v}\, dS \tag{3.1}$$

Thus the divergence of $\boldsymbol{v}$ measures the amount per unit volume that leaves the point $\boldsymbol{x}_0$. This definition is independent of coordinates, so the divergence is a tensor.

For a scalar field $\phi(\boldsymbol{x})$ there is an analogous quantity, the GRADIENT of ϕ, defined by

$$\text{grad}\, \phi = \lim_{\text{Vol}\to 0} \frac{1}{\text{Vol}} \int_S \boldsymbol{n}\, \phi\, dS \tag{3.2}$$

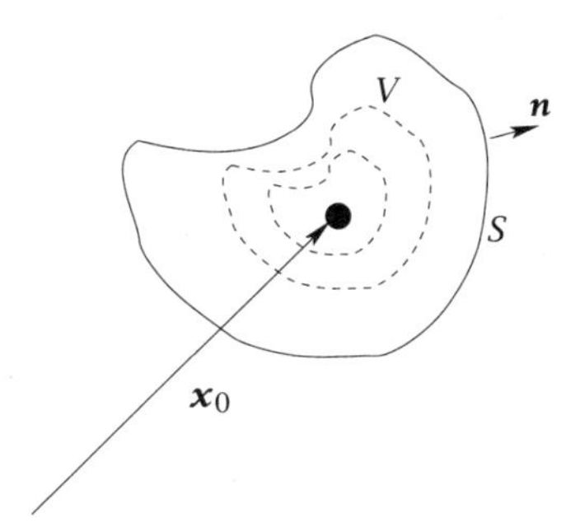

Figure 3.1: Volume V shrinking to zero size around a point $\boldsymbol{x}_0$.

Given a unit vector $\boldsymbol{s}$, the quantity $\boldsymbol{s} \cdot \text{grad}\,\phi$ is the derivative of $\boldsymbol{v}$ along the $\boldsymbol{s}$ direction, i.e., the DIRECTIONAL DERIVATIVE. The gradient of ϕ is a vector whose direction shows the direction of the maximum change in ϕ and whose magnitude is the magnitude of that change.

The final important operation, the CURL, measures the rotation of a vector field $\boldsymbol{v}$ at a point. It is defined by

$$\text{curl}\,\boldsymbol{v} = \lim_{\text{Vol}\to 0} \frac{1}{\text{Vol}} \int_S \boldsymbol{n} \times \boldsymbol{v}\, dS \tag{3.3}$$

Because of the cross product involved in its definition, the curl is a pseudovector.

The above definitions of div, grad, and curl are independent of coordinate system and illustrate the concepts underlying them, but to actually work with these operators we need coordinate systems. All three of the above operations can be expressed in terms of the GRADIENT operator, $\boldsymbol{\nabla}$, also called "nabla" or "del." It is also sometimes denoted

$$\frac{\partial}{\partial \boldsymbol{x}}$$

In Cartesian coordinates, it is given by

$$\boldsymbol{\nabla} = \sum_{i=1}^{3} \boldsymbol{e}_i \frac{\partial}{\partial x_i}$$

or in Cartesian tensor notation

$$\boldsymbol{\nabla} = \boldsymbol{e}_i \frac{\partial}{\partial x_i}$$

or simply $\frac{\partial}{\partial x_i}$. The presence of the basis vector $\boldsymbol{e}_i$ is implied by the unrepeated index i. The divergence, gradient, and curl operators are then given by

$$\text{div}\,\boldsymbol{v} = \boldsymbol{\nabla} \cdot v = \frac{\partial v_i}{\partial x_i}$$

$$\text{grad}\,\phi = \boldsymbol{\nabla}\phi = \frac{\partial \phi}{\partial x_i}$$

$$\text{curl}\,\boldsymbol{v} = \boldsymbol{\nabla} \times \boldsymbol{v} = \epsilon_{ijk}\frac{\partial v_k}{\partial x_j}$$

Another extremely important operator is the LAPLACIAN operator div grad, given by

$$\text{div grad} = \boldsymbol{\nabla} \cdot \boldsymbol{\nabla} = \frac{\partial^2}{\partial x_i \partial x_i}$$

The most common notation for the Laplacian operator is ∇^2. Unfortunately, this notation is somewhat misleading, implying that the operator is grad grad rather than div grad. Some literature uses the symbol Δ for the operator. We follow engineering convention and use ∇^2.

3.2.2 The Gradient Operator in Non-Cartesian Coordinates

In many applications, Cartesian coordinates are not the most practical for solving a problem.[2] We are familiar with cylindrical and spherical coordinate systems, but there are many others, including bipolar and parabolic systems. We consider here only orthogonal coordinate systems; the basis vectors may change from point to point, but at each point they are mutually orthogonal. We denote an arbitrary set of orthogonal coordinates by u_1, u_2, u_3 and the (orthonormal) base vectors by $\boldsymbol{e}_{u_1}, \boldsymbol{e}_{u_2}, \boldsymbol{e}_{u_3}$. The most important distinction between Cartesian and other coordinate systems is the actual distance traversed in moving from one coordinate line to another. For example, in Cartesian coordinates $(x_1, x_2, x_3) = (x, y, z)$, the distance between the coordinate lines $y = 1$ and $y = 2$, keeping x and z fixed, is always 1. But in cylindrical coordinates, $(u_1, u_2, u_3) = (r, \theta, z)$, the distance traveled going from $\theta = 1$ to $\theta = 2$ (at constant r and z) depends on r! This dependence is quantified in the SCALE FACTORS for a coordinate system, defined by

$$h_i = \sqrt{\left(\frac{\partial x_1}{\partial u_i}\right)^2 + \left(\frac{\partial x_2}{\partial u_i}\right)^2 + \left(\frac{\partial x_3}{\partial u_i}\right)^2}$$

[2] Appendix A of Bird, Stewart, and Lightfoot (2002) contains a great deal of useful information about this topic. Tensor analysis is not restricted to orthogonal coordinate systems; if you want to learn about tensor analysis in general coordinates, some good references are Aris (1962); Block (1978); Simmonds (1994); Bird, Armstrong, and Hassager (1987).

This quantity determines the distance traversed in moving along the u_i coordinate curve. For example, in cylindrical coordinates, it is easy to compute that $h_1 = 1, h_2 = r, h_3 = 1$. The distance covered in moving from θ to $\theta + d\theta$ is $h_2 d\theta = r d\theta$. If we let $\boldsymbol{g}_i = \frac{1}{h_i}\boldsymbol{e}_{u_i}$ (scale the basis vector by the scale factor), then we can write the basis vectors in terms of the Cartesian basis

$$\boldsymbol{g}_i = \frac{1}{h_i^2}\sum_{j=1}^{3}\frac{\partial x_j}{\partial u_i}\boldsymbol{e}_j$$

Note that despite the notation, the number h_i is *not* a component of a vector but rather is a property of the particular coordinate system under consideration.

For any orthogonal coordinate system, we can now write the gradient operator as

$$\nabla = \boldsymbol{g}_i \frac{\partial}{\partial u_i}$$

(summation implied). In general, the $\boldsymbol{g}_i$ depend on position. The importance of this fact becomes clear when we consider operators like the Laplacian

$$\begin{aligned}\nabla \cdot \nabla &= \boldsymbol{g}_i \frac{\partial}{\partial u_i} \cdot \boldsymbol{g}_j \frac{\partial}{\partial u_j} \\ &= \delta_{ij}\frac{1}{h_i h_j}\frac{\partial^2}{\partial u_i \partial u_j} + \left(\boldsymbol{g}_i \cdot \frac{\partial \boldsymbol{g}_j}{\partial u_i}\right)\frac{\partial}{\partial u_j}\end{aligned}$$

The second term in this expression does not appear in Cartesian coordinates, where the base vectors are independent of position. In terms of the scale factors, the derivative of a basis vector with respect to position can be written as follows

$$\frac{\partial \boldsymbol{g}_j}{\partial u_k} = -\frac{1}{h_j}\boldsymbol{g}_j \frac{\partial h_j}{\partial u_k} + \frac{1}{h_j}\left(\boldsymbol{g}_k \frac{h_k}{h_j}\frac{\partial h_k}{\partial u_j} - \delta_{jk}\sum_{i=1}^{3}\boldsymbol{g}_i \frac{\partial h_j}{\partial u_i}\right)$$

Summation is not implied, as $\frac{\partial \boldsymbol{g}_j}{\partial u_k}$ is not a component of a tensor.

Example 3.1: Gradient (del) and Laplacian operators in polar (cylindrical) coordinates

(a) Without referring to Cartesian coordinates at all, derive a formula for the gradient (del) operator in polar coordinates shown in Figure 3.2 so that one obtains for the differential of a scalar function ϕ

$$d\phi = \nabla\phi \cdot d\boldsymbol{x}$$

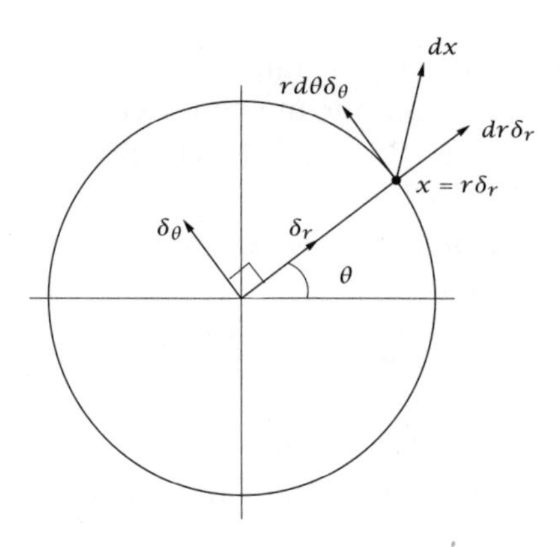

Figure 3.2: Polar coordinates (r, θ) and unit vectors $\boldsymbol{e}_r$ and $\boldsymbol{e}_\theta$.

in which $d\boldsymbol{x}$ is the differential of the position vector in polar coordinates.

(b) Using this formula for $\boldsymbol{\nabla}$, derive the formula for the Laplacian in polar coordinates.

(c) Finally check these two results by relating them to Cartesian coordinates using the h_i and $\boldsymbol{g}_i$ formulas given previously.

Solution

(a) As shown in Figure 3.2 we have for the differential of position

$$d\boldsymbol{x} = dr\boldsymbol{e}_r + rd\theta\boldsymbol{e}_\theta$$

From the definition of partial derivative, we have the formula for the total differential of an arbitrary function $\phi(r, \theta)$

$$d\phi = \frac{\partial\phi}{\partial r}dr + \frac{\partial\phi}{\partial\theta}d\theta$$

We substitute $\boldsymbol{\nabla}\phi = \boldsymbol{e}_r a_1 + \boldsymbol{e}_\theta a_2$ and solve for a_1, a_2, the two vector components of $\boldsymbol{\nabla}\phi$

$$d\phi = \boldsymbol{\nabla}\phi \cdot dx$$

$$\frac{\partial\phi}{\partial r}dr + \frac{\partial\phi}{\partial\theta}d\theta = (\boldsymbol{e}_r a_1 + \boldsymbol{e}_\theta a_2) \cdot (dr\boldsymbol{e}_r + rd\theta\boldsymbol{e}_\theta)$$

$$\frac{\partial\phi}{\partial r}dr + \frac{\partial\phi}{\partial\theta}d\theta = a_1 dr + a_2 rd\theta$$

Comparing the two sides, we have

$$a_1 = \frac{\partial \phi}{\partial r} \qquad a_2 = \frac{1}{r}\frac{\partial \phi}{\partial \theta}$$

which gives for ∇ in polar coordinates

$$\nabla \phi = \boldsymbol{e}_r \frac{\partial \phi}{\partial r} + \boldsymbol{e}_\theta \frac{1}{r}\frac{\partial \phi}{\partial \theta}$$

$$\nabla = \boldsymbol{e}_r \frac{\partial}{\partial r} + \boldsymbol{e}_\theta \frac{1}{r}\frac{\partial}{\partial \theta} \tag{3.4}$$

(b) Next we use the definition of the Laplacian to obtain

$$\begin{aligned}\nabla^2 &= \nabla \cdot \nabla \\ &= \left(\boldsymbol{e}_r \frac{\partial}{\partial r} + \boldsymbol{e}_\theta \frac{1}{r}\frac{\partial}{\partial \theta}\right) \cdot \left(\boldsymbol{e}_r \frac{\partial}{\partial r} + \boldsymbol{e}_\theta \frac{1}{r}\frac{\partial}{\partial \theta}\right)\end{aligned}$$

Taking the derivatives, and noting the dot product $\boldsymbol{e}_r \cdot \boldsymbol{e}_\theta = 0$ because the unit vectors are orthogonal, gives

$$\begin{aligned}\nabla^2 = \boldsymbol{e}_r \cdot \left(\frac{\partial \boldsymbol{e}_r}{\partial r}\frac{\partial}{\partial r} + \boldsymbol{e}_r \frac{\partial^2}{\partial r^2} + \frac{1}{r}\frac{\partial \boldsymbol{e}_\theta}{\partial r}\frac{\partial}{\partial \theta}\right) + \\ \frac{1}{r}\boldsymbol{e}_\theta \cdot \left(\frac{\partial \boldsymbol{e}_r}{\partial \theta}\frac{\partial}{\partial r} + \boldsymbol{e}_\theta \frac{1}{r}\frac{\partial^2}{\partial \theta^2} + \frac{1}{r}\frac{\partial \boldsymbol{e}_\theta}{\partial \theta}\frac{\partial}{\partial r}\right)\end{aligned}$$

Now we require the derivatives of the unit vectors with respect to (r, θ). As shown in Figure 3.2 these are given by (see also Exercise 3.2)

$$\frac{\partial \boldsymbol{e}_r}{\partial r} = 0 \qquad \frac{\partial \boldsymbol{e}_\theta}{\partial r} = 0 \qquad \frac{\partial \boldsymbol{e}_r}{\partial \theta} = \boldsymbol{e}_\theta \qquad \frac{\partial \boldsymbol{e}_\theta}{\partial \theta} = -\boldsymbol{e}_r \tag{3.5}$$

Substituting these derivatives into the previous result and collecting the nonzero terms gives

$$\nabla^2 = \frac{\partial^2}{\partial r^2} + \frac{1}{r}\frac{\partial}{\partial r} + \frac{1}{r^2}\frac{\partial^2}{\partial \theta^2}$$

Note that we can combine the first two terms for an equivalent form

$$\nabla^2 = \frac{1}{r}\frac{\partial}{\partial r}\left(r\frac{\partial}{\partial r}\right) + \frac{1}{r^2}\frac{\partial^2}{\partial \theta^2} \tag{3.6}$$

(c) The partial derivatives of the coordinates are

$$\frac{\partial x}{\partial r} = \cos\theta \qquad \frac{\partial y}{\partial r} = \sin\theta \qquad \frac{\partial x}{\partial \theta} = -r\sin\theta \qquad \frac{\partial y}{\partial \theta} = r\cos\theta$$

Substituting into the previously given formulas for h and g gives

$$h_1 = 1 \qquad h_2 = r$$

$$\boldsymbol{g}_1 = \cos\theta\,\boldsymbol{e}_x + \sin\theta\,\boldsymbol{e}_y \qquad \boldsymbol{g}_2 = \frac{1}{r^2}(-r\sin\theta\,\boldsymbol{e}_x + r\cos\theta\,\boldsymbol{e}_y)$$

$$\boldsymbol{g}_1 = \boldsymbol{e}_r \qquad \boldsymbol{g}_2 = \frac{1}{r}\boldsymbol{e}_\theta$$

We then have

$$\nabla = \boldsymbol{e}_r\frac{\partial}{\partial r} + \frac{1}{r}\boldsymbol{e}_\theta\frac{\partial}{\partial \theta}$$

which agrees with (3.4).

For the Laplacian, we require the derivatives of $\boldsymbol{g}_1, \boldsymbol{g}_2$

$$\frac{\partial \boldsymbol{g}_1}{\partial r} = 0 \qquad \frac{\partial \boldsymbol{g}_1}{\partial \theta} = \boldsymbol{e}_\theta \qquad \frac{\partial \boldsymbol{g}_2}{\partial r} = -\frac{1}{r^2}\boldsymbol{e}_\theta \qquad \frac{\partial \boldsymbol{g}_2}{\partial \theta} = -\frac{1}{r}\boldsymbol{e}_r$$

The formula for the Laplacian then gives

$$\nabla^2 = \frac{1}{h_1^2}\frac{\partial^2}{\partial r^2} + \frac{1}{h_2^2}\frac{\partial^2}{\partial \theta^2} + \boldsymbol{g}_1 \cdot \left(\frac{\partial \boldsymbol{g}_1}{\partial r}\frac{\partial}{\partial r} + \frac{\partial \boldsymbol{g}_2}{\partial r}\frac{\partial}{\partial \theta}\right) + \boldsymbol{g}_2 \cdot \left(\frac{\partial \boldsymbol{g}_1}{\partial \theta}\frac{\partial}{\partial r} + \frac{\partial \boldsymbol{g}_2}{\partial \theta}\frac{\partial}{\partial \theta}\right)$$

The $\boldsymbol{g}_1$ term vanishes upon substituting the various derivatives, and the $\boldsymbol{g}_2$ term produces the additional term $(1/r)\partial/\partial r$ giving

$$\begin{aligned}\nabla^2 &= \frac{\partial^2}{\partial r^2} + \frac{1}{r^2}\frac{\partial^2}{\partial \theta^2} + \frac{1}{r}\frac{\partial}{\partial r} \\ &= \frac{1}{r}\frac{\partial}{\partial r}\left(r\frac{\partial}{\partial r}\right) + \frac{1}{r^2}\frac{\partial^2}{\partial \theta^2}\end{aligned}$$

which agrees with (3.6). □

Table 3.1 collects expressions for the gradient and Laplacian operators in Cartesian, cylindrical, and spherical coordinate systems. The convention used for the angles θ and ϕ in spherical coordinates are shown in Figure 3.3.

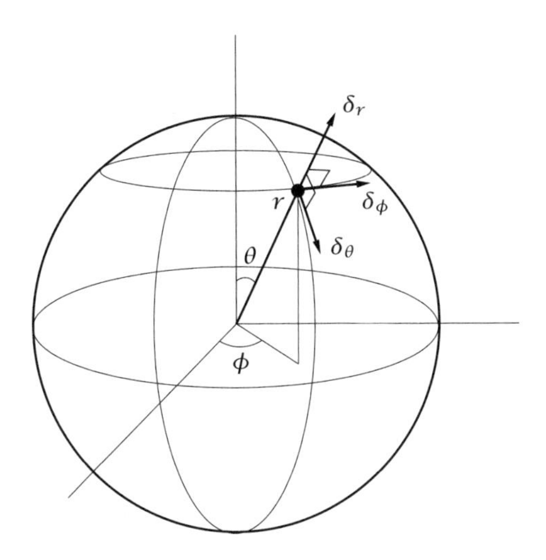

Figure 3.3: The orthonormal unit vectors in spherical coordinates.

Cartesian	$\boldsymbol{\nabla} = \boldsymbol{e}_x \frac{\partial}{\partial x} + \boldsymbol{e}_y \frac{\partial}{\partial y} + \boldsymbol{e}_z \frac{\partial}{\partial z}$ $\nabla^2 = \frac{\partial^2}{\partial x^2} + \frac{\partial^2}{\partial y^2} + \frac{\partial^2}{\partial z^2}$
Cylindrical	$\boldsymbol{\nabla} = \boldsymbol{e}_r \frac{\partial}{\partial r} + \boldsymbol{e}_\theta \frac{1}{r} \frac{\partial}{\partial \theta} + \boldsymbol{e}_z \frac{\partial}{\partial z}$ $\nabla^2 = \frac{1}{r} \frac{\partial}{\partial r}\left(r \frac{\partial}{\partial r}\right) + \frac{1}{r^2} \frac{\partial^2}{\partial \theta^2} + \frac{\partial^2}{\partial z^2}$
Spherical	$\boldsymbol{\nabla} = \boldsymbol{e}_r \frac{\partial}{\partial r} + \boldsymbol{e}_\theta \frac{1}{r} \frac{\partial}{\partial \theta} + \boldsymbol{e}_\phi \frac{1}{r \sin\theta} \frac{\partial}{\partial \phi}$ $\nabla^2 = \frac{1}{r^2} \frac{\partial}{\partial r}\left(r^2 \frac{\partial}{\partial r}\right) + \frac{1}{r^2 \sin\theta} \frac{\partial}{\partial \theta}\left(\sin\theta \frac{\partial}{\partial \theta}\right) + \frac{1}{r^2 \sin^2\theta} \frac{\partial^2}{\partial \phi^2}$

Table 3.1: Gradient and Laplacian operators in Cartesian, cylindrical, and spherical coordinates.

3.2.3 The Divergence Theorem

The divergence theorem concerns the integral of the divergence of a vector field $\boldsymbol{v}(\boldsymbol{x})$ in a region V. It is central to many aspects of the derivation and solution of partial differential equations. For example, many partial differential equations arise from conservation laws. These are often most easily stated in integral form, i.e., as conservation of some quantity over a finite volume of space. It is often useful, however, to have representations of the same laws that apply at each point in the volume and the divergence theorem plays a key role in development of these.

To illustrate the arguments underlying the divergence theorem without digressing too far, we will prove a limited version of it. Consider the two-dimensional "volume" V_A shown in Figure 3.4, whose "surface" consists of three pieces, S_1, S_2, and S_3 and whose outward unit normal is $\boldsymbol{n}_A$. In this domain

$$
\begin{aligned}
\int_{V_A} \nabla \cdot \boldsymbol{v}\, dV &= \int_{V_A} \left(\frac{\partial v_x}{\partial x} + \frac{\partial v_y}{\partial y} \right) dV \\
&= \int_0^{y_1} \int_0^{x_c(y)} \frac{\partial v_x}{\partial x}\, dx\, dy + \int_0^{x_1} \int_0^{y_c(x)} \frac{\partial v_y}{\partial y}\, dy\, dx \\
&= \int_0^{y_1} (v_x(x_c, y) - v_x(0, y))\, dy \\
&+ \int_0^{x_1} (v_y(x, y_c) - v_y(x, 0)\, dx \\
&= \int_0^{x_1} -v_y(x, 0)\, dx + \int_0^{y_1} -v_x(0, y))\, dy \\
&+ \int_0^{x_1} v_y(x, y_c)\, dx + \int_0^{y_1} v_x(x_c, y)\, dy
\end{aligned}
$$

Since on S_1, $\boldsymbol{n}_A = -\boldsymbol{e}_y$ and S_2, $\boldsymbol{n}_A = -\boldsymbol{e}_x$, the first two terms in the last expression can be simplified

$$
\begin{aligned}
\int_{V_A} \nabla \cdot \boldsymbol{v}\, dV &= \int_{S_1} \boldsymbol{n}_A \cdot \boldsymbol{v}\, dS + \int_{S_2} \boldsymbol{n}_A \cdot \boldsymbol{v}\, dS \\
&+ \int_0^{x_1} v_y(x, y_c)\, dx + \int_0^{y_1} v_x(x_c, y)\, dy
\end{aligned} \tag{3.7}
$$

To simplify the remaining two terms, observe that they both correspond to integrals along the surface (a curve in two dimensions) S_3. They can be combined into one by converting the second term into an integral over x, noting that $dy = \frac{dy_c}{dx} dx$ and changing the limits of

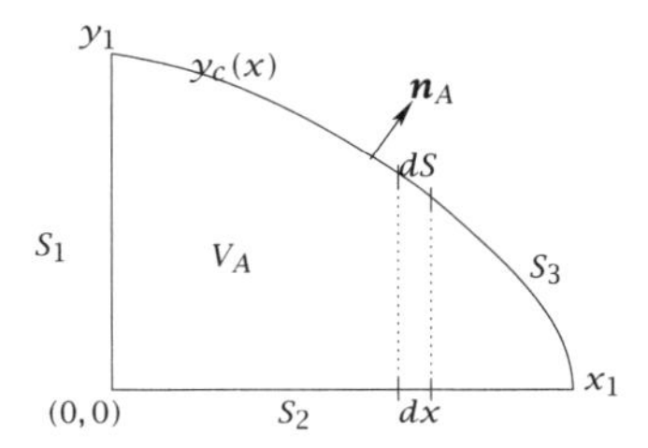

Figure 3.4: A two-dimensional volume for evaluation of the integral of the divergence. Differential elements dS and dx are also shown.

integration appropriately

$$\begin{aligned}
&\int_0^{x_1} v_y(x, y_c)\, dx + \int_0^{y_1} v_x(x_c, y)\, dy \\
&= \int_0^{x_1} v_y(x, y_c)\, dx + \int_{x|_{y=0}}^{x|_{y_c=y_1}} v_x(x_c, y)\frac{dy_c}{dx}\, dx \\
&= \int_0^{x_1} \left(v_y(x, y_c(x)) - v_x(x, y_c(x))\frac{dy_c}{dx}\right) dx \\
&= \int_0^{x_1} \left(-\frac{dy_c}{dx}\boldsymbol{e}_x + \boldsymbol{e}_y\right) \cdot \boldsymbol{v}\, dx
\end{aligned}$$

On S_3 the normal vector can be written

$$\boldsymbol{n}_A = \frac{1}{\sqrt{1 + \left(\frac{dy_c}{dx}\right)^2}} \left(-\frac{dy_c}{dx}\boldsymbol{e}_x + \boldsymbol{e}_y\right)$$

and $dS = \sqrt{1 + \left(\frac{dy_c}{dx}\right)^2}\, dx$ (see Figure 3.4), so this integral becomes

$$\int_{S_3} \boldsymbol{n}_A \cdot \boldsymbol{v}\, dS$$

Combining this result with (3.7), we find that

$$\int_{V_A} \nabla \cdot \boldsymbol{v}\, dV = \int_{S_1} \boldsymbol{n}_A \cdot \boldsymbol{v}\, dS + \int_{S_2} \boldsymbol{n}_A \cdot \boldsymbol{v}\, dS + \int_{S_3} \boldsymbol{n}_A \cdot \boldsymbol{v}\, dS \tag{3.8}$$

Finally, consider what happens if we extend the integral to the larger domain V that includes both V_A and a contiguous subdomain V_B, with

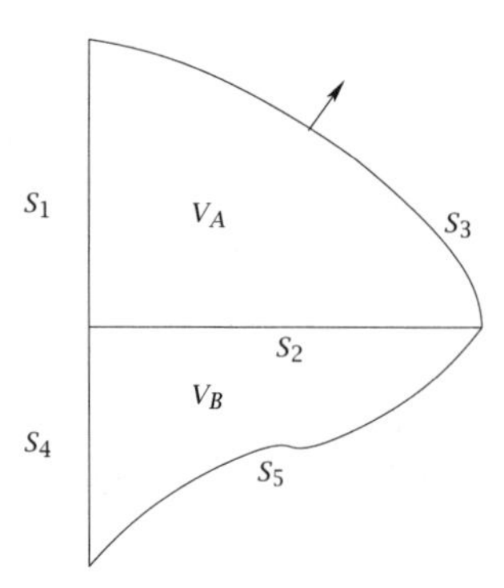

Figure 3.5: Two contiguous subdomains.

normal vector $\boldsymbol{n}_B$, as shown in Figure 3.5. By the same arguments given above,

$$\int_{V_B} \nabla \cdot \boldsymbol{v} \, dV = \int_{S_2} \boldsymbol{n}_B \cdot \boldsymbol{v} \, dS + \int_{S_4} \boldsymbol{n}_B \cdot \boldsymbol{v} \, dS + \int_{S_5} \boldsymbol{n}_B \cdot \boldsymbol{v} \, dS \tag{3.9}$$

The two domains V_A and V_2 share one side S_2; on this side $\boldsymbol{n}_A = -\boldsymbol{n}_B$. Adding (3.8) and (3.9) and recognizing that the integrals over S_2 cancel—anything leaving V_A via S_2 is entering V_B—we have that

$$\int_V \nabla \cdot \boldsymbol{v} \, dV = \int_S \boldsymbol{n} \cdot \boldsymbol{v} \, dS \tag{3.10}$$

where S is the boundary of the entire domain and $\boldsymbol{n}$ its outward unit normal. By piecing together domains like these and repeating the arguments used here, (3.10) can be seen to hold for any closed domain on the plane.

Equation (3.10) is the divergence theorem. By extending these elementary arguments it can be shown to be valid for arbitrary bounded domains in an arbitrary number of dimensions. In one dimension it reduces to the FUNDAMENTAL THEOREM OF INTEGRAL CALCULUS: $\int_a^b \frac{df}{dx} \, dx = f(b) - f(a)$. It is extremely important in a wide variety of contexts as it relates behavior in the interior of a domain to behavior on its boundary. Finally, one can see that the definition of the divergence operator mirrors this result for an infinitesimal domain.

In Cartesian tensor notation, the divergence theorem is

$$\int_V \frac{\partial v_i}{\partial x_i} \, dV = \int_S n_i v_i \, dS$$

By replacing the vector v_i by a scalar ϕ or by a second-order tensor τ_{ij} in this expression, the related results can be found (now expressed in boldface notation)

$$\int_V \nabla\phi \, dV = \int_S \boldsymbol{n}\phi \, dS \tag{3.11}$$

$$\int_V \nabla \cdot \boldsymbol{\tau} \, dV = \int_S \boldsymbol{n} \cdot \boldsymbol{\tau} \, dS \tag{3.12}$$

A closely related result is the multidimensional version of LEIBNIZ'S RULE, which we state without proof here. Consider the time derivative of an integral over a volume that is moving with time, e.g., a fluid element in a velocity field. If a point on the boundary is moving with a velocity $\boldsymbol{q}(\boldsymbol{x})$, then Leibniz's rule states that

$$\frac{d}{dt}\int_{V(t)} m(x,t) \, dV = \int_V \frac{\partial m}{\partial t} dV + \int_S m\boldsymbol{n} \cdot \boldsymbol{q} \, dS$$

The second term in this formula appears only if the volume is moving or changing shape with time and represents the net amount that is swept into V because of the motion of its boundaries.

Example 3.2: The divergence theorem and conservation laws

Conservation laws can be written for many quantities. Important examples include mass, energy, chemical species, and probability. Consider a quantity A that satisfies a conservation law in some arbitrary region of space V with boundary S and outward unit normal $\boldsymbol{n}$. The density (amount per unit volume) of A is ρ_A and the flux (amount per unit area per unit time) is $\boldsymbol{F}_A$. We allow for the possibility that A is created or destroyed within the volume, with rate R_A having units of amount of A per unit volume per unit time. If A is a chemical species, then R_A is a volumetric reaction rate of production of A. The conservation law for A can thus be written for the domain V as follows

$$\frac{d}{dt}\int_V \rho_A \, dV = -\int_S \boldsymbol{n} \cdot \boldsymbol{F}_A \, dS + \int_V R_A \, dV \tag{3.13}$$

The left-hand side is the rate of accumulation of A in the domain. The first term on the right-hand side is the net rate of entry of A into the domain across its boundary and the final term is the net rate of production of A via sources or sinks of A within the domain. Use the divergence theorem to write a conservation statement for A that is valid at every point in the domain.

Solution

The divergence theorem allows the lone surface integral to be recast as a volume integral

$$\int_S \boldsymbol{n} \cdot \boldsymbol{F}_A \, dS = \int_V \nabla \cdot \boldsymbol{F}_A \, dV$$

Furthermore, because V is time independent

$$\frac{d}{dt} \int_V \rho_A \, dV = \int_V \frac{\partial \rho_A}{\partial t} \, dV$$

Substituting these two equations into (3.13) yields

$$\int_V \frac{\partial \rho_A}{\partial t} \, dV = -\int_V \nabla \cdot \boldsymbol{F}_A \, dV + \int_V R_A \, dV$$

Since all terms in this equation are volume integrals, they can be combined

$$\int_V \left[\frac{\partial \rho_A}{\partial t} + \nabla \cdot \boldsymbol{F}_A \, dV - R_A \right] dV$$

Because the volume V is arbitrary, the only way that this equation can be satisfied in general is if its integrand vanishes at every point within V. That is

$$\frac{\partial \rho_A}{\partial t} = -\nabla \cdot \boldsymbol{F}_A + R_A \tag{3.14}$$

This is the general pointwise statement of the conservation law for A.

To be more specific, let A be a chemical species. Its molar density, or concentration, will be denoted c_A. Chemical species are transported by molecular diffusion and flow; if the species is dilute the flux of A can be written $F_A = c_A \boldsymbol{v} - D_A \nabla c_A$, where $\boldsymbol{v}$ is the velocity field for the fluid in which A is dissolved, and D_A is the diffusivity of the species. Now (3.14) becomes

$$\frac{\partial c_A}{\partial t} = -\nabla \cdot (c_A \boldsymbol{v}) + D_A \nabla^2 c_A + R_A \tag{3.15}$$

This is a partial differential equation for spatial and temporal distribution of a chemical species. If U and L are characteristic scales for the fluid velocity $\boldsymbol{v}$ and domain size, respectively, then the relative importance of convection and diffusion is estimated by the PECLET NUMBER $\mathrm{Pe} = UL/D_A$. □

3.2.4 Further Integral Relations and Adjoints of Multidimensional Differential Operators

GREEN'S IDENTITIES are special cases of the divergence theorem that are useful for working with integrals over quantities involving differential operators other than the divergence. Green's first identity is the divergence theorem for the case where $\boldsymbol{v}$ is replaced by $u\nabla v$, where u and v are now scalars

$$\int_V (\nabla u \cdot \nabla v + u\nabla^2 v)\, dV = \int_S u\nabla v \cdot \boldsymbol{n}\, dS \tag{3.16}$$

Green's second identity comes writing Green's first identity with u and v exchanged and subtracting this expression from Green's first identity as written above

$$\int_V (u\nabla^2 v - v\nabla^2 u)\, dV = \int_S (u\nabla v - v\nabla u) \cdot \boldsymbol{n}\, dS \tag{3.17}$$

Finally, GREEN'S FORMULA comes from replacing $\boldsymbol{v}$ in the original expression by $u\boldsymbol{v}$ where u is a scalar and $\boldsymbol{v}$ a vector

$$\int_V (\nabla u \cdot \boldsymbol{v} + u\nabla \cdot \boldsymbol{v})\, dV = \int_S u\boldsymbol{v} \cdot \boldsymbol{n}\, dS \tag{3.18}$$

In one dimension, Green's formula reduces to the expression for integration by parts.

The above theorems all deal with the divergence and its closest relatives, the gradient and the Laplacian. The final results are instead for the curl. In two dimensions, $\nabla \times \boldsymbol{v}$ reduces to $(\frac{\partial v_y}{\partial x} - \frac{\partial v_x}{\partial y})\boldsymbol{e}_3$. GREEN'S THEOREM shows how the integral of this over an area A can be reduced to an integral over the (closed) boundary curve C

$$\int_A \left(\frac{\partial v_y}{\partial x} - \frac{\partial v_x}{\partial y}\right) dA = \int_C (v_x\, dx + v_y\, dy)$$

The proof of this result closely follows what we did above with the divergence theorem. STOKES'S THEOREM is more general, applying to any bounded orientable ("two-sided") curved surface A floating in three dimensions, with boundary curve C

$$\int_A \boldsymbol{n} \cdot (\nabla \times \boldsymbol{v})\, dA = \int_C \boldsymbol{v} \cdot \boldsymbol{t}\, dC$$

Here $\boldsymbol{t}$ is the unit vector *tangent* to the boundary C, pointing in the direction in which the integration around C is being performed. The orientability condition precludes surfaces like a Möbius strip. The vector $\boldsymbol{n}$ is a unit normal vector to the surface A. Since the surface does not enclose a three-dimensional volume, however, inward and outward are undefined, the direction of $\boldsymbol{n}$ cannot be defined as it has been in earlier theorems. Here $\boldsymbol{n}$ is determined from the right-hand rule, based on the direction of the integration path for C. For example, if S were a region on a sheet of paper, then $\boldsymbol{n}$ would point upward out of the paper if the integration path around C is counterclockwise.

One important application of the above results is in the determination of the adjoints to div, grad, and curl. First, we define the relevant inner products. Let

$$(u, v) = \int_V uv dV$$

if u and v are (real) scalars, and

$$(\boldsymbol{u}, \boldsymbol{v}) = \int_V \boldsymbol{u} \cdot \boldsymbol{v} dV$$

if they are vectors. In our earlier discussion of adjoints, we used integration by parts to help us compute them; in multiple dimensions, Green's formula and identities are the appropriate replacements. For example, using Green's formula, (3.18), we can easily find that, with $u(S) = 0$ (Dirichlet boundary conditions)

$$(\boldsymbol{\nabla} u, \boldsymbol{v}) = -(u, \boldsymbol{\nabla} \cdot \boldsymbol{v})$$

Thus the adjoint of grad (with Dirichlet boundary conditions) is -div. Similarly, rearranging Green's second identity we find that

$$(\nabla^2 u, v) = -\int_S (u\boldsymbol{\nabla} v - v\boldsymbol{\nabla} u) \cdot \boldsymbol{n} dS + (u, \nabla^2 v)$$

If we impose the same boundary conditions on u and v, then $u\boldsymbol{\nabla} v = v\boldsymbol{\nabla} u$ on the boundary. Thus the boundary term vanishes, leaving

$$(\nabla^2 u, v) = (u, \nabla^2 v)$$

Therefore, the Laplacian operator is *always* self-adjoint. This fact has important implications for the solution of partial differential equations that involve the Laplacian.

3.3 Linear Partial Differential Equations: Properties and Solution Techniques

3.3.1 Classification and Canonical Forms for Second-Order Partial Differential Equations

Many general properties of partial differential equations can be introduced with this second-order equation in two dimensions

$$au_{xx} + 2bu_{xy} + cu_{yy} = f(x, y, u, u_x, u_y) \tag{3.19}$$

where $x, y \in \mathbb{R}$ and $u_x = \frac{\partial u}{\partial x}$ etc. For the moment x and y are not necessarily position variables—they are simply the independent variables for the problem. The coefficients a, b, and c are real and constant, though the latter restriction can be relaxed. Now consider the question of whether there exists a change of independent variables

$$\begin{aligned} \xi &= \xi_x x + \xi_y y \\ \eta &= \eta_x x + \eta_y y \end{aligned}$$

that can simplify the left-hand-side of this equation. Here $\xi_x, \xi_y, \eta_x, \eta_y$ are constants and $\xi_x\eta_y - \xi_y\eta_x$ must be nonzero for the coordinate transformation to be invertible. Applying the chain rule yields that

$$\begin{aligned} \left(a\xi_x^2 + 2b\xi_x\xi_y + c\xi_y^2\right) u_{\xi\xi} + \\ \left(a\xi_x\eta_x + b(\xi_x\eta_y + \xi_y\eta_x) + c\xi_y\eta_y\right) u_{\xi\eta} + \\ \left(a\eta_x^2 + 2b\eta_x\eta_y + c\eta_y^2\right) u_{\eta\eta} = g(\xi, \eta, u, u_\xi, u_\eta) \end{aligned} \tag{3.20}$$

If $b^2 - ac > 0$, then (3.19) is said to be HYPERBOLIC[3]. In this case, we can find real constants $\xi_x, \xi_y, \eta_x, \eta_y$ such that the coefficients multiplying $u_{\xi\xi}$ and $u_{\eta\eta}$ in (3.20) vanish, leaving the simpler differential equation

$$u_{\xi\eta} = g \tag{3.21}$$

This is the *canonical*, or simplest, form for a hyperbolic partial differential equation. Lines ξ = constant and η = constant are called CHARACTERISTICS for the equation. The WAVE EQUATION

$$u_{tt} - c^2 u_{xx} = 0 \tag{3.22}$$

[3]The nomenclature introduced in this section arises from an analogy with conic sections defined by the equation $ax^2 + 2bxy + cy^2 + dx + ey + f = 0$. If they exist, real solutions to this equation are hyperbolas, ellipses, or parabolas, depending on whether $b^2 - ac$ is positive, negative, or zero.

has this form, with $\xi = x - ct$, $\eta = x + ct$. We present the general solution to this equation in Section 3.3.6.

If $b^2 - ac < 0$, then (3.19) is ELLIPTIC. No real coefficients $\xi_x, \xi_y, \eta_x, \eta_y$ exist that will make the coefficients of $u_{\xi\xi}$ and $u_{\eta\eta}$ vanish. Instead, one finds complex conjugate characteristics $\xi = \xi_R + i\xi_I$, $\eta = \xi_R - i\xi_I$. All is not lost, however. Using $\xi' = \xi_R$ and $\eta' = \xi_I$ as new coordinates, the coefficient of $u_{\xi'\eta'}$ can be made to vanish, leading to the canonical form

$$u_{\xi'\xi'} + u_{\eta'\eta'} = g \tag{3.23}$$

The left-hand side of this equation is the two-dimensional Laplacian operator. At steady state, (3.15) above reduces to this form in two spatial dimensions. If g is only a function of x and y, this equation is called the POISSON EQUATION. If $g = 0$, it is called the LAPLACE EQUATION.

The borderline case $b^2 - ac = 0$ leads to the PARABOLIC equation

$$u_{\eta\eta} = g \tag{3.24}$$

The standard example of a parabolic equation is the transient species conservation equation, (3.15) in one spatial dimension, which we can write

$$u_t + v u_x = D u_{xx} + R_A$$

The Schrödinger equation is also parabolic. Elliptic and parabolic equations are treated extensively in the sections below.

The classification of partial differential equations into these categories plays an important role in the mathematical theory of existence of solutions for given boundary conditions. Fortunately, the physical settings commonly encountered by engineers generally lead to well-posed mathematical problems for which we do not need to worry about these more abstract issues. Therefore we now proceed to the presentation of classical solution approaches, many of which are insensitive to the type of equation encountered.

3.3.2 Separation of Variables and Eigenfunction Expansion with Equations involving ∇^2

The technique of SEPARATION OF VARIABLES is perhaps the most familiar classical technique for solving linear partial differential equations. It arises in problems in transport, electrostatics, quantum mechanics, and many other applications. The technique is based on the superposition property of linear problems (Section 2.2.1) as well as the following several assumptions:

1. a solution $u(x_1, x_2, x_3, \ldots)$ to a PDE with independent variables x_i can be written $u = X(x_1)Y(x_2)Z(x_3)\ldots$.
2. The boundaries of the domain are coordinate surfaces and the boundary conditions for the PDE can also be written in the above form.
3. A distinct ODE can be derived from the original PDE for each function $X, Y, Z, \ldots$.
4. Using superposition, a solution satisfying the boundary conditions can be constructed from an infinite series of solutions to these ODEs. This condition implies that separation of variables is primarily useful for equations involving self-adjoint partial differential operators such as the Laplacian, in which case eigenfunctions of various Sturm-Liouville problems provide bases for representing the solutions. Consider a problem with three independent variables and two of them, say x_2 and x_3, lead to Sturm-Liouville problems with eigenfunctions $Y_k(x_2)$ and $Z_l(x_3)$, $k = 0, 1, 2, \ldots, l = 0, 1, 2, \ldots$. The basis functions for the $x_2 - x_3$ direction are thus $Y_k(x_2)Z_l(x_3)$. The solutions to the problem in the inhomogeneous direction are then coefficients in the series and the total solution has this form

$$u(x_1, x_2, x_3) = \sum_{k=1}^{\infty}\sum_{l=1}^{\infty} X_{kl}(x_1)\left\{Y_k(x_2)Z_l(x_3)\right\}$$

We illustrate the method with several examples.

Example 3.3: Steady-state temperature distribution in a circular cylinder

Consider a circular cylinder with unit radius and an imposed temperature profile $u_s(\theta)$ on its surface. The steady-state temperature profile $u(r, \theta)$ is a solution to LAPLACE'S EQUATION

$$\nabla^2 u = 0 \tag{3.25}$$

in polar coordinates

$$\frac{1}{r}\frac{\partial}{\partial r}r\frac{\partial u}{\partial r} + \frac{1}{r^2}\frac{\partial^2 u}{\partial \theta^2} = 0$$

with u bounded at the origin and satisfying $u(1,\theta) = u_s(\theta)$. As described above, seek a solution $u(r,\theta) = R(r)\Theta(\theta)$.

Solution

Plugging into the equation and simplifying yields

$$\frac{r}{R}\left(rR'\right)' = -\frac{\Theta''}{\Theta}$$

where $R' = dR/dr$ and $\Theta' = d\Theta/d\theta$. Notice that the LHS of the equation is a function of r only and the RHS a function of θ. The only way for the two sides to be equal is for them both to equal a constant, c. This observation gives us a pair of ODEs

$$r\left(rR'\right)' - cR = 0 \tag{3.26}$$

$$\Theta'' + c\Theta = 0 \tag{3.27}$$

The constant c is as yet unspecified.

Equation (3.26) satisfies periodic boundary conditions $\Theta(\theta) = \Theta(\theta + 2\pi), \Theta'(\theta) = \Theta'(\theta + 2\pi)$; it is a Sturm-Liouville eigenvalue problem with eigenvalue c. This has solutions $\Theta_k(\theta) = e^{ik\theta}$ for all integers k with the corresponding eigenvalue $c = k^2$. So in fact we have found not a single solution, but a family of solutions; a basis for functions in the θ direction.

Now consider the equation for $R(r)$, setting $c = k^2$. A little manipulation puts the equation in this form

$$r^2R'' + rR' + k^2R = 0$$

This is a Cauchy-Euler equation, with k as a parameter and solutions $R_k = A_k r^k + B_k r^{-k}$. To satisfy the boundedness condition at $r = 0$, only the solution with a positive exponent must remain, so $R_k = a_k r^{|k|}$. Since every integer k gives a solution, we can use the superposition principle to write

$$u(r,\theta) = \sum_{k=-\infty}^{\infty} a_k r^{|k|} e^{ik\theta}$$

This is a Fourier series, using the Sturm-Liouville eigenfunctions $\Theta_k(\theta)$ as basis functions. The coefficients a_k come from the boundary condition. At $r = 1$,

$$u(1,\theta) = \sum_{k=-\infty}^{\infty} a_k e^{ik\theta} = u_s(\theta)$$

We can extract the coefficients a_k from this formula by using the orthogonality of the Sturm-Liouville basis functions: take inner products (in θ) of both sides with the basis function $e^{il\theta}$

$$\left(\sum_{k=-\infty}^{\infty} a_k e^{ik\theta}, e^{il\theta}\right) = \left(u_s(\theta), e^{il\theta}\right)$$

Letting $c_k = (u_s(\theta), e^{il\theta})/2\pi$, this process simply gives us that $a_k = c_k$. That is, the (known) Fourier coefficients of the boundary temperature determine the Fourier coefficients in the cylinder, so

$$u(r,\theta) = \sum_{k=-\infty}^{\infty} c_k r^{|k|} e^{ik\theta} \qquad \square$$

Example 3.4: Transient diffusion in a slab

The transient diffusion of heat or a chemical species in one direction is governed by the transient diffusion equation, also called the heat equation

$$\frac{\partial u}{\partial t} = D\frac{\partial^2 u}{\partial x^2} \tag{3.28}$$

Consider the initial and boundary conditions $u(x,0) = 0$, $u(0,t) = 0$, $u(\ell, t \geq 0) = u_\ell$, i.e., the initial concentration in the domain $0 < x < \ell$ is zero and at $t = 0$ the right end of the domain is exposed to a known concentration $u = u_\ell$. Seek a separation of variables solution $u(x,t) = X(x)T(t)$.

Solution

Using the form $u(x,t) = X(x)T(t)$, (3.28) becomes

$$XT' = DX''T$$

where again $'$ denotes the derivative of a function with respect to its independent variable. Rearranging yields

$$\frac{1}{D}\frac{T'}{T} = \frac{X''}{X}$$

Observing that this expression equates a function of t to a function of x we again conclude that each side of it must be constant

$$T' = cDT \tag{3.29}$$

$$X'' = cX \tag{3.30}$$

The second of these contains a Sturm-Liouville operator and must satisfy boundary conditions $X(0) = 0$, $X(\ell) = u_\ell$. These boundary conditions, however, are not homogeneous, so the problem as written is not a Sturm-Liouville problem.

A simple change of variable solves this problem. We let $u = u_s(x) + v(x,t)$ and choose u_s to satisfy the inhomogeneous boundary conditions at $x = 0$ and $x = \ell$, in which case v satisfies homogeneous boundary conditions $v(0,t) = v(\ell,t) = 0$. A particularly convenient choice is $u_s = u_\ell \frac{x}{\ell}$, which is the steady-state solution to this problem. Thus $v(x,t)$ is the deviation from the steady state. Substituting into (3.28) and observing that $\partial u_s/\partial t = \partial^2 u_s/\partial x^2 = 0$ yields

$$\frac{\partial v}{\partial t} = D\frac{\partial^2 v}{\partial x^2}$$

with $v(x,0) = -u_s$, $v(0,t) = 0$, $v(\ell, t \geq 0) = 0$. Now letting $v(x,t) = X(x)T(t)$ and repeating the above steps we find that the problem for X *is* a true Sturm-Liouville problem, including the homogeneous boundary conditions $X(0) = X(\ell) = 0$. The eigenvalues are $c = -k^2$, where now $k = n\pi/\ell$ for positive integer n and the eigenfunctions are $\sin\frac{n\pi x}{\ell}$. Equation (3.29) is an initial-value problem. Its solutions, parametrized by n, are

$$T_n(t) = T_n(0)e^{-\frac{n^2\pi^2}{\ell^2}Dt}$$

so the overall solution again has the Fourier series form

$$v(x,t) = \sum_{n=1}^{\infty} T_n(0)e^{-\frac{n^2\pi^2}{\ell^2}Dt}\sin\frac{n\pi x}{\ell} \tag{3.31}$$

The initial conditions $T_n(0)$ are determined from the initial condition $v(x,0) = -u_s$ by setting $t = 0$ in (3.31) and taking its inner product with basis function $\sin\frac{m\pi x}{\ell}$

$$\left(-u_\ell\frac{x}{\ell}, \sin\frac{m\pi x}{\ell}\right) = \left(\sum_{n=1}^{\infty} T_n(0)\sin\frac{n\pi x}{\ell}, \sin\frac{m\pi x}{\ell}\right)$$

Thus

$$T_m(0) = \frac{\int_0^\ell -u_\ell\frac{x}{\ell}\sin\frac{m\pi x}{\ell}\,dx}{\int_0^\ell \sin\frac{m\pi x}{\ell}\sin\frac{m\pi x}{\ell}\,dx} = (-1)^m\frac{2u_\ell}{m\pi}$$

The final exact solution is thus

$$u(x,t) = u_\ell\frac{x}{\ell} + \sum_{n=1}^{\infty}(-1)^n\frac{2u_\ell}{n\pi}e^{-\frac{n^2\pi^2}{\ell^2}Dt}\sin\frac{n\pi x}{\ell} \tag{3.32}$$

At short times $t \ll \ell^2/D$, this series converges very slowly because of the n^{-1} decay of the Fourier coefficients $T_n(0)$ of the initial condition. In this situation, alternate approaches that approximate the domain as semi-infinite are more appropriate because diffusion has only had time to spread the heat or solute over a short distance from the boundary. See Exercises 3.23 and 3.36. As t increases, the exponential decay term becomes smaller and the series converges more rapidly.

□

With these two examples, one can see a pattern emerging. Separation of variables leads to at least one direction that presents a Sturm-Liouville problem whose eigenfunctions are a useful basis for representing the solution. In the second example, a change of variable was required to find a direction with the homogeneous boundary conditions required of a Sturm-Liouville eigenvalue problem. The following example extends this idea.

Example 3.5: Steady-state diffusion in a square domain

Solve Laplace's equation $\nabla^2 u = 0$ in a unit square domain $0 < x < 1, 0 < y < 1$, with boundary conditions $u = 200$ on $x = 0$ and $y = 0$, $u = 300$ on $x = 1$ and $u = 500$ on $y = 1$, as shown in Figure 3.6(a).

Solution

As stated, there are no homogeneous directions. Now we split the solution into three pieces: $u(x,y) = U(x,y) + V(x,y) + W(x,y)$, where U, V, and W all satisfy Laplace's equation, but with conveniently chosen boundary conditions that sum to the boundary conditions for the original problem, as illustrated in Figure 3.6(b). The problem for U is trivial because all the boundaries have the same value of 200; thus $U = 200$. The problem for V has homogeneous boundary conditions at $y = 0$ and $x = 1$, while that for W has homogeneous boundary conditions at $x = 0$ and $x = 1$, aside from a multiplicative constant, it is just a $\pi/2$ rotation of the problem for V. The solution to the W problem (to within a multiplicative constant) is Exercise 3.32. From it the solution to the V problem can be found so the solution for $u = U + V + W$ is complete.

□

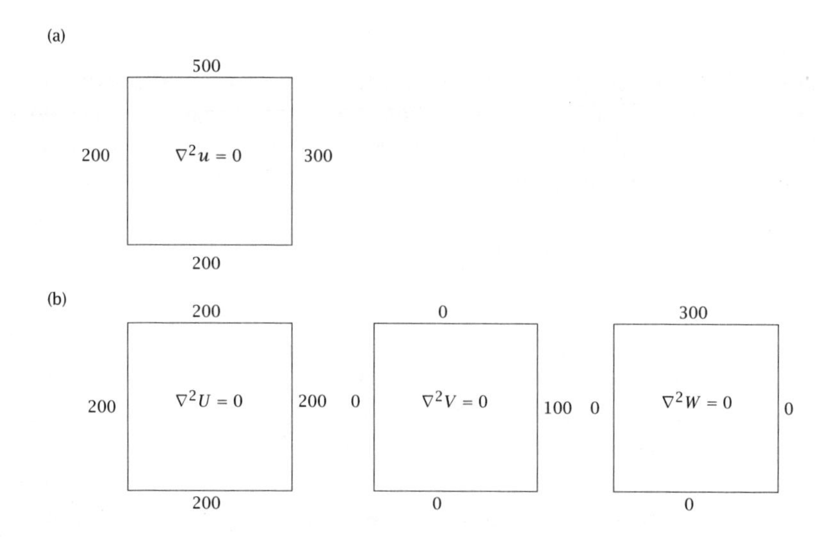

Figure 3.6: Laplace's equation in a square domain. (a) Original problem. (b) Three subproblems whose solutions sum to the solution of the original problem.

Example 3.6: Eigenfunction expansion for an inhomogeneous problem

Solve the Poisson equation

$$u_{xx} + u_{yy} = f(x, y)$$

in a unit square with Dirichlet boundary conditions, which models a steady-state distribution given a source $f(x, y)$ distributed within the domain.

Solution

Separation of variables does not work for this problem (try it), but a version of eigenfunction expansion does. Think of this problem as a linear algebra problem $Lu = f$. Here L is self-adjoint, so the solutions to the eigenvalue problem $Lw + \lambda w = 0$ form an orthogonal basis and allow us to diagonalize L. We can express u and f in this basis, and since L becomes diagonal we can easily solve for u.

To perform this procedure in the present case, we need to solve

$$w_{xx} + w_{yy} + \lambda w = 0$$

in the unit square with $w = 0$ on the boundary. We can solve this problem by separation of variables: it gives Sturm-Liouville problems in both x and y, and yields eigenfunctions $w_{mn}(x,y) = \sin m\pi x \sin n\pi y$ with (real) eigenvalues $\lambda_{mn} = \pi^2(m^2+n^2)$, for all integer pairs mn. Now to solve the Poisson equation, we express u and f in terms of the eigenfunctions

$$u(x,y) = \sum_{m=1}^{\infty}\sum_{m=1}^{\infty} u_{mn}w_{mn}(x,y)$$

$$f(x,y) = \sum_{m=1}^{\infty}\sum_{m=1}^{\infty} f_{mn}w_{mn}(x,y)$$

Since f is known, $f_{mn} = (f, w_{mn})/(w_{mn}, w_{mn})$, where the inner product in this case is just the integral over the square. Now since $w_{xx} + w_{yy} = \lambda w$, we can write $\lambda_{mn}u_{mn} = f_{mn}$, which we can solve immediately to give $u_{mn} = f_{mn}/\lambda_{mn}$, so

$$u(x,y) = \sum_{m=1}^{\infty}\sum_{m=1}^{\infty} \frac{f_{mn}}{-\lambda_{mn}} \sin m\pi x \sin n\pi y \qquad \square$$

In some situations, a separation of variable solution can be obtained via multiple approaches. For example, the Laplacian operator in polar coordinates can be written

$$\nabla^2 = L_r + \frac{1}{r^2}L_\theta + L_z$$

where

$$L_r = \frac{1}{r}\frac{\partial}{\partial r}\left(r\frac{\partial}{\partial r}\right) \qquad L_\theta = \frac{\partial^2}{\partial \theta^2} \qquad L_z = \frac{\partial^2}{\partial z^2}$$

Given appropriate homogeneous boundary conditions, all three of these are Sturm-Liouville operators, so depending on the boundary conditions, there may by the possibility of more than one method of solution. The following example illustrates this situation.

Example 3.7: Steady diffusion in a cylinder: eigenfunction expansion and multiple solution approaches

Consider Laplace's equation in a cylindrical domain with boundary conditions $u(r, z=0) = 1$, $u(r=1, z) = 0$, $u(r, z=1) = 0$. That is, the bottom is heated, and the top and side are cooled. Solve this equation in *two* different ways:

(a) Using basis functions that depend on r.

(b) Using basis functions that depend on z.

Solution

We could proceed by seeking a solution $v(r,z) = R(r)Z(z)$ as above. Instead we will directly impose a Fourier series form for the solution based on the eigenfunctions.

(a) For the current problem there is no θ-dependence and we first seek a solution that uses basis functions in the r-direction., i.e., eigenfunctions of L_r. This is a singular Sturm-Liouville operator ($p(r) = r$) so only boundedness is required at the origin, and the boundary condition at $r = 1$ is homogeneous. Referring back to Example 2.8, we recognize that the eigenfunctions of L_r are the Bessel functions of order zero so we can seek a solution

$$u(r,z) = \sum_{n=1}^{\infty} u_n(z) J_0(\sqrt{\lambda_n} r)$$

where $\sqrt{\lambda_n} = 2.4, 5.5, 8.7, 11.8, \ldots$. To simplify notation, let $k_n = \sqrt{\lambda_n}$. Substituting this solution form into Laplace's equation and using the fact that $L_r J_0(k_n r) = -k_n^2 J_0(k_n r)$ yields that

$$\frac{d^2 u_n}{dz^2} - k^2 u_n = 0$$

Because of the bounded domain, it is convenient to represent the solution to this problem as

$$u_n(z) = a_n \cosh k_n z + b_n \sinh k_n z$$

so

$$u(r,z) = \sum_{n=1}^{\infty} (a_n \cosh k_n z + b_n \sinh k_n z) J_0(k_n r) \tag{3.33}$$

At $z = 0$, $u = 1$. Taking the inner product, i.e., weighted integral from $r = 0$ to $r = 1$ of (3.33), evaluated at $z = 0$, with $J_0(k_m r)$ leads to

$$a_n = \frac{(1, J_0(k_m r))_w}{(J_0(k_m r), J_0(k_m r))_w} \tag{3.34}$$

Evaluation of these and related integrals is facilitated by the following general results for Bessel functions with integer n and arbitrary k, l

$$\frac{d}{dx}[x^n J_n(kx)] = x^n k J_{n-1}(kx) \tag{3.35}$$

$$\frac{d}{dx}[x^n Y_n(kx)] = x^n k Y_{n-1}(kx) \tag{3.36}$$

$$x\frac{d}{dx}[J_n(kx)] + nkJ_n(kx) = xkJ_{n-1}(kx) \tag{3.37}$$

$$x\frac{d}{dx}[Y_n(kx)] + nkY_n(kx) = xkY_{n-1}(kx) \tag{3.38}$$

$$x\frac{d}{dx}[J_n(kx)] - nkJ_n(kx) = -xkJ_{n+1}(kx) \tag{3.39}$$

$$x\frac{d}{dx}[Y_n(kx)] - nkY_n(kx) = xkY_{n+1}(kx) \tag{3.40}$$

$$J_{-n}(kx) = (-1)^n J_n(kx) \tag{3.41}$$

$$Y_{-n}(kx) = (-1)^n Y_n(kx) \tag{3.42}$$

$$\int_0^1 xJ_n(kx)J_n(lx)\,dx = \begin{cases} 0 & k \neq l \\ \frac{1}{2}J_{n+1}^2(k) & k = l, J_n(k) = 0, n > -1 \end{cases} \tag{3.43}$$

Using the first and last of these expressions, one can find that

$$(1, J_0(k_m r))_w = \int_0^1 J_0(k_m r) r\,dr = \frac{1}{k_m} J_1(k_m) \tag{3.44}$$

$$(J_0(k_m r), J_0(k_m r))_w = \int_0^1 J_0^2(k_m r) r\,dr = \frac{1}{2} J_1^2(k_m) \tag{3.45}$$

The boundary condition $u = 0$ at $z = 1$ requires that

$$b_n = -\frac{a_n \cosh k_n}{\sinh k_n}$$

Using these results, the solution is

$$u(r,z) = \sum_{n=1}^{\infty} a_n \left(\cosh k_n z - \frac{\cosh k_n}{\sinh k_n} \sinh k_n z \right) J_0(k_n r)$$

with a_n given by (3.34), (3.44), and (3.45).

(b) Alternately, we can seek a solution using eigenfunctions of L_z. The boundary conditions at $z = 0$ and $z = 1$ are not homogeneous but this is easily addressed by letting $u = (1-z)+v$. Now $\nabla^2 v = 0$ (because $\nabla^2(1-z) = 0$) with the homogeneous boundary conditions $v(r,0) = v(r,1) = 0$. Now, however, $v(1,z) = -(1-z)$. We could proceed by seeking a solution $v(r,z) = R(r)Z(z)$ as above. Instead we will directly impose a Fourier series form for the solution based on the eigenfunctions $\sin n\pi z$ of L_z

$$v(r,z) = \sum_{n=1}^{\infty} v_n(r) \sin n\pi z$$

Substituting this solution form into Laplace's equation leads to

$$\sum_{n=1}^{\infty} \left(\frac{1}{r}\frac{d}{dr} r \frac{dv_n}{dr} - n^2\pi^2 v_n(r) \right) \sin n\pi z = 0$$

Taking the inner product of this equation with $\sin m\pi z$, invoking orthogonality, and changing m to n yields

$$\frac{1}{r}\frac{d}{dr} r \frac{dv_n}{dr} - n^2\pi^2 v_n(r) = 0$$

This is called the MODIFIED BESSEL EQUATION OF ORDER ZERO. It differs from Bessel's equation by the sign in front of the second term. Its solution can be found by the method of Frobenius; the general solution is

$$v_n(r) = a_n I_0(n^2\pi^2 r) + b_n K_0(n^2\pi^2 r)$$

The functions I_0 and K_0 are the MODIFIED BESSEL FUNCTIONS of order zero; they are shown in Table 2.3. The function K_0 has a logarithmic singularity at the origin, so for boundedness we require that $b_n = 0$. The coefficients a_n are found by imposing the boundary condition at $r = 1$ and again taking the inner product with an eigenfunction

$$a_n = \frac{(-(1-z), \sin n\pi z)}{\frac{1}{2} I_0(n^2\pi^2)} = \frac{-2}{n\pi I_0(n^2\pi^2)}$$

The solution in final form is

$$v(r,z) = \sum_{n=1}^{\infty} a_n I_0(n^2\pi^2 r) \sin n\pi z \qquad \square$$

In spherical coordinates, the Laplacian operator can be written

$$\nabla^2 = \left(L_r u + \frac{1}{r^2} L_\theta u + \frac{1}{r^2 \sin^2 \theta} L_\phi u \right)$$

where

$$L_r = \frac{1}{r^2} \frac{\partial}{\partial r} r^2 \frac{\partial}{\partial r} \qquad L_\theta = \frac{1}{\sin \theta} \frac{\partial}{\partial \theta} \sin \theta \frac{\partial}{\partial \theta} \qquad L_\phi = \frac{\partial^2}{\partial \phi^2}$$

It often is useful to rewrite the first of these in this form

$$L_r f = \frac{1}{r} \frac{\partial^2}{\partial r^2} (r f) \tag{3.46}$$

Accordingly, the introduction of a new variable $g = rf$ often is useful in problems in spherical coordinates.

Example 3.8: Transient diffusion from a sphere

Consider the transient diffusion of a chemical species out of a sphere with radius R into uniform surroundings where the species concentration is zero. This problem satisfies

$$\frac{\partial u}{\partial t} = D L_r u$$

with $u(r, 0) = u_0, u(R, t > 0) = 0$.

Solution

Spherically symmetric problems like this can be solved using the eigenfunctions of

$$L_r \Phi + \lambda \Phi = 0$$

This is the SPHERICAL BESSEL'S EQUATION

$$\frac{d}{dx} x^2 \frac{dy}{dx} + \left(m^2 x^2 - n(n+1) \right) y = 0$$

in the specific case $\lambda = m^2$ and $n = 0$. Its solutions are the SPHERICAL BESSEL FUNCTIONS of order zero, which are simply

$$f(x) = a \frac{\sin mx}{x} + b \frac{\cos mx}{x}$$

These functions are orthogonal with respect to an inner product with weight function $w(r) = r^2$. This factor arises naturally in the differential volume element in spherical coordinates. The eigenvalues m^2, and coefficients a and b are determined as usual by the (homogeneous/boundedness) boundary conditions. For example, for diffusion in a sphere,boundedness at the origin will require that $b = 0$. □

Example 3.9: Temperature field around a sphere in a linear gradient

Consider the steady-state temperature field $T(r, \theta)$ that surrounds a sphere of radius R, e.g., a spherical inclusion in a solid material, exposed to a temperature gradient aligned along the z axis. We take the thermal conductivity of the sphere to be small enough that there is no heat flux into it. Thus we are solving

$$0 = L_r T + \frac{1}{r^2} L_\eta T$$

with boundary conditions

$$\begin{aligned} \frac{\partial T}{\partial r} &= 0, & r &= R \\ \nabla T &\to G\boldsymbol{e}_z, & r &\to \infty \end{aligned}$$

Solution

Axisymmetric diffusion problems involving the Laplacian in spherical geometries are naturally treated by expansion in the eigenfunctions of

$$L_\theta \Phi + \lambda \Phi = 0$$

If we make the substitution $\eta = \cos\theta$, L_θ becomes

$$L_\eta = \frac{d}{d\eta}\left(1 - \eta^2\right)\frac{d}{d\eta}$$

and the eigenvalue problem can be written as

$$\left(1 - z^2\right)\frac{d^2\Phi}{dz^2} - 2z\frac{d\Phi}{dz} + \lambda\Phi = 0$$

This is Legendre's differential equation, see Example 2.9. Its eigenvalues are $\lambda = n(n+1)$ for nonnegative integers n and its eigenfunctions are the Legendre polynomials $P_n(\eta)$.

Substituting the solution form

$$T(r, \eta) = \sum_{n=0}^{\infty} T_n(r) P_n(\eta) \tag{3.47}$$

into the governing equation, recalling that $L_\eta P_n = -n(n+1)P_n$, and using the orthogonality of the Legendre polynomials yields that

$$r^2 L_r T_n - n(n+1)T_n = 0$$

Rewriting this as

$$r^2\frac{d^2T_n}{dr^2} + 2r\frac{dT_n}{dr} - n(n+1)T_n = 0 \tag{3.48}$$

we recognize it as a Cauchy-Euler equation with solution

$$T_n(r) = a_n r^n + b_n r^{-(n+1)} \tag{3.49}$$

Consider first the boundary condition at infinity. We can rewrite this as $T \to Gr\eta + T_\infty = rGP_1(\eta) + T_\infty$, where T_∞ is arbitrary; we have not specified the temperature anywhere, only its gradient. Comparing this form to the series solution (3.47), we see that $a_0 = T_\infty$, $a_1 = G$, and $a_n = 0$ for $n > 1$. At $r = R$

$$\begin{aligned}
0 &= \frac{\partial T}{\partial r} \\
&= \sum_{n=0}^{\infty} \frac{dT_n}{dr} P_n(\eta) \\
&= \sum_{n=0}^{\infty} \left(na_n R^{n-1} - (n+1)b_n R^{-(n+1)-1}\right) P_n(\eta)
\end{aligned}$$

Because of the orthogonality of the $P_n(\eta)$, this sum must vanish term by term

$$na_n R^{n-1} - (n+1)b_n R^{-(n+1)-1} = 0$$

Using the known values of a_n

$$\begin{aligned}
n = 0: &\qquad b_0 = 0 \\
n = 1: &\qquad a_1 = G = 2b_1 R^{-3} \Rightarrow b_1 = \frac{GR^3}{2} \\
n > 1: &\qquad a_n = 0 = (n+1)b_n R^{-(n+1)-1} \Rightarrow b_n = 0
\end{aligned}$$

The final result is

$$T(r,\theta) = T_\infty + G\left(r + \frac{R^3}{2r^2}\right)\cos\theta \tag{3.50}$$

□

Example 3.10: Domain perturbation: heat conduction around a near-sphere

Consider the problem of heat conduction outside an object described in spherical coordinates by

$$r = R(\theta) = 1 + \epsilon P_2(\cos\theta)$$

where $P_2(x) = (3x^2 - 1)/2$ is the quadratic Legendre polynomial. This shape is slightly elongated at the poles and narrower at the equator than is a sphere, but has the same surface area. Use a regular perturbation approach based on the smallness of the deviation of the surface from spherical.

Solution

This example illustrates the technique of DOMAIN PERTURBATION. This approach is applicable to problems where the possibly unknown boundary shape is a small perturbation from a shape for which a closed form (e.g. separation of variables or Fourier transform) solution can be obtained. This approach is sometimes also used in numerical solution approaches to simplify the domain shape. In the present example, the choice of the Legendre polynomial simplifies the calculation but the solution procedure would be similar, but more tedious, with a more complicated surface shape, as long as the deviation from a sphere is uniformly small.

The equation and boundary conditions are

$$\nabla^2 T = 0, \; T(r = R) = 1, \; T \to 0 \text{ as } r \to \infty$$

Because the boundary is not a constant-coordinate surface, separation of variables (in spherical coordinates) cannot be used to find an exact solution. Nevertheless, a perturbation approach can be used to impose an asymptotically exact boundary condition at $r = 1$. This is done by expanding the boundary condition in a Taylor series around $r = 1$:

$$\begin{aligned} 1 &= T(r = R(\theta)) \\ &\sim T(1,\theta) + \left.\frac{\partial T}{\partial r}\right|_{r=1} (R(\theta) - 1) + \frac{1}{2}\left.\frac{\partial^2 T}{\partial r^2}\right|_{r=1} (R(\theta) - 1)^2 + O(\epsilon^3) \end{aligned}$$

Inserting the particular expression for the boundary shape:

$$1 \sim T(1,\theta) + \left.\frac{\partial T}{\partial r}\right|_{r=1} \epsilon P_2(\cos\theta) + \frac{1}{2}\left.\frac{\partial^2 T}{\partial r^2}\right|_{r=1} \epsilon^2 P_2(\cos\theta)^2 + O(\epsilon^3)$$

Note that this boundary condition is imposed at $r = 1$, allowing the use of separation of variables. There is no indication that a singular perturbation approach is necessary, so we posit a regular perturbation expansion $T(r,\theta) \sim T_0(r,\theta) + \epsilon T_1(r,\theta) + \epsilon^2 T_2(r,\theta)$. The governing equation at each order is simply Laplace's equation, with boundary condition at $r = 1$ at each order

$$\begin{aligned}
\epsilon^0 &: \quad T_0 = 1 \\
\epsilon^1 &: \quad T_1 = -P_2(\cos\theta)\tfrac{\partial T_0}{\partial r} \\
\epsilon^2 &: \quad T_2 = -P_2(\cos\theta)\tfrac{\partial T_1}{\partial r} - \tfrac{1}{2}P_2(\cos\theta)^2\tfrac{\partial^2 T_0}{\partial r^2}.
\end{aligned}$$

Using the fact that axisymmetric decaying solutions to Laplace's equation in spherical coordinates are given by

$$\sum_{i=0}^{\infty} c_i \frac{P_i(\cos\theta)}{r^{i+1}}$$

we find that the solutions at each order are:

$$\begin{aligned}
T_0(r,\theta) &= \frac{1}{r} \\
T_1(r,\theta) &= \frac{P_2(\cos\theta)}{r^3} \\
T_2(r,\theta) &= \frac{2}{5}\frac{1}{r} + \frac{4}{7}\frac{P_2(\cos\theta)}{r^3} + \frac{36}{35}\frac{P_4(\cos\theta)}{r^5}
\end{aligned}$$

Given these solutions, we can find that the dimensionless heat flux from the object is

$$Q = -2\pi \int_0^{\pi} \left.\frac{\partial T}{\partial r}\right|_{r=1} \sin\theta \, d\theta = 4\pi(1 + \frac{2}{5}\epsilon^2)$$

where $Q = 1$ corresponds to the heat flux from a sphere. Thus the change in heat flux from the sphere is proportional to the square of the deviation of the surface from spherical. Notice that the entire solution procedure is valid, and the heat flux the same if $\epsilon < 0$, so the object is actually a slightly flattened sphere. Therefore both prolate and oblate deviations from a spherical shape increase the heat flux. □

3.3.3 Laplace's Equation, Spherical Harmonics, and the Hydrogen Atom

Schrödinger's equation for the wave function $\Psi(\boldsymbol{x}, t)$ of a particle exposed to a potential energy field $V(\boldsymbol{x})$ is

$$i\frac{\partial \Psi}{\partial t} = -\nabla^2\Psi + V(\boldsymbol{x})\Psi \tag{3.51}$$

We will consider the case of a spherically symmetric potential $V(\boldsymbol{x}) = V(r)$, whose form we will specify later, so it is natural to work in spherical coordinates. The solutions of this equation have a very rich structure that encompasses many features of systems with spherical symmetry.

In contrast to the previous couple examples, we will allow the separation of variables procedure to again guide us. To begin, let $\Psi(\boldsymbol{x}, t) = f(t)\psi(\boldsymbol{x})$—the temporal and spatial variables are separated but not (yet) the individual coordinate directions. Inserting this form into (3.51) and rearranging yields

$$i\frac{\frac{df}{dt}}{f} = \frac{-\nabla^2\psi + V\psi}{\psi} = E$$

where E is a constant. Thus

$$i\frac{df}{dt} = Ef \tag{3.52}$$

$$(-\nabla^2 + V(r))\psi = E\psi \tag{3.53}$$

The solution to (3.52) is

$$f(t) = f_0 e^{-iEt}$$

Equation (3.53) has the form of an eigenvalue problem where the eigenvalue E is a dimensionless energy. This must be real so that Ψ does not vanish at past or future times. Now, since $L_\phi = \partial^2/\partial\phi^2$ with periodic boundary conditions, we let $\psi(r, \eta, \phi) = u(r, \eta)e^{im\phi}$ for any integer m. As above, we have let $\eta = \cos\theta$. Equation (3.53) becomes

$$\left(-L_r - \frac{1}{r^2}L_\eta + \frac{m^2}{r^2(1 - \eta^2)} + V(r)\right)u = Eu \tag{3.54}$$

We now write $u(r, \eta) = R(r)P(\eta)$. Substitution into (3.54) and rearrangement to group terms dependent only on r and η yields

$$\frac{1}{R}r^2 L_r R + r^2(E - V(r)) = -\frac{1}{P}L_\eta P + \frac{m^2}{1 - \eta^2} = c$$

Therefore

$$r^2 L_r R + r^2(E - V(r))R - cR = 0 \tag{3.55}$$

$$L_\eta P - \frac{m^2}{1 - \eta^2}P + cP = 0 \tag{3.56}$$

Equation (3.56) describes the angular behavior of the solutions. For $m = 0$, it reduces to Legendre's differential equation—we know that for boundedness at $\eta = \pm 1$ ($\theta = 0$ and π), that $c = l(l+1)$, with l a whole number. For $m \neq 0$ it is called the ASSOCIATED LEGENDRE DIFFERENTIAL EQUATION. Seeking a power series solution reveals that this equation has bounded solutions in $-1 \leq \eta \leq 1$ only if $m = -l, -(l-1), -(l-2), \ldots, 0, 1, 2, \ldots, l$. These solutions are the ASSOCIATED LEGENDRE POLYNOMIALS

$$P_{lm}(\eta) = \left(1 - \eta^2\right)^{m/2} \frac{d^m}{d\eta^m} P_l(\eta), \qquad m \geq 0 \tag{3.57}$$

and $P_{l,-m} = P_{lm}$.

Recapitulating, the products $P_{lm}(\cos\theta)e^{im\phi}$ contain the angular dependence of the solution. Suitably normalized and denoted $Y_{lm}(\theta, \phi)$, these products are called SURFACE SPHERICAL HARMONICS, or sometimes just SPHERICAL HARMONICS; they are the eigenfunctions of the angular part of the Laplacian

$$\left(L_\theta + \frac{1}{\sin^2\theta} L_\phi\right) Y_{lm} + l(l+1) Y_{lm} = 0 \tag{3.58}$$

Each eigenvalue l has $l+1$ corresponding eigenfunctions Y_{lm} with $m = 0, 1, \ldots l$. The normalized functions have the form

$$Y_{lm}(\theta, \phi) = (-1)^{(m+|m|)/2} \sqrt{\frac{2l+1}{4\pi} \frac{(l-m)!}{(l+m)!}} P_{lm}(\cos\theta) e^{im\phi} \tag{3.59}$$

and satisfy orthonormality with respect to integration over the surface of the unit sphere

$$\int_0^{2\pi} \int_0^{\pi} Y_{lm}(\theta, \phi) \bar{Y}_{np}(\cos\theta, \phi) \sin\theta \, d\theta \, d\phi = \delta_{ln} \delta_{mp} \tag{3.60}$$

The functions Y_{lm} for $l = 4$ are shown in Figure 3.7. Surface spherical harmonics are widely used to represent functions on the surface of a sphere.

Returning to (3.55) for the r-dependence, consider first the case $E = V(r) = 0$, in which (3.53) becomes the Laplace equation $\nabla^2 \psi = 0$. Equation (3.55) and its solution reduce to (3.48) and (3.49), respectively, with n replaced by l. Thus the general solution to $\nabla^2 \psi = 0$, expressed in spherical coordinates, is

$$\psi(r, \theta, \phi) = \sum_{l=0}^{\infty} \sum_{m=0}^{l} \left(a_{lm} r^l + b_{lm} r^{-(l+1)}\right) Y_{lm}(\theta, \phi) \tag{3.61}$$

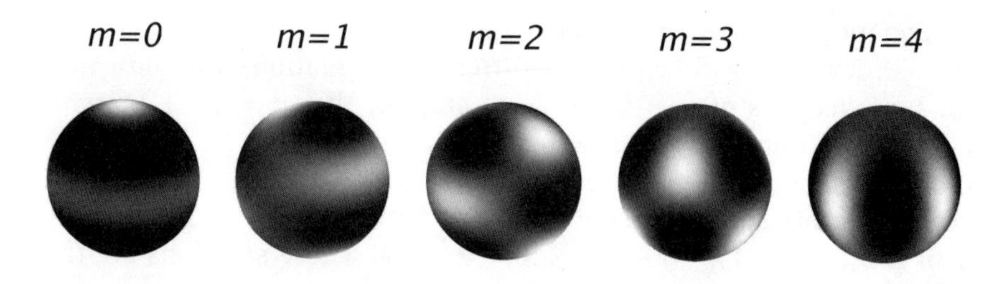

Figure 3.7: From left to right, real parts of the surface spherical harmonics $Y_{40}, Y_{41}, Y_{42}, Y_{43}, Y_{44}$.

Equation (3.50) is a particular case of this solution. Terms $r^l Y_{lm}(\theta, \phi)$ and $r^{-(l+1)} Y_{lm}(\theta, \phi)$ are called the growing and decaying SOLID SPHERICAL HARMONICS, respectively.

Now consider the case of an electron "orbiting" a proton—a hydrogen atom—where the potential energy is the Coulomb potential

$$V(\boldsymbol{x}) = -\frac{1}{r}$$

As boundary conditions, we require that Ψ is bounded at $r = 0$ and that it vanishes as $r \to \infty$. If the latter condition is not satisfied, the electron is not bound to the proton and we do not have an atom. Equation (3.46) motivates the substitution $w(r) = rR(r)$ into (3.55), yielding

$$\frac{d^2 w}{dr^2} + \left[E - \frac{1}{r} - \frac{l(l+1)}{r^2}\right] w = 0$$

As $r \to \infty$ we can approximate this as $w'' + Ew = 0$, suggesting that we seek a solution $w(r) = F(r)e^{-\beta r}$, where $\beta = \sqrt{-E}$. This result indicates that $E < 0$ for a bound electron. Without going into the details (with which we are now largely familiar), seeking a Frobenius solution $F(r) = r^\alpha g(r)$ and requiring that $Fe^{-\beta r} \to 0$ as $r \to \infty$ leads to $\alpha = l+1$ and requires that $g(r)$ be a truncated power series, i.e., a polynomial. Inspecting the recursion relation for the power series, one finds in close analogy to the results in Chapter 2 regarding Legendre and other orthogonal polynomials that it will truncate at degree n' if

$$\sqrt{-E}(l + n') = 1$$

The solutions, which we denote $R_{n'}$ can be written in terms of ASSOCIATED LAGUERRE POLYNOMIALS (Merzbacher, 1970; Winter, 1979).

Defining $n = l + n'$ as the PRINCIPAL QUANTUM NUMBER this becomes

$$E = -\frac{1}{n^2}$$

This expression determines the eigenvalues of (3.53) and describes very well the energy levels of a hydrogen atom. The eigenfunctions $\psi_{lmn'}(\boldsymbol{x}) = Y_{lm}(\theta, \phi)R_{n'}(r)$ of (3.53) are characterized by l, $m = -l, \ldots, l$, called the ANGULAR MOMENTUM QUANTUM NUMBERS, and n', called the RADIAL QUANTUM NUMBER. Since the eigenvalues E depend only on $l + n'$, various combinations of l and n' have the same energy. The same is true for all eigenfunctions with the same m. The s, p, d, and f atomic orbitals correspond to $l = 0, 1, 2$, and 3, respectively. Since $E < 0$, when $n = 1$ only $l = 0$ states, s orbitals, can exist. This is the *ground state* or lowest-energy state of the hydrogen atom. When $n = 2$, both $l = 0$ (s orbitals) and $l = 1$ (p orbitals) can exist, and so on. Thus we see in this analysis the basic features of the electronic structure of atoms.

3.3.4 Applications of the Fourier Transform to PDEs

In Section 2.4.1 we saw that functions in a finite domain could be represented as a trigonometric Fourier series

$$f(x) = \sum_{k=-\infty}^{\infty} c_k e^{ikx} \qquad c_k = \frac{(f, e^{ikx})}{(e^{ikx}, e^{ikx})}$$

The FOURIER TRANSFORM generalizes this idea to an unbounded domain. First some definitions: the Fourier transform $\hat{f}(k)$ of a function $f(x)$ is given by

$$\hat{f}(k) = \int_{-\infty}^{\infty} f(x)e^{-ikx}dx = F\{f(x)\}$$

This is the analogue of the expression for c_k in a bounded domain; because periodicity is no longer required over a finite interval, k can be any real number rather than needing to be an integer. The INVERSE FOURIER TRANSFORM is the analogue of the Fourier series representation of f

$$f(x) = \frac{1}{2\pi}\int_{-\infty}^{\infty} \hat{f}(k)e^{ikx}dk = F^{-1}\left\{\hat{f}(k)\right\}$$

These operations are mappings from "x-space" to "k-space" and vice versa. Here are some useful properties of Fourier transforms, which are easily derived from its definition:

1. Derivative property

$$F\left\{\frac{df(x)}{dx}\right\} = ikF\{f(x)\} = ik\hat{f}(k) \tag{3.62}$$

2. Integral property

$$F\left\{\int_{x_0}^{x} f(x)dx\right\} = \frac{1}{ik}\hat{f}(k) + c\delta(k) \tag{3.63}$$

where c depends on the lower limit x_0 of the integration.

3. Shift in x

$$F\{f(x-a)\} = e^{-ika}\hat{f}(k) \tag{3.64}$$

4. Shift in k

$$F\left\{e^{ilx}f(x)\right\} = \hat{f}(k-l) \tag{3.65}$$

5. Scaling

$$F\{f(\alpha x)\} = \frac{1}{|\alpha|}\hat{f}\left(\frac{k}{\alpha}\right) \tag{3.66}$$

where α is a real scalar.

6. Behavior upon exchanging variables: if $\hat{f}(k)$ is the Fourier transform of $f(x)$, then $\hat{f}(x) = 2\pi f(-k)$. This property is useful for extending the usefulness of lists or tables of transforms, like the one in the following paragraph.

7. Convolution theorem: the CONVOLUTION of two functions G and h is

$$u(x) = \int_{-\infty}^{\infty} G(x-\xi)h(\xi)\,d\xi = \int_{-\infty}^{\infty} G(\xi)h(x-\xi)\,d\xi$$

This is often written $u = G * h$. The CONVOLUTION THEOREM states that

$$F\{G*h\} = \widehat{G*h} = \hat{G}(k)\hat{h}(k) \tag{3.67}$$

A convolution in x-space is a product in k-space.

These properties will help us to solve PDEs.

Fourier transforms of some important functions are:

1. $f(x) = \delta(x) \iff \hat{f}(k) = 1$. A spike localized in space has equal components at every wavelength.

2. $\hat{f}(k) = 2\pi\delta(k) \iff f(x) = 1$. Conversely, a spike located at zero wavenumber is smeared all over space.

3. $\hat{f}(k) = 2\pi\delta(k-l) \iff f(x) = e^{ilx}$. A spike at $k = l$ corresponds to a sinusoid of wavenumber l.

4. $f(x) = 1, -L < x < L$ and zero elsewhere $\iff \hat{f}(k) = (2\sin kL)/k$.

5.
$$f(x) = \frac{1}{1+x^2} \Leftrightarrow \hat{f}(k) = \pi e^{-|k|}$$

6.
$$f(x) = e^{-b|x|} \quad (b > 0) \Leftrightarrow \hat{f}(k) = \frac{2b}{b^2+k^2} \qquad b > 0$$

7.
$$f(k) = e^{-ak^2}, (a > 0) \iff f(x) = \frac{1}{2\sqrt{\pi a}} e^{-x^2/4a}$$

   The Fourier transform of a Gaussian is a Gaussian. If a is large, the function decays very quickly as $|k|$ increases, so the Gaussian in k-space is very localized. Because a appears in the denominator in x-space, however, the function is very spread out in x. The opposite is true if a is small, with the balance holding at $a = 1/2$. Here and here only is the spread of the function the same in k and x. As $a \to 0$, $\frac{1}{2\sqrt{\pi a}} e^{-x^2/4a} \to \delta(x)$, in which case this property reduces to the first result on the list: $f(x) = \delta(x) \iff \hat{f}(k) = 1$.

Example 3.11: Derivation of a Fourier transform formula

Let $f(x) = e^{-b|x|}$, with $b > 0$. Find its Fourier transform.

Solution

$$\begin{aligned}
\hat{f}(k) &= \int_{-\infty}^{\infty} e^{-b|x|} e^{-ikx}\, dx \\
&= \int_{-\infty}^{0} e^{(b-ik)x}\, dx + \int_{0}^{\infty} e^{(-b-ik)x}\, dx \\
&= \frac{1}{b-ik} + \frac{1}{b+ik} \\
&= \frac{2b}{b^2+k^2}
\end{aligned}$$

□

We illustrate the use of Fourier transforms to solve PDEs with examples.

Example 3.12: Transient diffusion in an unbounded domain: one and multiple dimensions

(a) Consider transient diffusion

$$u_t = Du_{xx} \tag{3.68}$$

in the one-dimensional infinite domain $(-\infty, \infty)$ with initial condition $u(x, 0) = u_0(x)$, where $u_0(x)$ is known but otherwise arbitrary. Use the Fourier transform in x to find the solution.

(b) Extend this result to three directions, with initial condition $u(\boldsymbol{x}, 0) = u_0(\boldsymbol{x})$. Do so by first considering a δ-function initial condition $\delta(\boldsymbol{x}) = \delta(x)\delta(y)\delta(z)$ and noting that it can be incorporated into the governing equation as a point source in space and time

$$u_t - D\left(u_{xx} + u_{yy} + u_{zz}\right) = \delta(t)\delta(\boldsymbol{x}) \tag{3.69}$$

Solution

(a) Taking the Fourier transform of the equation and applying the derivative property yields

$$F\left\{u_t = Du_{xx}\right\} \implies \hat{u}_t(k, t) = D(ik)^2\hat{u}(k, t) = -k^2 D\hat{u}$$

This gives us an ODE for each value of k, with initial condition $\hat{u}_0(k)$. The solution is simply $\hat{u}(k, t) = \hat{u}_0(k)e^{-Dk^2 t}$. The inverse Fourier transform puts this back in physical space. Consider the evolution of a delta function, whose Fourier transform is simply $\hat{u}(k) = 1$. Now

$$\begin{aligned} u(x, t) &= F^{-1}\left\{e^{-Dtk^2}\right\} \\ &= \frac{1}{2\sqrt{\pi Dt}}e^{-x^2/4Dt} \end{aligned} \tag{3.70}$$

Thus at any time, the temperature field that starts as a delta function is a Gaussian distribution, with height $\sim 1/2\sqrt{\pi Dt}$ and width $\sim \sqrt{4Dt}$. An important extension of this result comes from the observation that *any* function can be written as a superposition of delta functions

$$u_0(x) = \int_{-\infty}^{\infty} u_0(\xi)\delta(x - \xi)\, d\xi \tag{3.71}$$

Thus the solution can be written as the superpositions of the Gaussians induced by these delta functions

$$u(x,t) = \frac{1}{2\sqrt{\pi D t}} \int_{-\infty}^{\infty} u_0(\xi) e^{-(x-\xi)^2/4Dt}\, d\xi \tag{3.72}$$

(b) Now the THREE-DIMENSIONAL FOURIER TRANSFORM will be introduced

$$\begin{aligned}\hat{f}(k_x, k_y, k_z) &= F_{3D}\{f(x,y,z)\} \\ &= \iiint_{-\infty}^{\infty} f(x,y,z) e^{-ik_x x} e^{-ik_y y} e^{-ik_z z}\, dx\, dy\, dz\end{aligned}$$

Here Fourier transforms have been applied in all three spatial coordinate directions. Defining the WAVEVECTOR $\boldsymbol{k} = (k_x, k_y, k_z)$, this can be written

$$\hat{f}(\mathbf{k}) = \iiint_{-\infty}^{\infty} f(\mathbf{x}) e^{-i\mathbf{k}\cdot\mathbf{x}}\, d\mathbf{x}$$

Similarly

$$f(\mathbf{x}) = F_{3D}^{-1}\left\{\hat{f}(\mathbf{k})\right\} = \frac{1}{(2\pi)^3} \iiint_{-\infty}^{\infty} \hat{f}(\mathbf{k}) e^{i\mathbf{k}\cdot\mathbf{x}}\, d\mathbf{k}$$

The results presented above for one-dimensional transforms can be used to generate formulas for multidimensional transforms. For example

$$\begin{aligned}F_{3D}\{\nabla f\} &= i\mathbf{k}\hat{f}(\mathbf{k}) \\ F_{3D}\{\nabla\cdot\boldsymbol{v}\} &= i\mathbf{k}\cdot\hat{\boldsymbol{v}}(\mathbf{k}) \\ F_{3D}\left\{\nabla^2 f\right\} &= -k^2\hat{f}, \quad k^2 = k_x^2 + k_y^2 + k_z^2 \\ F_{3D}\left\{\iiint_{-\infty}^{\infty} G(\mathbf{x}-\boldsymbol{\xi}) h(\boldsymbol{\xi})\, d\boldsymbol{\xi}\right\} &= \hat{G}(\mathbf{k})\hat{h}(\mathbf{k})\end{aligned}$$

Taking the three-dimensional Fourier transform of (3.69) yields

$$\hat{u}_t - k^2 D\hat{u} = \delta(t)$$

which is easily solved (by Laplace transform in time for example) to yield

$$\hat{u}(t) = e^{-k^2 Dt}$$

Applying the inverse three-dimensional transform formula results in

$$\begin{aligned} u(\boldsymbol{x},t) &= \frac{1}{(2\pi)^3}\iiint_{-\infty}^{\infty} e^{-k^2Dt}e^{i\boldsymbol{k}\cdot\boldsymbol{x}}\,d\boldsymbol{k} \\ &= \frac{1}{2\pi}\int_{-\infty}^{\infty} e^{-k_x^2Dt}e^{ik_x x}dk_x \\ &\times \frac{1}{2\pi}\int_{-\infty}^{\infty} e^{-k_y^2Dt}e^{ik_y y}dk_y\,\frac{1}{2\pi}\int_{-\infty}^{\infty} e^{-k_z^2Dt}e^{ik_z z}dk_z \\ &= \frac{1}{2\sqrt{\pi Dt}}e^{-x^2/4Dt}\frac{1}{2\sqrt{\pi Dt}}e^{-y^2/4Dt}\frac{1}{2\sqrt{\pi Dt}}e^{-z^2/4Dt} \\ &= \frac{1}{(2\sqrt{\pi Dt})^3}e^{-r^2/4Dt} \end{aligned}$$

where $r = |x|$. Using this result, (3.72) generalizes to an arbitrary three-dimensional initial condition $u_0(\boldsymbol{x})$

$$u(\boldsymbol{x}) = \frac{1}{(2\sqrt{\pi Dt})^3}\iiint_{-\infty}^{\infty} u_0(\boldsymbol{\xi})e^{-|\boldsymbol{x}-\boldsymbol{\xi}|^2/4Dt}\,d\boldsymbol{\xi}$$

□

Example 3.13: Steady diffusion from a wall with an imposed concentration profile

Solve the steady state diffusion or heat conduction problem

$$u_{xx} + u_{yy} = 0$$

in the half-plane $-\infty < x < \infty$, $0 < y < \infty$, with boundary conditions $u(x,0) = u_0(x)$ and $u(x,y)$ bounded as $y \to \infty$.

Solution

Based on our experience with the previous example, we begin by considering the boundary condition $u(x,0) = \delta(x)$. Taking the Fourier transform of the equation and boundary condition in the x-direction (the problem is not unbounded in y) yields

$$-k^2\hat{u}(y) + \hat{u}_{yy}(y) = 0, \qquad \hat{u}(0) = F\{\delta(x)\} = 1$$

Requiring that the solution be bounded as $y \to \infty$, this has the solution

$$\hat{u}(y) = \hat{u}(0)e^{-|k|y} \tag{3.73}$$

where $\hat{u}_0 = 1$. Now the inverse transform of this solution must be found. Recall that from the point of view of the Fourier transform and its inverse, the variable y is a constant (we are only considering Fourier transforms involving the x-coordinate). Therefore we can combine the following results from above: $\hat{f}(k) = \pi e^{-|k|} \Leftrightarrow f(x) = \frac{1}{1+x^2}$ and $\hat{f}(\alpha x) = \frac{1}{\alpha}\hat{f}(k/\alpha)$. Letting $y = \frac{1}{\alpha}$, we have that

$$\hat{f}(ky) = e^{-|k|y} \Rightarrow f(x) = \frac{1}{\pi}\frac{1}{y}\frac{1}{1+\frac{x^2}{y^2}} = \frac{1}{\pi}\frac{y}{x^2+y^2}$$

Given this solution, we can use (3.71) and the superposition principle for linear problems to determine that, given an arbitrary boundary condition $u(x,0) = u_0(x)$, the solution is

$$u(x,y) = \frac{1}{\pi}\int_{-\infty}^{\infty} u_0(\xi)\frac{y}{(x-\xi)^2+y^2}\,d\xi$$

Observe that (3.73) has the form of a convolution $\hat{G}(k)\hat{h}(k)$, with $\hat{G}(k) = u_0(k)$ and $\hat{h}(k) = e^{-|k|y}$, i.e, it is a convolution. Thus, the solution arises directly from the convolution theorem

$$u(x,y) = F^{-1}\left\{\hat{G}(k)\hat{h}(k)\right\} = \frac{1}{\pi}\int_{-\infty}^{\infty} u_0(\xi)\frac{y}{(x-\xi)^2+y^2}\,d\xi \qquad \square$$

3.3.5 Green's Functions and Boundary-Value Problems

Overview

The transient diffusion problem we solved in Example 3.12 gave us an example of a GREEN'S FUNCTION, a solution to a differential equation with a point source forcing.[4] We saw in that example that the solutions for an arbitrary initial distribution $u(x,0) = u_0(x)$ could be written as a convolution of u_0 and the Green's function. Exercise 3.36 extends that result. In the present section we will develop the basic theory of Green's functions, with a particular focus on boundary value problems.

Consider a linear boundary value problem

$$Lu = f(\boldsymbol{x}) \tag{3.74}$$

[4]In quantum mechanics in particular, a Green's function for a transient problem like this one is called a PROPAGATOR, since it propagates a δ-function initial condition forward in time.

with specified boundary conditions that may be inhomogeneous in general. The Green's function $G(\boldsymbol{x}, \boldsymbol{x}_0)$ for the operator L is the solution to

$$LG = \delta(\boldsymbol{x} - \boldsymbol{x}_0) \tag{3.75}$$

For example, $G(\boldsymbol{x}, \boldsymbol{x}_0)$ is the solution for a point source placed at an arbitrary position $\boldsymbol{x}_0$ within the domain of interest. The discussion below reveals what boundary conditions G should satisfy. For the present, we will consider Green's functions for self-adjoint problems and as a specific initial example will consider Sturm-Liouville operators. Recall (2.33) from Section 2.4.2

$$\begin{aligned}
(Lu, v)_w &= \int_a^b \frac{1}{w(x)}\left(\frac{d}{dx}\left[p(x)\frac{du}{dx}\right] + r(x)u\right) v\, w\, dx \\
&= p(b)\left(u'(b)v(b) - u(b)v'(b)\right) \\
&\quad - p(a)\left(u'(a)v(a) - u(a)v'(a)\right) \\
&\quad + \int_a^b u\frac{1}{w(x)}\left(\frac{d}{dx}\left[p(x)\frac{dv}{dx}\right] + r(x)v\right) w\, dx
\end{aligned}$$

Letting $v(x) = G(x, x_0)$, this becomes

$$\begin{aligned}
(Lu, G)_w &= (u, LG)_w \\
&\quad + p(b)\left(u'(b)G(b, x_0) - u(b)v'(b, x_0)\right) \\
&\quad - p(a)\left(u'(a)G(a, x_0) - u(a)G'(a, x_0)\right)
\end{aligned}$$

Applying (3.74) and (3.75) in the two inner products gives us that

$$\begin{aligned}
(f, G)_w &= (u, \delta(x - x_0))_w \\
&\quad + p(b)\left(u'(b)G(b, x_0) - u(b)G'(b, x_0)\right) \\
&\quad - p(a)\left(u'(a)G(a, x_0) - u(a)G'(a, x_0)\right)
\end{aligned}$$

The inner product $(u, \delta(x - x_0))_w$ evaluates to $u(x_0)w(x_0)$, so rearranging leads to

$$\begin{aligned}
u(x_0) = \frac{1}{w(x_0)}\Bigg[&\int_a^b f(x)G(x, x_0)w(x)\, dx \\
&- p(b)(u'(b)G(b, x_0) - u(b)G'(b, x_0)) \\
&+ p(a)(u'(a)G(a, x_0) - u(a)G'(a, x_0))\Bigg]
\end{aligned}$$

Finally, we specify boundary conditions. For example, we can set inhomogeneous Dirichlet boundary conditions $u(a) = u_a, u(b) = u_b$. Because we are not specifying homogeneous boundary conditions here, the operator L is said to be FORMALLY self-adjoint; true self-adjointness for a differential operator requires that we impose homogeneous boundary conditions such that the boundary terms vanish. In this case $u'(a)$ and $u'(b)$ are not specified. If we require G to satisfy *homogeneous* Dirichlet boundary conditions $G(a,x_0) = G(b,x_0) = 0$, however, then the unknown boundary values u' do not appear and we arrive at a solution for u in terms of f, G and the boundary conditions

$$u(x_0) = \frac{1}{w(x_0)}\left[\int_a^b f(x)G(x,x_0)w(x)\,dx + p(x)u(x)G'(x,x_0)\Big|_{x=a}^{x=b}\right] \quad (3.76)$$

Therefore, given the solution $G(x,x_0)$ to the problem $LG = \delta(x-x_0)$, $G(a,x_0) = G(b,x_0) = 0$, we can find the solution to $Lu = f$ for any f through (3.76). Note that (3.76) is closely analogous to the solution $x = A^{-1}b$ of the algebraic problem $Ax = b$, with G playing the role of A^{-1}. Example 2.15 shows a derivation of this formula for a specific problem. Because that example already imposes homogeneous Dirichlet boundary conditions, reworking it with $u(x) = G(x,x_0)$ and $f(x) = \delta(x-x_0)$ would directly yield the Green's function for the Dirichlet problem.

The above discussion focused on a Sturm-Liouville problem, which is formally self-adjoint. For a *non*-self-adjoint operator, the Green's function for the adjoint operator satisfies

$$L^*G^*(\boldsymbol{x},\boldsymbol{x}_1) = \delta(\boldsymbol{x}-\boldsymbol{x}_1) \quad (3.77)$$

along with appropriate homogeneous boundary conditions. In general, the position of the source is arbitrary, which is why we let its position here be $\boldsymbol{x}_1$, which is generally distinct from $\boldsymbol{x}_0$. From the definition of the adjoint

$$(LG, G^*) = (G, L^*G^*)$$

(we have chosen homogeneous boundary conditions on G and G^* so the boundary terms vanish), and inserting (3.75) and (3.77), yields

$$(\delta(\boldsymbol{x}-\boldsymbol{x}_0), G^*(\boldsymbol{x},\boldsymbol{x}_1)) = (G(\boldsymbol{x},\boldsymbol{x}_0), \delta(\boldsymbol{x}-\boldsymbol{x}_1))$$

This reduces to simply

$$G^*(\boldsymbol{x}_0,\boldsymbol{x}_1) = G(\boldsymbol{x}_1,\boldsymbol{x}_0) \quad (3.78)$$

This result is the analog of the matrix adjoint result $A^*_{ij} = A_{ji}$. For a (formally) self-adjoint operator it becomes the analog of the result for a symmetric matrix: $G(\boldsymbol{x}_0, \boldsymbol{x}_1) = G(\boldsymbol{x}_1, \boldsymbol{x}_0)$. A specific case of this result arose in Example 2.15. For complex functions, (3.78) simply becomes $G^*(\boldsymbol{x}_0, \boldsymbol{x}_1) = \bar{G}(\boldsymbol{x}_1, \boldsymbol{x}_0)$.

Green's Function Solution to the Poisson Equation

In multiple dimensions, Green's identities provide the foundation for developing solutions based on Green's functions. We focus here on the solution of the Poisson equation[5]

$$-\nabla^2 u = f(\boldsymbol{x})$$

with boundary conditions specified below. The Green's function of interest here satisfies

$$-\nabla^2 G(\boldsymbol{x}, \boldsymbol{x}_0) = \delta(\boldsymbol{x} - \boldsymbol{x}_0) \tag{3.79}$$

Green's second identity, (3.17), with v replaced by G, is

$$\begin{aligned}\int_V (u(\boldsymbol{x})\nabla^2 G(\boldsymbol{x}, \boldsymbol{x}_0) - G(\boldsymbol{x}, \boldsymbol{x}_0)\nabla^2 u(\boldsymbol{x}))\, dV(\boldsymbol{x}) \\ = \int_S (u(\boldsymbol{x})\boldsymbol{\nabla} G(\boldsymbol{x}, \boldsymbol{x}_0) - G(\boldsymbol{x}, \boldsymbol{x}_0)\boldsymbol{\nabla} u(\boldsymbol{x})) \cdot \boldsymbol{n}\, dS(\boldsymbol{x})\end{aligned}$$

where we have written the differential volume and surface elements as explicit functions of $\boldsymbol{x}$ to remind us that it is the independent variable. Inserting (3.3.5) and (3.79) and evaluating the integral containing the δ-function yields

$$\begin{aligned}u(\boldsymbol{x}_0) = \int_V G(\boldsymbol{x}, \boldsymbol{x}_0) f(\boldsymbol{x})\, dV(\boldsymbol{x}) \\ - \int_S (u(\boldsymbol{x})\boldsymbol{\nabla} G(\boldsymbol{x}, \boldsymbol{x}_0) - G(\boldsymbol{x}, \boldsymbol{x}_0)\boldsymbol{\nabla} u(\boldsymbol{x})) \cdot \boldsymbol{n}\, dS(\boldsymbol{x})\end{aligned} \tag{3.80}$$

If u satisfies Dirichlet boundary conditions $u = u_S$ on S, then requiring that $G = 0$ on S yields a solution for u

$$u(\boldsymbol{x}_0) = \int_V G(\boldsymbol{x}, \boldsymbol{x}_0) f(\boldsymbol{x})\, dV(\boldsymbol{x}) - \int_S u_S(\boldsymbol{x}) \frac{\partial G}{\partial n}(\boldsymbol{x}, \boldsymbol{x}_0)\, dS(\boldsymbol{x}) \tag{3.81}$$

[5]We put a negative sign in front of the Laplacian here so that physically, the term $f(\boldsymbol{x})$ represents a *source* of heat, chemical species, etc., and thus the Green's function represents a *point* source. Some authors do not use the negative sign.

where $\partial G/\partial n = \boldsymbol{n} \cdot \boldsymbol{\nabla} G$. A Green's function satisfying homogeneous Dirichlet boundary conditions is sometimes called a GREEN'S FUNCTION OF THE FIRST KIND. If u satisfies Neumann boundary conditions $\partial u/\partial n = j_S$, then we apply homogeneous Neumann boundary conditions $\partial G/\partial n = 0$ to the Green's function, in which case the solution for u is

$$u(\boldsymbol{x}_0) = \int_V G(\boldsymbol{x}, \boldsymbol{x}_0) f(\boldsymbol{x})\, dV(\boldsymbol{x}) + \int_S G(\boldsymbol{x}, \boldsymbol{x}_0) j_S\, dS(\boldsymbol{x}) \tag{3.82}$$

and G is a GREEN'S FUNCTION OF THE SECOND KIND.

Evaluating the solutions (3.81) or (3.82) requires us to determine the solution to $-\nabla^2 G = \delta(\boldsymbol{x} - \boldsymbol{x}_0)$ with the appropriate boundary conditions. To do this, it is useful to let G be written as the sum of two parts: $G = G_\infty + G_B$. In this sum, G_∞ is called the FREE-SPACE GREEN'S FUNCTION. It is a solution to the equation $Lu = \delta$ in an unbounded domain, and contains the singular behavior induced by the presence of the point source. The boundary correction G_B satisfies $LG_B = 0$ (the singular behavior is contained in G_∞), and is determined by the requirement that G satisfy specific boundary conditions on S. We will find G_∞ and G_B for $L = -\nabla^2$ in two dimensions.

For the purpose of obtaining the free-space Green's function, we will place the source at the origin: $\boldsymbol{x}_0 = 0$. Because the δ-function has no angular dependence, we will seek a two-dimensional solution to $-\nabla^2 G_\infty = \delta(\boldsymbol{x})$ that is only a function of r. Therefore, at every point in the domain except the origin, $G_\infty(r)$ satisfies the equation

$$\frac{1}{r}\frac{d}{dr} r \frac{dG_\infty}{dr} = 0$$

The solution to this is simple

$$G_\infty(r) = c_1 \ln r + c_2$$

We set $c_2 = 0$; any constant component of the solution can be incorporated into G_B. To find c_1 we first integrate the equation $-\nabla^2 G_\infty = \delta(\boldsymbol{x})$ over any volume V (area in this case) containing the origin

$$-\int_V \nabla^2 G_\infty\, dV = \int_V \delta(\boldsymbol{x})\, dV = 1$$

Recalling that $\nabla^2 = \boldsymbol{\nabla} \cdot \boldsymbol{\nabla}$ and applying the divergence theorem to the left-hand side of this expression yields that

$$-\int_S \frac{\partial G_\infty}{\partial n}\, dS = 1$$

The integral is simple to evaluate if we let V be a circle of radius ϵ surrounding the origin, in which case

$$-\int_S \frac{\partial G_\infty}{\partial n}\, dS = -\int_S \frac{c_1}{r} r\, d\theta = 1$$

Therefore $c_1 = -\frac{1}{2\pi}$. Letting $r = |\boldsymbol{x} - \boldsymbol{x}_0|$, the free-space Green's function for $-\nabla^2$ in two dimensions becomes

$$G_\infty(\boldsymbol{x}, \boldsymbol{x}_0) = G_\infty(\boldsymbol{x} - \boldsymbol{x}_0) = \frac{-1}{2\pi} \ln |\boldsymbol{x} - \boldsymbol{x}_0| \tag{3.83}$$

To determine G_B, the shape of the domain and the boundary conditions must be specified. We will take the domain to be the half-plane $-\infty < x < \infty$, $0 < y < \infty$ and seek a solution that vanishes as $y \to \infty$. In the case of Dirichlet boundary conditions, G_B satisfies

$$-\nabla^2 G_B(\boldsymbol{x}, \boldsymbol{x}_0) = 0$$

with $G_B = -G_\infty$ on $y = 0$. We can solve this problem using the "method of images." Since G_∞ represents the field due to a point source at the position $\boldsymbol{x}_0 = (x_0, y_0)$, if we place a point *sink* (an "image" or "reflection" of the source) at $\boldsymbol{x}_{0I} = (x_0, -y_0)$, symmetry shows us that the field due to the source-sink combination will be zero at $y = 0$ (Figure 3.8). Therefore we set

$$G_B(\boldsymbol{x}, \boldsymbol{x}_0) = -\frac{-1}{2\pi} \ln |\boldsymbol{x} - \boldsymbol{x}_{0I}|$$

This satisfies $\nabla^2 G_B = 0$ in $y > 0$ because the sink is in the image region $y < 0$. Thus the total Green's function is given by

$$\begin{aligned} G(\boldsymbol{x}, \boldsymbol{x}_0) &= \frac{-1}{2\pi} \ln |\boldsymbol{x} - \boldsymbol{x}_0| - \frac{-1}{2\pi} \ln |\boldsymbol{x} - \boldsymbol{x}_{0I}| \\ &= \frac{-1}{2\pi} \ln \frac{|\boldsymbol{x} - \boldsymbol{x}_0|}{|\boldsymbol{x} - \boldsymbol{x}_{0I}|} \end{aligned}$$

Finally, the solution, (3.81), becomes

$$\begin{aligned} u(x_0, y_0) &= \frac{-1}{2\pi} \int_0^\infty \int_{-\infty}^\infty \ln \frac{|\boldsymbol{x} - \boldsymbol{x}_0|}{|\boldsymbol{x} - \boldsymbol{x}_{0I}|} f(x, y)\, dx\, dy \\ &\quad + \frac{y_0}{\pi} \int_{-\infty}^\infty \frac{u_S(x)}{(x - x_0)^2 + y_0^2}\, dx \end{aligned}$$

If $f(x, y) = 0$, this solution reduces to what we found using Fourier transforms in Example 3.13. For the solution with Neumann boundary

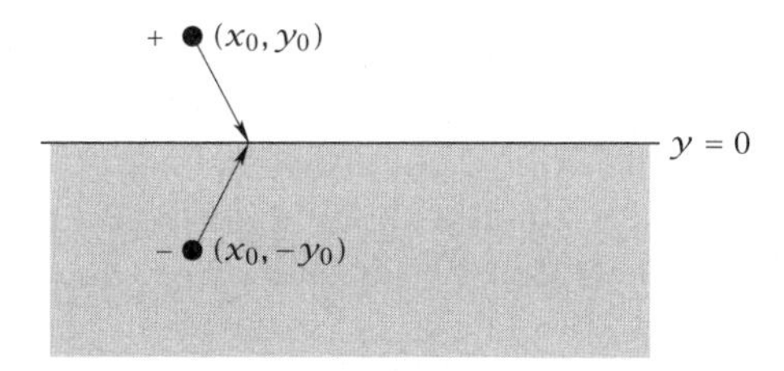

Figure 3.8: A source (as indicated by the +) in the physical domain at position (x_0, y_0) and an image sink (−) at $(x_0, -y_0)$. The shaded region is outside the physical domain. Because source and sink have equal magnitude and opposite sign, and are the same distance from the plane $y = 0$, the fields due to them cancel out on that line.

conditions, (3.82), G_B would have to satisfy $\partial G_B/\partial n = -\partial G_\infty/\partial n$ on $y = 0$. In this case G_B is the field due to an image *source* rather than a sink at position $\boldsymbol{x}_{0I}$.

The simple geometry used here required only one "image point" to satisfy the boundary conditions. Nevertheless, the geometry does not need to be much more complicated to require many or even an infinite number of image points. The infinite strip, $-\infty < x < \infty, 0 < y < 1$, requires an infinite number of image points since the image point we use to satisfy, say, the boundary condition at $y = 0$ will change the field at $y = 1$, which must be compensated by another image point, and so on *ad infinitum.* As a practical matter, often using one image for each of the two boundaries provides an adequate approximation.

Boundary Integral Formulation of the Laplace Equation

Equations (3.81) and (3.82) require the availability of the solution for the Green's function with the appropriate boundary conditions and in the domain of interest. In some cases, as we saw above, this solution is available in closed form but often it is not. To address this situation, we step back to (3.80). In developing this equation, the boundary conditions on G have not yet been specified. For example, it is valid if we let $G = G_\infty$, which has a simple closed-form solution, (3.83). Using this choice and letting $f(\boldsymbol{x}) = 0$ so that we are considering the Laplace equation, (3.80) becomes

$$u(\boldsymbol{x}_0) = \int_S \left(G_\infty(\boldsymbol{x}, \boldsymbol{x}_0) \frac{\partial u(\boldsymbol{x})}{\partial n} - u(\boldsymbol{x}) \frac{\partial G_\infty(\boldsymbol{x}, \boldsymbol{x}_0)}{\partial n} \right) dS(\boldsymbol{x}) \qquad (3.84)$$

Above, we have taken $\boldsymbol{x}_0$ to be a point within the domain. If instead we take it to be on the boundary itself, we would have a self-consistent INTEGRAL EQUATION for the boundary values of u and $\partial u/\partial n$. The solution to this equation could then be inserted into (3.84) to find the solution at any point within the domain. We will derive this equation for the case where the domain is the interior of a bounded volume and the boundary of the volume is smooth.

There is an important subtlety in doing this, which arises from the fact that $\partial G_\infty/\partial n$ changes sign as x_0 crosses from one side of the boundary to the other. Consider a vertical boundary defined by the line $x = 0$ (with the outward normal pointing to the right) and let $\boldsymbol{x}_0 = (x_0, y)$. Taking the limit $x_0 \to 0$ corresponds to approaching the point $(0, y)$ on the boundary

$$\begin{aligned}
\lim_{x_0\to 0} \frac{\partial G_\infty(\boldsymbol{x},\boldsymbol{x}_0)}{\partial n} &= \lim_{x_0\to 0} \frac{\partial G_\infty(\boldsymbol{x},\boldsymbol{x}_0)}{\partial x} \\
&= \lim_{x_0\to 0} -\frac{1}{2\pi}\frac{x_0}{x_0^2+y^2} \\
&= -\frac{1}{2}\,\mathrm{sgn}(x_0)\delta(y)
\end{aligned}$$

where the last step is accomplished by recognizing $\frac{|x_0|}{\pi(x_0^2+y^2)}$ as a delta family, see Section 2.2.5. Thus this term is singular as $\boldsymbol{x}_0$ approaches the boundary, and the sign depends on the side from which it approaches. Using this result, and recalling that here $\boldsymbol{x}_0$ is approaching the boundary from the left (interior)

$$\lim_{\boldsymbol{x}_0\to S}\int_S u(\boldsymbol{x})\frac{\partial G_\infty(\boldsymbol{x},\boldsymbol{x}_0)}{\partial n}\,dS(\boldsymbol{x}) = \int_S u(\boldsymbol{x})\frac{\partial G_\infty(\boldsymbol{x},\boldsymbol{x}_0)}{\partial n}\,dS(\boldsymbol{x}) - \frac{1}{2}u(\boldsymbol{x}_0) \tag{3.85}$$

Here the integral on the boundary must be evaluated in the sense of its CAUCHY PRINCIPAL VALUE

$$\lim_{\epsilon\to 0}\int_{S-S_\epsilon} u(\boldsymbol{x})\frac{\partial G_\infty(\boldsymbol{x},\boldsymbol{x}_0)}{\partial n}\,dS(\boldsymbol{x})$$

where S_ϵ is the portion of S within a tiny radius ϵ of $\boldsymbol{x}_0$.

Finally, inserting (3.85) into (3.84) yields that for points $\boldsymbol{x}_0$ on the boundary

$$\frac{1}{2}u(\boldsymbol{x}_0) = \int_S \left(G_\infty(\boldsymbol{x},\boldsymbol{x}_0)\frac{\partial u(\boldsymbol{x})}{\partial n} - u(\boldsymbol{x})\frac{\partial G_\infty(\boldsymbol{x},\boldsymbol{x}_0)}{\partial n}\right) dS(\boldsymbol{x}) \tag{3.86}$$

If Dirichlet boundary conditions $u = g$ are imposed, then the left-hand side and the second integral is known and the boundary values of $\partial u/\partial n$ are determined by the solution of this equation. If Neumann boundary conditions are imposed, then the boundary values of u are the unknowns. If u is imposed on some part of the boundary, and $\partial u/\partial n$ on the remainder, then $\partial u/\partial n$ is an unknown on the part of the boundary where u is imposed and vice versa.

Closed form solutions to (3.86) can be obtained in special cases, but its importance goes beyond these. On a fundamental level, it shows that Laplace's equation, a partial differential equation, can be reformulated as an integral equation whose domain is the boundary of the original domain. On a practical level, it forms the basis of an important computational approach to solving the Laplace equation and related problems, the BOUNDARY ELEMENT METHOD. In this approach, the integrals in (3.86) are discretized, leading to a system of linear algebraic equations whose unknowns are values of u and $\partial u/\partial n$ at points on the boundary.

3.3.6 Characteristics and D'Alembert's Solution to the Wave Equation

The wave equation

$$u_{tt} = c^2\nabla^2 u \tag{3.87}$$

governs wave propagation in many physical contexts, including electromagnetic waves (light), vibrations of strings and membranes, and sound propagation. In one spatial dimension, the equation is

$$u_{tt} = c^2 u_{xx} \tag{3.88}$$

which was introduced in Section 3.3.1 as an archetypal hyperbolic equation. Following the change of variable procedure introduced there, we find that $\xi = x - ct$ and $\eta = x + ct$. Rewriting (3.88) in these coordinates yields (3.21) with $g = 0$

$$\frac{\partial^2 u}{\partial \xi \partial \eta} = 0$$

We can easily integrate this twice to find the general solution of the wave equation

$$u(x,t) = F_1(\xi) + F_2(\eta) = F_1(x - ct) + F_2(x + ct)$$

It says that any solution is a superposition of a right-moving and a left-moving wave. Usually, we want to understand the wave equation as an initial-value problem, so we look at two cases of initial conditions and then combine them to get a general result.

First, consider the initial condition $u(x,0) = u_0(x), u_t(x,0) = 0$. This condition corresponds to a plucked string: the string is pulled to a stationary shape, held, then released. There is an initial deformation, but no initial velocity. At $t = 0$ the above general solution and its time derivative become

$$\begin{aligned} u(x,0) &= u_0(x) = F_1(x) + F_2(x) \\ u_t(x,0) &= 0 = -cF_1' + cF_2' \end{aligned}$$

The latter equation integrates to yield $F_1 = F_2$, and using this fact in the first equation gives $F_1 = F_2 = \frac{1}{2}u_0$. Thus the solution for these initial conditions is

$$u(x,t) = \frac{1}{2}u_0(x - ct) + \frac{1}{2}u_0(x + ct)$$

The initial condition splits immediately into two identical waves, one traveling to the right and one to the left. These waves have the same shape, but half the amplitude, of the initial condition. In contrast to the parabolic heat equation $u_t = u_{xx}$, which smooths discontinuous initial conditions as illustrated in Example 3.12, no smoothing occurs in the wave equation. If an initial condition contains a discontinuity at a point x, this will simply propagate along the characteristic directions $\xi =$ constant, $\eta =$ constant.

Now consider a struck string rather than a plucked one. The initial condition is $u(x,0) = 0, u_t(x,0) = v_0(x) \neq 0$. There is no initial deformation, but there is an initial velocity. Now at $t = 0$ we have

$$\begin{aligned} u(x,0) &= 0 = F_1(x) + F_2(x) \\ u_t(x,0) &= v_0(x) = -cF_1' + cF_2' \end{aligned}$$

This tells us that $F_1 = -F_2$ and that $v_0 = 2cF_2'$. We can integrate this to find that

$$F_2(x + ct) = \frac{1}{2c}\int_0^{x+ct} v_0(\xi)\, d\xi$$

Similarly

$$F_1(x - ct) = -\frac{1}{2c}\int_0^{x-ct} v_0(\xi)\, d\xi$$

The solution is $F_1 + F_2$, which is

$$u(x,t) = \frac{1}{2c}\int_{x-ct}^{x+ct} v_0(\xi)\, d\xi$$

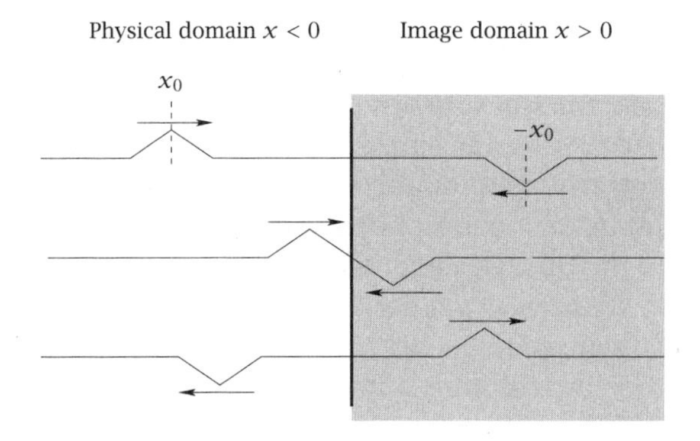

Figure 3.9: An initially right-traveling wave in the domain $x < 0$ reflecting across a wall where $u = 0$, as solved using superposition of a left-traveling "image" with opposite sign.

The complete solution to the initial-value problem is the sum of the above two cases. This is D'Alembert's solution

$$u(x,t) = \frac{1}{2}u_0(x - ct) + \frac{1}{2}u_0(x + ct) + \frac{1}{2c}\int_{x-ct}^{x+ct} v_0(\xi)\, d\xi$$

We have only considered the very simplest hyperbolic equation here. For example, if the coefficients a, b, c depend on position, then the characteristics are curved. The references contain extensive information about more complex hyperbolic problems.

Because the wave equation is linear, we can superpose multiple solutions to form another solution. As an application of this fact, imagine a pulse traveling rightward toward a boundary at $x = 0$, at which the boundary condition is $u = 0$. At time $t = 0$, the pulse is centered at $x = x_0$. To understand this situation, recall Figure 3.8 and the "method of images" analysis of Section 3.3.5. Applying the same idea here, we place an "image" pulse of the same shape but opposite sign at the position $x = -x_0$ (which is outside the physical domain) and make it move leftward as shown in Figure 3.9. Now the real and image pulses will eventually overlap, and by symmetry they will satisfy $u = 0$ at $x = 0$. Once the "image" pulse enters the physical domain, it is no longer an image, but a component of the true solution. The implication of this construction is that when a wave hits a boundary where no deformation is allowed, it reflects but with a change of sign. What happens if the boundary condition is $u_x = 0$?

3.3.7 Laplace Transform Methods

Next we illustrate the solution of several linear PDEs with Laplace transforms. For a user with some experience, Laplace transforms are probably the most powerful method for solving linear, low-dimensional PDEs in closed form. After taking the Laplace transform of a PDE, usually with respect to the time variable, the result is a linear ODE in the transform function. We can often solve this ODE. To perform the inverse transform, we then require some inverse formulas for transforms with singularities. We develop these inverse formulas next and then solve some example PDEs. Let the transform function

$$\overline{f}(s) = \frac{\overline{p}(s)}{\overline{q}(s)}$$

have singularities at the zeros of $\overline{q}(s)$, which is assumed to have m simple zeros [6]

$$\overline{q}(s) = 0 \qquad s = s_1, s_2, \ldots, s_m$$

The inverse of this Laplace transform is given by the following formula

$$f(t) = \sum_{n=1}^{m} a_n e^{s_n t} \qquad a_n = \frac{\overline{p}(s_n)}{\overline{q}'(s_n)} \tag{3.89}$$

which is usually called the Heaviside expansion theorem. When $\overline{p}(s)$ and $\overline{q}(s)$ are polynomials, the coefficients a_n can be derived using partial fractions. But the result applies to more general cases as we require in the two examples below, where $\overline{q}(s) = \sinh\sqrt{k+s}$ and $\overline{q}(s) = \sinh s$.[7]

When the zeros of $\overline{q}(s)$ are higher than first order, $f(t)$ is a linear combination of products of polynomials and exponentials of time, and the coefficients are more complex. Let the zero s_n have order r_n, $n = 1, 2, \ldots m$. Then the inverse is given by

$$f(t) = \sum_{n=1}^{m} e^{s_n t} \sum_{i=1}^{r_n} a_{ni} t^{i-1} \tag{3.90}$$

[6] The singularities of complex-valued functions are poles, branch points, and essential singularities (Levinson and Redheffer, 1970). The *order* of a zero is the smallest integer i such that $q(s_n)/(s-s_n)^i$ is nonzero, and a simple zero is a first-order zero. So we are assuming here that the function $\overline{f}(s)$ has m simple poles.

[7] We are in good company. Heaviside also used the expansion for the case of $\overline{q}(s) = \sinh xs$ (Vallarta, 1926)(Heaviside, 1899, p. 88).

The coefficients a_{ni}, for $i = 1, \ldots, r_n$, $n = 1, 2, \ldots m$, are given by

$$a_{ni} = \frac{\phi^{(r_n - i)}(s_n)}{(r_n - 1)!(i-1)!}$$

in which

$$\phi(s) = (s - s_n)^{r_n} \frac{\overline{p}(s)}{\overline{q}(s)}$$

and $\phi^{(i)}(s_n)$ denotes the ith derivative of $\phi(s)$ evaluated at $s = s_n$. For students with a background in complex variables, Exercise A.2 provides some hints to establish (3.90) (and hence also (3.89)), which requires inverting the Laplace transform by performing the contour integral (2.7).

Next we use Laplace transforms to solve the reaction-diffusion equation and the wave equation. We will see that the transform in both problems has only simple zeros and we will use (3.89) for calculating the inverse.

Example 3.14: Reaction and diffusion in a membrane

The following model describes diffusion through a membrane in which component A decomposes by a first-order reaction. The membrane initially has zero concentration of A. At $t = 0$ the concentration at the side of the membrane at $x = 0$ is abruptly raised to concentration c_{A_0} and the other side is maintained at zero concentration.

$$\text{PDE} \quad \frac{\partial c_A}{\partial t} = D_A \frac{\partial^2 c_A}{\partial x^2} - K c_A$$

$$\begin{array}{lll} \text{BC1} & c_A(0,t) = c_{A_0} & t > 0 \\ \text{BC2} & c_A(L,t) = 0 & t > 0 \\ \text{IC} & c_A(x,0) = 0 & 0 < x < L \end{array}$$

(a) Define the dimensionless variables

$$c = \frac{c_A}{c_{A_0}} \qquad z = \frac{x}{L} \qquad \tau = \frac{tD_A}{L^2} \qquad k = \frac{KL^2}{D_A}$$

and show that the model reduces to

$$\text{PDE} \quad \frac{\partial c}{\partial \tau} = \frac{\partial^2 c}{\partial z^2} - kc$$

$$\begin{array}{lll} \text{BC1} & c(0,\tau) = 1 & \tau > 0 \\ \text{BC2} & c(1,\tau) = 0 & \tau > 0 \\ \text{IC} & c(z,0) = 0 & 0 < z < 1 \end{array}$$

in which $k = KL^2/D_A$ is the only dimensionless parameter appearing in the problem. This dimensionless parameter is known as the Thiele number or Thiele modulus in the chemical reaction engineering literature (Rawlings and Ekerdt, 2012, p. 363). It indicates the ratio of the reaction rate to the diffusion rate.

(b) Take the Laplace transform of your model (also the boundary conditions). Solve the resulting differential equation and boundary conditions for $\overline{c}(z,s)$ and show that

$$\overline{c}(z,s) = \frac{\sinh(\sqrt{s+k}(1-z))}{s \sinh\sqrt{s+k}}$$

(c) Apply the final-value theorem to $\overline{c}(z,s)$ to find the steady-state solution $c_s(z)$.

(d) Take the limit of this solution as $k \to 0$ for the zero-reaction case. Does your solution satisfy the diffusion equation?

(e) Sketch the solution $c_s(z)$ for a range of k values and show the effect of reaction on the steady-state concentration profile.

(f) Let $\overline{p}(s) = \sinh\sqrt{s+k}(1-z)$ and $\overline{q}(s) = s\sinh\sqrt{s+k}$, and find the zeros s_n of $\overline{q}(s)$. Also find the value of $\overline{p}(s_n)/\overline{q}'(s_n)$ at the zeros of $\overline{q}(s)$. The following formulas may be helpful: $\cosh(iu) = \cos(u)$, $\sinh iu = i \sin u$.

(g) Invert the transform and find $c(z,t)$. Check that the solution satisfies the PDE and boundary conditions.

Solution

(a) Inserting the defined dimensionless variables in the PDE gives

$$\frac{D_A c_{A0}}{L^2}\frac{\partial c}{\partial \tau} = D_A \frac{c_{A0}}{L^2}\frac{\partial^2 c}{\partial z^2} - K c_{A0} c$$

and rearranging gives

$$\frac{\partial c}{\partial \tau} = \frac{\partial^2 c}{\partial z^2} - \frac{KL^2}{D_A} c$$

$$\frac{\partial c}{\partial \tau} = \frac{\partial^2 c}{\partial z^2} - kc$$

Inserting the dimensionless variables in the boundary and initial conditions gives

$$c_{A_0}c(z,\tau) = c_{A_0} \qquad z = 0, \quad \tau > 0$$
$$c_{A_0}c(z,\tau) = 0 \qquad zL = L, \quad \tau > 0$$
$$c_{A_0}c(z,\tau) = 0 \qquad 0 < zL < L, \quad \tau = 0$$

Simplifying these expression gives

$$c(z,\tau) = 1 \qquad z = 0, \quad \tau > 0$$
$$c(z,\tau) = 0 \qquad z = 1, \quad \tau > 0$$
$$c(z,\tau) = 0 \qquad 0 < z < 1, \quad \tau = 0$$

(b) Taking the Laplace transform of the PDE and BCs gives

$$s\overline{c}(z,s) - c(z,0) = \frac{d^2\overline{c}}{dz^2} - k\overline{c}$$
$$\frac{d^2\overline{c}}{dz^2} - (k+s)\overline{c} = 0$$
$$\overline{c}(1,s) = 0 \qquad \overline{c}(0,s) = \frac{1}{s}$$

The solution of the ODE can be written

$$\overline{c}(z,s) = a\cosh\sqrt{s+k}(1-z) + b\sinh\sqrt{s+k}(1-z)$$

and we use the two BCs to find the constants a and b. We have

$$0 = a \qquad \frac{1}{s} = b\sinh\sqrt{s+k}$$

so we have

$$b = \frac{1}{s\sinh\sqrt{s+k}}$$

which gives for the Laplace transform of the solution

$$\overline{c}(z,s) = \frac{\sinh\sqrt{s+k}(1-z)}{s\sinh\sqrt{s+k}}$$

(c) Applying the final-value theorem gives

$$
\begin{aligned}
c_s(z) &= \lim_{s\to 0} s\overline{c}(z,s) \\
&= \lim_{s\to 0} s \frac{\sinh\sqrt{s+k}(1-z)}{s\sinh\sqrt{s+k}} \\
&= \lim_{s\to 0} \frac{\sinh\sqrt{s+k}(1-z)}{\sinh\sqrt{s+k}} \\
c_s(z) &= \frac{\sinh\sqrt{k}(1-z)}{\sinh\sqrt{k}}
\end{aligned}
$$

(d) Using the fact that $\sinh x \approx x$ for small x gives

$$
\begin{aligned}
\lim_{k\to 0} c_s(z) &= \frac{\sqrt{k}(1-z)}{\sqrt{k}} \\
&= 1 - z
\end{aligned}
$$

Yes, the solution satisfies the steady-state diffusion equation and boundary conditions

$$
\frac{d^2 c_s(z)}{dz^2} = 0 \qquad c_s(0) = 1 \qquad c_s(1) = 0
$$

(e) The concentration profile $c_s(z)$ versus z for a variety of rate constant k are given in Figure 3.10. We see that a large reaction rate constant prevents species A from diffusing very far into the membrane.

(f) Since the zeros of $\sin u$ are $u = \pm n\pi$, $n = 0, 1, 2, \ldots$, the zeros of $\sinh u$ are $u = \pm n\pi i$, $n = 0, 1, 2, \ldots$.[8] The zeros of $\sinh\sqrt{k+s}$ are given by $s_n = -(n^2\pi^2 + k)$, and for these roots, we have that $\sqrt{s_n + k} = n\pi i$, in which we choose the positive square root. Therefore the zeros of the denominator $\overline{q}(s)$ are given by

$$
s = \{0, -(n^2\pi^2 + k)\}, \qquad n = 0, 1, 2, \ldots
$$

These are simple zeros so the inversion formula in (3.89) is applicable. Differentiating $\overline{q}(s)$ and evaluating $\overline{q}'(s)$ at the zeros

[8]See Exercise 3.48 for a proof that these are the only zeros of $\sin u$ for $u \in \mathbb{C}$.

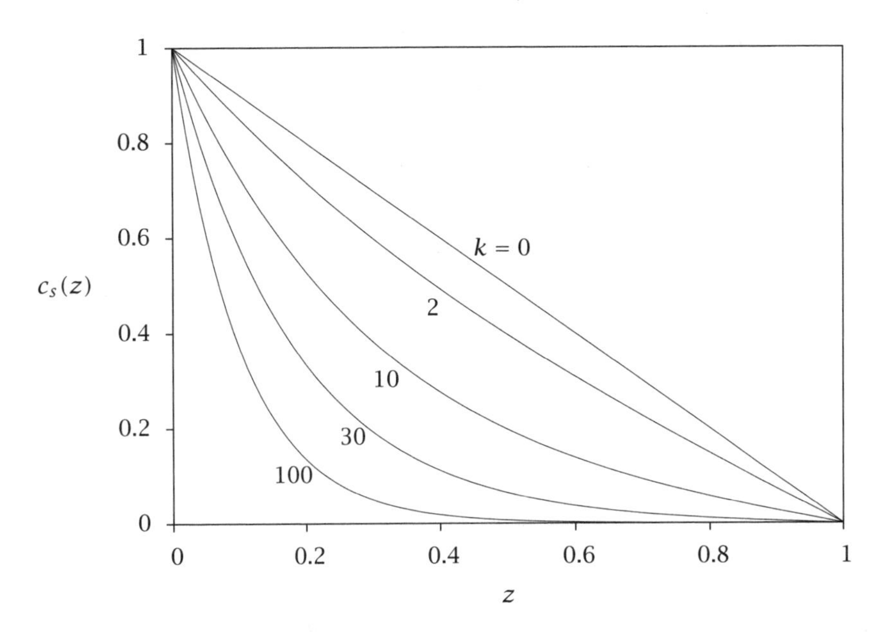

Figure 3.10: Concentration versus membrane penetration distance for different reaction rate constants.

gives

$$\overline{q}'(s) = \sinh\sqrt{s+k} + \frac{s\cosh\sqrt{s+k}}{2\sqrt{s+k}}$$

$$\overline{q}'(0) = \sinh\sqrt{k}$$

$$\overline{q}'(-(n^2\pi^2+k)) = -\frac{(n^2\pi^2+k)(-1)^n}{2n\pi i}$$

Evaluating $\overline{p}(s)$ at the zeros gives

$$\overline{p}(0) = \sinh\sqrt{k}(1-z) \qquad \overline{p}(-(n^2\pi^2+k)) = i\sin n\pi(1-z)$$

(g) Putting these terms together in (3.89) gives

$$c(z,\tau) = \frac{\sinh\sqrt{k}(1-z)}{\sinh\sqrt{k}} + \sum_{n=0}^{\infty}(-1)^n\frac{2\pi n}{n^2\pi^2+k}\ \sin(n\pi(1-z))\ e^{-(n^2\pi^2+k)\tau}$$

Noticing the $n = 0$ term vanishes, we can rewrite the solution as

$$c(z,\tau) = \frac{\sinh\sqrt{k}(1-z)}{\sinh\sqrt{k}} - 2\sum_{n=1}^{\infty}\frac{(-1)^{n+1}\pi n}{n^2\pi^2+k}\sin(n\pi(1-z))\,e^{-(n^2\pi^2+k)\tau}$$

Compare also to entry 34 in Table A.1. □

Example 3.15: Solving the wave equation

Revisit the wave equation $u_{tt} = c^2 u_{xx}$ on $x \in [0,1]$ for a string with fixed ends $u(0,t) = u(1,t) = 0$, and the plucked string initial condition, $u(x,0) = u_0(x)$, $u_t(x,0) = 0$. Solve this equation using the Laplace transform. Compare the solution to D'Alembert's solution. Which form do you prefer and why?

Solution

First we define $\tau = ct$ to remove the velocity c and simplify our work. The problem is now

$$\begin{aligned} u_{\tau\tau} &= u_{xx} \\ u(x,0) = u_0(x), \quad u_\tau(x,0) &= 0 \qquad x \in (0,1) \\ u(0,t) = 0, \quad u(1,t) &= 0 \qquad \tau \geq 0 \end{aligned}$$

Taking the Laplace transform with respect to the time variable gives

$$\overline{u}(x,s)_{xx} - s^2\overline{u}(x,s) = -su_0(x)$$

with transformed boundary conditions

$$\overline{u}(0,s) = \overline{u}(1,s) = 0$$

We obtain a second-order nonhomogeneous differential equation for the transform. We already have solved this problem in Chapter 2, and obtained the Green's function. The solution is therefore

$$\overline{u}(x,s) = \int_0^1 \overline{G}(x,\xi,s)u_0(\xi)d\xi$$

in which

$$\overline{G}(x,\xi,s) = \begin{cases} -\dfrac{\sinh(s\xi)\sinh(s(1-x))}{\sinh s}, & \xi < x \\ -\dfrac{\sinh(sx)\sinh(s(1-\xi))}{\sinh s}, & \xi > x \end{cases}$$

Notice that, as we expect, $\overline{G}(x, \xi, s)$ is symmetric in (x, ξ) because the second-order boundary-value problem is self-adjoint. Next we invert $\overline{G}(x, \xi, s)$. We require a Laplace inverse for the following form

$$\overline{f}(s) = \frac{\sinh(as)\sinh(bs)}{\sinh s}$$

Notice that $\sinh s$ has simple zeros at $s_n = n\pi i$ with n an integer. We use the formula given in (3.89) to obtain

$$p(s_n) = \sinh(n\pi ai)\sinh(n\pi bi) = -\sin(n\pi a)\sin(n\pi b)$$
$$q'(s_n) = \cosh(n\pi i) = (-1)^n$$

Therefore the inverse is

$$f(\tau) = \sum_{n=-\infty}^{\infty} (-1)^{n+1}\sin(n\pi a)\sin(n\pi b)e^{in\pi\tau}$$

Substituting $e^{in\pi\tau} = \cos(n\pi\tau) + i\sin(n\pi\tau)$ and combining terms gives

$$f(\tau) = 2\sum_{n=1}^{\infty}(-1)^{n+1}\sin(n\pi a)\sin(n\pi b)\cos n\pi\tau$$

Notice that the function is now real valued as it must be. Using this result to invert the Green's function gives

$$G(x, \xi, \tau) = 2\sum_{n=1}^{\infty}\begin{cases}(-1)^n\sin(n\pi\xi)\sin(n\pi(1-x))\cos(n\pi\tau) & \xi < x\\ (-1)^n\sin(n\pi x)\sin(n\pi(1-\xi))\cos(n\pi\tau) & \xi > x\end{cases}$$

But noticing that $\sin(n\pi(1-\xi)) = (-1)^{n+1}\sin(n\pi\xi)$ reduces this to

$$G(x, \xi, \tau) = 2\sum_{n=1}^{\infty}\sin(n\pi\xi)\sin(n\pi x)\cos(n\pi\tau)$$

Substituting this into the solution gives

$$u(x, t) = 2\sum_{n=1}^{\infty}\sin(n\pi x)\cos(n\pi t)\int_0^1 u_0(\xi)\sin(n\pi\xi)d\xi$$

Defining the Fourier coefficients representing the initial condition

$$a_n = 2\int_0^1 u_0(\xi)\sin(n\pi\xi)d\xi$$

we have finally[9]

$$u(x, \tau) = \sum_{n=1}^{\infty} a_n \sin(n\pi x) \cos(n\pi\tau) \tag{3.91}$$

Returning to the original time variable with the substitution $\tau = ct$ gives

$$u(x, t) = \sum_{n=1}^{\infty} a_n \sin(n\pi x) \cos(n\pi ct)$$

Notice that this solution does not resemble D'Alembert's solution. The Laplace transform has provided a Fourier series representation of the solution to the wave equation. It is easy to see that the solution satisfies the wave equation. Taking two x derivatives gives $u_{xx} = -(n\pi)^2 u$; similarly, taking two τ derivatives gives $u_{\tau\tau} = -c^2(n\pi)^2 u$ so $u_{tt} = c^2 u_{xx}$ and the solution satisfies the wave equation. The zero boundary conditions are satisfied because all the sine terms vanish at $x = 0, 1$. The initial condition is satisfied because of the Fourier series representation of $u_0(x)$. We see immediately that the solution is periodic (in time) with period $T = 2/c$ since all the cosine terms have this period. The Fourier series solution is also convenient if we wish to analyze the frequency content of the solution, which is often a quantity of interest when modeling sound propagation.

D'Alembert's solution, on the other hand, provides the nice structural insight that the solution splits into two waves traveling in opposite directions. But then we also require the additional insight from the method of images to enforce zero boundary conditions and extend the solution to the (x, t) values where $x - ct < 0$ or $x + ct > 1$, for which $u_0(x - ct)$ or $u_0(x + ct)$ is not defined. □

3.4 Numerical Solution of Initial-Boundary-Value Problems: Discretization and Numerical Stability

Chapter 2 introduced numerical methods for solving initial-value and boundary-value problems. These approaches will be combined here to solve initial-boundary-value problems

$$\frac{\partial u}{\partial t} + Lu = f(\boldsymbol{x}), \quad Bu(S, t) = h, \quad u(\boldsymbol{x}, 0) = u_0(\boldsymbol{x})$$

[9]If we knew enough about the problem to propose a solution of this form, we could arrive at this answer more quickly. The value of the Laplace transform here is that it is *prescriptive.* You do not have to know (or guess) the structure of the solution to apply the method.

where L is a differential operator that contains all the spatial derivatives, S is the boundary of the domain, and B is an operator that determines the boundary conditions. To treat this problem the spatial-dependence will be discretized using the approaches of Chapter 2 to yield a set of ordinary differential equations in the form of a normal initial-value problem. Then the time-integration approaches also introduced in Chapter 2 can be used. This approach is sometimes called the METHOD OF LINES. We will see that a central issue in this approach is the numerical stability of time integration, which is now closely coupled to the spatial discretization (Press, Teukolsky, Vetterling, and Flannery, 1992; Strang, 1986).

Any of the methods introduced in Section 2.9 can be used for spatial discretization. In the weighted residual formulation for one spatial dimension, we look for an approximate solution $u_N(x,t)$; a truncated (discretized) series of basis (trial) functions $\phi_j(x)$; the difference now is that we allow the coefficients in the series to depend on time. That is

$$u_N(x,t) = \sum_{j=1}^{N} c_j(t)\phi_j(x)$$

Note the similarity of this expression to those arising in the separation of variables technique. We assume for the moment that the basis functions satisfy the boundary conditions and define the residual or error by

$$R = \frac{\partial u_N}{\partial t} + Lu_N - f(x)$$

The residual is now forced to be orthogonal to the set of N test functions ψ_i; that is $(R,\psi) = 0, i = 1,2,\ldots,N$. In the Galerkin method the test functions equal the trial functions so this condition becomes

$$\sum_{j=1}^{N} (\phi_j,\phi_i)\dot{c}_j + \sum_{j=1} N(L\phi_j,\phi_i)c_j = (f,\phi_i), i = 1,2,\ldots,N$$

If we let $M_{ij} = (\phi_j,\phi_i)$, $A_{ij} = (L\phi_j,\phi_i)$, and $b_i = (f,\phi_i)$, we can write the weighted residual conditions as

$$\dot{c} = -M^{-1}Ac + M^{-1}b$$

This is a set of linear ODEs (an initial-value problem) for the vector of coefficients c in the series for u_N. We have reduced a partial differential equation to a system of ordinary differential equations.

If the Galerkin tau method is used, then there are only $N - N_{bc}$ ordinary differential equations, where N_{bc} is the number of boundary conditions. The boundary conditions add N_{bc} algebraic equations. Typically, these can be explicitly solved for the last N_{bc} values of c and the formulas substituted into the ODEs.

A similar result arises if we use the collocation approach. Now we replace the spatial derivative operators in L by their matrix approximations, the collocation differentiation operators. This yields

$$\dot{u}(x_i) + L^N_{ij} u(x_j) = f(x_i) \text{ for } x_i \text{ in the interior of the domain}$$
$$u(x_i) = u_c(x_i) \text{ on the boundaries}$$

Here L^N is the matrix operator obtained by inserting the collocation differentiation operators.

In both Galerkin and collocation approaches, the PDE has been reduced to a system of ODEs. In principle, we know how to solve these. In practice, though, there are numerical stability considerations that arise because the matrices derive from the approximation of derivative operators.

3.4.1 Numerical Stability Analysis for the Diffusion Equation

To get an initial idea of the stability issues we face when numerically solving PDEs, we look at the diffusion equation in one dimension,

$$u_t = D u_{xx}$$

in an unbounded domain. Taking the Fourier transform of this equation gives $\hat{u}_t(k) = -k^2 D \hat{u}(k)$, for all real values of k. This is a system of linear ODEs with eigenvalues $\lambda = -Dk^2$. If we want spatial resolution of wavelengths as short as $2\pi/k_{max}$, an explicit Euler scheme would require $\Delta t < -2/\lambda_{max} = 2/(Dk^2_{\max})$ to ensure stability. Defining $\ell_{\min} = \frac{2\pi}{k_{max}}$ as the smallest wavelength resolved, we can rewrite this stability limit as

$$\frac{\Delta t D}{\ell^2_{\min}} < \left(2\pi^2\right)^{-1}$$

This result shows that, to within a numerical constant, the time step for explicit Euler must be shorter than the time scale for diffusion over a distance $\ell_{\min}$.

A similar result holds when finite element or finite difference methods are applied. For simplicity, we will consider a finite difference approximation to the diffusion equation, using the central difference formula (2.95):

$$\frac{du_j}{dt} = D\frac{u_{j-1} - 2u_j + u_{j+1}}{h^2}$$

where h is the spacing between mesh points x_j and $u_j = u(x_j)$. Recall from Chapter 2 that the finite element discretization using hat functions leads to an identical form for the second derivative. The forward Euler approximation to this ODE is

$$u_j^{(n+1)} = u_j^{(n)} + \frac{D\Delta t}{h^2}\left(u_{j-1}^{(n)} - 2u_j^{(n)} + u_{j+1}^{(n)}\right)$$

Following an approach initially developed by von Neumann, we will seek a spatially periodic solution to this equation: $u_j^{(n)} = e^{ikx_j} = e^{ikjh}$ where k is arbitrary. (The full solution is a superposition over all k.) In a periodic or unbounded domain, this yields an exact solution to the discretized problem—in a bounded domain it works very well when $kL \gg 1$ where L is the domain size. Substituting into the equation above gives

$$\begin{aligned} u_j^{(n+1)} &= \left\{1 - \frac{2D\Delta t}{h^2} + \frac{2D\Delta t}{h^2}\cos kh\right\} e^{ikjh} \\ &= Gu_j^{(n)} \end{aligned}$$

Here $G = \left\{1 - \frac{2D\Delta t}{h^2} + \frac{2D\Delta t}{h^2}\cos kh\right\}$ is the growth factor, which for numerical stability must satisfy $|G| < 1$. When $k = 0$, $G = 1$, which makes physical sense because $k = 0$ corresponds to a constant function, which does not decay by diffusion (there are no gradients). As k increases, G decreases, taking on its most negative value when $kh = \pi$. To maintain stability at this value of k requires that

$$\frac{2D\Delta t}{h^2} < 1 \tag{3.92}$$

Indeed, one common indication of numerical instability in a solution is the observation of "sawtooth" patterns with a length scale close to h.

Equation (3.92) is the key result of numerical stability theory for parabolic differential equations and is sometimes called the diffusive Courant-Friedrichs-Levy (CFL) condition[10]. It mirrors the result we found above using Fourier transforms: to within a constant, the time

[10]The true CFL condition was derived for convection problems and is given in the following section.

step Δt must be smaller than the time h^2/D required for diffusion between two mesh points or across one element. The maximum time step scales as the square of the minimum element size h. This can be a *very* severe restriction on the time step if high spatial resolution is required, as in problems with boundary layers.

This severe stability restriction means that for problems where diffusion is important (Peclet number is not high), implicit integration techniques are almost always used. The second-order Adams-Moulton method (AM2) is popular. For the finite difference approach used here, AM2 becomes

$$\begin{aligned} u_j^{(n+1)} = u_j^{(n)} &+ \frac{1}{2}\frac{D\Delta t}{h^2}\left(u_{j-1}^{(n)} - 2u_j^{(n)} + u_{j+1}^{(n)}\right) \\ &+ \frac{1}{2}\frac{D\Delta t}{h^2}\left(u_{j-1}^{(n+1)} - 2u_j^{(n+1)} + u_{j+1}^{(n+1)}\right) \end{aligned}$$

This is called the CRANK-NICHOLSON method. The linear system that must be solved at each time step is tridiagonal so it can be factored quickly.

3.4.2 Numerical Stability Analysis for the Convection Equation

We just considered diffusion, so it makes sense now to look at convection. The transient convection equation in one dimension (also called the FIRST-ORDER WAVE EQUATION)

$$u_t + v u_x = 0 \tag{3.93}$$

where v is a constant velocity. The Fourier transform of this is $\hat{u}_t = -ikv\hat{u}$. Now the eigenvalue is purely imaginary. Recall that imaginary eigenvalues pose a problem for many time-integration schemes; many, including forward Euler and RK2, are *never* stable for problems with imaginary eigenvalues.

Using the central difference formula

$$\frac{\partial u}{\partial x} \approx \frac{u(x_{j+1}) - u(x_{j-1})}{2h}$$

Equation (3.93) becomes

$$\frac{du_j}{dt} = -\frac{v\Delta t}{2h}(u_{j+1}^{(n)} - u_{j-1}^{(n)})$$

The same right-hand side arises in the finite element approximation with hat functions. The forward Euler approximation is

$$u_j^{(n+1)} = u_j^{(n)} - \frac{v\Delta t}{2h}\left(u_{j+1}^{(n)} - u_{j-1}^{(n)}\right) \tag{3.94}$$

which is sometimes called the forward-time center-space (FTCS) discretization. It is first-order accurate in time and second-order accurate in space.

To analyze stability we again seek a solution $c_j^n = e^{ikx_j} = e^{ikjh}$. This analysis shows that $G = 1 + i\frac{c_w \Delta t}{h}\sin kh$, which *always* has magnitude greater than one when $k \neq 0$. Forward Euler will never work for the convection equation, as we guessed from the Fourier analysis above, which revealed that the eigenvalues of the convection operator are imaginary. The same conclusion holds for the wave equation.

There are number of possible resolutions to this problem. One is to use a different approximation for the spatial derivative. In particular, we might expect that we should only use information from "upwind" when computing the solution at the next time step—after all, in the physical problem, convection carries the value of u downstream, so the approximate solution $u_j^{(n+1)}$ should ideally only be determined by values upstream of it. Applying this idea, we replace the central difference above by a forward or "upwind" difference. For $v > 0$ the forward Euler approximation becomes

$$u_j^{(n+1)} = u_j^{(n)} - \frac{v\Delta t}{h}\left(u_j^{(n)} - u_{j-1}^{(n)}\right)$$

This gives the growth factor

$$G = 1 - \frac{v\Delta t}{h}\left(1 - e^{-ikh}\right)$$

The stability condition $|G| < 1$ will hold if

$$\frac{v\Delta t}{h} < 1$$

Defining $C = \frac{v\Delta t}{h}$ as the COURANT NUMBER, the stability condition becomes

$$C < C_{\max} \tag{3.95}$$

where in this case $C_{\max} = 1$. This is the COURANT-FRIEDRICHS-LEVY CONDITION, often simply called the COURANT CONDITION. Physically, it tells us that the time step must be smaller than the time it takes for convection at speed v over one mesh unit h. By replacing the central difference, which has second-order accuracy in space, with an upwind difference, we have lost an order in spatial accuracy but have gained stability. And anyway, the method is still first order in time. One small complication of this method is that for problems where the velocity can change sign, it is necessary to take care that the appropriate upwind difference is used. If downwind differencing is used, the approximation is always unstable.

Stability also can be gained without use of upwind differences. The LAX-FRIEDRICHS method is a simple modification to the FTCS discretization where the present value at point x_j is replaced by the average of the values at points $j+1$ and $j-1$

$$u_j^{(n+1)} = \frac{1}{2}\left(u_{j+1}^{(n)} + u_{j-1}^{(n)}\right) - \frac{v\Delta t}{2h}\left(u_{j+1}^{(n)} - u_{j-1}^{(n)}\right) \tag{3.96}$$

By applying the average, this change effectively introduces a small amount of smoothing or "numerical diffusion" into the time-integration process. This can be seen explicitly by rewriting (3.96) so that it has the form of (3.94) with an additional remainder term that indicates the difference between the two methods

$$u_j^{(n+1)} = u_j^{(n)} - \frac{v\Delta t}{2h}\left(u_{j+1}^{(n)} - u_{j-1}^{(n)}\right) + \frac{1}{2}\left(u_{j-1}^{(n)} - 2u_j^{(n)} + u_{j+1}^{(n)}\right) \tag{3.97}$$

The remainder term has very nearly the form of a central difference approximation to the second-derivative operator, and in fact this expression is precisely the FTCS approximation to a convection-diffusion equation with artificial or numerical diffusivity $\frac{h^2}{2\Delta t}$

$$u_t + vu_x = \frac{h^2}{2\Delta t}u_{xx}$$

This diffusion term is enough to stabilize the method: using the von Neumann analysis the stability criterion is found to be very similar to what we found for the upwind scheme but is now insensitive to the sign of v

$$\frac{|v|\,\Delta t}{h} = |C| < 1$$

All the methods developed so far for the convection equation are first order in time, so even if the stability condition is satisfied, the solution may not be very accurate. The LAX-WENDROFF scheme builds on the Lax-Friedrichs scheme to yield second-order accuracy. Let $u_{j\pm1/2}^{(n+1/2)}$ be intermediate values at the midpoint $t+\Delta t/2$ of the time step and on "half-mesh points" $x_j \pm h/2$. Lax-Friedrichs, (3.96) is used to generate these intermediate values

$$u_{j\pm1/2}^{(n+1/2)} = \frac{1}{2}\left(u_{j\pm1}^{(n)} + u_j^{(n)}\right) \mp \frac{v\Delta t}{2h}\left(u_{j\pm1}^{(n)} - u_j^{(n)}\right)$$

This solution is used in a modified FTCS step to generate the solution at time $t + \Delta t$

$$u_j^{(n+1)} = u_j^{(n)} - \frac{v\Delta t}{h}\left(u_{j+1/2}^{(n+1/2)} - u_{j-1/2}^{(n+1/2)}\right)$$

Eliminating the intermediate values, this can be rewritten in the more illuminating form

$$u_j^{(n+1)} = u_j^{(n)} - \frac{v\Delta t}{2h}\left(u_{j+1}^{(n)} - u_{j-1}^{(n)}\right) + \frac{(v\Delta t)^2}{2h^2}\left(u_{j-1}^{(n)} - 2u_j^{(n)} + u_{j+1}^{(n)}\right) \tag{3.98}$$

This is almost identical to 3.97; the difference is that now the artificial diffusivity has the value $v^2\Delta t/2$. The stability condition is again $|C| < 1$ and now, since the method is second order in time, the time step can be set very close to the stability limit and still yield enough accuracy for many purposes. Lax-Wendroff and related methods are thus widely used.

Pure convection does not change the amplitude of an initial condition; convection only carries the initial condition downstream. In all of the methods described here, some amplitude damping occurs ($|G| < 1$) except precisely when $k = 0$ or $C = 1$. We care most about this damping when kh is small, corresponding to length scales that are large compared to the grid size, i.e., $|G|$ should be very close to unity for all length scales of interest. If we care about length scales close to h, then we have made h too big; h should always be chosen to be much smaller than the length scales over which the true solution varies. Taylor-expanding $|G|^2$ around $kh = 0$ yields

$$|G|^2 = 1 - (1 - C^2)(kh)^2 + O((kh)^4)$$

and

$$|G|^2 = 1 - C^2(1 - C^2)\frac{(kh)^4}{4} + O(kh)^6$$

for Lax-Friedrichs and Lax-Wendroff, respectively. The latter is substantially better, since the deviation from $|G|^2 = 1$ scales as $(kh)^4$ rather than $(kh)^2$.

3.4.3 Operator Splitting for Convection-Diffusion Problems

The cases above represent the low and high Peclet number limits of the general convection-diffusion equation

$$u_t + vu_x = Du_{xx}$$

A simple explicit scheme for this equation would use central differences for the diffusion term and Lax-Wendroff for the convection term

$$u_j^{(n+1)} = u_j^{(n)} + \frac{D\Delta t}{h^2}\left(u_{j-1}^{(n)} - 2u_j^{(n)} + u_{j+1}^{(n)}\right) - \frac{v\Delta t}{2h}\left(u_{j+1}^{(n)} - u_{j-1}^{(n)}\right) + \frac{(v\Delta t)^2}{2h^2}\left(u_{j-1}^{(n)} - 2u_j^{(n)} + u_{j+1}^{(n)}\right)$$

Unless the Peclet number is very large, the diffusive term controls the stability, because it leads to a growth factor that scales as h^{-2} rather than the h^{-1} from the convective term. We could use an implicit method on the whole problem, but this entails solution of a large non-self-adjoint matrix problem (at every time step if the problem is nonlinear). It would be preferable to use an implicit method only on the diffusive piece, which is self-adjoint. A popular solution is called OPERATOR SPLITTING; an explicit method is used for the convective terms and an implicit method for the diffusive ones. For example, Lax-Wendroff can be used for the convective terms and Crank-Nicholson for the diffusive. This is often executed in two steps:

1. The convective terms are applied, to give an intermediate solution $u_j^{(*)}$

$$u_j^{(*)} = u_j^{(n)} - \frac{v\Delta t}{2h}\left(u_{j+1}^{(n)} - u_{j-1}^{(n)}\right) + \frac{(v\Delta t)^2}{2h^2}\left(u_{j-1}^{(n)} - 2u_j^{(n)} + u_{j+1}^{(n)}\right)$$

2. Crank-Nicholson is applied, using the intermediate values instead of the values at step n

$$u_j^{(n+1)} = u_j^{(*)} + \frac{1}{2}\frac{D\Delta t}{h^2}\left(u_{j-1}^{(*)} - 2u_j^{(*)} + u_{j+1}^{(*)}\right) + \frac{1}{2}\frac{D\Delta t}{h^2}\left(u_{j-1}^{(n+1)} - 2u_j^{(n+1)} + u_{j+1}^{(n+1)}\right)$$

In methods like this, because the diffusion terms are evaluated implicitly, the stability limit is set by a Courant condition on the convective terms. In fact, one might also get away with an unstable (e.g., FTCS) method for the convection term, relying on the implicit treatment of the diffusion term to stabilize the overall result. There is not generally a good reason to do this.

3.5 Exercises

Exercise 3.1: Gradient formula from gradient definition

Consider a cubic volume with one corner at the origin and the opposite corner at $(x, y, z) = (\Delta x, \Delta y, \Delta z)$. In this case the integral definition of the gradient grad ϕ becomes

$$\lim_{\Delta x \to 0} \lim_{\Delta y \to 0} \lim_{\Delta z \to 0} \frac{1}{\Delta x \Delta y \Delta z} \int \boldsymbol{n}\phi \, dS$$

Because we are going to shrink the volume to zero, we can make the truncated Taylor-series approximation

$$\phi(x, y, z) = \phi(0, 0, 0) + x\frac{\partial \phi}{\partial x} + y\frac{\partial \phi}{\partial y} + z\frac{\partial \phi}{\partial z} + \cdots$$

where the derivatives are evaluated at the origin. Combine these to derive the formula $\boldsymbol{\nabla} = \sum_{i=1}^{3} \boldsymbol{e}_i \frac{\partial}{\partial x_i}$ (where $x_1 = 1, x_2 = y, x_3 = z$).

The same arguments hold for the other two terms, so

$$\text{grad}\ \phi = \boldsymbol{e}_x \frac{\partial \phi}{\partial x} + \boldsymbol{e}_y \frac{\partial \phi}{\partial y} + \boldsymbol{e}_z \frac{\partial \phi}{\partial z} = \boldsymbol{\nabla}\phi$$

Exercise 3.2: Derivatives of unit vectors in polar (cylindrical) coordinates

By taking limits in polar coordinates, derive the formulas for the derivatives of the unit vectors $\boldsymbol{e}_r, \boldsymbol{e}_\theta$

$$\frac{\partial \boldsymbol{e}_r}{\partial r} = 0 \qquad \frac{\partial \boldsymbol{e}_\theta}{\partial r} = 0 \qquad \frac{\partial \boldsymbol{e}_r}{\partial \theta} = \boldsymbol{e}_\theta \qquad \frac{\partial \boldsymbol{e}_\theta}{\partial \theta} = -\boldsymbol{e}_r$$

Do not refer to Cartesian coordinates in your derivation.

Exercise 3.3: Divergence of the flux in polar coordinates

Derive an expression for the divergence of a flux in polar coordinates, $\boldsymbol{\nabla} \cdot q$, in which q is an arbitrary vector. Do not use Cartesian coordinates in your derivation.

Hint: the answer is

$$\boldsymbol{\nabla} \cdot \boldsymbol{q} = \frac{1}{r}\frac{\partial}{\partial r}(r q_r) + \frac{1}{r}\frac{\partial q_\theta}{\partial \theta}$$

Exercise 3.4: Gradient and Laplacian in spherical coordinates

Repeat Example 3.1 and find expressions for $\boldsymbol{\nabla}$ and ∇^2 for spherical coordinates shown in Figure 3.3. Do not refer to Cartesian coordinates in your derivation. Then derive the result using the h and g formulas provided in the text. Which method do you prefer and why?

Hint: the answers are in Table 3.1.

Exercise 3.5: Fundamental identities in vector calculus

Using Cartesian tensor notation, derive the following identities (here $\boldsymbol{u}, \boldsymbol{v}$, and $\boldsymbol{w}$ are vectors and ϕ is a scalar).

(a) $\boldsymbol{\nabla} \cdot \boldsymbol{\nabla} \times \boldsymbol{u} = 0$

(b) $\boldsymbol{\nabla} \times \boldsymbol{\nabla}\phi = 0$

(c) $\boldsymbol{\nabla} \cdot (\boldsymbol{u}\boldsymbol{u}) = \boldsymbol{u} \cdot \boldsymbol{\nabla}\boldsymbol{u} + \boldsymbol{u}\boldsymbol{\nabla} \cdot \boldsymbol{u}$

Exercise 3.6: Cross-product identities

(a) Verify that $\epsilon_{ijk}\epsilon_{klm} = \delta_{il}\delta_{jm} - \delta_{im}\delta_{jl}$. Now use this to derive the following results.

(b) $\nabla \times \nabla \times \boldsymbol{u} = \nabla\nabla \cdot \boldsymbol{u} - \nabla \cdot \nabla \boldsymbol{u}$

(c) $\nabla \times (\boldsymbol{v} \times \boldsymbol{w}) = (\boldsymbol{w} \cdot \nabla)\boldsymbol{v} - \boldsymbol{w}(\nabla \cdot \boldsymbol{v}) - (\boldsymbol{v} \cdot \nabla)\boldsymbol{w} + (\boldsymbol{v}\nabla \cdot \boldsymbol{w})$

(d) $(\boldsymbol{u} \times \boldsymbol{v}) \times \boldsymbol{w} = (\boldsymbol{u} \cdot \boldsymbol{w})\boldsymbol{v} - (\boldsymbol{v} \cdot \boldsymbol{w})\boldsymbol{u}$

(e) $(\boldsymbol{u} \times \boldsymbol{v}) \times \boldsymbol{w} = (\boldsymbol{v}\boldsymbol{u} - \boldsymbol{u}\boldsymbol{v}) \cdot \boldsymbol{w}$

(f) $\boldsymbol{v} \times (\nabla \times \boldsymbol{v}) = \nabla(\frac{1}{2}\, \|\boldsymbol{v}\|^2) - (\boldsymbol{v} \cdot \nabla)\boldsymbol{v}$

Exercise 3.7: A special case of Leibniz's rule

Derive Leibniz's rule for the special case where the volume V is a cube whose size is constant but is moving with velocity $\boldsymbol{q}$. In other words, explicitly show that the contribution from the motion of V becomes $\int_S m\boldsymbol{q} \cdot \boldsymbol{n} dA$.

Exercise 3.8: Adjoint of curl

Find the adjoint of the curl operator with Dirichlet boundary conditions.

Exercise 3.9: Volume as surface integral

(a) If $\mathbf{A}$ is a constant vector and $r = \|\boldsymbol{x}\|$, then show using Cartesian tensor notation that

$$\mathbf{A} \cdot \nabla \left(\frac{1}{r}\right) = -\frac{\mathbf{A} \cdot \boldsymbol{x}}{r^3}$$

and

$$\nabla(\mathbf{A} \cdot \boldsymbol{x}) = \mathbf{A}$$

(b) Show that

$$\nabla \cdot \boldsymbol{x} = 3$$

(c) Use this result and the divergence theorem to derive a formula for the total volume $T = \int_V dV$ of a region V in terms of an integral over the surface S of the volume.

Exercise 3.10: Curl theorem

Use the divergence theorem and results of vector algebra to show that

$$\int_V \nabla \times \boldsymbol{v}\, dV = \int_S \boldsymbol{n} \times \boldsymbol{v}\, dS$$

Exercise 3.11: Poisson equation in a no-flux domain

Consider the Poisson equation

$$\nabla^2 u = f$$

in a volume T with (no-flux) boundary condition $\boldsymbol{n} \cdot \nabla u = 0$ on the boundary S of T. $\boldsymbol{n}$ is the outward unit normal vector on the boundary.

(a) Use the divergence theorem to show that a necessary condition for the existence of a solution to this problem is

$$\int_T f \, dV = 0$$

(b) If $f = \nabla \cdot \boldsymbol{v}$ for some vector $\boldsymbol{v}$, what condition must $\boldsymbol{v}$ satisfy for the result of part (a) to be satisfied?

Exercise 3.12: Helmholtz decomposition

Under rather general conditions it is possible to write a vector field $\boldsymbol{q}(\boldsymbol{x})$ as

$$\boldsymbol{q} = \nabla \phi + \nabla \times \boldsymbol{v}$$

Using the results of Problem 3.5, find independent equations for ϕ and $\boldsymbol{v}$ in terms of $\boldsymbol{q}$. Find ϕ and $\boldsymbol{v}$ for the case $\boldsymbol{q} = x_1 \boldsymbol{e}_1 + x_1^2 \boldsymbol{e}_2$.

Exercise 3.13: The Stokes equations for viscous flow

The Stokes equations for the velocity $\boldsymbol{u}$ and pressure p in a viscous flow driven by a body force $\boldsymbol{f}$ are

$$\nabla^2 \boldsymbol{u} - \nabla p = -\boldsymbol{f}$$

$$\nabla \cdot \boldsymbol{u} = 0$$

These equations can be written in matrix-vector form as

where

$$\mathbf{A} = \begin{bmatrix} \nabla^2 & -\nabla \\ \nabla \cdot & 0 \end{bmatrix} \qquad \mathbf{U} = \begin{bmatrix} \boldsymbol{u} \\ p \end{bmatrix} \qquad \mathbf{F} = \begin{bmatrix} \boldsymbol{f} \\ 0 \end{bmatrix}$$

If $\boldsymbol{u} = 0$ on the boundary S of the flow domain V, show that the Stokes operator $\mathbf{A}$ is self-adjoint. That is, if

$$\mathbf{V} = \begin{bmatrix} \boldsymbol{v} \\ q \end{bmatrix}$$

then

$$(\mathbf{AU}, \mathbf{V}) = (\mathbf{U}, \mathbf{AV})$$

where the inner product is given by

$$(\mathbf{U}, \mathbf{V}) = \int_V \boldsymbol{u} \cdot \boldsymbol{v} \, dV + \int_V pq \, dV$$

Exercise 3.14: Differentiating functions of a matrix and matrix determinant

Derive the following two differentiation formulas.

(a) Use the polynomial expansion of a matrix function to show that

$$\frac{d}{dt} f(A) = g(A) \frac{d}{dt} A$$

in which $g(\cdot) = d/d(\cdot) f(\cdot)$ is the usual derivative of the scalar function $f(\cdot)$. For the special case of $f(A) = \ln A$ for A nonsingular, we obtain

$$\frac{d}{dt} \ln A = A^{-1} \frac{d}{dt} A$$

(b) For nonsingular A, differentiate (1.35) with respect to scalar t and use the result of the previous part to show that

$$\frac{d}{dt}\det A = \det(A)\,\mathrm{tr}\left(A^{-1}\frac{d}{dt}A\right) \tag{3.99}$$

Exercise 3.15: Euler expansion formula

Let coordinates $\boldsymbol{u}$ represent the reference position of a point in a deformable continuum, and $\boldsymbol{x}$ be the position at time t so $\boldsymbol{x}(0) = \boldsymbol{u}$. We have that $\frac{d}{dt}\boldsymbol{x} = \boldsymbol{v}$, the velocity of the continuum. We know that the size of the volume element when transforming coordinates is given by

$$dV_{\boldsymbol{u}} = \det\left(\frac{\partial \boldsymbol{x}}{\partial \boldsymbol{u}}\right)dV_{\boldsymbol{x}}$$

in which $\det(\frac{\partial \boldsymbol{x}}{\partial \boldsymbol{u}})$ is the determinant of the Jacobian matrix of the transformation. Assuming the Jacobian is nonsingular, use the matrix differentiation formula of the previous exercise to establish that

$$\frac{d}{dt}\det\left(\frac{\partial \boldsymbol{x}}{\partial \boldsymbol{u}}\right) = \det\left(\frac{\partial \boldsymbol{x}}{\partial \boldsymbol{u}}\right)(\nabla \cdot \boldsymbol{v})$$

which is known as the Euler expansion (or dilation) formula.

Exercise 3.16: Temperature profile in tube flow

Read Example 12.2-2 in (Bird et al., 2002, p.384). Check the following points.

(a) Substitute Ψ from Equation 12.2-21 into 12.2-23. Then exchange the order of integration, and show that the inner integral can be performed. Then, make a change of variable to obtain

$$\Theta(\chi,\lambda) = \frac{\sqrt[3]{9\lambda}}{\Gamma(\frac{2}{3})}\left[e^{-\chi^3} - \chi\int_{\chi^3}^{\infty} t^{-1/3}\,e^{-t}\,dt\right]$$

which is equivalent to 12.2-24.

(b) Evaluate the derivatives $\left(\frac{\partial\Theta}{\partial\chi}\right)_\lambda \quad \left(\frac{\partial\Theta}{\partial\lambda}\right)_\chi \quad \left(\frac{\partial^2\Theta}{\partial\chi^2}\right)_\lambda$

(c) Verify that the temperature profile in (a) satisfies the differential equation in Equation 12.2-13. Use the chain rule and the results from (b).

(d) What is the numerical value of $\Gamma(2/3)$?

Exercise 3.17: The error function and some useful integrals

The error function is defined by

$$\mathrm{erf}\,(z) = \frac{2}{\sqrt{\pi}}\int_0^z e^{-t^2}dt \qquad z > 0$$

Note that

$$\int_0^\infty e^{-t^2}dt = \frac{\sqrt{\pi}}{2}$$

The complementary error function defined by

$$\mathrm{erfc}(z) = 1 - \mathrm{erf}\,(z)$$

$$\mathrm{erfc}(z) = \frac{2}{\sqrt{\pi}}\int_z^\infty e^{-t^2}dt \qquad z > 0$$

(a) Sketch the error function and the complementary error function.

(b) Consider the function $f(x)$

$$f(x) = \int_0^\infty e^{-t^2} \cos(2tx)dt$$

Differentiate $f(x)$ and then integrate by parts to show that f satisfies the differential equation

$$\frac{df}{dx} + 2xf(x) = 0$$

What is the initial condition for this ODE?

(c) Solve the ODE and show that

$$f(x) = \frac{\sqrt{\pi}}{2} e^{-x^2}$$

(d) Let $t = au$ and $x = b/(2a)$ and show that

$$\frac{\sqrt{\pi}}{2a} \exp\left(-\frac{b^2}{4a^2}\right) = \int_0^\infty e^{-a^2u^2} \cos(bu)du$$

(e) Integrate the previous equation with respect to b on the interval $[0, \beta]$. Change the order of integration and show finally that

$$\frac{\pi}{2} \operatorname{erf}\left(\frac{\beta}{2a}\right) = \int_0^\infty e^{-a^2u^2} \frac{\sin(\beta u)}{u} du$$

Exercise 3.18: Other useful integrals

Differentiate the following function with respect to x

$$\frac{\sqrt{\pi}}{4a}\left[e^{2ab} \operatorname{erf}\,(ax + \frac{b}{x}) + e^{-2ab} \operatorname{erf}\,(ax - \frac{b}{x})\right]$$

and derive the indefinite integral (Abramowitz and Stegun, 1970, p. 304)

$$\int e^{-a^2x^2 - \frac{b^2}{x^2}} dx = \frac{\sqrt{\pi}}{4a}\left[e^{2ab} \operatorname{erf}\,(ax + \frac{b}{x}) + e^{-2ab} \operatorname{erf}\,(ax - \frac{b}{x})\right] + \text{const.} \qquad a \neq 0$$

Use the indefinite integral to derive the definite integral

$$\int_0^x e^{-a^2x^2 - \frac{b^2}{x^2}} dx = \frac{\sqrt{\pi}}{4a}\left[e^{-2ab} \operatorname{erfc}(\frac{b}{x} - ax) - e^{2ab} \operatorname{erfc}(\frac{b}{x} + ax)\right] \qquad a \neq 0, b \geq 0 \tag{3.100}$$

From this result, show that

$$\int_0^\infty e^{-x^2 - \frac{b^2}{x^2}} dx = \frac{\sqrt{\pi}}{2} e^{-2b} \qquad b \geq 0 \tag{3.101}$$

This integral arises in transport problems in semi-infinite domains (see Exercises 3.19 and 3.23).

Exercise 3.19: Some useful Laplace transforms

The following Laplace transform pairs are useful for solving transient heat-conduction and diffusion equations (see Exercises 3.23 and 5.6).

$\overline{f}(s)$	$f(t)$
$\dfrac{e^{-k\sqrt{s}}}{s}$, $k>0$	$\operatorname{erfc}\left(\dfrac{k}{2\sqrt{t}}\right)$
$e^{-k\sqrt{s}}$, $k>0$	$\dfrac{k}{2\sqrt{\pi t^3}}e^{-\frac{k^2}{4t}}$
$\dfrac{e^{-k\sqrt{s}}}{\sqrt{s}}$, $k>0$	$\dfrac{1}{\sqrt{\pi t}}e^{-\frac{k^2}{4t}}$

(a) Establish the first entry by taking the Laplace transform of the function

$$f(t) = \operatorname{erfc}\left(\frac{k}{2\sqrt{t}}\right)$$

Use the definition of the Laplace transform, switch the order of integration, and use Equation 3.101.

(b) Establish the second entry by differentiating the first $f(t)$ with respect to t.

(c) Establish the third entry by differentiating the second $\overline{f}(s)$ with respect to s.

Exercise 3.20: A transform pair for reaction-diffusion problems

The following Laplace transform pair is useful in solving problems with simultaneous diffusion and first-order reaction (Carslaw and Jaeger, 1959, p. 496)

$$\overline{f}(s) = \frac{e^{-k\sqrt{s}}}{(s-\alpha)\sqrt{s}} \qquad k>0$$

$$f(t) = \frac{1}{2\sqrt{\alpha}}e^{\alpha t}\left\{e^{-k\sqrt{\alpha}}\operatorname{erfc}\left(\frac{k}{2\sqrt{t}} - \sqrt{\alpha t}\right) - e^{k\sqrt{\alpha}}\operatorname{erfc}\left(\frac{k}{2\sqrt{t}} + \sqrt{\alpha t}\right)\right\}$$

Derive this result by using the convolution theorem and the last entry in the table in Exercise 3.19. You will also require the integral (3.100).

Exercise 3.21: Integral representations of K_0

The following integral representation of K_0 proves useful in applying Laplace transforms to solve the diffusion equation

$$K_0(x) = \frac{1}{2}\int_0^\infty t^{-1}e^{-(x^2t+\frac{1}{4t})}dt \tag{3.102}$$

The following argument provides a derivation.

(a) Denote the integral by

$$f_0(x) = \frac{1}{2}\int_0^\infty t^{-1}e^{-(x^2t+\frac{1}{4t})}dt$$

Differentiate with respect to x and show

$$\frac{1}{x}\frac{d}{dx}\left(x\frac{df_0}{dx}\right) = 2x^2\int_0^\infty te^{-(x^2t+\frac{1}{4t})}dt - 2\int_0^\infty e^{-(x^2t+\frac{1}{4t})}dt$$

(b) Next use integration by parts to show

$$\int_0^\infty e^{-(x^2t+\frac{1}{4t})}dt = x^2\int_0^\infty te^{-(x^2t+\frac{1}{4t})}dt - \frac{1}{4}\int_0^\infty t^{-1}e^{-(x^2t+\frac{1}{4t})}dt$$

Substitute this result into the previous equation and show that f_0 satisfies the Bessel equation

$$\frac{1}{x}\frac{d}{dx}\left(x\frac{df_0}{dx}\right) - f_0(x) = 0$$

Therefore f_0 is of the form

$$f_0(x) = a_1 I_0(x) + a_2 K_0(x)$$

with some constants a_1, a_2.

(c) Given the integral defining $f_0(x)$, what value does $f_0(x)$ approach for large x? Use this fact to deduce the value of a_1.

(d) Next use l'Hôpital's rule to show that

$$\lim_{x\to 0}\frac{f_0(x)}{\ln(x)} = -1$$

It is known that $K_0(x) \approx -\ln(x)$ as $x \to 0$ (see (Abramowitz and Stegun, 1970, p. 375)), so we conclude that $a_2 = 1$ and $f_0(x) = K_0(x)$.

Exercise 3.22: More useful Laplace transforms

Use the integral representations of the modified Bessel function K_0 derived in Exercise 3.21 to derive the following Laplace transform pairs.

$\overline{f}(s)$	$f(t)$
$K_0(k\sqrt{s}),\ k > 0$	$\frac{1}{2t}e^{-\frac{k^2}{4t}}$
$\frac{1}{\sqrt{s}}K_1(k\sqrt{s}),\ k > 0$	$\frac{1}{k}e^{-\frac{k^2}{4t}}$

These transforms are also useful in solving transient heat-conduction and diffusion equations (see Exercise 5.9).

Exercise 3.23: Time-dependent heating of a semi-infinite slab

Consider a slab of infinite thickness, density ρ, heat capacity $\hat{C}_p$, and thermal conductivity k with a surface at $x = 0$. The boundary conditions are

$$T(x,0) = T_0 \qquad x > 0$$
$$T(0,t) = T_1 \qquad t > 0$$

(a) Define the following scaled variables

$$\Theta = \frac{T - T_0}{T_1 - T_0} \qquad \tau = \frac{kt}{\rho\hat{C}_P}$$

Show that the energy equation reduces to

$$\frac{\partial \Theta}{\partial \tau} = \frac{\partial^2 \Theta}{\partial x^2}$$

with boundary conditions

$$\Theta(x, 0) = 0 \qquad x > 0$$
$$\Theta(0, \tau) = 1 \qquad \tau > 0$$

Notice that there are no parameters in the problem, but there is also no natural length scale for this problem.

(b) Take the Laplace transform of the PDE and show that

$$\overline{\Theta}(x, s) = \frac{e^{-x\sqrt{s}}}{s}$$

What assumptions did you make?

(c) Take the inverse transform using Exercise 3.19 to obtain

$$\Theta(x, \tau) = \text{erfc}\left(\frac{x}{2\sqrt{\tau}}\right)$$

Plot $\Theta(x, \tau)$ as a function of x on $0 \le x \le 10$ for $\tau = [0.01, 0.1, 1, 10, 100, 1000]$.

(d) Show that the proposed solution satisfies the PDE and BCs.

Exercise 3.24: Partial fraction expansion

We often teach inversion of Laplace transforms by so-called partial fraction expansion. For example, given

$$\overline{f}(s) = \frac{1}{(s-a)(s-b)(s-c)} \qquad a \neq b \neq c$$

Note that $a \neq b \neq c$ ensures a, b, and c are simple zeros of the denominator polynomial. The function $\overline{f}(s)$ is first written as a summation of simpler fractions

$$\frac{1}{(s-a)(s-b)(s-c)} = \frac{A}{s-a} + \frac{B}{s-b} + \frac{C}{s-c} \tag{3.103}$$

and the coefficients A, B, and C are determined. Then the inverse is simply

$$f(t) = Ae^{at} + Be^{bt} + Ce^{ct}$$

(a) Determine A, B, and C in the partial expansion approach and determine $f(t)$.

(b) Apply (3.89) with $\overline{p}(s) = 1$ and $\overline{q}(s) = (s-a)(s-b)(s-c)$, and find $f(t)$ using (3.89). Which method do you prefer and why? Notice that (3.89) can be applied when the denominator $\overline{q}(s)$ is more general than a polynomial as shown in Example 3.14.

Exercise 3.25: Transient heat conduction in a finite slab

Consider the transient heat conduction equation

$$\rho \hat{C}_P \frac{\partial T}{\partial t} = -\nabla \cdot (-k\nabla T)$$

We have a one-dimensional slab with ends located at $x = \pm L$. The slab is initially at uniform temperature T_0. Just after $t = 0$, the two ends are immediately raised to temperature T_1 and held at this temperature. We wish to find the transient solution $T(x,t)$ for this problem.

(a) Write the PDE and (three) boundary conditions for this situation, i.e., conditions at $x = L$, $x = -L$, and $t = 0$. How many parameters appear in this problem?

(b) Choose nondimensional temperature, spatial position, and time variables as follows

$$\Theta = \frac{T - T_0}{T_1 - T_0} \qquad z = \frac{x}{L} \qquad \tau = \frac{tk}{\rho \hat{C}_P L^2}$$

Express the PDE and BCs in these nondimensional variables. How many parameters appear in this problem?

(c) Take the Laplace transform of the PDE, apply the boundary conditions, and show that

$$\overline{\Theta}(z,s) = \frac{\cosh(\sqrt{s}z)}{s \cosh \sqrt{s}}$$

(d) For what s values in the complex plane is $\overline{\Theta}(z,s)$ singular?

(e) Invert the transform and find $\Theta(z,\tau)$.
Hint: the answer is

$$\Theta(z,\tau) = 1 - 2\sum_{n=0}^{\infty} \frac{(-1)^n}{(n+1/2)\pi} \cos((n+1/2)\pi z)\, e^{-((n+1/2)\pi)^2 \tau} \tag{3.104}$$

(f) Show that $\Theta(z,\tau)$ satisfies the PDE and boundary conditions at $z = \pm 1$. Does the solution satisfy the initial condition? How would you check this?

(g) What is the steady state, $\Theta_s(z)$, i.e., take the limit of $\Theta(z,\tau)$ as $\tau \to \infty$.

Exercise 3.26: Heat conduction in a cylinder and a sphere

Let's change the body in Exercise 3.25 from a slab to a cylinder and a sphere and see what happens. Again assume the body is initially at uniform temperature T_0. Just after $t = 0$, the outer boundary at $r = R$ is immediately raised to temperature T_1 and held at this temperature. We wish to find the transient solution $T(r,t)$ for these problems.

(a) Write the PDE and (three) boundary conditions for the **cylindrical** body, i.e., conditions at $r = R$, $r = 0$, and $t = 0$. How many parameters appear in this problem?

(b) Choose nondimensional temperature, radial position, and time variables as follows

$$\Theta = \frac{T - T_0}{T_1 - T_0} \qquad \xi = \frac{r}{R} \qquad \tau = \frac{tk}{\rho \hat{C}_P R^2}$$

Express the PDE and BCs in these nondimensional variables. How many parameters appear in this problem?

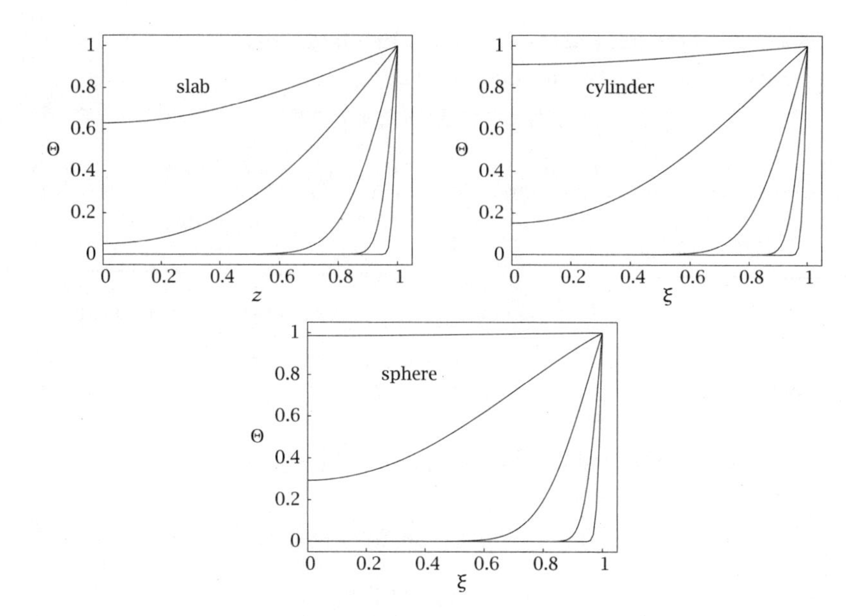

Figure 3.11: Transient heating of slab, cylinder, and sphere from (3.104), (3.105), and (3.106). Dimensionless temperature $\Theta(\xi, \tau)$ versus ξ at $\tau = 10^{-4}, 10^{-3}, 10^{-2}, 0.1, 0.5$.

(c) Take the Laplace transform of the PDE, apply the boundary conditions and find $\overline{\Theta}(\xi, s)$ for the cylinder. You do not need to invert this transform.

(d) Write the PDE and (three) boundary conditions for the **spherical** body, i.e., conditions at $r = R$, $r = 0$, and $t = 0$. How many parameters appear in this problem?

(e) Choose the same nondimensional temperature, radial position, and time variables as follows

$$\Theta = \frac{T - T_0}{T_1 - T_0} \qquad \xi = \frac{r}{R} \qquad \tau = \frac{tk}{\rho \hat{C}_P R^2}$$

Express the PDE and BCs in these nondimensional variables. How many parameters appear in this problem?

(f) Take the Laplace transform of the PDE, apply the boundary conditions and find $\overline{\Theta}(\xi, s)$ for the sphere. You do not need to invert this transform.

Exercise 3.27: Transient solutions for slab, cylinder, and sphere

We wish to plot and compare the temperature profile $\Theta(\xi, \tau)$ versus ξ at different τ for the slab, cylinder, and sphere geometries.

(a) The transform for the cylinder is given by

$$\overline{\Theta}(\xi, s) = \frac{I_0(\sqrt{s}\xi)}{sI_0(\sqrt{s})}$$

Find the zeros of the denominator.
Hint: you may want to use the Bessel function relations $I_0(r) = J_0(ir)$ and $I_1(r) = J_1(ir)/i$ (Abramowitz and Stegun, 1970, p. 375, 9.6.3.).

(b) Use (3.89) and show that the inverse is given by

$$\Theta(\xi, \tau) = 1 - 2\sum_{n=1}^{\infty} \frac{1}{\alpha_n J_1(\alpha_n)} J_0(\alpha_n \xi)\, e^{-\alpha_n^2 \tau} \tag{3.105}$$

(c) The transform for the sphere is

$$\overline{\Theta}(\xi, s) = \frac{\sinh(\sqrt{s}\xi)}{\xi s \sinh\sqrt{s}}$$

Find the zeros of the denominator $\sinh\sqrt{s}$. Note that the denominator has a *double* zero at $s = 0$ because both s and $\sinh\sqrt{s}$ vanish at $s = 0$.

(d) Because of the double zero, we cannot use the inversion formula (3.89), which assumes simple zeros. But notice the following fact. If the Laplace transforms $\overline{f}(s)$ and $\overline{g}(s)$ satisfy

$$\overline{g}(s) = \frac{\overline{f}(s)}{s}$$

then their inverse transforms satisfy

$$g(t) = \int_0^t f(t')dt'$$

Therefore define

$$\overline{f}(\xi, s) = s\overline{\Theta}(\xi, s) = \frac{\sinh(\sqrt{s}\xi)}{\xi \sinh\sqrt{s}}$$

Use (3.89) to invert this transform and show

$$f(\xi, \tau) = 2\sum_{n=1}^{\infty} \frac{(-1)^{n+1}}{\xi}(n\pi)\sin(n\pi\xi)e^{-n^2\pi^2\tau}$$

(e) Perform the time integral and show that

$$\Theta(\xi, \tau) = 2\sum_{n=1}^{\infty} \frac{(-1)^{n+1}}{n\pi\xi}\sin(n\pi\xi)[1 - e^{-n^2\pi^2\tau}]$$

Notice that the following series is the Fourier sine series of the linear function ξ (Selby, 1973, p. 480)

$$\frac{2}{\pi}\sum_{n=1}^{\infty} \frac{(-1)^{n+1}}{n}\sin(n\pi\xi) = \xi$$

so we have

$$\Theta(\xi, \tau) = 1 - 2\sum_{n=1}^{\infty} \frac{(-1)^{n+1}}{n\pi\xi}\sin(n\pi\xi)\, e^{-n^2\pi^2\tau} \tag{3.106}$$

(f) Make plots of the temperature profile $\Theta(\xi, \tau)$ versus ξ at several τ for the slab, cylinder, and sphere geometries. Your results should resemble Figure 3.11.

Which geometry heats up the quickest? The slowest? Give a physical explanation for these results.

Exercise 3.28: Transient diffusion in a sphere by separation of variables

Revisit the transient diffusion problem in the spherical geometry described in Exercises 3.26 and 3.27. Solve it by separation of variables, using the information in Example 3.8.

Exercise 3.29: Fourier series

Find the Fourier series coefficients for the function $f(x) = 1$ on the interval $x \in [-\pi/2, \pi/2]$ using the odd cosine terms $\{\cos x, \cos 3x, \cos 5x, \ldots\}$

$$f(x) = \sum_{n=0}^{\infty} a_n \cos(2n+1)x$$

Use this result to check the initial condition of (3.104) in Exercise 3.25.
Hint: first establish the orthogonality property $\int_{-\pi/2}^{\pi/2} \cos nx \cos mx dx = \frac{\pi}{2}\delta_{mn}$, $n, m = 1, 2, \ldots$. Then obtain the a_n by taking inner products as discussed in Section 2.4.1.

Exercise 3.30: Plancherel's formula

Plancherel's formula states that

$$2\pi \int_{-\infty}^{\infty} |f(x)|^2 \, dx = \int_{-\infty}^{\infty} |\hat{f}(k)|^2 \, dk$$

Begin with the expression on the left and from it derive the expression on the right. In general, both $f(x)$ and $\hat{f}(k)$ can be complex. Hint: $\int_{-\infty}^{\infty} e^{i(k-l)x} \, dx = 2\pi\delta(k-l)$.

Exercise 3.31: Green's function for a fourth-order problem

(a) Use the Fourier transform technique to solve the ordinary differential equation

$$\frac{d^4G}{d\xi^4} - 2\frac{d^2G}{d\xi^2} + G = \delta(\xi - x)$$

Use a computer algebra program or a math handbook to perform the integral that is required to find $G(\xi)$.

(b) The function G from the previous problem is the Green's function for the ordinary differential operator $L = \frac{d^4}{dx^4} - 2\frac{d^2}{dx^2} + 1$. Use this Green's function to solve $Lu = f(x), u(x) \to 0$ as $x \to \pm\infty$, where $f(x) = 1$ if $0 < x < 1$ and 0 elsewhere. Use numerical integration to approximate the solution for $|x| < 10$.

Exercise 3.32: A square with one heated wall

Solve $\nabla^2 u = 0$ in a unit square domain $0 < x < 1, 0 < y < 1$, with boundary conditions $u = 0$ on $x = 0$, $x = 1$ and $y = 0$, and $u = 1$ on $y = 1$.

Exercise 3.33: Separation of variables for the wave equation

Use separation of variables to solve

$$u_{tt} = c^2 u_{xx}$$

with the following boundary conditions

$$u(x,0) = \begin{cases} x, & x < 1/2 \\ 1-x, & x > 1/2 \end{cases}, \qquad u(0,t) = u(1,t) = 0, \qquad u_t(x,0) = 0$$

Show that your solution can be put in the D'Alembert form $u = F_1(x - ct) + F_2(x + ct)$.

Exercise 3.34: Separation of variables for a partially heated sphere

Use separation of variables to solve for the steady-state temperature distribution in a sphere whose bottom half the surface temperature is kept at $T = 0$, and whose top half is at $T = 1$. Use the transformation $X = \cos\phi$ to convert the equation in the polar angle direction to Legendre's equation. Note that the eigenvalues of Legendre's equation are $\lambda = n(n+1)$ for positive integers n. The corresponding eigenfunction is the Legendre polynomial $P_n(X)$. Explicitly find the first four terms in the expansion. Laplace's equation in spherical coordinates (r, θ, ϕ) is

$$\frac{1}{r^2}\frac{\partial}{\partial r}\left(r^2\frac{\partial T}{\partial r}\right) + \frac{1}{r^2 \sin\phi}\frac{\partial}{\partial \phi}\left(\sin\phi\frac{\partial T}{\partial \phi}\right) + \frac{1}{r^2\sin^2\phi}\frac{\partial^2 T}{\partial \theta^2} = 0$$

Exercise 3.35: The Helmholtz equation

Consider the wave equation

$$u_{tt} = c^2\nabla^2 u$$

in the domain $y > 0$, $\infty < x < \infty$, with boundary condition $u(x, y = 0, t) = f(x)e^{i\omega t}$. This equation governs sound emanating from a vibrating wall.

1. By assuming a solution of the form $u(x,y,t) = v(x,y)e^{i\omega t}$, show that the equation can be reduced to

$$-\omega^2 v - c^2\nabla^2 v = 0$$

   with boundary condition $v(x,0) = f(x)$. This is the HELMHOLTZ EQUATION.

2. Find the Green's function $G = G_\infty + G_B$ for this operator with the appropriate boundary conditions, using the fact that the Bessel function $Y_0(r) \sim \frac{2}{\pi}\ln r$ as $r \to 0$.

3. Use the Green's function to solve for $u(x,y,t)$.

Exercise 3.36: Transient diffusion via Green's function and similarity solution approaches

In Example 3.12 we found that the transient diffusion problem

$$G_t = DG_{xx}$$

has solution

$$G(x,t,\xi,\tau) = \frac{1}{2\sqrt{\pi D(t-\tau)}}e^{-(x-\xi)^2/4D(t-\tau)}$$

We have changed notation here to emphasize that this solution is the Green's function for transient diffusion with delta function source term $f(x,t) = \delta(x-\xi)\delta(t-\tau)$.

(a) Use this result, along with a result analogous to (3.71), to find the solution to the initial-value problem

$$u_t = Du_{xx} + f(x,t)$$

in $-\infty < x < \infty$ and initial condition $u(x,0) = u_0(x)$.

(b) Now consider the case $f = 0$ in the domain $x > 0$ with boundary condition $u(0,t) = 0$ and initial condition $u(x > 0, 0) = 1$. Use an image or symmetry argument to convert this into a problem in the unbounded domain, where you can apply (3.72). The information in Exercise 3.17 may be useful. This solution is found by Laplace transforms in Exercise 3.23.

(c) Solve this problem again by the method of SIMILARITY SOLUTION. That is, observe that the only length scale in the problem is the combination $\eta = x/2\sqrt{Dt}$ (the factor of 2 is arbitrary but convenient), and seek a solution

$$u(x,t) = u(\eta)$$

Substitution of this form into the governing equation and application of the chain rule leads to an ordinary differential equation.

Exercise 3.37: Schrödinger equation in a circular domain

The wave function for the "quantum corral"—an arrangement of atoms on a surface designed to localize electrons—is governed by the Schrödinger equation

$$i\frac{\partial \psi}{\partial t} = \nabla^2 \psi$$

Use separation of variables to find the general (bounded) axisymmetric solution to this problem in a circular domain with $\psi = 0$ at $r = 1$. Hint: if you assume exponential growth or decay in time, the spatial dependence will be determined by the so-called *modified* Bessel equation. Use the properties of solutions to this equation to show that there are no nontrivial solutions that are exponentially growing or decaying in time, thus concluding that the time dependence must be oscillatory.

Exercise 3.38: Temperature profile with wavy boundary temperature

Solve the steady-state heat-conduction problem

$$u_{xx} + u_{yy} = 0$$

in the half-plane $-\infty < x < \infty$, $0 < y < \infty$, with boundary conditions $u(x,0) = A + B\cos\alpha x = A + \frac{B}{2}(e^{i\alpha x} + e^{-i\alpha x})$ and $u(x,y)$ bounded as $y \to \infty$. Use the Fourier transform in the x direction. How far does the wavy temperature variation imposed at the boundary penetrate out into the material?

Exercise 3.39: Domain perturbation analysis of diffusion in a wavy-walled slab

Solve $\nabla^2 T = 0$ in the wavy-walled domain shown in Figure 3.12. The top surface is at $y = 1$, the left and right boundaries are $x = 0$ and $x = L$, respectively, and the bottom surface is $y = \epsilon\cos 2\pi x/L$, where $\epsilon \ll 1$. Find the solution to $O(\epsilon)$ using domain perturbation.

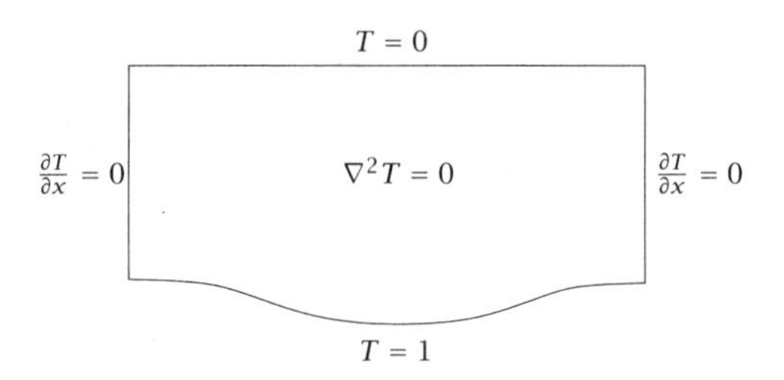

Figure 3.12: Wavy-walled domain.

Exercise 3.40: Fourier transform for solving heat conduction in a strip

Solve the steady-state heat-conduction problem

$$u_{xx} + u_{yy} = 0$$

in the infinite strip $-\infty < x < \infty$, $0 < y < 1$, with boundary conditions $u(x,0) = u_0(x), u(x,1) = u_1(x)$. Use the Fourier transform in the x direction to get an ordinary differential equation and boundary conditions for $\hat{u}(k,y)$.

Exercise 3.41: Separation of variables and Laplace's equation for a wedge

Use separation of variables to solve Laplace's equation in the wedge $0 < \theta < \alpha, 0 < r < 1$, with boundary conditions $u(r,0) = 0, u(r,\alpha) = 50, u(1,\theta) = 0$.

Exercise 3.42: Laplace's equation in a wedge

Again consider Laplace's equation in a wedge, but now fix the wedge angle at $\alpha = \pi/4$. Use the method of images to find the Green's function for this domain—where should the images be, and what should their signs be? A well-drawn picture showing the positions and signs of the images is sufficient. The first two images are shown. They don't completely solve the problem because each messes up the boundary condition on the side of the wedge further from it.

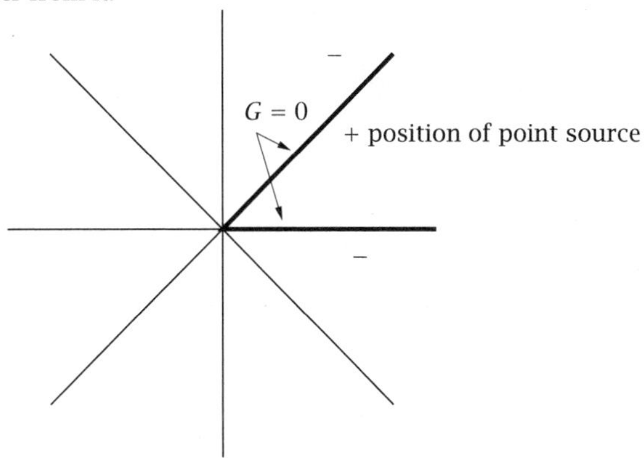

Exercise 3.43: D'Alembert form of the wave equation

(a) By looking for solutions of the form $u(x,t) = U(x - at)$ where a is to be determined, show that the general solution of the WAVE EQUATION

$$\frac{\partial^2 u}{\partial t^2} = c^2 \frac{\partial^2 u}{\partial x^2}$$

is

$$u(x,t) = f(x - ct) + g(x + ct)$$

where f and g are arbitrary.

(b) Use this solution to find the solution with initial condition $u(x,0) = w(x)$, $\partial u/\partial t(x,0) = 0$ in an unbounded domain. Pick a shape for $w(x)$ and sketch the solution $u(x,t)$.

Exercise 3.44: Heat equation in a semi-infinite domain

The solution to the heat equation

$$u_t = Du_{xx}$$

subject to the initial condition $u(x,0) = u_0(x)$ is

$$u(x,t) = \int_{-\infty}^{\infty} u_0(\xi) \frac{1}{2\sqrt{\pi D t}} e^{-\frac{(x-\xi)^2}{4Dt}}\, d\xi$$

Use this solution and an argument based on images to find the analogous solution for the same problem, but in the domain $x > 0$, with boundary condition $u(0,t) = 0$ and with initial condition $u(x > 0, t = 0) = u_+(x)$.

Exercise 3.45: Vibrating beam

The transverse vibrations of an elastic beam satisfy the equation

$$u_{tt} + \kappa u_{xxxx} = 0$$

where $\kappa > 0$. Use separation of variables to find $u(x,t)$ subject to initial condition $u(x,0) = u_0(x), u_t(x,0) = 0$ and boundary conditions $u(0,t) = u(L,t) = 0$, $u_{xx}(0,t) = u_{xx}(L,t) = 0$. These conditions correspond to a beam given an initial deformation and held fixed at its ends $x = 0$ and $x = L$. Hint: the equation $\alpha^4 = c$ has solutions $\alpha = \pm c^{1/4}$ and $\alpha = \pm i c^{1/4}$, where $c^{1/4}$ is the real positive fourth root of c.

Exercise 3.46: Convection and reaction with a point source

Use the Fourier transform and its properties to find a solution that vanishes at $\pm\infty$ for the ordinary differential equation

$$\frac{du}{dx} = -au + \delta(x)$$

where $a > 0$. Recall that $F^{-1}\left\{\frac{1}{a^2+k^2}\right\} = \frac{1}{2a} e^{-a|x|}$.

Exercise 3.47: Green's function for Laplacian operator

(a) Find the free space Green's function G_∞ for the Laplacian operator in three dimensions. It is spherically symmetric.

(b) Show that if u is a solution to Laplace's equation, then so is ∇u, as well as $\boldsymbol{c} \cdot \nabla u$, where $\boldsymbol{c}$ is any constant vector.

(c) Show that $E_{ij} \frac{\partial}{\partial x_i} \frac{\partial}{\partial x_j} u$ is also a solution, for any constant tensor $\boldsymbol{E}$.

Exercise 3.48: Zeros of sine, cosine, and exponential in the complex plane

We extend the definition of the exponential to complex argument $z \in \mathbb{C}$ as follows

$$e^z = e^{x+iy} = e^x(\cos y + i \sin y) \qquad z \in \mathbb{C}, \quad x, y \in \mathbb{R}$$

in which we take the usual definitions of real-valued $e^x, \cos y, \sin y$ for $x, y \in \mathbb{R}$. We extend the sine and cosine to complex arguments in terms of the exponential

$$\cos z = \frac{e^{iz} + e^{-iz}}{2} \qquad \sin z = \frac{e^{iz} - e^{-iz}}{2i}$$

Given these definitions, find all the zeros of the following functions in the complex plane

(a) The function $\sin z$.
Hint: using the definition of sine, convert the zeros of $\sin z$ to solutions of the equation $e^{2iz} = 1$. Substitute $z = x + iy$ and find all solutions $x, y \in \mathbb{R}$. Notice that all the zeros in the complex plane are only the usual ones on the real axis.

(b) The function $\cos z$. (Answer: only the usual ones on the real axis.)

(c) The function e^z. (Answer: no zeros in $\mathbb{C}$.)

Exercise 3.49: A Laplace transform inverse

The Laplace inverse for the following transform has been used in solving the wave equation

$$\overline{f}(s) = \frac{\sinh(as)\sinh(bs)}{\sinh s} \qquad a, b \in \mathbb{R}$$

Find $f(t)$, and note that your solution should be real valued, i.e., the imaginary number i should not appear anywhere in your final expression for $f(t)$.

Exercise 3.50: Wave equation with struck string

Revisit Example 3.15 and the wave equation $u_{\tau\tau} = u_{xx}$ on $x \in [0, 1]$ for a string with fixed ends $u(0, \tau) = u(1, \tau) = 0$, but the plucked string initial condition, $u(x, 0) = u_0(x)$, $u_\tau(x, 0) = v(x)$. Here there is zero initial deformation, but a nonzero initial velocity.

(a) Solve this equation using the Laplace transform.

(b) Note that this initial condition requires an inverse Laplace transform for

$$\overline{f}(s) = \frac{\sinh(as)\sinh(bs)}{s \sinh s}$$

Show that this inverse is given by

$$f(\tau) = 2 \sum_{n=1}^{\infty} \frac{(-1)^{n+1}}{n\pi} \sin(n\pi a)\ \sin(n\pi b)\ \sin(n\pi\tau) \tag{3.107}$$

(c) Denote the Fourier coefficients for the initial velocity $v(x)$ as b_n

$$v(x) = \sum_{n=1}^{\infty} b_n \sin(n\pi x)$$

(d) Next consider the mixed initial condition $u(x,0) = u_0(x)$ and $u_\tau(x,0) = v(x)$. Let a_n denote the Fourier coefficients of $u_0(x)$ as in Example 3.15. Show that the solution for the mixed initial condition is

$$u(x,\tau) = \sum_{n=1}^{\infty} \sin(n\pi x)\left(a_n \cos(n\pi\tau) + \frac{b_n}{n\pi}\sin(n\pi\tau)\right)$$

Exercise 3.51: Wave equation with triangle wave initial condition

The wave equation is useful to describe propagation of sound and vibration of strings and membranes. Consider the wave equation $u_{tt} = c^2 u_{zz}$ on $z \in [0,L]$ for a string with fixed ends $u(0,t) = u(L,t) = 0$, and the "plucked" string initial condition, i.e., fixed arbitrary position and zero velocity at $t = 0$, $u(z,0) = u_0(z)$, $u_t(z,0) = 0$. The constant c is known as the wave speed.

(a) First rescale time and position as $\tau = (c/L)t$, $x = z/L$, to remove the parameters c and L and simplify your work. Show that the rescaled problem is

$$u_{\tau\tau} = u_{xx}$$
$$u(x,0) = u_0(x), \quad u_\tau(x,0) = 0 \qquad x \in (0,1)$$
$$u(0,t) = 0, \quad u(1,t) = 0 \qquad \tau \geq 0$$

(b) Consider the solution (3.91) given in Example 3.15. Establish that the solution $u(x,\tau)$ satisfies the wave equation, both boundary conditions, and the initial condition. Establish that the solution $u(x,\tau)$ is periodic in time. What is the period?

(c) Consider the string's initial condition to be the triangle function depicted in Figure 2.32 with $a = 0.1$. Given the Fourier coefficients for this triangle function from Exercise 2.10, plot your solution at the following times on separate plots:

1. $\tau = 0, 0.0175, 0.035, 0.0525, 0.07$
2. $\tau = 0.45, 0.48, 0.49, 0.495, 0.4975, 0.50$
3. $\tau = 0.50, 0.5025, 0.505, 0.51, 0.52, 0.55$
4. $\tau = 0.90, 0.95, 1.00, 1.05, 1.10$
5. $\tau = 1.90, 1.95, 2.00, 2.05, 2.10$

Provide a physical description (comprehensible to the general public) of what is happening as the wave equation evolves forward in time. In particular, explain what the initial condition does just after $\tau = 0$. Explain what happens when waves arrive at the boundaries $x = 0, 1$?

Exercise 3.52: Numerical solution of the heat equation

(a) Write (and run) a program to use Chebyshev collocation to solve the heat equation

$$u_t = u_{xx}$$

with boundary conditions

$$\begin{aligned} u(x,0) &= 0, & & 0 < x < 1, \\ u(0,t) &= 0, \quad u(1,t) = 1, & & 0 < t \end{aligned}$$

Use the AM2 method and compare your solutions at a number of values of N at different times. Approximately how long does it take for the temperature at $x = 0.9$ to reach 0.5?.

(b) How many terms in the exact solution are needed to find the time at which $u(0.9) = 0.5$? (to five percent precision)?

Exercise 3.53: Propagation of a reaction front

Using the Chebyshev collocation technique for spatial discretization and the Adams-Moulton time integration, write a MATLAB or Octave program to solve the transient reaction-diffusion problem)

$$\frac{\partial T}{\partial t} = \alpha \frac{\partial^2 T}{\partial x^2} + T - T^3$$

$$T(-1,0) = 1 \qquad T(1,t) = -1 \qquad T(x,0) = -1$$

using $\alpha = 0.05$. Perform simulations for a long enough time that the solution reaches a steady state, and perform convergence checks to verify that your spatial and temporal discretizations are adequate, i.e., that the solution does not change much when the resolution is increased).

Exercise 3.54

Use von Neumann stability analysis to find the growth factor and the stability (Courant) condition for the Lax-Wendroff method, (3.98).

Bibliography

M. Abramowitz and I. A. Stegun. *Handbook of Mathematical Functions.* National Bureau of Standards, Washington, D.C., 1970.

R. Aris. *Vectors, Tensors, and the Basic Equations of Fluid Mechanics.* Dover Publications Inc., New York, 1962.

R. B. Bird, R. C. Armstrong, and O. Hassager. *Dynamics of Polymeric Liquids, Vol. 1, Fluid Dynamics.* Wiley, New York, second edition, 1987.

R. B. Bird, W. E. Stewart, and E. N. Lightfoot. *Transport Phenomena.* John Wiley & Sons, New York, second edition, 2002.

H. D. Block. *Tensor Analysis.* Charles E. Merrill Books, Inc., Columbus, Ohio, 1978.

C. Canuto, M. Y. Hussaini, A. Quarteroni, and T. A. Zang. *Spectral Methods: Fundamentals in Single Domains.* Springer-Verlag, Berlin, 2006.

H. S. Carslaw and J. C. Jaeger. *Conduction of Heat in Solids.* Oxford University Press, Oxford, second edition, 1959.

R. Courant. *Methods of Mathematical Physics. Volume II. Partial Differential Equations.* Wiley, New York, 1962.

W. M. Deen. *Analysis of Transport Phenomena.* Topics in chemical engineering. Oxford University Press, Inc., New York, second edition, 2011.

M. D. Greenberg. *Foundations of Applied Mathematics.* Prentice-Hall, New Jersey, 1978.

O. Heaviside. *Electromagnetic Theory*, volume II. The Electrician Printing and Publishing Company, London, 1899.

N. Levinson and R. M. Redheffer. *Complex Variables.* Holden Day, Oakland, CA, 1970.

E. Merzbacher. *Quantum Mechanics.* John Wiley and Sons, New York, second edition, 1970.

J. Ockendon, S. Howison, A. Lacey, and A. Movchan. *Applied Partial Differential Equations, Revised Edition.* Cambridge University Press, Cambridge, 2003.

T. A. Osswald and J. P. Hernández-Ortiz. *Polymer Processing: Modeling and Simulation.* Hanser, Munich, 2006.

C. Pozrikidis. *Introduction to Theoretical and Computational Fluid Dynamics.* Oxford University Press, New York, 1997.

W. H. Press, S. A. Teukolsky, W. T. Vetterling, and B. T. Flannery. *Numerical Recipes in C: The Art of Scientific Computing.* Cambridge University Press, Cambridge, 1992.

A. Prosperetti. *Advanced Mathematics for Applications.* Cambridge University Press, Cambridge, 1980.

J. B. Rawlings and J. G. Ekerdt. *Chemical Reactor Analysis and Design Fundamentals.* Nob Hill Publishing, Madison, WI, second edition, 2012.

M. Renardy and R. C. Rogers. *An Introduction to Partial Differential Equations.* Springer-Verlag, New York, 1992.

S. M. Selby. *CRC Standard Mathematical Tables.* CRC Press, twenty-first edition, 1973.

J. J. Simmonds. *A Brief on Tensor Analysis.* Springer, New York, second edition, 1994.

I. Stakgold. *Green's Functions and Boundary Value Problems.* John Wiley & Sons, New York, second edition, 1998.

G. Strang. *Introduction to Applied Mathematics.* Wellesley-Cambridge Press, Wellesley, MA, 1986.

M. S. Vallarta. Heaviside's proof of his expansion theorem. *Trans. A. I. E. E.*, pages 429–434, February 1926.

R. Winter. *Quantum Physics.* Wadsworth, Belmont, CA, 1979.

4

Probability, Random Variables, and Estimation

4.1 Introduction and the Axioms of Probability

For those engineers familiar with only deterministic models, we now make a big transition to random or stochastic models in the final two chapters of the text. Why? The motivation for including stochastic models is simple: they have proven highly useful in many fields of science and engineering. Moreover, even basic scientific *literacy* demands reasonable familiarity with stochastic methods. Students who have been exposed to primarily deterministic descriptions of physical processes sometimes initially regard stochastic methods as mysterious, vague, and difficult. We hope to change this perception, remove any mystery, and perhaps even make these methods easy to understand and enjoyable to use. To achieve this goal, we must maintain a clear separation between the physical process, and the stochastic model we choose to represent it, and the mathematical reasoning we use to make deductions about the stochastic model. Ignoring this separation and calling upon physical intuition in place of mathematical deduction invariably creates the confusion and mystery that we are trying to avoid.

Probability is the branch of mathematics that provides the inference engine that allows us to derive correct consequences from our starting assumptions. The starting assumptions are stated in terms of undefinable notions, such as outcomes and events. This should not cause any alarm, because this is the same pattern in all fields of mathematics, such as geometry, where the undefinable notions are point, line, plane, and so forth. Since human intuition about geometry is quite strong, however, the undefinable starting notions of geometry are taken in stride without much thought. Exposure to games of chance may provide the same human intuition about probability's undefinable starting terms.

We start with the set or space of possible outcomes which we denote by $\mathcal{I}$. Let $\mathcal{A}$ and $\mathcal{B}$ be events, which are subsets of $\mathcal{I}$. We use the empty set $\varnothing$ to denote an impossible event. Let $\mathcal{A} \cup \mathcal{B}$ denote the event "either $\mathcal{A}$ or $\mathcal{B}$," and let $\mathcal{A} \cap \mathcal{B}$ denote the event "both $\mathcal{A}$ and $\mathcal{B}$." The close analogy with the set operations of union and intersection is intentional, and helpful. To each event $\mathcal{A} \subseteq \mathcal{I}$, we can assign a probability to that event, denoted $\Pr(\mathcal{A})$. The three axioms of probability can then be stated as follows.

I. (Nonnegativity) $\Pr(\mathcal{A}) \geq 0$ for all $\mathcal{A} \in \mathcal{I}$.

II. (Normalization) $\Pr(\mathcal{I}) = 1$.

III. (Finite additivity) $\Pr(\mathcal{A} \cup \mathcal{B}) = \Pr(\mathcal{A}) + \Pr(\mathcal{B})$ for all $\mathcal{A}, \mathcal{B} \in \mathcal{I}$ satisfying $\mathcal{A} \cap \mathcal{B} = \varnothing$.

These three axioms, due to Kolmogorov (1933), are the source from which all probabilistic deductions follow. It may seem surprising at first that these three axioms are sufficient. In fact, we'll see soon that we do require a modified third axiom to handle infinitely many sets. First we state a few immediate consequences of these axioms. Exercise 4.1 provides several more. When $\mathcal{A} \cap \mathcal{B} = \varnothing$ we say that events $\mathcal{A}$ and $\mathcal{B}$ are mutually exclusive, or pairwise disjoint. We use the symbol $\mathcal{A} \setminus \mathcal{B}$ to denote the events in set $\mathcal{A}$ that are not events in set $\mathcal{B}$, or, equivalently, the events in set $\mathcal{A}$ with the events in $\mathcal{B}$ removed. The set $\overline{\mathcal{A}}$ is then defined to be $\mathcal{I} \setminus A$, i.e., $\overline{\mathcal{A}}$ is the set of all events that are not events in $\mathcal{A}$. We say that two events $\mathcal{A}$ and $\mathcal{B}$ are independent if $\Pr(\mathcal{A} \cap \mathcal{B}) = \Pr(\mathcal{A})\Pr(\mathcal{B})$.

Some of the important immediate consequences of the axioms are the following

$$\Pr(\varnothing) = 0$$
$$\Pr(\mathcal{A}) + \Pr(\overline{\mathcal{A}}) = 1$$
$$\Pr(\mathcal{A}) \leq 1$$
$$\text{If } \mathcal{B} \subseteq \mathcal{A}, \text{ then } \Pr(\mathcal{B}) \leq \Pr(\mathcal{A})$$
$$\Pr(\mathcal{A} \cup \mathcal{B}) = \Pr(\mathcal{A}) + \Pr(\mathcal{B}) - \Pr(\mathcal{A} \cap \mathcal{B})$$

Proof. To establish the first result, note that $\mathcal{A} \cup \varnothing = \mathcal{A}$ and $\mathcal{A} \cap \varnothing = \varnothing$ for all $\mathcal{A} \in \mathcal{I}$, and apply the *third axiom* to obtain $\Pr(\mathcal{A} \cup \varnothing) = \Pr(\mathcal{A}) = \Pr(\mathcal{A}) + \Pr(\varnothing)$. Rearranging this last equality gives the first result.

To establish the second result note that from the definition of $\overline{\mathcal{A}}$, we have that $A \cup \overline{\mathcal{A}} = \mathcal{I}$ and $A \cap \overline{\mathcal{A}} = \emptyset$. Applying the third axiom then gives $\Pr(\mathcal{A} \cup \overline{\mathcal{A}}) = \Pr(\mathcal{I}) = \Pr(\mathcal{A}) + \Pr(\overline{\mathcal{A}})$, and applying the *second axiom* then gives the second result. Using this second result and the *first axiom*, then gives the third result.[1] To obtain the fourth result, note that if $B \subseteq A$, then $\mathcal{A}$ can be expressed as $A = B \cup (\mathcal{A} \cap \overline{\mathcal{B}})$, with $B \cap (\mathcal{A} \cap \overline{\mathcal{B}}) = \emptyset$. Applying the third axiom then gives $\Pr(\mathcal{A}) = \Pr(\mathcal{B}) + \Pr(\mathcal{A} \cap \overline{\mathcal{B}})$, and applying the first axiom gives $\Pr(\mathcal{A}) \geq \Pr(\mathcal{B})$.

To obtain the fifth result, we express both $\mathcal{A} \cup \mathcal{B}$ and $\mathcal{B}$ as the union of mutually exclusive events: $\mathcal{A} \cup \mathcal{B} = \mathcal{A} \cup (\overline{\mathcal{A}} \cap \mathcal{B})$ with $\mathcal{A} \cap (\overline{\mathcal{A}} \cap \mathcal{B}) = \emptyset$, and $\mathcal{B} = (\mathcal{A} \cap \mathcal{B}) \cup (\overline{\mathcal{A}} \cap \mathcal{B})$ with $(\mathcal{A} \cap \mathcal{B}) \cap (\overline{\mathcal{A}} \cap \mathcal{B}) = \emptyset$. Applying the third axiom to both gives

$$\Pr(\mathcal{A} \cup \mathcal{B}) = \Pr(\mathcal{A}) + \Pr(\overline{\mathcal{A}} \cap \mathcal{B}) \qquad \Pr(\mathcal{B}) = \Pr(\mathcal{A} \cap \mathcal{B}) + \Pr(\overline{\mathcal{A}} \cap \mathcal{B})$$

Solving the second equation for $\Pr(\overline{\mathcal{A}} \cap \mathcal{B})$ and substituting into the first gives the fifth result, which is known as the addition law of probability. Also note that due to the first result, the probability of two mutually exclusive events is zero. ∎

4.2 Random Variables and the Probability Density Function

Next we introduce the concept of an experiment and a random variable. An experiment is the set of all outcomes $\mathcal{I}$, the subsets $\mathcal{F} \subseteq \mathcal{I}$ that are the events of interest, and the probabilities assigned to these events. A random variable is a function that assigns a number to the possible outcomes of the experiment, $X(\omega)$, $\omega \in \mathcal{I}$. For an experiment with a finite number of outcomes, such as rolling a die, the situation is simple. We can enumerate all outcomes to obtain $\mathcal{I} = \{1, 2, 3, 4, 5, 6\}$, and the events, $\mathcal{F}$, can be taken as all subsets of $\mathcal{I}$. The set $\mathcal{F}$ obviously contains the six different possible outcomes of the die roll, $\{1\}, \{2\}, \{3\}, \{4\}, \{5\}, \{6\} \in \mathcal{F}$. But the random variable is a different idea. We may choose to assign the integers $1, 2, \ldots, 6$ to the different events. But we may choose instead to assign the values 1 to the events corresponding to an even number showing on the die, and 0 to the events corresponding to an odd number showing on the die. In the first case we have the simple assignment

$$X(\omega) = \omega, \quad \omega = 1, 2, 3, 4, 5, 6$$

[1] Notice that we have used all three axioms to reach this point.

and in the second case, we have the assignment

$$X(\omega) = \begin{cases} 0, & \omega = 1, 3, 5 \\ 1, & \omega = 2, 4, 6 \end{cases}$$

The experiment is the same in both cases, but we have chosen different random variables to reflect potentially different goals in our physical modeling that led to this random process.

The situation becomes considerably more complex when we have an experiment with *uncountably* many outcomes, which is the case when we require real-valued random variables. For example, if we measure the temperature in a reactor, and want to model the reactor as a random process, the random variable of interest $X(\omega)$ assigns a (positive, real) value to each outcome of the experiment $\omega \in \mathcal{I}$. If we let $\mathcal{I} = \mathbb{R}$, for example, it's not immediately clear what we should allow for the subsets $\mathcal{F}$. If we allow only the individual points on the real number line, we do not obtain a rich enough set of events to be useful, i.e., the probability of achieving *exactly* some real-valued temperature T is *zero* for all $T \in \mathbb{R}$. The events corresponding to infinite sets of points, e.g., $a \leq T \leq b$ with $a < b \in \mathbb{R}$, are the ones that have *nonzero* probability. If we try to allow *all* subsets of the real number line, however, we obtain a set that is so large that we cannot satisfy the axioms of probability. Probabilists have found a satisfactory resolution to this issue in which the events are chosen as all intervals $[a, b]$, for all $a, b \in \mathbb{R}$, and all *countable* intersections and unions of all such intervals. Moreover, we modify the third axiom of probability to cover additivity of countably infinitely many sets

III'. (Countable additivity) Let $\mathcal{A}_i \in \mathcal{I}, i = 1, 2, 3, \ldots$ be a countable set of mutually exclusive events. Then

$$\Pr(\mathcal{A}_1 \cup A_2 \cup \cdots) = \Pr(\mathcal{A}_1) + \Pr(\mathcal{A}_2) + \cdots$$

We can then assign probabilities to these events, $\Pr(\mathcal{A}), \mathcal{A} \in \mathcal{F}$ satisfying the axioms. The random variable $X(\omega)$ is then a mapping from $\omega \in \mathcal{I}$ to $\mathbb{R}$, and we have well-defined probabilities for the events $\Pr(X(\omega) \leq x) = \Pr(\omega : X(\omega) \leq x)$ for all $x \in \mathbb{R}$. At this point we have all the foundational elements that we require to develop the stochastic methods of most use in science and engineering. The interested reader may wish to consult Papoulis (1984, pp.22–27) and Thomasian (1969, pp.320–322) for further discussion of these issues.

We let ξ be a random variable taking values in the field of real numbers, and the function $F_\xi(x)$ denote the (cumulative) PROBABILITY DISTRIBUTION FUNCTION of the random variable so that

$$F_\xi(x) = \Pr(\xi \leq x)$$

i.e., we say that $F_\xi(x)$ is the probability that the random variable ξ takes on a value less than or equal to x. The function F_ξ is a nonnegative, nondecreasing function and has the following properties due to the axioms of probability

$$F_\xi(x_1) \leq F_\xi(x_2) \quad \text{if } x_1 < x_2$$

$$\lim_{x \to -\infty} F_\xi(x) = 0 \qquad \lim_{x \to \infty} F_\xi(x) = 1$$

We next define the PROBABILITY DENSITY FUNCTION, denoted $p_\xi(x)$, such that

$$F_\xi(x) = \int_{-\infty}^{x} p_\xi(s)ds, \qquad -\infty < x < \infty \tag{4.1}$$

We can allow discontinuous F_ξ if we are willing to accept generalized functions (delta functions and the like) for p_ξ. Also, we can define the density function for discrete as well as continuous random variables if we allow delta functions. Alternatively, we can replace the integral in (4.1) with a sum over a discrete density function. The random variable may be a coin toss or a dice game, which takes on values from a discrete set contrasted to a temperature or concentration measurement, which takes on values from a continuous set. The density function has the following properties

$$p_\xi(x) \geq 0 \qquad \int_{-\infty}^{\infty} p_\xi(x)dx = 1$$

and the interpretation in terms of probability

$$\Pr(x_1 \leq \xi \leq x_2) = \int_{x_1}^{x_2} p_\xi(x)dx$$

The MEAN OR EXPECTATION of a random variable ξ is defined as

$$\mathcal{E}(\xi) = \int_{-\infty}^{\infty} x p_\xi(x)dx \tag{4.2}$$

The MOMENTS of a random variable are defined by

$$\mathcal{E}(\xi^n) = \int_{-\infty}^{\infty} x^n p_\xi(x)dx$$

and the mean is the first moment. Moments of ξ about the mean are defined by

$$\mathcal{E}((\xi - \mathcal{E}(\xi))^n) = \int_{-\infty}^{\infty} (x - \mathcal{E}(\xi))^n p_\xi(x) dx$$

The VARIANCE is defined as the second moment about the mean

$$\begin{aligned} \text{var}(\xi) &= \mathcal{E}((\xi - \mathcal{E}(\xi))^2) \\ &= \mathcal{E}(\xi^2 - 2\xi\mathcal{E}(\xi) + \mathcal{E}^2(\xi)) \\ &= \mathcal{E}(\xi^2) - 2\mathcal{E}^2(\xi) + \mathcal{E}^2(\xi) \\ &= \mathcal{E}(\xi^2) - \mathcal{E}^2(\xi) \end{aligned}$$

The standard deviation is the square root of the variance

$$\sigma(\xi) = (\text{var}(\xi))^{1/2}$$

Normal distribution. The normal or Gaussian distribution is ubiquitous in applications. It is characterized by its mean, m, and variance, σ^2, and is given by

$$p_\xi(x) = \frac{1}{\sqrt{2\pi\sigma^2}} \exp\left(-\frac{1}{2}\frac{(x-m)^2}{\sigma^2}\right) \tag{4.3}$$

We proceed to check that the mean of this distribution is indeed m and the variance is σ^2 as claimed, and that the density is normalized so that its integral is one. We require the definite integral formulas

$$\boxed{\begin{aligned} \int_{-\infty}^{\infty} e^{-x^2} dx &= \sqrt{\pi} && (4.4) \\ \int_{-\infty}^{\infty} x e^{-x^2} dx &= 0 && (4.5) \\ \int_{-\infty}^{\infty} x^2 e^{-x^2} dx &= \frac{\sqrt{\pi}}{2} && (4.6) \end{aligned}}$$

The first formula may also be familiar from the error function in transport phenomena

$$\text{erf}\,(x) = \frac{2}{\sqrt{\pi}} \int_0^x e^{-u^2} du \qquad \text{erf}\,(\infty) = 1$$

The second integral follows because the function e^{-x^2} is even and the function x is odd. The third formula may also be familiar from the gamma function, defined by (Abramowitz and Stegun, 1970, p.255–260)

$$\Gamma(n) = \int_0^{\infty} t^{n-1} e^{-t} dt = (n-1)!$$

Changing the variable of integration using $t = x^2$ gives

$$\int_{-\infty}^{\infty} x^2 e^{-x^2} dx = 2\int_0^{\infty} x^2 e^{-x^2} dx$$
$$= \int_0^{\infty} t^{1/2} e^{-t} dt$$
$$= \Gamma(3/2) = \frac{\sqrt{\pi}}{2}$$

We calculate the integral of the normal density as follows

$$\int_{-\infty}^{\infty} p_\xi(x)dx = \frac{1}{\sqrt{2\pi\sigma^2}} \int_{-\infty}^{\infty} \exp\left(-\frac{1}{2}\frac{(x-m)^2}{\sigma^2}\right) dx$$

Define the change of variable

$$u = \frac{1}{\sqrt{2}}\left(\frac{x-m}{\sigma}\right)$$

which gives

$$\int_{-\infty}^{\infty} p_\xi(x)dx = \frac{1}{\sqrt{\pi}} \int_{-\infty}^{\infty} \exp\left(-u^2\right) du = 1$$

from (4.4) and the proposed normal density does have unit area. Computing the mean gives

$$\mathcal{E}(\xi) = \frac{1}{\sqrt{2\pi\sigma^2}} \int_{-\infty}^{\infty} x \exp\left(-\frac{1}{2}\frac{(x-m)^2}{\sigma^2}\right) dx$$

using the same change of variables as before yields

$$\mathcal{E}(\xi) = \frac{1}{\sqrt{\pi}} \int_{-\infty}^{\infty} (\sqrt{2}u\sigma + m)e^{-u^2} du$$

The first term in the integral is zero from (4.5), and the second term produces

$$\mathcal{E}(\xi) = m$$

as claimed. Finally the definition of the variance of ξ gives

$$\text{var}(\xi) = \frac{1}{\sqrt{2\pi\sigma^2}} \int_{-\infty}^{\infty} (x-m)^2 \exp\left(-\frac{1}{2}\frac{(x-m)^2}{\sigma^2}\right) dx$$

Changing the variable of integration as before gives

$$\text{var}(\xi) = \frac{2}{\sqrt{\pi}}\sigma^2 \int_{-\infty}^{\infty} u^2 e^{-u^2} du$$

and from (4.6)

$$\text{var}(\xi) = \sigma^2$$

Shorthand notation for the random variable ξ having a normal distribution with mean m and variance σ^2 is

$$\xi \sim N(m, \sigma^2)$$

In order to collect a more useful set of integration facts for manipulating normal distributions, we can derive the following integrals by changing the variable of integration in (4.4)–(4.6). For $x, a \in \mathbb{R}, a > 0$

$$\boxed{\begin{aligned} \int_{-\infty}^{\infty} e^{-\frac{1}{2}x^2/a} dx &= \sqrt{2\pi}\sqrt{a} \\ \int_{-\infty}^{\infty} x e^{-\frac{1}{2}x^2/a} dx &= 0 \\ \int_{-\infty}^{\infty} x^2 e^{-\frac{1}{2}x^2/a} dx &= \sqrt{2\pi} a^{3/2} \end{aligned}}$$

Figure 4.1 shows the normal distribution with a mean of one and variances of 1/2, 1, and 2. Notice that a large variance implies that the random variable is likely to take on large values. As the variance shrinks to zero, the probability density becomes a delta function and the random variable approaches a deterministic value.

Characteristic function. It is often convenient to handle the algebra of density functions, particularly normal densities, by using a close relative of the Fourier transform of the density function rather than the density itself. The transform, which we denote as $\varphi_\xi(u)$, is known as the characteristic function in the probability and statistics literature. It is defined by

$$\begin{aligned} \varphi_\xi(t) &= \mathcal{E}(e^{it\xi}) \\ &= \int_{-\infty}^{\infty} e^{itx} p_\xi(x) dx \end{aligned}$$

where we again assume that any random variable ξ of interest has a density $p_\xi(\cdot)$. Note the sign convention with a positive sign chosen on the imaginary unit i. Hence, under this convention, the conjugate of the characteristic function $\overline{\varphi_\xi}(t)$ is the Fourier transform of the density. The characteristic function has a one-to-one correspondence with the density function, which can be seen from the inverse transform formula

$$p_\xi(x) = \frac{1}{2\pi} \int_{-\infty}^{\infty} e^{-itx} \varphi_\xi(t) dt$$

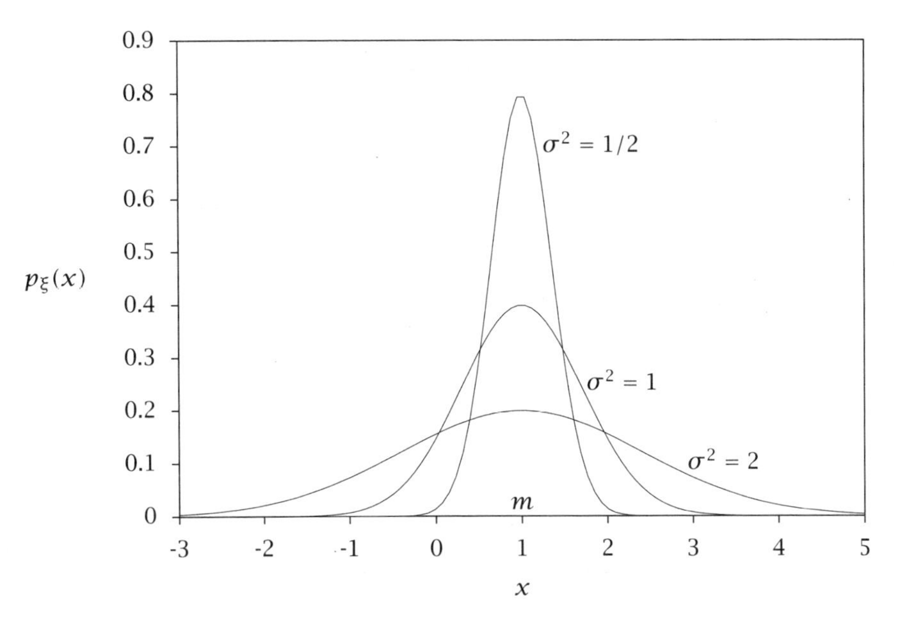

Figure 4.1: Normal distribution, with probability density $p_\xi(x) = (1/\sqrt{2\pi\sigma^2})\exp(-(1/2)(x-m)^2/\sigma^2)$. Mean is one and variances are 1/2, 1, and 2.

Again note the sign difference from the usual inverse Fourier transform. Note that multiplying a random variable by a constant $\eta = a\xi$ gives

$$\varphi_\eta(t) = \mathcal{E}(e^{ita\xi}) = \varphi_\xi(at) \tag{4.7}$$

Adding independent random variables $\eta = \xi_1 + \xi_2$ gives

$$\begin{aligned}
\varphi_\eta(t) &= \mathcal{E}(e^{it(x_1+x_2)}) \\
&= \iint_{-\infty}^{\infty} e^{it(x_1+x_2)} p_{\xi_1,\xi_2}(x_1,x_2)dx_1dx_2 \\
&= \int_{-\infty}^{\infty} e^{itx_1} p_{\xi_1}(x_1)dx_1 \int_{-\infty}^{\infty} e^{itx_2} p_{\xi_2}(x_2)dx_2 \\
\varphi_\eta(t) &= \varphi_{\xi_1}(t)\varphi_{\xi_2}(t)
\end{aligned} \tag{4.8}$$

We next compute the characteristic function of the normal distribution.

Example 4.1: Characteristic function of the normal density

Show the characteristic function of the normal density is

$$\varphi_\xi(t) = \exp\left(itm - \frac{1}{2}t^2\sigma^2\right)$$

Solution

The definition of the characteristic function and the normal density give

$$\varphi_\xi(t) = \frac{1}{\sqrt{2\pi\sigma^2}} \int_{-\infty}^{\infty} e^{-itx} e^{-(1/2)(x-m)^2/\sigma^2} dx$$

Changing the variable of integration to $z = x - m$ gives

$$\begin{aligned}
\varphi_\xi(t) &= \frac{1}{\sqrt{2\pi\sigma^2}} e^{itm} \int_{-\infty}^{\infty} e^{-itz} e^{-(1/2)z^2/\sigma^2} dz \\
&= \frac{2}{\sqrt{2\pi\sigma^2}} e^{itm} \int_{0}^{\infty} e^{-(1/2)z^2/\sigma^2} \cos tz dz \\
\varphi_\xi(t) &= e^{itm - t^2\sigma^2/2}
\end{aligned}$$

in which we used the definite integral

$$\int_0^{\infty} e^{-a^2x^2} \cos bx dx = \frac{\sqrt{\pi}}{2a} e^{-b^2/(4a^2)} \qquad a \neq 0$$

on the last line. Exercise 4.49 discusses how to derive this definite integral. Note also that the integral with the $\sin tz$ term vanished because sine is an odd function. □

4.3 Multivariate Density Functions

In applications we usually do not have a single random variable but a collection of them. We group these variables together in a vector and let random variable ξ now take on values in $\mathbb{R}^n$. Proceeding analogously to the single variable case, the JOINT DISTRIBUTION FUNCTION $F_\xi(x)$ is defined so that

$$F_\xi(x) = \Pr(\xi \leq x)$$

in which the vector inequality is defined to be the n corresponding scalar inequalities for the components. Note that $F_\xi(x)$ remains a *scalar*-valued function taking values in the interval $[0, 1]$

$$F_\xi(x) : \mathbb{R}^n \to [0, 1]$$

Also, as in the single variable case, we define the JOINT DENSITY FUNCTION, denoted $p_\xi(x) : \mathbb{R}^n \to \mathbb{R}_{\geq 0}$ such that

$$F_\xi(x) = \int_{-\infty}^{x_n} \cdots \int_{-\infty}^{x_1} p_\xi(s) ds_1 \cdots ds_n$$

or, provided the derivatives exist,

$$p_\xi(x) = \frac{\partial^n}{\partial x_1 \partial x_2 \cdots \partial x_n} F_\xi(x) \tag{4.9}$$

As in the scalar case, the probability that the n-dimensional random variable ξ takes on values between a and b is given by

$$\Pr(a \leq \xi \leq b) = \int_{a_n}^{b_n} \cdots \int_{a_1}^{b_1} p_\xi(x) dx_1 \cdots dx_n$$

Mean and covariance. The mean of the vector-valued random variable ξ is simply the vector-valued integral

$$\mathcal{E}(\xi) = \int_{-\infty}^{\infty} x p_\xi(x) dx \tag{4.10}$$

Writing out this integral in terms of its components we have

$$\mathcal{E}(\xi) = \begin{bmatrix} \int_{-\infty}^{\infty} \cdots \int x_1 p_\xi(x) dx_1 \dots dx_n \\ \int_{-\infty}^{\infty} \cdots \int x_2 p_\xi(x) dx_1 \dots dx_n \\ \vdots \\ \int_{-\infty}^{\infty} \cdots \int x_n p_\xi(x) dx_1 \dots dx_n \end{bmatrix}$$

The covariance of two scalar random variables ξ, η is defined as

$$\text{cov}(\xi, \eta) = \mathcal{E}\left((\xi - \mathcal{E}(\xi)) (\eta - \mathcal{E}(\eta)) \right)$$

The covariance matrix, C, of the vector-valued random variable ξ with components $\xi_i, i = 1, \dots n$ is defined as

$$C_{ij} = \text{cov}(\xi_i, \xi_j)$$

$$C = \begin{bmatrix} \text{var}(\xi_1) & \text{cov}(\xi_1, \xi_2) & \cdots & \text{cov}(\xi_1, \xi_n) \\ \text{cov}(\xi_2, \xi_1) & \text{var}(\xi_2) & \cdots & \text{cov}(\xi_2, \xi_n) \\ \vdots & \vdots & \ddots & \vdots \\ \text{cov}(\xi_n, \xi_1) & \text{cov}(\xi_n, \xi_2) & \cdots & \text{var}(\xi_n) \end{bmatrix}$$

Again, writing out the integrals in terms of the components gives

$$C_{ij} = \int_{-\infty}^{\infty} \cdots \int (x_i - \mathcal{E}(\xi_i))(x_j - \mathcal{E}(\xi_j)) p_\xi(x) dx_1 \ldots dx_n \tag{4.11}$$

Notice that $C_{ij} = C_{ji}$, so C is symmetric and has positive elements on the diagonal. We often express this definition of the variance with the convenient shorthand

$$\text{var}(\xi) = C = \mathcal{E}((\xi - \mathcal{E}(\xi))(\xi - \mathcal{E}(\xi))^T)$$

Notice that the vector outer product xx^T appears here, which is an $n \times n$ matrix, rather than the usual inner or dot product $x^T x$, which is a scalar.

Marginal density functions. We often are interested in only some subset of the random variables in a problem. Consider two vectors of random variables, $\xi \in \mathbb{R}^n$ and $\eta \in \mathbb{R}^m$. We can consider the joint distribution of both of these random variables $p_{\xi,\eta}(x, y)$ or we may only be interested in the ξ variables, in which case we can integrate out the m η variables to obtain the marginal density of ξ

$$p_\xi(x) = \int_{-\infty}^{\infty} \cdots \int p_{\xi,\eta}(x, y) dy_1 \cdots dy_m$$

Analogously, to produce the marginal density of η we use

$$p_\eta(y) = \int_{-\infty}^{\infty} \cdots \int p_{\xi,\eta}(x, y) dx_1 \cdots dx_n$$

4.3.1 Multivariate normal density

We define the multivariate normal density of the random variable $\xi \in \mathbb{R}^n$ as

$$p_\xi(x) = \frac{1}{(2\pi)^{n/2}(\det P)^{1/2}} \exp\left[-\frac{1}{2}(x - m)^T P^{-1}(x - m) \right] \tag{4.12}$$

in which $m \in \mathbb{R}^n$ is the mean and $P \in \mathbb{R}^{n \times n}$ is a real, symmetric, positive definite matrix. We show subsequently that P is the covariance matrix of ξ. The notation $\det P$ denotes the determinant of P. The multivariate normal density is well defined only for $P > 0$. The singular, or degenerate, case $P \geq 0$ is discussed subsequently. Shorthand notation for the random variable ξ having a normal distribution with mean m and covariance P is

$$\xi \sim N(m, P)$$

We also find it convenient to define the notation

$$n(x, m, P) = \frac{1}{(2\pi)^{n/2}(\det P)^{1/2}} \exp\left[-\frac{1}{2}(x-m)^T P^{-1}(x-m)\right] \quad (4.13)$$

so that we can write compactly for the normal with mean m and covariance P

$$p_\xi(x) = n(x, m, P)$$

in place of (4.12). The matrix P is real and symmetric. Figure 4.2 displays a multivariate normal for

$$P^{-1} = \begin{bmatrix} 3.5 & 2.5 \\ 2.5 & 4.0 \end{bmatrix}$$

As displayed in Figure 4.2, lines of constant probability in the multivariate normal are lines of constant

$$(x - m)^T P^{-1}(x - m)$$

To understand the geometry of lines of constant probability (ellipses in two dimensions, ellipsoids or hyperellipsoids in three or more dimensions) we examine the eigenvalues and eigenvectors of a positive definite matrix A as shown in Figure 4.3. Each eigenvector of A points along one of the axes of the ellipse. The eigenvalues show us how stretched the ellipse is in each eigenvector direction.

If we want to put simple bounds on the ellipse, then we draw a box around it as shown in Figure 4.3. Notice that the box contains much more area than the corresponding ellipse and we have lost the correlation between the elements of x. This loss of information means we can put different tangent ellipses of quite different shapes inside the same box. The size of the bounding box is given by

$$\text{length of } i\text{th side} = \sqrt{b\widetilde{A}_{ii}}$$

in which

$$\widetilde{A}_{ii} = (i, i) \text{ element of } A^{-1}$$

See Exercise 4.15 for a derivation of the size of the bounding box. Figure 4.3 displays these results: the eigenvectors are aligned with the

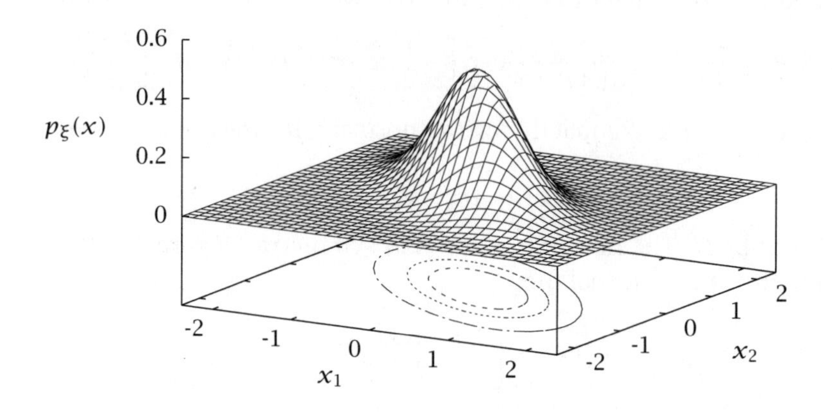

Figure 4.2: Multivariate normal for $n = 2$. The contour lines show ellipses containing 95, 75, and 50 percent probability.

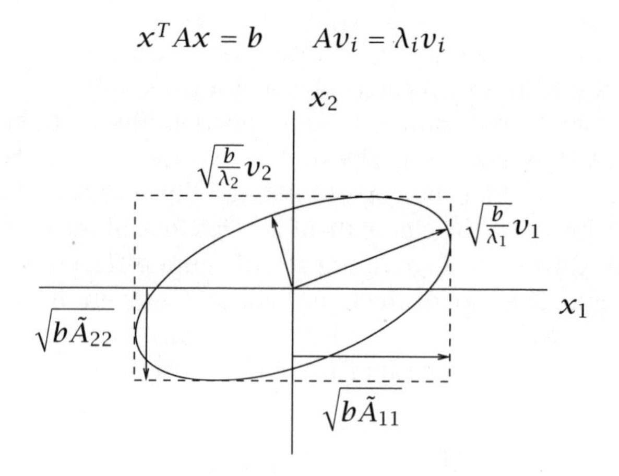

Figure 4.3: The geometry of quadratic form $x^T Ax = b$.

ellipse axes and the eigenvalues scale the lengths. The lengths of the sides of the box that is tangent to the ellipse are proportional to the square root of the diagonal elements of A^{-1}.

Example 4.2: The mean and covariance of the multivariate normal

Assume the random variable ξ is distributed normally as in (4.12)

$$p_\xi(x) = \frac{1}{(2\pi)^{n/2}(\det P)^{1/2}} \exp\left[-\frac{1}{2}(x-m)^T P^{-1}(x-m) \right]$$

1. Establish the following facts of integration. For $z \in \mathbb{R}^n$ with $A \in \mathbb{R}^{n \times n}, A > 0$

$$\int_{-\infty}^{\infty} \exp\left[-\frac{1}{2} z^T A^{-1} z \right] dz = (2\pi)^{n/2}(\det A)^{1/2} \quad \text{(scalar)} \tag{4.14}$$

$$\int_{-\infty}^{\infty} z \exp\left[-\frac{1}{2} z^T A^{-1} z \right] dz = 0 \quad \text{(n-vector)} \tag{4.15}$$

$$\int_{-\infty}^{\infty} z z^T \exp\left[-\frac{1}{2} z^T A^{-1} z \right] dz = (2\pi)^{n/2}(\det A)^{1/2} A \quad \text{($n \times n$-matrix)} \tag{4.16}$$

2. Show that the first and second integrals, and the definition of mean, (4.10), lead to

$$\mathcal{E}(\xi) = m$$

Show that the second and third integrals, and the definition of covariance, (4.11), lead to

$$C = P$$

So we have established that vector m and matrix P in (4.12) are the mean and covariance, respectively, of the normally distributed random variable ξ.

Solution

1. To compute the integrals, we first note that because A is real and symmetric, there exists a factorization

$$A = Q\Lambda Q^T \qquad A^{-1} = Q\Lambda^{-1}Q^T$$

in which Λ is a diagonal matrix containing the eigenvalues of A and Q is real and orthogonal. To establish the first integral, we use the variable transformation $z = Qx$ and change the variable of integration in (4.14)

$$\begin{aligned}
\int_{-\infty}^{\infty} \exp\left[-\frac{1}{2}z^T A^{-1} z\right] dz &= \int_{-\infty}^{\infty} \exp\left[-\frac{1}{2}x^T \Lambda^{-1} x\right] |\det Q|\, dx \\
&= \int_{-\infty}^{\infty} \exp\left[-\frac{1}{2}\sum_{i=1}^{n} x_i^2 / \lambda_i\right] dx \\
&= \prod_{i=1}^{n} \int_{-\infty}^{\infty} e^{-\frac{1}{2}x_i^2/\lambda_i} dx_i
\end{aligned}$$

in which $|\det Q| = 1$ because $QQ^T = I$, which makes $\det(QQ^T) = (\det Q)^2 = 1$ so $\det Q = \pm 1$. Performing the integrals gives

$$\begin{aligned}
\int_{-\infty}^{\infty} \exp\left[-\frac{1}{2}z^T A^{-1} z\right] dz &= \prod_{i=1}^{n} \sqrt{2\pi}\sqrt{\lambda_i} \\
&= (2\pi)^{n/2}\left(\prod_{i=1}^{n} \lambda_i\right)^{1/2} \\
&= (2\pi)^{n/2}(\det A)^{1/2}
\end{aligned}$$

and we have established the first result.

To establish the second integral, use the variable transformation $z = Qx$ to obtain

$$\int_{-\infty}^{\infty} z \exp\left[-\frac{1}{2}z^T A^{-1} z\right] dz = Q \int_{-\infty}^{\infty} x \exp\left[-\frac{1}{2}x^T \Lambda^{-1} x\right] dx$$

Notice that the ith element of this vector equation is of the form

$$Q \int_{-\infty}^{\infty} x_i \exp\left[-\frac{1}{2}x^T \Lambda^{-1} x\right] dx =$$
$$Q \int_{-\infty}^{\infty} x_i e^{-\frac{1}{2}x_i^2/\lambda_i} dx_i \prod_{k=1, k\neq i}^{n} \int_{-\infty}^{\infty} e^{-\frac{1}{2}x_k^2/\lambda_k} dx_k = 0$$

This integral vanishes because of the first term in the product. Since the integral vanishes for each element i, the vector of integrals is therefore zero.

To establish the third integral, we again use the variable transformation $z = Qx$ and change the variable of integration in (4.16)

$$\int_{-\infty}^{\infty} zz^T \exp\left[-\frac{1}{2}z^T A^{-1} z\right] dz =$$
$$Q\left(\int_{-\infty}^{\infty} xx^T \exp\left[-\frac{1}{2}x^T \Lambda^{-1} x\right] \det Q dx\right) Q^T =$$
$$Q\left(\int_{-\infty}^{\infty} xx^T \exp\left[-\frac{1}{2}\sum_{i=1}^{n} x_i^2/\lambda_i\right] dx\right) Q^T = QVQ^T \quad (4.17)$$

in which, again, $\det Q = 1$, and the V matrix is defined to be the integral on the right-hand side. Examining the components of V we note that if $i \neq j$ then the integral is of the form

$$V_{ij} = \int_{-\infty}^{\infty} x_i e^{-\frac{1}{2}x_i^2/\lambda_i} dx_i \int_{-\infty}^{\infty} x_j e^{-\frac{1}{2}x_j^2/\lambda_j} dx_j$$
$$\prod_{k=1,k\neq i,j}^{n} \int_{-\infty}^{\infty} e^{-\frac{1}{2}x_k^2/\lambda_k} dx_k = 0 \qquad i \neq j$$

The off-diagonal integrals vanish because of the odd functions in the integrands for the x_i and x_j integrals. The diagonal terms, on the other hand, contain even integrands and they do not vanish

$$V_{ii} = \int_{-\infty}^{\infty} x_i^2 e^{-\frac{1}{2}x_i^2/\lambda_i} dx_i \prod_{k=1,k\neq i}^{n} \int_{-\infty}^{\infty} e^{-\frac{1}{2}x_k^2/\lambda_k} dx_k$$

Evaluating these integrals gives

$$V_{ii} = \sqrt{2\pi}\lambda_i^{3/2} \prod_{k=1,k\neq i}^{n} \sqrt{2\pi}\sqrt{\lambda_k}$$
$$= (2\pi)^{n/2}\left(\prod_{k=1}^{n} \lambda_k\right)^{1/2} \lambda_i$$
$$V_{ii} = (2\pi)^{n/2}(\det A)^{1/2}\lambda_i$$

Substituting this result into (4.17) gives

$$\int_{-\infty}^{\infty} zz^T \exp\left[-\frac{1}{2}z^T A^{-1} z\right] dz = QVQ^T$$
$$= (2\pi)^{n/2}(\det A)^{1/2} Q \begin{bmatrix} \lambda_1 & & \\ & \ddots & \\ & & \lambda_n \end{bmatrix} Q^T$$
$$= (2\pi)^{n/2}(\det A)^{1/2} Q\Lambda Q^T = (2\pi)^{n/2}(\det A)^{1/2} A$$

and we have established the integral result of interest.

2. Using the probability density of the multivariate normal and the definition of the mean give

$$\mathcal{E}(\xi) = \int_{-\infty}^{\infty} x p_\xi(x) dx$$

$$\mathcal{E}(\xi) = \frac{1}{(2\pi)^{n/2} (\det P)^{1/2}} \int_{-\infty}^{\infty} x \exp\left[-\frac{1}{2}(x-m)^T P^{-1}(x-m)\right] dx$$

Changing the variable of integration to $z = x - m$ gives

$$\mathcal{E}(\xi) = \frac{1}{(2\pi)^{n/2} (\det P)^{1/2}} \int_{-\infty}^{\infty} (m+z) \exp\left[-\frac{1}{2} z^T P^{-1} z\right] dz$$

$$\mathcal{E}(\xi) = m$$

in which the integral with m produces unity by (4.14) and the integral involving z vanishes because the integrand is odd.

Next using the probability density of the multivariate normal, the definition of the covariance, and changing the variable of integration give

$$\begin{aligned}
C &= \int_{-\infty}^{\infty} (x - \mathcal{E}(\xi))(x - \mathcal{E}(\xi))^T p_\xi(x) dx \\
&= \frac{1}{(2\pi)^{n/2} (\det P)^{1/2}} \int_{-\infty}^{\infty} (x-m)(x-m)^T \\
&\qquad \exp\left[-\frac{1}{2}(x-m)^T P^{-1}(x-m)\right] dx \\
&= \frac{1}{(2\pi)^{n/2} (\det P)^{1/2}} \int_{-\infty}^{\infty} z z^T \exp\left[-\frac{1}{2} z^T P^{-1} z\right] dz \\
&= \frac{1}{(2\pi)^{n/2} (\det P)^{1/2}} (2\pi)^{n/2} (\det P)^{1/2} P \\
C &= P
\end{aligned}$$

□

Characteristic function of multivariate density. The characteristic function of an n-dimensional multivariate random variable ξ, is defined as

$$\varphi_\xi(t) = \int_{-\infty}^{\infty} e^{it^T x} p_\xi(x) dx$$

in which t is now an n-dimensional variable. The inverse transform is now

$$p_\xi(x) = \frac{1}{(2\pi)^n} \int_{-\infty}^{\infty} e^{-it^T x} \varphi_\xi(t) dt$$

Note that if one has the characteristic function of the entire random variable vector available, one can easily compute the characteristic function of any marginal distribution. We simply set the components of the t vector to zero for any variables we wish to integrate over to create the marginal. To illustrate the idea, assume we have a joint density for two (vector-valued) random variables ξ and η, $p_{\xi,\eta}(x, y)$, and its characteristic function $\varphi(t_x, t_y)$

$$\varphi(t_x, t_y) = \iint_{-\infty}^{\infty} \exp\left(i \begin{bmatrix} t_x^T & t_y^T \end{bmatrix} \begin{bmatrix} x \\ y \end{bmatrix} \right) p_{\xi,\eta}(x, y) dx dy$$

If we are interested in the characteristic function of η's marginal, $\varphi_\eta(t_y)$, we set $t_x = 0$ in the joint characteristic function to obtain it

$$\begin{aligned}
\varphi_{\xi,\eta}(0, t_y) &= \iint_{-\infty}^{\infty} \exp\left(i \begin{bmatrix} 0 & t_y^T \end{bmatrix} \begin{bmatrix} x \\ y \end{bmatrix} \right) p_{\xi,\eta}(x, y) dx dy \\
&= \int_{-\infty}^{\infty} e^{it_y^T y} \int_{-\infty}^{\infty} p_{\xi,\eta}(x, y) dx dy \\
&= \int_{-\infty}^{\infty} e^{it_y^T y} p_\eta(y) dy \\
&= \varphi_\eta(t_y)
\end{aligned}$$

Example 4.3: Characteristic function of the multivariate normal

Show that the characteristic function of the multivariate normal $\xi \sim N(m, P)$ is given by

$$\varphi_\xi(t) = e^{it^T m - (1/2) t^T P t}$$

Solution

From the definition of the characteristic function we are required to evaluate the integral

$$\varphi_\xi(t) = \int_{-\infty}^{\infty} \frac{e^{it^T x} e^{-(1/2)(x-m)^T P^{-1} (x-m)}}{(2\pi)^{n/2} (\det P)^{1/2}} dx$$

Changing the variable of integration to $z = x - m$ gives

$$\varphi_\xi(t) = \frac{e^{it^T m}}{(2\pi)^{n/2} (\det P)^{1/2}} \int_{-\infty}^{\infty} e^{it^T z} e^{-(1/2) z^T P^{-1} z} dz$$

Since P is positive definite, by Theorem 1.16 it can be factored as $P = Q\Lambda Q^T$ so $P^{-1} = Q\Lambda^{-1}Q^T$, and changing the variable of integration to $w = Q^T z$ in the integral gives

$$\int_{-\infty}^{\infty} e^{it^T z} e^{-(1/2)z^T P^{-1} z} dz = \int_{-\infty}^{\infty} e^{it^T Q w} e^{-(1/2)w^T \Lambda^{-1} w} dw$$

after noting that $\det(Q) = \pm 1$ since Q is orthogonal. Denoting $t^T Q = \begin{bmatrix} b_1 & b_2 & \cdots & b_n \end{bmatrix}$ gives

$$\begin{aligned}
\int_{-\infty}^{\infty} e^{it^T z} e^{-(1/2)z^T P^{-1} z} dz &= \prod_{j=1}^{n} \int_{-\infty}^{\infty} e^{ib_j w_j} e^{-(1/2)w_j^2/\lambda_j} dw_j \\
&= \prod_{j=1}^{n} \sqrt{\pi}\sqrt{2\lambda_j} e^{-(1/2)b_j^2 \lambda_j} \\
&= (2\pi)^{n/2} (\det P)^{1/2} \exp\Big(-(1/2) \sum_{j=1}^{n} b_j^2 \lambda_j \Big)
\end{aligned}$$

in which we used (4.95) to evaluate the integral. Noting that $\sum_{j=1}^{n} b_j^2 \lambda_j = t^T Q \Lambda Q^T t = t^T P t$ gives

$$\int_{-\infty}^{\infty} e^{it^T z} e^{-(1/2)z^T P^{-1} z} dz = (2\pi)^{n/2} (\det P)^{1/2} e^{-(1/2)t^T P t} \tag{4.18}$$

Substituting this result into the characteristic function gives

$$\varphi_\xi(t) = e^{it^T m - (1/2)t^T P t}$$

which is the desired result. □

Example 4.4: Marginal normal density

Given that ξ and η are jointly, normally distributed with mean and covariance

$$m = \begin{bmatrix} m_x \\ m_y \end{bmatrix} \qquad P = \begin{bmatrix} P_x & P_{xy} \\ P_{yx} & P_y \end{bmatrix}$$

show that the marginal density of ξ is normal with the following parameters

$$\xi \sim N(m_x, P_x) \tag{4.19}$$

Solution

Method 1. As a first approach to establish (4.19), we could directly integrate the y variables. Let $\bar{x} = x - m_x$ and $\bar{y} = y - m_y$, and n_x and n_y be the dimension of the ξ and η variables, respectively, and $n = n_x + n_y$. Then the definition of the marginal density gives

$$p_\xi(x) = \frac{1}{(2\pi)^{n/2}(\det P)^{1/2}} \int_{-\infty}^{\infty} \exp\left[-\frac{1}{2}\begin{pmatrix}\bar{x}\\ \bar{y}\end{pmatrix}^T \begin{bmatrix}P_x & P_{xy}\\ P_{yx} & P_y\end{bmatrix}^{-1}\begin{pmatrix}\bar{x}\\ \bar{y}\end{pmatrix}\right] d\bar{y}$$

If we follow this approach, we'll also need to use the matrix inversion lemma. This is left as an exercise for the interested reader.

Method 2. In the second approach, we use the previously derived results of characteristic function of the multivariate normal and its marginals. First the characteristic function of the joint density is given by

$$\varphi_{\xi,\eta}(t_x, t_y) = \exp\left(i\begin{bmatrix}t_x^T & t_y^T\end{bmatrix}\begin{bmatrix}m_x\\ m_y\end{bmatrix} - (1/2)\begin{bmatrix}t_x\\ t_y\end{bmatrix}^T P \begin{bmatrix}t_x\\ t_y\end{bmatrix}\right)$$

Setting $t_y = 0$ to compute the characteristic function of ξ's marginal gives

$$\begin{aligned}\varphi_\xi(t_x) &= \varphi_{\xi,\eta}(t_x, 0)\\ &= \exp\left(i\begin{bmatrix}t_x^T & 0\end{bmatrix}\begin{bmatrix}m_x\\ m_y\end{bmatrix} - (1/2)\begin{bmatrix}t_x\\ 0\end{bmatrix}^T P \begin{bmatrix}t_x\\ 0\end{bmatrix}\right)\\ &= e^{it_x^T m_x - (1/2)t_x^T P_x t_x}\end{aligned}$$

But notice that this last expression is the characteristic function of a normal with mean m_x and covariance P_x, so inverting this result back to the densities gives

$$p_\xi(x) = \frac{1}{(2\pi)^{n_x/2}(\det P_x)^{1/2}} e^{-(1/2)(x-m_x)^T P_x^{-1}(x-m_x)}$$

or $\xi \sim N(m_x, P_x)$.

Summarizing, since we have already performed the required integrals to derive the characteristic function of the normal, the second approach saves significant time and algebraic manipulation. It pays off to do the required integrals one time, "store" them in the characteristic function, and then reuse them whenever possible, such as here when deriving marginals. □

4.3.2 Functions of random variables.

In many applications we need to know how the density of a random variable is related to the density of a function of that random variable. Let $f : \mathbb{R}^n \to \mathbb{R}^n$ be a mapping of the random variable ξ into the random variable η, and assume that the inverse mapping also exists

$$\eta = f(\xi), \qquad \xi = f^{-1}(\eta)$$

Given the density of $\xi, p_\xi(x)$, we wish to compute the density of η, $p_\eta(y)$, induced by the function f. Let $\mathbb{X}$ denote an arbitrary region of the field of the random variable ξ, and define the set $\mathbb{Y}$ as the transform of this set under the function f

$$\mathbb{Y} = \{y | y = f(x), x \in \mathbb{X}\}$$

Then we seek a function $p_\eta(y)$ such that

$$\int_{\mathbb{X}} p_\xi(x)dx = \int_{\mathbb{Y}} p_\eta(y)dy \tag{4.20}$$

for every admissible set $\mathbb{X}$. Using the rules of calculus for transforming a variable of integration we can write

$$\int_{\mathbb{X}} p_\xi(x)dx = \int_{\mathbb{Y}} p_\xi(f^{-1}(y)) \left| \det \left(\frac{\partial f^{-1}(y)}{\partial y} \right) \right| dy \tag{4.21}$$

in which $\left| \det(\partial f^{-1}(y)/\partial y) \right|$ is the absolute value of the determinant of the Jacobian matrix of the transformation from η to ξ.[2] Subtracting (4.21) from (4.20) gives

$$\int_{\mathbb{Y}} \left(p_\eta(y) - p_\xi(f^{-1}(y)) \left| \det \left(\frac{\partial f^{-1}(y)}{\partial y} \right) \right| \right) dy = 0 \tag{4.22}$$

Because (4.22) must be true for any set $\mathbb{Y}$, we conclude (a proof by contradiction is immediate)[3]

$$\boxed{p_\eta(y) = p_\xi(f^{-1}(y)) \left| \det \left(\frac{\partial f^{-1}(y)}{\partial y} \right) \right|} \tag{4.23}$$

---

[2]See Appendix A for various notations for derivatives with respect to vectors.

[3]Some care should be exercised if one has generalized functions in mind for the probability density.

Example 4.5: Nonlinear transformation

Find the density function of the random variable η under the transformation $\eta = \xi^3$ for ξ normally distributed $\xi \sim N(m, \sigma^2)$.

Solution

The transformation is invertible and we have that $\xi = \eta^{1/3}$. Taking the derivative gives $d\xi/d\eta = (1/3)\eta^{-2/3}$, and using (4.23) gives

$$p_\eta(y) = \frac{1}{3\sqrt{2\pi}\sigma} y^{-2/3} \exp\left(-(1/2)(y^{1/3} - m)^2/\sigma^2\right) \qquad \square$$

Noninvertible transformations. Given random variables ξ having n components $\xi = (\xi_1, \xi_2, \ldots \xi_n)$ with joint density p_ξ and k random variables $\eta = (\eta_1, \eta_2, \ldots, \eta_k)$ defined by the transformation $\eta = f(\xi)$

$$\eta_1 = f_1(\xi) \quad \eta_2 = f_2(\xi) \quad \cdots \quad \eta_k = f_k(\xi)$$

We wish to find p_η in terms of p_ξ. Consider the region generated in $\mathbb{R}^n$ by the vector inequality

$$f(x) \le c$$

Call this region $\mathbb{X}(c)$, which is by definition

$$\mathbb{X}(c) = \{x | f(x) \le c\}$$

Note that $\mathbb{X}$ is not necessarily simply connected. The (cumulative) probability distribution (not density) for η then satisfies

$$F_\eta(y) = \int_{\mathbb{X}(y)} p_\xi(x) dx \tag{4.24}$$

If the density p_η is of interest, it can be obtained by differentiating F_η.

Example 4.6: Maximum of two random variables

Given two independent random variables, ξ_1, ξ_2 and the new random variable defined by the noninvertible, nonlinear transformation

$$\eta = \max(\xi_1, \xi_2)$$

Show that η's density is given by

$$p_\eta(y) = p_{\xi_1}(y) \int_{-\infty}^{y} p_{\xi_2}(x) dx + p_{\xi_2}(y) \int_{-\infty}^{y} p_{\xi_1}(x) dx$$

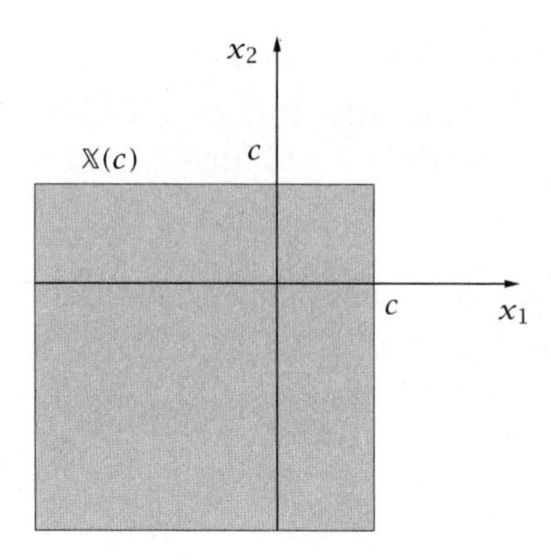

Figure 4.4: The region $\mathbb{X}(c)$ for $y = \max(x_1, x_2) \leq c$.

Solution

The region $\mathbb{X}(c)$ generated by the inequality $y = \max(x_1, x_2) \leq c$ is sketched in Figure 4.4. Applying (4.24) then gives

$$\begin{aligned}
F_\eta(y) &= \int_{-\infty}^{y} \int_{-\infty}^{y} p_\xi(x_1, x_2) dx_1 dx_2 \\
&= F_\xi(y, y) \\
&= F_{\xi_1}(y) F_{\xi_2}(y)
\end{aligned}$$

which has a clear physical interpretation. It says the probability that the *maximum* of two independent random variables is less than some value is equal to the probability that *both* random variables are less than that value. To obtain the density, we differentiate

$$\begin{aligned}
p_\eta(y) &= p_{\xi_1}(y) F_{\xi_2}(y) + F_{\xi_1}(y) p_{\xi_2}(y) \\
&= p_{\xi_1}(y) \int_{-\infty}^{y} p_{\xi_2}(x) dx + p_{\xi_2}(y) \int_{-\infty}^{y} p_{\xi_1}(x) dx \qquad \square
\end{aligned}$$

4.3.3 Statistical Independence and Correlation

From the definition of independence, two events $\mathcal{A}$ and $\mathcal{B}$ are independent if $\Pr(\mathcal{A} \cap \mathcal{B}) = \Pr(\mathcal{A}) \Pr(\mathcal{B})$. We translate this definition into an equivalent statement about probability distributions as follows. Given random variables ξ, η, let event $\mathcal{A}$ be $\xi \leq x$ and event $\mathcal{B}$ be $\eta \leq y$, then

$\mathcal{A} \cap \mathcal{B}$ is $\xi \leq x$ and $\eta \leq y$. By the definitions of joint and marginal probability distribution, these events have probabilities: $\Pr(\mathcal{A} \cap \mathcal{B}) = F_{\xi,\eta}(x,y)$, $\Pr(\mathcal{A}) = F_\xi(x)$, $\Pr(\mathcal{B}) = \mathcal{F}_\eta(y)$. So events $\mathcal{A}$ and $\mathcal{B}$ are independent if for the corresponding x and y, $F_{\xi,\eta}(x,y) = F_\xi(x)F_\eta(y)$. We say that the two *random variables* ξ, η are STATISTICALLY INDEPENDENT or simply independent if this relation holds for all x, y

$$F_{\xi,\eta}(x,y) = F_\xi(x)F_\eta(y), \qquad \text{all } x, y \tag{4.25}$$

See Exercise 4.2 for the proof that an equivalent condition for statistical independence can be stated in terms of the probability densities instead of distributions

$$p_{\xi,\eta}(x,y) = p_\xi(x)p_\eta(y), \qquad \text{all } x, y \tag{4.26}$$

provided that the densities are defined. We say two random variables, ξ and η, are UNCORRELATED if

$$\operatorname{cov}(\xi, \eta) = 0$$

Example 4.7: Independent implies uncorrelated

Prove that if ξ and η are statistically independent, then they are uncorrelated.

Solution

The definition of covariance and statistical independence gives

$$\begin{aligned}
\operatorname{cov}(\xi,\eta) &= \mathcal{E}((\xi - \mathcal{E}(\xi))(\eta - \mathcal{E}(\eta))) \\
&= \iint_{-\infty}^{\infty} (x - \mathcal{E}(\xi))(y - \mathcal{E}(\eta))p_{\xi,\eta}(x,y)dxdy \\
&= \int_{-\infty}^{\infty} (x - \mathcal{E}(\xi))p_\xi(x)dx \int_{-\infty}^{\infty} (y - \mathcal{E}(\eta))p_\eta(y)dy \\
&= 0
\end{aligned}$$

□

Example 4.8: Does uncorrelated imply independent?

Let ξ and η be jointly distributed random variables with probability density function

$$p_{\xi,\eta}(x,y) = \begin{cases} \frac{1}{4}[1 + xy(x^2 - y^2)], & |x| < 1, \quad |y| < 1, \\ 0, & \text{otherwise} \end{cases}$$

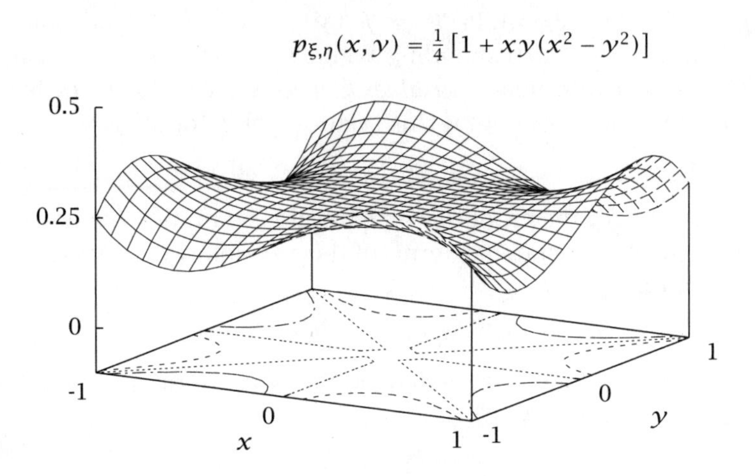

Figure 4.5: A joint density function for the two uncorrelated random variables in Example 4.8.

(a) Compute the marginals $p_\xi(x)$ and $p_\eta(y)$. Are ξ and η independent?

(b) Compute $\text{cov}(\xi, \eta)$. Are ξ and η uncorrelated?

(c) What is the relationship between independent and uncorrelated? Are your results on this example consistent with this relationship? Why or why not?

Solution

The joint density is shown in Figure 4.5.

(a) Direct integration of the joint density produces

$$p_\xi(x) = \frac{1}{2}, \quad |x| < 1 \qquad \mathcal{E}(\xi) = 0$$

$$p_\eta(y) = \frac{1}{2}, \quad |y| < 1 \qquad \mathcal{E}(\eta) = 0$$

and we see that both marginals are zero mean, uniform densities. Obviously ξ and η are not independent because the joint density is not the product of the marginals.

(b) Performing the double integral for the expectation of the product term gives

$$\begin{aligned} \mathcal{E}(\xi\eta) &= \iint_{-1}^{1} xy + (xy)^2(x^2 - y^2)dxdy \\ &= 0 \end{aligned}$$

and the covariance of ξ and η is therefore

$$\begin{aligned} \operatorname{cov}(\xi, \eta) &= \mathcal{E}(\xi\eta) - \mathcal{E}(\xi)\mathcal{E}(\eta) \\ &= 0 \end{aligned}$$

and ξ and η are uncorrelated.

(c) We know that independent implies uncorrelated. This example does not contradict that relationship. This example shows uncorrelated does not imply independent, in general, but see the next example for normals. □

Example 4.9: Independent and uncorrelated are equivalent for normals

If two random variables are jointly normally distributed,

$$\begin{bmatrix} \xi \\ \eta \end{bmatrix} \sim N\left(\begin{bmatrix} m_x \\ m_y \end{bmatrix}, \begin{bmatrix} P_x & P_{xy} \\ P_{yx} & P_y \end{bmatrix} \right)$$

Prove ξ and η are statistically independent if and only if ξ and η are uncorrelated, or, equivalently, P is block diagonal.

Solution

We have shown already that independent implies uncorrelated for any density, so we now show that, *for normals*, uncorrelated implies independent. Given $\operatorname{cov}(\xi, \eta) = 0$, we have

$$P_{xy} = P_{yx}^T = 0 \qquad \det P = \det P_x \det P_y$$

so the density can be written

$$p_{\xi,\eta}(x,y) = \frac{1}{(2\pi)^{\frac{1}{2}(n_x+n_y)}\left(\det P_x \det P_y\right)^{1/2}} \exp\left(-\frac{1}{2}\begin{bmatrix}\bar{x}\\ \bar{y}\end{bmatrix}^T \begin{bmatrix}P_x & 0\\ 0 & P_y\end{bmatrix}^{-1}\begin{bmatrix}\bar{x}\\ \bar{y}\end{bmatrix}\right) \tag{4.27}$$

For any joint normal, we know that the marginals are simply

$$\xi \sim N(m_x, P_x) \qquad \eta \sim N(m_y, P_y)$$

so we have

$$p_\xi(x) = \frac{1}{(2\pi)^{n_x/2}(\det P_x)^{1/2}} \exp\left(-\frac{1}{2}\bar{x}^T P_x^{-1}\bar{x}\right)$$
$$p_\eta(y) = \frac{1}{(2\pi)^{n_y/2}(\det P_y)^{1/2}} \exp\left(-\frac{1}{2}\bar{y}^T P_y^{-1}\bar{y}\right)$$

Forming the product and combining terms gives

$$p_\xi(x)p_\eta(y) = \frac{1}{(2\pi)^{\frac{1}{2}(n_x+n_y)}\left(\det P_x \det P_y\right)^{1/2}} \exp\left(-\frac{1}{2}\begin{bmatrix}\bar{x}\\ \bar{y}\end{bmatrix}^T \begin{bmatrix}P_x^{-1} & 0\\ 0 & P_y^{-1}\end{bmatrix}\begin{bmatrix}\bar{x}\\ \bar{y}\end{bmatrix}\right)$$

Comparing this equation to (4.27), and using the inverse of a block-diagonal matrix, we have shown that ξ and η are statistically independent. □

4.4 Sampling

Let scalar random variable ξ have density p_ξ with mean m and variance P, and consider n independent samples of ξ, denoted $x_1, x_2, \ldots, x_n$. By independent samples, we mean that the joint density of the samples is the product of the marginals, which all are identical and equal to p_ξ

$$p_{x_1,\ldots,x_n}(z_1,\ldots,z_n) = p_{x_1}(z_1)\cdots p_{x_n}(z_n) = p_\xi(z_1)\cdots p_\xi(z_n)$$

4.4.1 Linear Transformation

The following facts about the linear transformations of random variables prove useful. Consider random variable $\xi \in \mathbb{R}^n$ with density p_ξ, and let $A \in \mathbb{R}^{m\times n}$ be a constant matrix. Then the following formulas give the mean and variance of random variable $\eta = A\xi$

$$\mathcal{E}(\eta) = A\mathcal{E}(\xi) \qquad \text{var}(\eta) = A\text{var}(\xi)A^T \tag{4.28}$$

We establish these formulas as follows. Using the definition of expectation, we have that

$$\begin{aligned}\mathcal{E}(\eta) &= \mathcal{E}(A\xi) \\ &= \int_{-\infty}^{\infty} Axp_\xi(x)dx \\ &= A\int_{-\infty}^{\infty} xp_\xi(x)dx \\ &= A\mathcal{E}(x)\end{aligned}$$

Using the definition of variance, we have that

$$\begin{aligned}\text{var}(\eta) &= \text{var}(A\xi) \\ &= \int_{-\infty}^{\infty} (Ax - \mathcal{E}(Ax))(Ax - \mathcal{E}(Ax))^T p_\xi(x)dx \\ &= A\int_{-\infty}^{\infty} (x - \mathcal{E}(x))(x - \mathcal{E}(x))^T p_\xi(x)dxA^T \\ &= A\text{var}(\xi)A^T\end{aligned}$$

With normals, we often wish to check if the variance is positive definite after a linear transformation. Let $P \in \mathbb{R}^{n\times n}$ be positive definite and $A \in \mathbb{R}^{m\times n}$ be an arbitrary matrix. The following result is often useful: $P > 0$ and A's rows linearly independent $\iff$ $APA^T > 0$. See also statement 5 in Section 1.4.4.

Singular or degenerate normal distributions. It is often convenient to extend the definition of the normal distribution to admit positive *semidefinite* covariance matrices. The distribution with a semidefinite covariance is known as a singular or degenerate normal distribution (Anderson, 2003, p. 30). Figure 4.6 shows a nearly singular normal distribution.

To see how the singular normal arises, let the scalar random variable ξ be distributed normally with zero mean and positive definite covari-

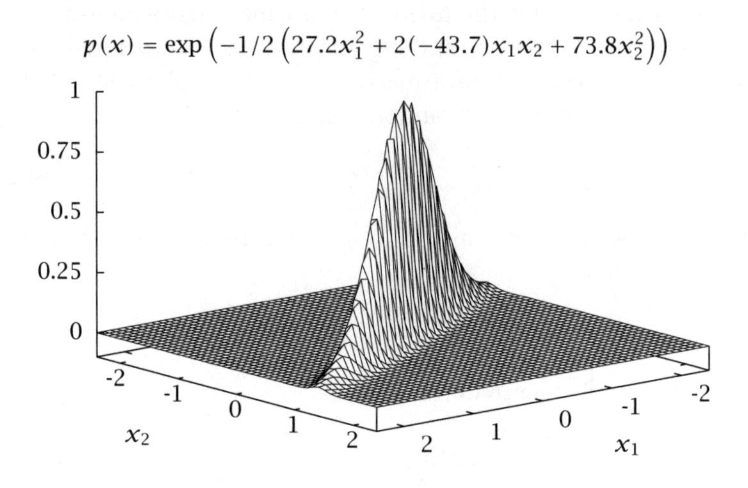

Figure 4.6: A nearly singular normal density in two dimensions.

ance, $\xi \sim N(0, P_x)$, and consider the simple linear transformation

$$\eta = A\xi \qquad A = \begin{bmatrix} 1 \\ 1 \end{bmatrix}$$

in which we have created two identical copies of ξ for the two components η_1 and η_2 of η. Now consider the density of η. If we try to use the standard formulas for transformation of a normal, we would have

$$\eta \sim N(0, P_y) \qquad P_y = AP_xA^T = \begin{bmatrix} P_x & P_x \\ P_x & P_x \end{bmatrix}$$

and P_y is singular since its rows are linearly dependent. Therefore one of the eigenvalues of P_y is zero, and P_y is positive semidefinite and not positive definite. Obviously we cannot use (4.12) for the density in this case because the inverse of P_y does not exist. To handle these cases, we first provide an interpretation that remains valid when the covariance matrix is singular and semidefinite.

Definition 4.10 (Density of a singular normal). A singular joint normal density of random variables (ξ_1, ξ_2), $\xi_1 \in \mathbb{R}^{n_1}$, $\xi_2 \in \mathbb{R}^{n_2}$, is denoted

$$\begin{bmatrix} \xi_1 \\ \xi_2 \end{bmatrix} \sim N\left[\begin{bmatrix} m_1 \\ m_2 \end{bmatrix}, \begin{bmatrix} \Lambda_1 & 0 \\ 0 & 0 \end{bmatrix} \right]$$

with $\Lambda_1 > 0$. The density is defined by

$$p_\xi(x_1, x_2) = \frac{1}{(2\pi)^{n_1/2}(\det \Lambda_1)^{1/2}} \exp\left[-\frac{1}{2}(x_1 - m_1)^T \Lambda_1^{-1} (x_1 - m_1) \right] \delta(x_2 - m_2) \tag{4.29}$$

In this limit, the "random" variable ξ_2 becomes deterministic and equal to its mean m_2. For the case $n_1 = 0$, we have the completely degenerate case in which $p_{\xi_2}(x_2) = \delta(x_2 - m_2)$, which describes the completely deterministic case $\xi_2 = m_2$, and there is no random component ξ_2. Notice that by performing the required integrals of (4.29) the two marginal densities are found to be

$$p_{\xi_1}(x_1) = \frac{1}{(2\pi)^{n_1/2}(\det \Lambda_1)^{1/2}} \exp\left[-\frac{1}{2}(x_1 - m_1)^T \Lambda_1^{-1} (x_1 - m_1) \right]$$
$$p_{\xi_2}(x_2) = \delta(x_2 - m_2)$$

Example 4.11: Computing a singular density

Consider again the motivating example with the unit normal scalar random variable $\xi \sim N(0, P_x), P_x = 1$ and the linear transformation

$$\eta = A\xi \qquad A = \begin{bmatrix} 1 \\ 1 \end{bmatrix}$$

Use Definition 4.10 to express the density p_η for this case, and draw a figure showing the appearance of p_η.

Solution

We first compute the eigenvalue decomposition of the semidefinite covariance P_y

$$P_y = AP_xA^T = AA^T = \begin{bmatrix} 1 & 1 \\ 1 & 1 \end{bmatrix}$$

and obtain

$$P_y = Q\Lambda Q^T \qquad Q = \frac{1}{\sqrt{2}} \begin{bmatrix} -1 & -1 \\ -1 & 1 \end{bmatrix} \qquad \Lambda = \begin{bmatrix} 2 & 0 \\ 0 & 0 \end{bmatrix}$$

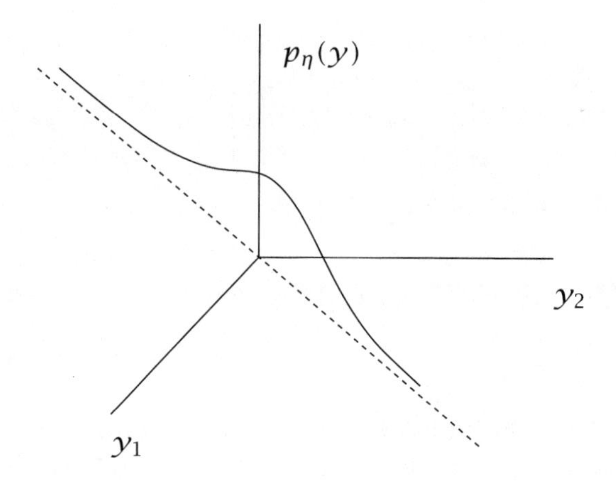

Figure 4.7: The singular normal resulting from $y = Ax$ with rank deficient A.

Next we define the invertible variable transformation

$$\zeta = Q^T \eta \qquad \eta = Q\zeta$$

and we can write the covariance of ζ, P_z, as

$$P_z = Q^T P_y Q = \Lambda = \begin{bmatrix} 2 & 0 \\ 0 & 0 \end{bmatrix}$$

which is in the form of Definition 4.10. Using that definition gives the density for ζ

$$p_\zeta(z_1, z_2) = \frac{1}{\sqrt{2\pi}\sqrt{2}} e^{-\frac{1}{2}(z_1^2/2)}\, \delta(z_2)$$

Finally transforming back to the variable η using

$$z_1 = -\frac{1}{\sqrt{2}}(y_1 + y_2) \qquad z_2 = -\frac{1}{\sqrt{2}}(y_1 - y_2)$$

and noting $\delta(ax) = (1/a)\delta(x)$ gives

$$p_\eta(y_1, y_2) = \frac{1}{\sqrt{2\pi}} \exp\left[-\frac{1}{2} \frac{(y_1 + y_2)^2}{4} \right] \delta(y_1 - y_2)$$

To draw a sketch, first we note that $p_\eta(y_1, y_2) = 0$ for $y_1 \neq y_2$ because of the delta function. So we have a singular normal defined in the plane, and the density is nonzero on the line $y_1 = y_2$. Therefore take a zero mean, unit variance normal defined on the y_1 axis, and rotate it by 45 degrees to the $y_1 = y_2$ line, and that is the joint density for $p_\eta(y_1, y_2)$. The result is shown in Figure 4.7. □

The expanded definition of normal distribution enables us to generalize the important result that the linear transformation of a normal is normal, so that it holds for *any* linear transformation, including rank-deficient transformations such as the A matrix given above in which the rows are not independent We state this result as the following theorem and defer the proof to Exercise 4.24.

Theorem 4.12 (Normal distributions under linear transformation). *Consider a normally distributed random variable $\xi \in \mathbb{R}^n$, $\xi \sim N(0, P_x)$, with semidefinite covariance $P_x \geq 0$ and an arbitrary linear transformation $A \in \mathbb{R}^{m \times n}$ and transformed random variable $\eta = A\xi$. Then $\eta \sim N(0, P_y)$ with $P_y = AP_xA^T \geq 0$.*

4.4.2 Sample Mean, Sample Variance, and Standard Error

Usually in applications we do not obtain nearly enough samples to obtain convergence to the entire density, and we settle for convergence to a few low-order moments of the distribution, such as the mean and variance. The SAMPLE MEAN is defined as

$$\hat{x}_n = \frac{1}{n} \sum_{i=1}^{n} x_i$$

and we expect this quantity to converge to ξ's mean as the number of samples increases. Indeed if we take expectations

$$\mathcal{E}(\hat{x}_n) = \mathcal{E}\left(\frac{1}{n} \sum_{i=1}^{n} x_i\right) = \frac{1}{n} \sum_{i=1}^{n} \mathcal{E}(x_i) = \frac{1}{n} \sum_{i=1}^{n} m = m$$

which means that the sample mean is an *unbiased* estimate of the mean of random variable ξ, for all values of n. An estimator's BIAS is defined to be the difference between the expectation of the estimator and the true value, and an estimator is termed UNBIASED if the bias is zero.

Next, toward defining an appropriate sample variance, we consider the sum of squares of the samples' differences from the sample mean

$S_n = \sum_{i=1}^n (x_i - \hat{x}_n)^2$, which can be rearranged as follows

$$\begin{aligned}
S_n &= \sum_{i=1}^n (x_i - \hat{x}_n)^2 \\
&= \sum_{i=1}^n ((x_i - m) - (\hat{x}_n - m))^2 \\
&= \sum_{i=1}^n \left((x_i - m)^2 - 2(x_i - m)(\hat{x}_n - m) + (\hat{x}_n - m)^2\right) \\
&= \left(\sum_{i=1}^n (x_i - m)^2\right) - 2n(\hat{x}_n - m)^2 + n(\hat{x}_n - m)^2 \\
S_n &= \left(\sum_{i=1}^n (x_i - m)^2\right) - n(\hat{x}_n - m)^2
\end{aligned}$$

Taking the expectation gives

$$\mathcal{E}(S_n) = \sum_{i=1}^n \text{var}(x_i) - n\text{var}(\hat{x}_n)$$

We know $\text{var}(x_i) = P$, for all $i = 1, \ldots, n$, and to compute the variance of $\hat{x}_n$, it is convenient to first determine the variance of vector X, obtained by stacking the samples together in a column vector $X = \begin{bmatrix} x_1 & x_2 & \cdots & x_n \end{bmatrix}^T$. Since the x_i are mutually independent, we have that $\text{var}(x_i, x_j) = P\delta_{ij}, i, j = 1, \ldots, n$ or in matrix form

$$\text{var}(X) = \begin{bmatrix} P & & & \\ & P & & \\ & & \ddots & \\ & & & P \end{bmatrix}$$

Using $\hat{x}_n = AX$ with $A = (1/n)\begin{bmatrix} 1 & 1 & \cdots & 1 \end{bmatrix}$ and the second part of (4.28) gives

$$\text{var}(\hat{x}_n) = A\text{var}(X)A^T = \frac{1}{n}P$$

Substituting these into the equation for expectation of S_n gives

$$\mathcal{E}(S_n) = nP - P = (n-1)P$$

So here we notice an interesting outcome; if we want to obtain an *unbiased* estimate of the variance, we should define the SAMPLE VARIANCE as $s_n = S_n/(n-1)$ to obtain

$$s_n = \frac{1}{n-1}\sum_{i=1}^{n}(x_i - \hat{x}_n)^2$$
$$\mathcal{E}(s_n) = P$$

This explains the somewhat mysterious definition of sample variance involving division of the sum of squares by $n-1$ instead of n, which one might have anticipated. We show later that division by n gives the maximum-likelihood estimate of the variance, which is also a good estimate because it converges to P as $n \to \infty$. Although the maximum-likelihood estimate is *not* an unbiased estimate for finite n, the bias decreases to zero as $n \to \infty$.

Standard error is the standard deviation of the sampling distribution of an estimator. For example, in the scalar case, if we consider the sample mean above to be an estimator of the mean, we have worked out that the variance of the sample mean is $\text{var}(\hat{x}_n) = (1/n)\text{var}(x) = (1/n)\sigma^2$, and the STANDARD ERROR OF THE MEAN is therefore

$$\text{SE}(\hat{x}_n) = \frac{\sigma}{\sqrt{n}}$$

When the standard deviation of the random variable being sampled is also unknown, people sometimes replace σ in the previous expression with an estimate of it, such as the square root of the sample variance $\sqrt{s_n}$. We then have

$$\text{SE}(\hat{x}_n) \approx \frac{\sqrt{s_n}}{\sqrt{n}}$$

This quantity does provide a rough measure of the uncertainty in $\hat{x}_n$ due to the finite sample size. But if we want to say something precise about the uncertainty in the sample mean as an estimate of mean, we must calculate a true confidence interval for that estimate. We show how to calculate confidence intervals in the discussion of maximum likelihood estimation in Section 4.7.

4.5 Central Limit Theorems

Central limit theorems are concerned with the following remarkable observation: if we have a set of n independent random variables x_i, $i = 1, 2, \ldots, n$, then, under fairly general conditions, the density p_y of their sum

$$y = x_1 + x_2 + \cdots + x_n$$

tends to a normal density as $n \to \infty$. We require only mild restrictions on how the x_i themselves are distributed for the sum y to tend to a normal. It is perhaps best to illustrate this observation with a concrete example.

Example 4.13: Sum of 10 uniformly distributed random variables

Consider 10 uniformly and independently distributed random variables, $x_1, x_2, \ldots, x_{10}$. Consider a new random variable y, which is the sum of the 10 x random variables

$$y = x_1 + x_2 + \cdots x_{10}$$

What is y's mean and variance? Draw samples of the 10 x_i random variables, and compute samples of y. Plot frequency distributions of x and y. Even though the 10 x random variables are uniformly distributed, and their probability distribution looks nothing like a normal distribution, discuss how well y is approximated by a normal.

Solution

The x random variables are distributed as $x \sim U(0,1)$, which means

$$p_x(x') = \begin{cases} 1 & x' \in [0,1] \\ 0 & \text{otherwise} \end{cases} \tag{4.30}$$

Computing the mean and variance gives

$$\mathcal{E}(x) = \int_0^1 x dx = \frac{1}{2}$$

$$\text{var}(x) = \int_0^1 (x - (1/2))^2 dx = \frac{1}{12}$$

If we stack the x variables in a vector

$$x = \begin{bmatrix} x_1 & x_2 & \cdots & x_{10} \end{bmatrix}^T$$

we can write the y random variable as the linear transformation of x

$$y = Ax \qquad A = \begin{bmatrix} 1 & 1 & \cdots & 1 \end{bmatrix}$$

We have that y's mean and variance are given by

$$\mathcal{E}(y) = A\mathcal{E}(x) = 5$$

$$\text{var}(y) = A\text{var}(x)A^T = \frac{5}{6}$$

So, if the central limit theorem is in force with only 10 random variables in the sum, we might expect y to be distributed as

$$y \sim N(5, 5/6)$$

A histogram of the 10,000 samples of x_1 and y are shown in Figures 4.8 and 4.9. It is clear that even 10 uniformly distributed x random variables produce nearly a normal distribution for their sum y. □

4.5.1 Identically distributed random variables

Consider n independent random variables, X_i, $i = 1, 2, \ldots, n$, each with identical distribution having mean μ and variance σ^2. We are interested in the distribution of the sum $S_n = X_1 + X_2 + \cdots + X_n$ as n becomes large. Since the X_i are independent, the mean and variance of S_n are given by

$$\begin{aligned}
\mathcal{E}(S_n) &= \sum_{i=1}^{n} \mathcal{E}(X_i) = n\mu \\
\operatorname{var}(S_n) &= \sum_{i=1}^{n} \operatorname{var}(X_i) = n\sigma^2
\end{aligned}$$

Since we want to take the limit as $n \to \infty$, we first rescale the sum to keep the mean and variance finite. Given the formulas for shifting mean and variance we choose $Z_n = (S_n - n\mu)/(\sqrt{n}\sigma)$ and obtain

$$\begin{aligned}
\mathcal{E}(Z_n) &= \frac{1}{\sqrt{n}\sigma}(\mathcal{E}(S_n) - n\mu) = 0 \\
\operatorname{var}(Z_n) &= \frac{1}{n\sigma^2}\operatorname{var}(S_n) = 1
\end{aligned}$$

Theorem 4.14 (De Moivre-Laplace central limit theorem). *Let $X_i, i = 1, 2, \ldots, n$ be independent and identically distributed with mean μ and variance σ^2, then Z_n tends to the standard normal $N(0, 1)$ as $n \to \infty$.*

Proof. In keeping with Laplace's approach to the problem, we shall use characteristic functions to establish this result. We shall find useful the following bound on the error in the Taylor series approximation of the exponential with a purely imaginary argument.

$$\left| e^{ix} - \sum_{m=0}^{n} \frac{(ix)^m}{m!} \right| \le \frac{|x|^{n+1}}{(n+1)!} \tag{4.31}$$

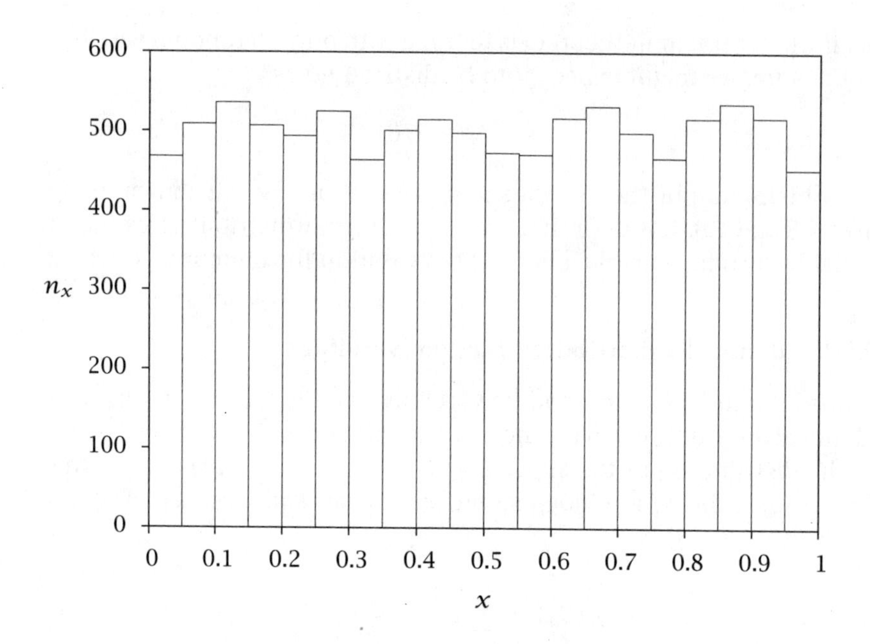

Figure 4.8: Histogram of 10,000 samples of uniformly distributed x.

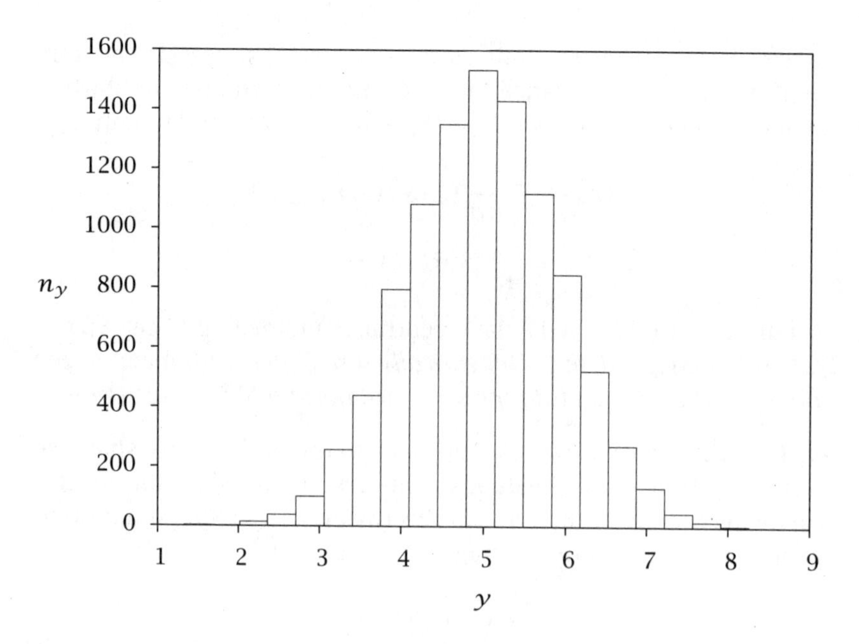

Figure 4.9: Histogram of 10,000 samples of $y = \sum_{i=1}^{10} x_i$.

This bound is simple to establish (see Exercise 4.53). We will use it with $n = 2$ stated in this form

$$e^{ix} = 1 + ix - \frac{x^2}{2!} + O(|x|^3) \tag{4.32}$$

in which $O(|x|^3)$ denotes that the size of the error term in (4.31) is less than some constant time $|x|^3$. We first show that the characteristic function of Z_n converges to the characteristic function of $N(0,1)$, which is $e^{-(1/2)t^2}$. Let $Y_i = (X_i - \mu)/\sigma$ so that the Y_i have zero mean and unit variance. We use (4.32) with argument e^{itx} and obtain a series expansion for Y_i's characteristic function

$$\begin{aligned}
\varphi_{Y_i}(t) &= \int_{-\infty}^{\infty} e^{itx} p_{Y_i}(x) dx \\
&= \int_{-\infty}^{\infty} (1 + itx - (1/2)t^2x^2 + |x|^3 O(|t|^3)) p_{Y_i}(x) dx \\
&= 1 + i\mathcal{E}(Y_i) - (1/2)t^2 \mathrm{var}(Y_i) + \mathcal{E}(|Y_i|^3) O(|t|^3) \\
&= 1 - (1/2)t^2 + O(|t|^3)
\end{aligned}$$

Notice that here we have assumed $\mathcal{E}(|Y_i|^3)$ is finite, so that it can be absorbed into the $O(|t|^3)$ term. Next, since $Z_n = (1/\sqrt{n}) \sum_{i=1}^{n} Y_i$, we have from (4.7) and (4.8) that

$$\varphi_{Z_n}(t) = (\varphi_{Y_i}(t/\sqrt{n}))^n = (1 - (1/2)t^2/n + O(t^3/n^{3/2}))^n$$

In taking the limit as $n \to \infty$ the last term is negligible and can be dropped to obtain

$$\lim_{n\to\infty} \varphi_{Z_n}(t) = \lim_{n\to\infty} (1 - (1/2)t^2/n)^n$$

Using the calculus result that $\lim_{x\to 0}(1 + ax)^{1/x} = e^a$ with $n = 1/x$ gives

$$\lim_{n\to\infty} \varphi_{Z_n}(t) = e^{-(1/2)t^2}$$

The final step, which unfortunately requires the most effort, is to show that if the characteristic function converges, then the random variable also converges (in distribution). Assuming this is true, we then have

$$\lim_{n\to\infty} Z_n \sim N(0,1)$$

and the result is established. ∎

This argument can be improved in several ways. For example, it is sufficient to assume only that the second moment is finite, not the third absolute moment $\mathcal{E}(|Y_i|^3)$ assumed here (Durrett, 2010, pp.114–116). And we have not justified the claim of convergence in distribution implied by convergence in characteristic function. However, the argument does nicely illustrate why characteristic functions prove so useful. In the next section we pursue a much more general approach that is not based on the characteristic function, so we content ourselves to leave this proof here.

4.5.2 Random variables with different distributions

The central limit theorem of de Moivre and Laplace is already a spectacular mathematical result. But as it stands, it is not a compelling reason to assume that unmodeled noise in a physical system would be well represented by a normal distribution. After all, how would we deduce that some unmodeled random effect in a physical system is the result of many different independent random causes, all of which have *identical* distributions? But the central limit theorem runs deeper. We next remove the assumption that the X_i are identically distributed. This version of the central limit theorem was developed by Lindeberg (1922). We consider the following conditions on the X_i variables.

Assumption 4.15 (Lindeberg conditions)**.** Consider independent random variables $X_i, i = 1, 2, \ldots, n$ satisfying $\mathcal{E}(X_i) = 0$ and $\mathrm{var}(X_i) = \sigma_i^2$, and let $s_n^2 = \sum_{i=1}^n \sigma_i^2$. The following two conditions hold as $n \to \infty$

(a) $s_n \to \infty$

(b) For every $\epsilon > 0$, $\dfrac{1}{s_n^2} \sum\limits_{k=1}^{n} \mathcal{E}(X_k^2; |X_k| > \epsilon s_n) \to 0$

The notation $\mathcal{E}(X_k^2; |X_k| > \epsilon s_n)$ is shorthand for taking expectations of the *truncated* random variable

$$\mathcal{E}(X^2; |X| > a) = \int_{-\infty}^{a} p_X(w) w^2 dw + \int_{a}^{\infty} p_X(w) w^2 dw$$

Notice that the definition implies that $\mathcal{E}(X^2; |X| > a) + \mathcal{E}(X^2; |X| \le a) = \mathrm{var}(X)$. Many sufficient conditions for the central limit theorem have been proposed over the years, but all were superseded by the Lindeberg conditions, which were also shown to be necessary (Feller, 1935; Lévy, 1935). For example, Exercise 4.55 shows that the identically distributed assumption of the de Moivre-Laplace central limit theorem is a special case of these conditions. Also, all *bounded* random variables satisfy these conditions. We have the following theorem with $\Phi(x)$ denoting the distribution function of the standard normal.

Theorem 4.16 (Lindeberg-Feller central limit theorem). *Consider independent random variables $X_i, i = 1, 2, \ldots, n$ with $\mathcal{E}(X_i) = 0$ and var(X_i) $= \sigma_i^2$ satisfying Assumption 4.15. The normalized sum $Z_n = S_n/s_n$ converges in distribution to the unit normal*

$$\lim_{n\to\infty} \sup_x \left| F_{Z_n}(x) - \Phi(x) \right| = 0$$

The proof of this theorem is given in Section 4.9.

4.5.3 Multidimensional central limit theorems

The central limit theorem (CLT) can be extended to vector-valued random variables, $X_i \in \mathbb{R}^d$. Consider first independent, identically distributed (IID) $X_i, i = 1, 2, \ldots, n$ random variables with $\mathcal{E}(X_i) = \mu$, and $\text{var}(X_i) = \Sigma$. We assume that $\Sigma > 0$ is positive definite. We have the following result.

Theorem 4.17 (Multivariate CLT—IID). *Let vector-valued random variables $X_i, i = 1, 2, \ldots, n$ be independent and identically distributed with $\mathcal{E}(X_i) = \mu$ and var(X_i) $= \Sigma$. The normalized sum $Z_n = (1/\sqrt{n}) \sum_{i=1}^n (X_i - \mu)$ converges in distribution to the normal $N(0, \Sigma)$.*

Again, the IID version is a special case of a more general version that assumes a generalization of the Lindeberg condition.

Theorem 4.18 (Multivariate CLT—Lindeberg-Feller). *Consider independent vector-valued random variables $X_i, i = 1, 2, \ldots, n$ with $\mathcal{E}(X_i) = \mu_i$ and var(X_i) $= \Sigma_i > 0$, and satisfying the following conditions as $n \to \infty$*

(a) $\sum_{i=1}^n \Sigma_i \to \Sigma$

(b) For every $\epsilon > 0$, $\sum_{i=1}^n \mathcal{E}(\, \|X_i\|^2 ; \|X_i\| > \epsilon) \to 0$

Then the sum $Z_n = \sum_{i=1}^n (X_i - \mu_i)$ converges in distribution to the normal $N(0, \Sigma)$.

See van der Vaart (1998, pp. 20-21) for further discussion of this case. Theorem 4.18 is the mathematical basis for the common physical assumption that noise in process measurements is often well modeled by a zero mean normal distribution. The variance often can be determined by examining samples of the measurement, which is an important part of the process modeling task that is often overlooked.

Finally, the history of the term "central limit theorem" is also interesting. Apparently coined by Polyá in 1920 (in German: zentraler Grenzwertsatz), he referred to the theorem as *central* to the theory of probability, a place of honor that it maintains to this day. But the word *central* can also be interpreted to mean the *center* of the normal distribution, where the distribution converges quickly as n increases compared to the tails of the distribution, where the convergence is much slower (Le Cam, 1986). Le Cam's article is highly recommended reading for anyone interested in the fascinating history of the central limit theorem.

4.6 Conditional Density Function and Bayes's Theorem

Let ξ and η be jointly distributed random variables with probability density $p_{\xi,\eta}(x,y)$. We seek the density function of ξ given that a specific realization y of η has been observed. We define the conditional density function as

$$p_{\xi|\eta}(x|y) = \frac{p_{\xi,\eta}(x,y)}{p_\eta(y)} \qquad p_\eta(y) \neq 0$$

Consider a roll of a single die in which η takes on values E or O to denote whether the outcome is even or odd and ξ is the integer value of the die. The 12 values of the joint density function are simply computed

$$\begin{aligned} p_{\xi,\eta}(1,\mathrm{E}) &= 0 & p_{\xi,\eta}(1,\mathrm{O}) &= 1/6 \\ p_{\xi,\eta}(2,\mathrm{E}) &= 1/6 & p_{\xi,\eta}(2,\mathrm{O}) &= 0 \\ p_{\xi,\eta}(3,\mathrm{E}) &= 0 & p_{\xi,\eta}(3,\mathrm{O}) &= 1/6 \\ p_{\xi,\eta}(4,\mathrm{E}) &= 1/6 & p_{\xi,\eta}(4,\mathrm{O}) &= 0 \\ p_{\xi,\eta}(5,\mathrm{E}) &= 0 & p_{\xi,\eta}(5,\mathrm{O}) &= 1/6 \\ p_{\xi,\eta}(6,\mathrm{E}) &= 1/6 & p_{\xi,\eta}(6,\mathrm{O}) &= 0 \end{aligned} \tag{4.33}$$

The marginal densities are then easily computed; we have for ξ

$$p_\xi(x) = \sum_{y=\mathrm{O}}^{\mathrm{E}} p_{\xi,\eta}(x,y)$$

which gives by summing across rows of (4.33)

$$p_\xi(x) = 1/6, \qquad x = 1,2,\ldots 6$$

Similarly, we have for η

$$p_\eta(y) = \sum_{x=1}^{6} p_{\xi,\eta}(x,y)$$

which gives by summing down the columns of (4.33)

$$p_\eta(y) = 1/2, \qquad y = \text{E}, \text{O}$$

These are both in accordance of our intuition on the rolling of the die: uniform probability for each value 1 to 6 and equal probability for an even or an odd outcome.

Now the conditional density is a different concept. The conditional density $p_{\xi|\eta}(x|y)$ tells us the density of x given that $\eta = y$ has been observed. So consider the value of this function

$$p_{\xi|\eta}(1|\text{O})$$

which tells us the probability that the die has a 1 given that we know that it is odd. We expect that the additional information on the die being odd causes us to revise our probability that it is 1 from 1/6 to 1/3. Applying the defining formula for conditional density indeed gives

$$p_{\xi|\eta}(1|\text{O}) = p_{\xi,\eta}(1,\text{O})/p_\eta(\text{O}) = \frac{1/6}{1/2} = 1/3$$

Consider the reverse question, the probability that we have an odd given that we observe a 1. The definition of conditional density gives

$$p_{\eta,\xi}(\text{O}|1) = p_{\eta,\xi}(\text{O},1)/p_\xi(1) = \frac{1/6}{1/6} = 1$$

i.e., we are sure the die is odd if it is 1. Notice that the arguments to the conditional density do not commute as they do in the joint density.

This fact leads to a famous result. Consider the definition of conditional density, which can be expressed as

$$p_{\xi,\eta}(x,y) = p_{\xi|\eta}(x|y)p_\eta(y)$$

or

$$p_{\eta,\xi}(y,x) = p_{\eta|\xi}(y|x)p_\xi(x)$$

Because $p_{\xi,\eta}(x,y) = p_{\eta,\xi}(y,x)$, we can equate the right-hand sides and deduce

$$p_{\xi|\eta}(x|y) = \frac{p_{\eta|\xi}(y|x)p_\xi(x)}{p_\eta(y)} \qquad p_\eta(y) \neq 0 \tag{4.34}$$

which is known as Bayes's theorem (Bayes, 1763). Notice that this result comes in handy whenever we wish to switch the variable that is known in the conditional density, which we will see is a key step in state estimation problems.

Example 4.19: Conditional normal density

Show that if ξ and η are jointly normally distributed as

$$\begin{bmatrix} \xi \\ \eta \end{bmatrix} \sim N\left(\begin{bmatrix} m_x \\ m_y \end{bmatrix}, \begin{bmatrix} P_x & P_{xy} \\ P_{yx} & P_y \end{bmatrix} \right)$$

then the conditional density of ξ given η is also normal

$$p_{\xi|\eta}(x|y) = n(x, m, P) \tag{4.35}$$

in which the mean and covariance are

$$m = m_x + P_{xy}P_y^{-1}(y - m_y) \qquad P = P_x - P_{xy}P_y^{-1}P_{yx} \tag{4.36}$$

Solution

The definition of conditional density gives

$$p_{\xi|\eta}(x|y) = \frac{p_{\xi,\eta}(x,y)}{p_\eta(y)}$$

Because (ξ, η) is jointly normal, we know from Example 4.4

$$p_\eta(y) = n(y, m_y, P_y)$$

and therefore

$$p_{\xi|\eta}(x|y) = \frac{n\left(\begin{bmatrix} x \\ y \end{bmatrix}, \begin{bmatrix} m_x \\ m_y \end{bmatrix}, \begin{bmatrix} P_x & P_{xy} \\ P_{yx} & P_y \end{bmatrix} \right)}{n(y, m_y, P_y)}$$

Substituting in the definition of the normal density from (4.13) gives

$$p_{\xi|\eta}(x|y) = \frac{(\det P_y)^{1/2}}{(2\pi)^{n_\xi/2} \det\left(\begin{bmatrix} P_x & P_{xy} \\ P_{yx} & P_y \end{bmatrix} \right)^{1/2}} \exp(-(1/2)a) \tag{4.37}$$

in which the argument of the exponent is

$$a = \begin{bmatrix} x - m_x \\ y - m_y \end{bmatrix}^T \begin{bmatrix} P_x & P_{xy} \\ P_{yx} & P_y \end{bmatrix}^{-1} \begin{bmatrix} x - m_x \\ y - m_y \end{bmatrix} - (y - m_y)^T P_y^{-1}(y - m_y) \tag{4.38}$$

If we use $P = P_x - P_{xy}P_y^{-1}P_{yx}$ as defined in (4.36) then we can use the partitioned matrix inversion formula to express the matrix inverse in (4.38) as

$$\begin{bmatrix} P_x & P_{xy} \\ P_{yx} & P_y \end{bmatrix}^{-1} = \begin{bmatrix} P^{-1} & -P^{-1}P_{xy}P_y^{-1} \\ -P_y^{-1}P_{yx}P^{-1} & P_y^{-1} + P_y^{-1}P_{yx}P^{-1}P_{xy}P_y^{-1} \end{bmatrix}$$

Substituting this expression into (4.38) and multiplying out terms yields

$$\begin{aligned} a = (x - m_x)^T P^{-1}(x - m_x) - 2(y - m_y)^T (P_y^{-1}P_{yx}P^{-1})(x - m_x) \\ + (y - m_y)^T (P_y^{-1}P_{yx}P^{-1}P_{xy}P_y^{-1})(y - m_y) \end{aligned}$$

which is the expansion of the following quadratic term

$$a = \left[(x - m_x) - P_{xy}P_y^{-1}(y - m_y)\right]^T P^{-1}\left[(x - m_x) - P_{xy}P_y^{-1}(y - m_y)\right]$$

in which we use the fact that $P_{xy} = P_{yx}^T$. Substituting (4.36) into this expression yields

$$a = (x - m)^T P^{-1}(x - m)$$

Finally noting that for the partitioned matrix

$$\det \begin{bmatrix} P_x & P_{xy} \\ P_{yx} & P_y \end{bmatrix} = \det P_y \det P$$

and substituting the two previous equations into (4.37) yields

$$\frac{n\left(\begin{bmatrix} x \\ y \end{bmatrix}, \begin{bmatrix} m_x \\ m_y \end{bmatrix}, \begin{bmatrix} P_x & P_{xy} \\ P_{yx} & P_y \end{bmatrix}\right)}{n(y, m_y, P_y)} = n(x, m, P) \qquad (4.39)$$

or

$$p_{\xi|\eta}(x|y) = n(x, m, P)$$

which is the desired result. □

Example 4.20: More normal conditional densities

Let the joint conditional of random variables (A, B) given C be a normal distribution with the following mean and variance

$$p_{A,B|C}(a, b|c) = n((a, b), m, P) \qquad (4.40)$$

$$m = \begin{bmatrix} m_a \\ m_b \end{bmatrix} \qquad P = \begin{bmatrix} P_a & P_{ab} \\ P_{ba} & P_b \end{bmatrix}$$

Show that the conditional density of A given B and C is also normal

$$p_{A|B,C}(a|b,c) = n(a, m, P) \tag{4.41}$$

with mean and variance given by

$$m = m_a + P_{ab}P_b^{-1}(b - m_b) \qquad P = P_a - P_{ab}P_b^{-1}P_{ba}$$

Solution

From the definition of joint density we have that

$$p_{A|B,C}(a|b,c) = \frac{p_{A,B,C}(a,b,c)}{p_{B,C}(b,c)}$$

Multiplying the top and bottom of the fraction by $p_C(c)$ yields

$$p_{A|B,C}(a|b,c) = \frac{p_{A,B,C}(a,b,c)}{p_C(c)} \frac{p_C(c)}{p_{B,C}(b,c)}$$

or

$$p_{A|B,C}(a|b,c) = \frac{p_{A,B|C}(a,b|c)}{p_{B|C}(b|c)}$$

Substituting the distribution given in (4.20) and using the result in Example 4.4 to integrate over a to obtain the marginal $p_{B|C}(b|c) = \int p_{A,B|C}(a,b|c)da$ yields

$$p_{A|B,C}(a|b,c) = \frac{n\left(\begin{bmatrix} a \\ b \end{bmatrix}, \begin{bmatrix} m_a \\ m_b \end{bmatrix}, \begin{bmatrix} P_a & P_{ab} \\ P_{ba} & P_b \end{bmatrix}\right)}{n(b, m_b, P_b)}$$

Now using (4.39) and (4.36) gives

$$p_{A|B,C}(a|b,c) = n(a, m, P)$$
$$m = m_a + P_{ab}P_b^{-1}(b - m_b) \qquad P = P_a - P_{ab}P_b^{-1}P_{ba}$$

and the result is established. □

4.7 Maximum-Likelihood Estimation

We now turn to one of the most basic problems in modeling: how to determine model parameters from experimental measurement. Finding methods to solve parameter estimation problems has had a significant impact on the development of mathematics, generally, and statistics, in particular. To get started we consider the simplest but arguably still one of the most important problems, determining the parameters in a *linear* model. Consider some set of environmental or predictor variables, x, that we wish to use to explain some response variables, y. The linear model means simply that $y = \theta x$ in which θ is a set of parameters that we wish to determine from measurements of y for given values of x. We often intend to use the identified parameters to make *predictions* about y's response to x values that we have not used in any previous experiment. We may use the identified model to optimize over the x variables to find the conditions that maximize the responses y. This approach may save considerable time and expense compared to the alternative of trial and error experimental adjustment of the x variables.

In addition to finding the "best" parameter estimate, we would also like to *quantify* our uncertainty in the estimate. Modeling the uncertainty in the data as a random variable with some fixed probability density is one of the key methods that we can use to solve this problem. Uncertainty in measurement leads to uncertainty in estimate, and stipulating the structure of the measurement uncertainty allows us to find (exactly in some cases) the uncertainty in the estimate. Because of the central limit theorem, our first choice for modeling uncertainty in measurement is the normal distribution. We then have the model

$$y = \theta x + e$$

in which e is assumed normal and zero mean. The effect of nonzero mean is assumed to be included in θ as additional parameters to be estimated.

The six canonical linear estimation problems. We next look at the six versions of this problem that result from assuming (i) y is a scalar or vector, (ii) θ is a vector or matrix, and (iii) whether we know the measurement error variance, or if it has to be estimated from the data. The variable x will be a vector throughout. The goal in each problem is the same: find the optimal parameter estimate by maximize the probability of the data, and quantify the estimate's uncertainty, for example, by determining confidence intervals. The first five estimation problems have analytical, closed-form solutions. Number six requires iterative, numerical solution for both the optimal parameter estimate and the measurement error covariance estimate.

4.7.1 Scalar Measurement y, Known Measurement Variance σ

We consider first the case in which y_i is a scalar measurement for n samples $i = 1, \ldots, n$, e_i is the measurement error (a random variable) for the ith sample, $\theta \in \mathbb{R}^{n_p}$ is a vector of n_p model parameters, and $x_i \in \mathbb{R}^{n_p}$ is the n_p vector of environmental conditions for the ith sample

$$y_i = x_i^T \theta + e_i \qquad e_i \sim N(0, \sigma^2) \tag{4.42}$$

Consider $n \geq 1$ samples. The probability density of the set of y_i samples for given values of θ and σ are obtained from the given normal distribution for the measurement error $e_i = y_i - x^T \theta$. We have that

$$p(y_1, y_2, \ldots, y_n | \theta, \sigma) = \frac{1}{(2\pi)^{n/2} \sigma^n} \exp\left(-\frac{1}{2\sigma^2} \sum_{i=1}^{n} (y_i - x_i^T \theta)^2 \right)$$

Taking logarithm gives

$$-\ln p(y_1, y_2, \ldots, y_n | \theta, \sigma) = \frac{n}{2} \ln 2\pi + n \ln \sigma + \frac{1}{2\sigma^2} \sum_{i=1}^{n} (y_i - x_i^T \theta)^2$$

This equation is easier to express if we first stack the y_i in a vector and x_i in a matrix as

$$y = \begin{bmatrix} y_1 \\ y_2 \\ \vdots \\ y_n \end{bmatrix} \qquad X = \begin{bmatrix} x_1^T \\ x_2^T \\ \vdots \\ x_n^T \end{bmatrix}$$

giving

$$-\ln p(y | \theta, \sigma) = \frac{n}{2} \ln 2\pi + n \ln \sigma + \frac{1}{2\sigma^2} (y - X\theta)^T (y - X\theta)$$

We define the the log of the likelihood as a function of the parameters θ and σ with the data y regarded as fixed values

$$-L(\theta, \sigma) = \frac{n}{2} \ln 2\pi + n \ln \sigma + \frac{1}{2\sigma^2} (y - X\theta)^T (y - X\theta) \tag{4.43}$$

Because we assume that we know the measurement error variance σ^2, the only unknown in this first estimation problem is θ. Therefore, to find the maximum-likelihood estimate, we maximize $L(\theta, \sigma)$ by differentiating with respect to θ and set the result to zero

$$\begin{aligned} \frac{dL(\theta, \sigma)}{d\theta} &= \frac{1}{\sigma^2} X^T (y - X\theta) \qquad (4.44) \\ 0 &= \frac{1}{\sigma^2} X^T (y - X\theta) \end{aligned}$$

Assuming that X has full column rank, we solve the last equation giving the familiar least-squares formula for the maximum-likelihood estimate

$$\boxed{\widehat{\theta} = (X^T X)^{-1} X^T y} \tag{4.45}$$

We delay a discussion of what to do when X does not have full column rank until Section 4.8. But we know from Chapter 1 that the optimal estimate is not unique, and we have a linear subspace of estimates that all are optimal. This situation is depicted in Figure 1.6(b).

Probability density of parameters and parameter confidence interval. The next item of interest is the probability density of the estimates. Let θ_0 be the parameter generating the measurements so the model is $y = X\theta_0 + e$. Then we have

$$\begin{aligned}
\widehat{\theta} &= (X^T X)^{-1} X^T y \\
&= (X^T X)^{-1} X^T (X\theta_0 + e) \qquad e \sim N(0, \sigma^2 I_n) \\
&= \theta_0 + (X^T X)^{-1} X^T e \\
\widehat{\theta} &= \theta_0 + (X^T X)^{-1} X^T e
\end{aligned}$$

Using the result on linear transformation of a normal, we have

$$\boxed{\widehat{\theta} \sim N(\theta_0, \sigma^2 (X^T X)^{-1})} \tag{4.46}$$

As shown in Exercise 4.21, for a random variable $\xi \in \mathbb{R}^{n_p}$ distributed as a multivariate normal with mean m and covariance P, the probability that ξ takes on value x inside the ellipse

$$(x - m)^T P^{-1} (x - m) \leq b$$

is given by

$$\alpha = \frac{\gamma(n_p/2, b/2)}{\Gamma(n_p/2)}$$

in which the complete and incomplete gamma functions are defined by (Abramowitz and Stegun, 1970, p.255–260)

$$\Gamma(n_p) = \int_0^\infty t^{n_p - 1} e^{-t} dt = (n_p - 1)! \qquad \gamma(n_p, x) = \int_0^x t^{n_p - 1} e^{-t} dt$$

The function χ^2 inverts this relationship so we have that

$$\chi^2(n_p, \alpha) = b$$

Therefore, given a confidence level α, the region of the multivariate normal containing probability α is obtained by substituting this relation for b into the equation defining the ellipse

$$(x - m)^T P^{-1}(x - m) \leq \chi^2(n_p, \alpha)$$

Finally, substituting in the values of the mean and covariance gives the following α-level elliptical confidence region for the maximum-likelihood estimate

$$\boxed{(\hat{\theta} - \theta_0)^T \left(\frac{X^T X}{\sigma^2}\right)(\hat{\theta} - \theta_0) \leq \chi^2(n_p, \alpha)} \tag{4.47}$$

For a large-dimensional parameter vector, the elliptical region is cumbersome to present. In these cases we may wish to approximate the confidence region with the smallest bounding box that contains the ellipse. As shown in Exercise 4.15, this box is given by

$$\boxed{\left|\hat{\theta} - \theta_0\right|_i \leq \left(\chi^2(n_p, \alpha)\sigma^2 (X^T X)^{-1}_{ii}\right)^{1/2}}$$

which is commonly reported as plus/minus limits with the following notation

$$\hat{\theta} = \theta_0 \pm c$$

in which

$$c_i = \left(\chi^2(n_p, \alpha)\sigma^2 (X^T X)^{-1}_{ii}\right)^{1/2}$$

Note that the parameter uncertainly interval does *not* depend on the measurement samples y_i when we know the measurement error variance. We can compute c before we do the experiment, based solely on the chosen x_i. Only $\hat{\theta}$ depends on the experiment. And if we do an increasing number of experiments, $X^T X = \sum_{i=1}^n x_i x_i^T$ increases linearly with the number of samples n, so the confidence interval c decreases as $n^{-1/2}$. So one method to reduce uncertainty in parameter estimates is to replicate experiments.

Marginal parameter estimates. Another way to condense the multivariate density is to compute its marginals. Since $\hat{\theta}$ is distributed as a normal in (4.46), we compute marginals as in (4.19) giving

$$\hat{\theta}_i \sim N((\theta_0)_i, \sigma^2 (X^T X)^{-1}_{ii})$$

We can then compute α-level confidence levels on each of the n_p univariate normals giving

$$\hat{\theta} = \theta_0 \pm m$$

in which

$$m_i = \left(\chi^2(1,\alpha)\sigma^2(X^TX)^{-1}_{ii}\right)^{1/2}$$

Notice that the m_i and c_i formulas are *different*. The first is the bounding box for the true multivariate α-level confidence region; the second is simply a collection of the α-level confidence intervals for all the marginals of the multivariate estimate. Let's call this latter region the "marginal box" to distinguish it from the bounding box. Students often ask, "Since it is difficult to present a high-dimensional ellipse, which of these two plus/minus results should be reported as *the* confidence interval in a research presentation?" This question has no satisfactory answer. The important point is to know and communicate *what* you are reporting. The bounding box certainly contains more than the α-level probability since it contains the true α-level region in its interior. The marginal box does *not* have this property. The interpretation of the marginal box is the same as the interpretation of any marginal density. If you obtained many samples of the parameter estimates from many datasets, the ith interval of the marginal box would contain an α-level fraction of all the different samples of the ith parameter estimate. No statement about the probability of the *jointly distributed* parameter estimate follows from this characterization. We include the following example to help clarify these distinctions.

Example 4.21: The confidence region, bounding box, and marginal box

Assume that the two-dimensional random variable ξ is distributed as $N(m, P)$ with

$$m = \begin{bmatrix} 1 \\ 2 \end{bmatrix} \qquad P = \begin{bmatrix} 2 & 3/4 \\ 3/4 & 1/2 \end{bmatrix}$$

(a) Plot the multivariate density.

(b) Compute and plot the two marginal densities, and their 95% confidence intervals.

(c) Compute the bounding box and the marginal box, and plot them along with the joint density 95% confidence ellipse.

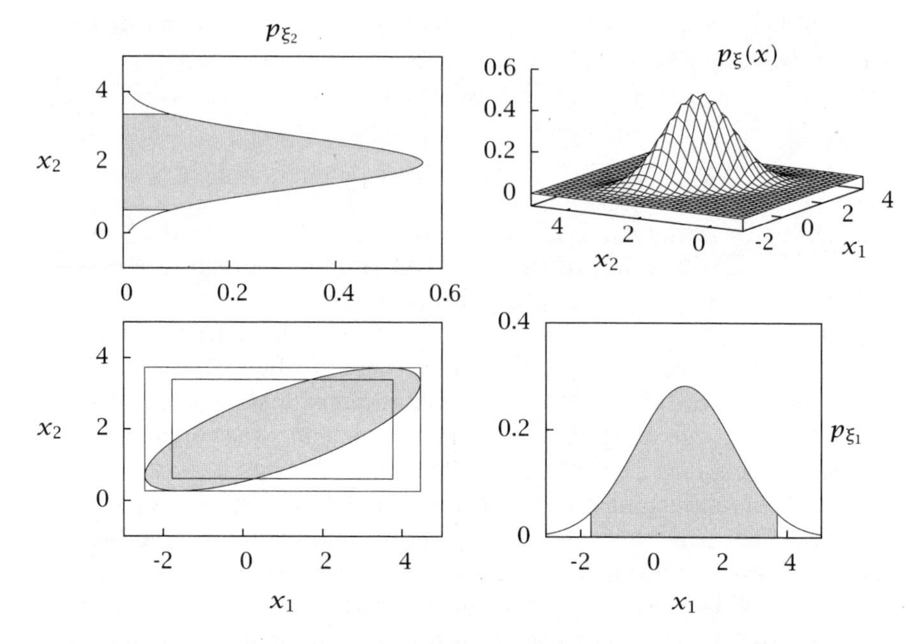

Figure 4.10: The multivariate normal density (top right). The two marginal densities and marginal 95% confidence regions (shaded) (top left and bottom right). The joint elliptical 95% confidence region (shaded), bounding box (outer), and the marginal box (inner) (bottom left).

(d) Take 1000 independent samples of ξ, and determine the number inside the ellipse, the bounding box, and the marginal box. Approximately what confidence levels can you assign to the bounding box and marginal box?

Solution

(a) The multivariate density is shown in the top right of Figure 4.10. The 95% confidence ellipse is given by

$$(x-m)^T P^{-1}(x-m) \leq \chi^2(2, 0.95) = 5.99$$

This ellipse is shown in the bottom left of Figure 4.10.

(b) The two marginals are

$$p_{\xi_1}(x_1) = \frac{1}{\sqrt{4\pi}} e^{-(1/4)(x_1-1)^2}$$
$$p_{\xi_2}(x_2) = \frac{1}{\sqrt{\pi}} e^{-(x_2-2)^2}$$

The marginal densities of ξ_1 and ξ_2 are shown in the bottom right and top left of Figure 4.10, respectively. The 95% interval for the two marginals are given by

$$\frac{(x_1-1)^2}{2} \leq \chi^2(1, 0.95) = 3.84 \qquad x_1 \in [-1.77, 3.77]$$
$$\frac{(x_2-2)^2}{(1/2)} \leq \chi^2(1, 0.95) = 3.84 \qquad x_2 \in [0.614, 3.39]$$

These intervals are shown as the shaded regions in the bottom right and top left of Figure 4.10.

(c) The ellipse's bounding box is given by

$$\frac{(x_1-1)^2}{2} \leq \chi^2(2, 0.95) = 5.99 \qquad x_1 \in [-2.46, 4.46]$$
$$\frac{(x_2-2)^2}{(1/2)} \leq \chi^2(2, 0.95) = 5.99 \qquad x_2 \in [0.269, 3.73]$$

The ellipse, bounding box, and marginal box are shown in the bottom left of Figure 4.10.

(d) Generating 1000 samples of ξ and counting the fraction of samples within each of the three regions gives

$$\text{ellipse} = 0.956$$
$$\text{bounding box} = 0.981 \quad \text{marginal box} = 0.920$$

□

4.7.2 Scalar Measurement y, Unknown Measurement Variance σ

We now consider the measurement error variance σ^2 to be unknown. We have the same model as in the previous section

$$y_i = x_i^T \theta + e_i \qquad e_i \sim N(0, \sigma^2)$$

When the measurement variance is unknown, we maximize the likelihood function given in (4.43) over *both* σ and θ and estimate both quantities from the data. The θ derivative is the same as in (4.44), and differentiating (4.43) with respect to σ gives

$$\frac{\partial L(\theta, \sigma)}{\partial \theta} = \frac{1}{\sigma^2} X^T (y - X\theta)$$

$$\frac{\partial L(\theta, \sigma)}{\partial \sigma} = -\frac{n}{\sigma} + \sigma^{-3} (y - X\theta)^T (y - X\theta)$$

Equating the derivatives to zero and solving simultaneously gives

$$\hat{\theta} = (X^T X)^{-1} X^T y \quad (4.48)$$

$$\widehat{\sigma^2} = \frac{1}{n} (y - X\hat{\theta})^T (y - X\hat{\theta}) \quad (4.49)$$

We see that the maximum-likelihood parameter estimate is unchanged from the known variance case, and the maximum-likelihood estimate of the variance is the mean of the square of the residual over the samples. Notice that the maximum-likelihood estimate of variance is close to but not equal to the sample variance s^2 given by the formula (for $n > n_p$)

$$s^2 = \frac{1}{n - n_p} (y - X\hat{\theta})^T (y - X\hat{\theta})$$

$$s^2 = \frac{n}{n - n_p} \widehat{\sigma^2}$$

We show subsequently that the sample variance is an unbiased estimate of σ^2 so the maximum-likelihood estimate of σ^2 is biased. But this bias is small for a large number of samples compared to parameters $n \gg n_p$.

Given the same result for $\hat{\theta}$ as in the previous problem, the probability density of $\hat{\theta}$ is unchanged from the previous problem. We next determine the probability density of $\widehat{\sigma^2}$. For this it is convenient to first consider the singular value decomposition of the X matrix. We assume that this $n \times n_p$ matrix has independent columns so the rank is n_p. As discussed in Chapter 1, a real $n \times n_p$ matrix with independent columns can be written as the product of orthogonal $n \times n$ matrix U and orthogonal $n_p \times n_p$ matrix V, and diagonal $n_p \times n_p$ matrix Σ

$$X = \begin{bmatrix} U_1 & U_2 \end{bmatrix} \begin{bmatrix} \Sigma \\ 0 \end{bmatrix} V^T$$

$$X = U_1 \Sigma V^T$$

in which the following relationships result from orthogonality

$$U_1^T U_1 = I_{n_p} \qquad U_2^T U_2 = I_{n-n_p} \qquad U_1^T U_2 = 0_{n_p \times n-n_p}$$
$$U_2^T U_1 = 0_{n-n_p \times n_p} \qquad U_1 U_1^T + U_2 U_2^T = I_n \qquad V^T V = V V^T = I_{n_p}$$

Using the singular value decomposition (SVD) for X, we find by substitution and orthogonality

$$\begin{aligned}(X^T X)^{-1} X^T &= V \Sigma^{-1} U_1^T \\ X(X^T X)^{-1} X^T &= U_1 U_1^T \\ I - X(X^T X)^{-1} X^T &= U_2 U_2^T\end{aligned}$$

These relations allow us to express the estimate and residual in terms of the measurement errors as

$$\begin{aligned}\hat{\theta} - \theta_0 &= V \Sigma^{-1} U_1^T e \\ y - X\hat{\theta} &= U_2 U_2^T e\end{aligned} \tag{4.50}$$

Using these relations we can express the following quadratic terms as

$$\begin{aligned}(\hat{\theta} - \theta_0)^T \left(X^T X\right)(\hat{\theta} - \theta_0) &= e^T U_1 U_1^T e \\ (y - X\hat{\theta})^T (y - X\hat{\theta}) &= e^T U_2 U_2^T e\end{aligned}$$

These relations provide an essential insight. The error e obviously affects both quadratic terms, but its effect in the sum of the squares of the residual (the sample variance) is through U_2 and its effect in the parameter estimate's distance from the true value is through U_1. Because these two matrices are orthogonal to each other, the effect of the measurement error is *independently* distributed in these two quadratic terms. We make this statement precise subsequently. First it is helpful to establish that the following two random variables, z_1, z_2 are statistically independent

$$z_1 = \frac{1}{\sigma} U_1^T e \qquad z_2 = \frac{1}{\sigma} U_2^T e$$

Given that $e \sim N(0, \sigma^2 I)$ and the result on linear transformation of a normal, the pair z_1, z_2 is distributed as

$$\begin{bmatrix} z_1 \\ z_2 \end{bmatrix} \sim N(0, P) \qquad P = \begin{bmatrix} I_{n_p} & 0 \\ 0 & I_{n-n_p} \end{bmatrix}$$

Since the pair is jointly normal and the covariance is diagonal, z_1 and z_2 are statistically independent. We also know that their quadratic products are distributed as chi-squared

$$z_1^T z_1 \sim \chi^2_{n_p} \qquad z_2^T z_2 \sim \chi^2_{n-n_p}$$

Exercise 4.33 discusses the chi-squared and chi densities, and also shows that the mean of χ^2_n is n.

From that fact we can deduce quickly the earlier claim that the sample variance is an unbiased estimate. Summarizing our results on sample variance thus far

$$\begin{aligned} s^2 &= \frac{1}{n-n_p}(y - X\hat{\theta})^T(y - X\hat{\theta}) \\ s^2 &= \frac{\sigma^2}{n-n_p}(z_2^T z_2) \end{aligned}$$

Taking expectation gives

$$\begin{aligned} \mathcal{E}(s^2) &= \frac{\sigma^2}{n-n_p}\mathcal{E}(z_2^T z_2) \\ &= \frac{\sigma^2}{n-n_p}\mathcal{E}(\chi^2_{n-n_p}) \\ \mathcal{E}(s^2) &= \sigma^2 \end{aligned}$$

and the result is established.

As shown in Exercise 4.3, if two random variables are statistically independent, then all functions of the two random variables are also statistically independent. Therefore we know that $z_1^T z_1$ and $z_2^T z_2$ are statistically independent. The ratio of two chi-squared, statistically independent random variables is defined as the F-distribution

$$\left(\frac{n-n_p}{n_p}\right)\frac{z_1^T z_1}{z_2^T z_2} \sim F(n_p, n-n_p)$$

The F-distribution can be shown to have density

$$p_F(z; n, m) = \frac{\Gamma(\frac{n+m}{2})}{\Gamma(\frac{n}{2})\Gamma(\frac{m}{2})}\frac{1}{z}\sqrt{\frac{(zn)^n m^m}{(zn+m)^{n+m}}} \qquad z \geq 0, \quad n, m \geq 1$$

Exercises 4.35 and 4.45 provide further discussion of the F-distribution. Substituting the definitions of z_1, z_2 in terms of $\hat{\theta}$ and $\widehat{\sigma^2}$ give

$$(\hat{\theta} - \theta_0)^T \left(\frac{X^T X}{\sigma^2}\right)(\hat{\theta} - \theta_0) \sim \chi^2_{n_p} \qquad n\left(\frac{\widehat{\sigma^2}}{\sigma^2}\right) \sim \chi^2_{n-n_p}$$

$$\left(\frac{n - n_p}{n n_p}\right) \frac{(\hat{\theta} - \theta_0)^T (X^T X)(\hat{\theta} - \theta_0)}{\widehat{\sigma^2}} \sim F(n_p, n - n_p) \tag{4.51}$$

This last distribution provides the basis for the confidence intervals on the parameter estimates. Summarizing our results so far, the densities for the parameter estimates and the measurement variance estimate are

$$\hat{\theta} \sim N(\theta_0, \sigma^2 (X^T X)^{-1}) \tag{4.52}$$

$$n\left(\frac{\widehat{\sigma^2}}{\sigma^2}\right) \sim \chi^2_{n-n_p} \tag{4.53}$$

Notice that these distributions are inadequate to construct confidence levels on the estimated parameter $\hat{\theta}$ because they both depend on the *unknown* measurement variance σ^2. One might be tempted to replace the unknown σ^2 in the normal density for $\hat{\theta}$ with the maximum-likelihood estimate $\widehat{\sigma^2}$ and obtain the confidence intervals for $\hat{\theta}$ from that density. That idea is in the right spirit, but is not quite correct. We obtain the correct confidence region by considering the distribution in (4.51). Notice that the ratio of the two quadratic terms has divided out the common term σ^2. We define the function $F(n, m, \alpha)$, $0 \le \alpha \le 1$ to return the argument of the cumulative F-distribution that achieves value α

$$\alpha = \int_0^{F(n,m,\alpha)} p_F(z; n, m) dz \tag{4.54}$$

Then the ellipsoidal confidence intervals for the parameter estimates follow from (4.51)

$$(\hat{\theta} - \theta_0)^T \left(\frac{X^T X}{\widehat{\sigma^2}}\right)(\hat{\theta} - \theta_0) \le \left(\frac{n n_p}{n - n_p}\right) F(n_p, n - n_p, \alpha)$$

We can use the sample variance s^2 in place of $\widehat{\sigma^2}$ in this formula to give a slightly simpler expression

$$(\hat{\theta} - \theta_0)^T \left(\frac{X^T X}{s^2}\right)(\hat{\theta} - \theta_0) \le n_p \, F(n_p, n - n_p, \alpha) \tag{4.55}$$

We can obtain the bounding box intervals as was done in the previous section

$$\hat{\theta} = \theta_0 \pm c$$

in which

$$c_i = \left(n_p F(n_p, n - n_p, \alpha)\, s^2 (X^T X)^{-1}_{ii}\right)^{1/2}$$

The significant difference between this and the previous case is that here the confidence interval c also depends on the measurements. The *size* as well as the center $(\hat{\theta})$ of the α-level confidence ellipse is therefore *random.* But the statistical interpretation remains the same; given many replicated experiments, the true parameter θ_0, which is not a random variable, will lie within the generated confidence ellipse for the experiment 95% of the time (for $\alpha = 0.95$). We have the same dependence as the previous case, the confidence interval c decreases with number of samples $n^{-1/2}$.

4.7.3 Vector of Measurements y, Different Parameters Corresponding to Different Measurements, Known Measurement Covariance R

We next consider the vector measurement case. This case arises frequently when identifying empirical linear models between a vector of input variables x and a vector of output or response variables y_i. We consider first the case in which each measurement type has its own vector of parameters describing it

$$\overset{1}{p\begin{bmatrix} y_1 \\ \vdots \\ y_p \end{bmatrix}_i} = \overset{q}{p\begin{bmatrix} \theta_1^T \\ \vdots \\ \theta_p^T \end{bmatrix}} \overset{1}{q\begin{bmatrix} \\ x_i \\ \\ \end{bmatrix}} + \overset{1}{p\begin{bmatrix} e_1 \\ \vdots \\ e_p \end{bmatrix}_i} \tag{4.56}$$

$$y_i = \Theta \quad x_i + e_i \qquad e_i \sim N(0, R)$$

The environmental variable x_i is assumed to have q components, $x_i \in \mathbb{R}^q$, and $\Theta \in \mathbb{R}^{p \times q}$, and we assume $q < n$. In this model we have $n_p = pq$ model parameters to estimate. Notice that this model is not restricted to only p independent versions of the model given by (4.42). The generalization allowed here comes from the covariance matrix R. To reduce this case to the (4.42), we would add the further restriction that $R = \sigma^2 I$. We will see that allowing the different measurements

$y_1, \ldots, y_p$ to be correlated does not prevent us from solving this estimation problem also in closed form. We continue to assume here (and assume throughout) that the different samples are independent (hence uncorrelated).

Consider $n \geq 1$ samples, $i = 1, \ldots, n$, and, given the deterministic variables Θ and the n x_i, we have for the probability density of the measurements

$$p(y_1, y_2, \ldots, y_n | \Theta, R) = \frac{1}{(2\pi)^{np/2}(\det R)^{n/2}} \exp\left(-\frac{1}{2} \sum_{i=1}^{n} (y_i - \Theta x_i)^T R^{-1} (y_i - \Theta x_i) \right)$$

or, by taking logarithm

$$-\ln p(y_1, y_2, \ldots, y_n | \Theta, R) = \frac{np}{2} \ln 2\pi + \frac{n}{2} \ln \det R + \frac{1}{2} \sum_{i=1}^{n} (y_i - \Theta x_i)^T R^{-1} (y_i - \Theta x_i)$$

We again define the log-likelihood as a function of the parameters Θ and R with the data y_i, $i = 1, 2, \ldots n$, regarded as fixed values

$$-L(\Theta, R) = \frac{np}{2} \ln 2\pi + \frac{n}{2} \ln \det R + \frac{1}{2} \sum_{i=1}^{n} (y_i - \Theta x_i)^T R^{-1} (y_i - \Theta x_i)$$

Since R is known, we take the derivative of L with respect to the matrix Θ. It is perhaps easiest to perform this derivative using component notation. Rewriting the expression for L in components gives

$$-L(\Theta, R) = \frac{np}{2} \ln 2\pi + \frac{n}{2} \ln \det R + \frac{1}{2} (y_{ir} - \Theta_{rj} x_{ij})^T R_{rs}^{-1} (y_{is} - \Theta_{sl} x_{il}) \tag{4.57}$$

in which we use the Einstein summation convention for repeated indices. Taking the derivative of scalar-valued function L with respect to Θ_{mn} gives a matrix derivative

$$\frac{\partial L}{\partial \Theta_{mn}} = \frac{1}{2} \Big(\delta_{rm} \delta_{jn} x_{ij} R_{rs}^{-1} (y_{is} - \Theta_{sl} x_{il}) + (y_{ir} - \Theta_{rj} x_{ij}) R_{rs}^{-1} \delta_{sm} \delta_{ln} x_{il} \Big)$$

Performing the sums over the deltas, noting R is symmetric, and collecting terms gives

$$\frac{\partial L}{\partial \Theta_{mn}} = R_{ms}^{-1} (y_{is} - \Theta_{sl} x_{il}) x_{ni}$$

If we convert this back to the vector/matrix notation of the problem statement we have

$$\frac{\partial L}{\partial \Theta} = \sum_{i=1}^{n} R^{-1}(y_i - \Theta x_i)x_i^T$$

Setting this matrix to zero and solving gives the maximum-likelihood estimate for the parameters Θ

$$\boxed{\hat{\Theta} = \left(\sum_i y_i x_i^T\right)\left(\sum_i x_i x_i^T\right)^{-1}}$$

in which we assume that the matrix $\sum_i x_i x_i^T$ has full rank. Again, we discuss what to do when this rank condition fails later in Section 4.8. Notice that the value of the measurement error covariance R is irrelevant in the estimation of Θ in this problem also. It is often convenient to arrange the variables so that the summation is performed by matrix operations. Arranging the data vectors in the following matrices

$$Y = \begin{bmatrix} y_1 & \cdots & y_n \end{bmatrix} \qquad X = \begin{bmatrix} x_1 & \cdots & x_n \end{bmatrix}$$

allows us to express the maximum-likelihood estimate as

$$\hat{\Theta} = YX^T \left(XX^T\right)^{-1} \tag{4.58}$$

Next we determine the probability density of the estimated parameter $\hat{\Theta}$. We denote the parameter value generating the data as Θ_0, so the measurements are given by

$$Y = \Theta_0 X + E \qquad E = \begin{bmatrix} e_1 & \cdots & e_n \end{bmatrix}$$

The estimate and its transpose are therefore

$$\begin{aligned} \hat{\Theta} &= \Theta_0 + EX^T(XX^T)^{-1} \\ \hat{\Theta}^T &= \Theta_0^T + (XX^T)^{-1}XE^T \end{aligned}$$

We find the transpose convenient because we now wish to stack the matrix $\hat{\Theta}^T$ in a vector giving

$$\hat{\Theta}^T = \begin{bmatrix} \theta_1 & \theta_2 & \cdots & \theta_p \end{bmatrix} \qquad \text{vec}\hat{\Theta}^T = \begin{bmatrix} \theta_1 \\ \theta_2 \\ \vdots \\ \theta_p \end{bmatrix}$$

Applying the vec operator to both sides of the transposed form of the parameter estimates gives

$$\text{vec}\widehat{\Theta}^T = \text{vec}\Theta_0^T + (I \otimes (XX^T)^{-1}X)\text{vec}E^T$$

From the definition of E we see

$$\text{vec}E^T = \begin{bmatrix} e_{1,1} \\ e_{1,2} \\ \vdots \\ e_{1,n} \\ \vdots \\ e_{p,1} \\ e_{p,2} \\ \vdots \\ e_{p,n} \end{bmatrix} \qquad e_{j,i} \; j\text{th measurement, } i\text{th sample}$$

Given this arrangement of these normally distributed random variables, we have for the density

$$\text{vec}E^T \sim N(0, P)$$

in which

$$P = \begin{bmatrix} R_{11} & & & R_{1p} & & \\ & \ddots & \cdots & & \ddots & \\ & & R_{11} & & & R_{1p} \\ & \vdots & & \ddots & \vdots & \\ R_{p1} & & & R_{pp} & & \\ & \ddots & \cdots & & \ddots & \\ & & R_{p1} & & & R_{pp} \end{bmatrix}$$

$$P = R \otimes I$$

Using the result on linear transformation of a normal, we have

$$\boxed{\text{vec}\widehat{\Theta}^T - \text{vec}\Theta_0^T \sim N(0, S) \qquad S = R \otimes (XX^T)^{-1}} \tag{4.59}$$

in which

$$S = (I \otimes (XX^T)^{-1}X)(R \otimes I)(I \otimes (XX^T)^{-1}X)^T$$

Using the Kronecker product formulas from Section 1.5.3, we can simplify this covariance as follows

$$\begin{aligned} S &= (I \otimes (XX^T)^{-1}X)(R \otimes I)(I \otimes (XX^T)^{-1}X)^T \\ &= (R \otimes (XX^T)^{-1}X)(I \otimes X^T(XX^T)^{-1}) \\ S &= R \otimes (XX^T)^{-1} \end{aligned}$$

Equation (4.59), with this result for S, is the matrix analog of the vector result in (4.46).

Given the normal density, the elliptical confidence region for $\mathrm{vec}\widehat{\Theta}^T$ can be found as in Section 4.7.1

$$\boxed{(\mathrm{vec}\widehat{\Theta}^T - \mathrm{vec}\Theta_0^T)^T\, S^{-1}(\mathrm{vec}\widehat{\Theta}^T - \mathrm{vec}\Theta_0^T) \;\le \chi^2(n_p, \alpha)} \qquad (4.60)$$

Interlude

Let's put the tools of orthogonality and Kronecker products to good use and prove a fundamental result in statistics, namely that the sample mean and sample variance from a normal distribution are statistically independent.

Theorem 4.22 (Mean and variance of samples from a normal). *Let $x_i \in \mathbb{R}^p, i = 1, \ldots, n$ be n independent samples from $N(\mu, \Sigma)$. Define the sample mean and the maximum-likelihood estimate of the variance as*

$$\overline{x} = \frac{1}{n}\sum_{i=1}^{n} x_i \qquad \widehat{\Sigma} = \frac{1}{n}\sum_{i=1}^{n}(x_i - \overline{x})(x_i - \overline{x})^T$$

Then $\overline{x}$ is distributed as $N(\mu, (1/n)\Sigma)$ and independently of $\widehat{\Sigma}$, and $n\widehat{\Sigma}$ is distributed as $\sum_{i=1}^{n-1} z_i z_i^T$ in which the z_i are distributed independently and identically as $N(0, \Sigma)$

Proof. Stack the n x_i vectors next to each other in a matrix

$$X = \begin{bmatrix} x_1 & x_2 & \cdots & x_n \end{bmatrix}$$

We next construct an orthogonal transformation of this matrix. Let 1 be $1/\sqrt{n}$ times an n-vector of ones so that $X1 = \sqrt{n}\,\overline{x}$. Next consider the null space of 1^T. From the fundamental theorem of linear algebra, that is an $n-1$ dimensional space. Collect an orthonormal basis in the

$n \times (n-1)$ matrix B_{n-1}. Then construct the following orthogonal B matrix

$$B = \begin{bmatrix} B_{n-1}^T \\ 1^T \end{bmatrix} \qquad B^T = \begin{bmatrix} B_{n-1} & 1 \end{bmatrix}$$
$$BB^T = B^T B = I$$

Define the transformed random variables

$$\begin{bmatrix} z_1 & z_2 & \cdots & z_n \end{bmatrix} = \begin{bmatrix} x_1 & x_2 & \cdots & x_n \end{bmatrix} B^T$$
$$Z = XB^T$$

in which $z_n = \sqrt{n}\,\overline{x}$. The samples x_i are distributed as

$$\text{vec}X = \begin{bmatrix} x_1 \\ \vdots \\ x_n \end{bmatrix} \sim N\left(\begin{bmatrix} \mu \\ \vdots \\ \mu \end{bmatrix}, \begin{bmatrix} \Sigma & & \\ & \ddots & \\ & & \Sigma \end{bmatrix}\right)$$

or in more compact notation

$$\text{vec}X \sim N(\sqrt{n}1 \otimes \mu, I \otimes \Sigma)$$

The transformation gives for Z

$$\text{vec}Z = (B \otimes I)\text{vec}X \qquad \text{vec}Z \sim N(m, P)$$

in which

$$m = \sqrt{n}(B \otimes I)(1 \otimes \mu) \qquad P = (B \otimes I)(I \otimes \Sigma)(B \otimes I)^T$$

Rearranging these expressions gives

$$m = \sqrt{n}B1 \otimes \mu \qquad P = (BB^T \otimes \Sigma)$$

From the orthogonality relations we have

$$B1 = \begin{bmatrix} 0 \\ \vdots \\ 0 \\ 1 \end{bmatrix} \qquad BB^T = I$$

so

$$\text{vec}Z = \begin{bmatrix} z_1 \\ \vdots \\ z_{n-1} \\ z_n \end{bmatrix} \sim N\left(\begin{bmatrix} 0 \\ \vdots \\ 0 \\ \sqrt{n}\mu \end{bmatrix}, \begin{bmatrix} \Sigma & & \\ & \ddots & \\ & & \Sigma & \\ & & & \Sigma \end{bmatrix}\right)$$

From the covariance we conclude that the variables $z_1, z_2, \ldots, z_n$ are statistically independent. Computing $\hat{\Sigma}$ gives

$$\begin{aligned}
\hat{\Sigma} &= \frac{1}{n}\sum_{i=1}^{n}(x_i - \overline{x})(x_i - \overline{x})^T \\
&= \frac{1}{n}\sum_{i=1}^{n}(x_i x_i^T - \overline{x}\overline{x}^T) \\
&= \frac{1}{n}(XX^T - n\overline{x}\overline{x}^T) \\
&= \frac{1}{n}(ZB^T BZ^T - z_n z_n^T) \\
&= \frac{1}{n}\left(ZZ^T - z_n z_n^T\right) \\
\hat{\Sigma} &= \frac{1}{n}\sum_{i=1}^{n-1} z_i z_i^T
\end{aligned}$$

which establishes the stated distribution for $\hat{\Sigma}$. Since $\hat{\Sigma}$ is a function of only $z_1, \ldots z_{n-1}$ and $\overline{x}$ is a function of only z_n, $\hat{\Sigma}$ and $\overline{x}$ are independent. Since $\overline{x} = z_n/\sqrt{n}$, we have that $\overline{x} \sim N(\mu, (1/n)\Sigma)$, and the theorem is proved. ∎

This result is established in many statistical texts using a dazzling variety of arguments, some bordering on the mystical. The proof given above is a compact expression of a standard method given by Anderson (2003, p. 77).

4.7.4 Vector of Measurements y, Different Parameters Corresponding to Different Measurements, Unknown Measurement Covariance R

When R is also unknown, we maximize L in (4.57) with respect to both Θ and R. The Θ derivative has been given previously. Differentiating (4.57) with respect to R is facilitated by using the following fact about the trace of a matrix product

$$\text{tr}(AB) = \text{tr}(BA)$$

which follows immediately from the definition of trace and expressing the matrix product in components

$$\text{tr}(AB) = A_{ij}B_{ji} = B_{ji}A_{ij} = \text{tr}(BA)$$

Using this result twice on a product of three matrices gives

$$\mathrm{tr}(ABC) = \mathrm{tr}(BCA) = \mathrm{tr}(CAB)$$

This identity allows us to rewrite the following scalar term

$$\begin{aligned}(y_i - \Theta x_i)^T R^{-1}(y_i - \Theta x_i) &= \mathrm{tr}\left((y_i - \Theta x_i)^T R^{-1}(y_i - \Theta x_i)\right) \\ &= \mathrm{tr}\left(R^{-1}(y_i - \Theta x_i)(y_i - \Theta x_i)^T\right)\end{aligned}$$

Next we use the following fact in differentiating the trace of a function of a matrix

$$\frac{d\mathrm{tr}(f(A))}{dA} = g(A^T) \qquad g(x) = \frac{df(x)}{dx}$$

in which g is the usual scalar derivative of the scalar function f. See Exercise 4.4 for a derivation of this fact. Applying this result and using the fact that R is symmetric gives

$$\frac{d(\mathrm{tr}(R^{-1}C))}{dR} = -R^{-2}C$$

The derivative of the determinant and the log of the determinant are (see Exercise 4.5 for a derivation)

$$\frac{d\det A}{dA} = (A^{-1})^T \det A \qquad \frac{d\ln\det A}{dA} = (A^{-1})^T$$

The R derivative of (4.57) is therefore

$$\frac{\partial L(\Theta, R)}{\partial R} = -\frac{n}{2}R^{-1} + \frac{1}{2}R^{-2}\sum_i (y_i - \Theta x_i)(y_i - \Theta x_i)^T$$

Setting this matrix equation to zero, using the estimate of Θ, and solving gives the maximum-likelihood estimates for this problem

$$\boxed{\begin{aligned}\hat{\Theta} &= \left(\sum_i y_i x_i^T\right)\left(\sum_i x_i x_i^T\right)^{-1} \\ \hat{R} &= \frac{1}{n}\sum_i (y_i - \hat{\Theta}x_i)(y_i - \hat{\Theta}x_i)^T\end{aligned}}$$

The estimate $\hat{R}$ is an unbiased estimate of the measurement variance R. The distribution for $n\hat{R}$ can be shown to be a Wishart distribution (see Exercise 4.51), which is a generalization of the χ^2 distribution to the multivariate case (Wishart, 1928). The Wishart distribution can be shown to be (Anderson, 2003, pp. 252-255)

$$p_W(W) = \frac{(\det W)^{\frac{n-p-1}{2}}}{2^{\frac{np}{2}} (\det R)^{\frac{n}{2}} \Gamma_p(\frac{n}{2})} e^{-\frac{1}{2}\mathrm{tr}(R^{-1}W)} \tag{4.61}$$

in which Γ_p is the multivariate gamma function defined by

$$\Gamma_p(z) = \pi^{p(p-1)/4} \prod_{i=1}^{p} \Gamma(z - \frac{1}{2}(i-1))$$

Note that the argument of the probability density $p_W(\cdot)$ is a positive definite matrix W. The probability is zero for W not positive definite.

4.7.5 Vector of Measurements y, Same Parameters for all Measurements, Known Measurement Covariance R

Next we consider the case in which the different measurement types are affected by the same set of parameters. The model is

$$\begin{bmatrix} y_1 \\ y_2 \\ \vdots \\ y_p \end{bmatrix}_i = \begin{bmatrix} x_1^T \\ x_2^T \\ \vdots \\ x_p^T \end{bmatrix}_i \begin{bmatrix} \\ \theta \\ \\ \end{bmatrix} + \begin{bmatrix} e_1 \\ e_2 \\ \vdots \\ e_p \end{bmatrix}_i$$

$$y_i \quad = \quad X_i \quad \theta \quad + \quad e_i \qquad\qquad e_i \sim N(0, R)$$

In this model, all of the different components of the measurement $y_1, y_2, \ldots, y_p$ are affected by the same, single vector of parameters θ. Consider $n \geq 1$ samples, $i = 1, \ldots, n$, and, given the deterministic variables θ and the n X_i, we have for the probability density of the measurements

$$p(y_1, y_2, \ldots, y_n | \theta, R) =$$

$$\frac{1}{(2\pi)^{np/2} (\det R)^{n/2}} \exp\Big(-\frac{1}{2} \sum_{i=1}^{n} (y_i - X_i\theta)^T R^{-1} (y_i - X_i\theta) \Big)$$

$$-L(\theta, R) =$$

$$\frac{np}{2} \ln 2\pi + \frac{n}{2} \ln \det R + \frac{1}{2} \sum_{i=1}^{n} (y_i - X_i\theta)^T R^{-1} (y_i - X_i\theta) \tag{4.62}$$

Taking the derivative with respect to θ gives

$$\begin{aligned}\frac{\partial L(\theta, R)}{\partial \theta} &= \frac{1}{2}\sum_{i=1}^{n} 2X_i^T R^{-1} y_i - 2X_i^T R^{-1} X_i \theta \\ &= \sum_{i=1}^{n} X_i^T R^{-1}(y_i - X_i\theta)\end{aligned}$$

Setting this vector equation to zero and solving for θ gives the maximum-likelihood estimate

$$\boxed{\hat{\theta} = \left(\sum_i X_i^T R^{-1} X_i\right)^{-1} \sum_i X_i^T R^{-1} y_i} \tag{4.63}$$

In this problem, it can make sense to estimate θ with a single sample ($n = 1$) if we can choose the number of measurements p significantly larger than the number of parameters n_p. For a single sample, the parameter estimate formula is

$$\hat{\theta} = \left(X^T R^{-1} X\right)^{-1} X^T R^{-1} y \tag{4.64}$$

which is the solution of a weighted least-squares problem using R^{-1} as the weight. Compare this expression to (4.45).

Notice also that this is the first estimation problem for which the maximum-likelihood estimate of the parameter $\hat{\theta}$ depends on the covariance of the measurement error R. We see next that this dependence prevents us from solving the final estimation problem in closed form.

We next calculate the probability density of the estimate. We denote the parameter value generating the data as θ_0, so the measurements are given by

$$y_i = X_i\theta_0 + e_i$$

and substituting this result into the estimate equation gives

$$\hat{\theta} = \theta_0 + \left(\sum_i X_i^T R^{-1} X_i\right)^{-1} \sum_i X_i^T R^{-1} e_i$$

$$\hat{\theta} = \theta_0 + \left(\sum_i X_i^T R^{-1} X_i\right)^{-1} \begin{bmatrix} X_1^T R^{-1} & \cdots & X_n^T R^{-1}\end{bmatrix} \begin{bmatrix} e_1 \\ \vdots \\ e_n \end{bmatrix}$$

Using the result on linear transformation of a normal, we have

$$\boxed{\hat{\theta} - \theta_0 \sim N(0, S)} \tag{4.65}$$

in which

$$S = \left(\sum_i X_i^T R^{-1} X_i\right)^{-1} \begin{bmatrix} X_1^T R^{-1} & \cdots & X_n^T R^{-1} \end{bmatrix} \cdot \begin{bmatrix} R & & \\ & \ddots & \\ & & R \end{bmatrix} \begin{bmatrix} R^{-1} X_1 \\ \vdots \\ R^{-1} X_n \end{bmatrix} \left(\sum_i X_i^T R^{-1} X_i\right)^{-1}$$

$$S = \left(\sum_i X_i^T R^{-1} X_i\right)^{-1}$$

Given the normal density, we can compute the elliptical confidence region as in Section 4.7.1

$$\boxed{(\hat{\theta} - \theta_0)^T\, S^{-1}\, (\hat{\theta} - \theta_0) \leq \chi^2(n_p, \alpha)} \tag{4.66}$$

The bounding box intervals follow as in Section 4.7.1. Notice that whenever the variance of the measurement errors are known, the maximum-likelihood estimate is normally distributed and the elliptical confidence intervals are given by $\chi^2(n_p, \alpha)$.

4.7.6 Vector of Measurements y, Same Parameters for all Measurements, Unknown Measurement Covariance R

The final case is the one that arises most often in mechanistic modeling of chemical and biological experiments. To determine the unknown R, we maximize $L(\theta, R)$ over R in addition to θ. Using the results of Section 4.7.4 we can take the derivative of (4.62) with respect to R giving

$$\frac{\partial L(\theta, R)}{\partial R} = -\frac{n}{2} R^{-1} + \frac{1}{2} R^{-2} \sum_i (y_i - X_i\theta)(y_i - X_i\theta)^T$$

Setting this result to zero and using the result of the previous section gives the following set of necessary conditions for the maximum-likelihood estimates

$$\hat{\theta} = \left(\sum_i X_i^T \hat{R}^{-1} X_i\right)^{-1} \sum_i X_i^T \hat{R}^{-1} y_i \tag{4.67}$$

$$\hat{R} = \frac{1}{n} \sum_i (y_i - X_i\hat{\theta})(y_i - X_i\hat{\theta})^T \tag{4.68}$$

These are two sets of nonlinear equations in the unknowns $\hat{\theta}$ and $\hat{R}$, which must be solved numerically. One simple solution strategy is to first estimate the parameter $\hat{\theta}_0$ with (4.67) using an initial guess for the covariance such as $\hat{R}_0 = I$. One then estimates the iterate $\hat{R}_1$ by substituting $\hat{\theta}_0$ into (4.68), and the process is repeated. If this iteration procedure converges, then one has found the maximum-likelihood estimates by solving a sequence of standard estimation problems. But there is no guarantee that this procedure converges. One may find that a crude initial guess like $\hat{R}_0 = I$ lies outside the region of convergence of the iteration procedure.

Maximum-Likelihood and Bayesian Estimation

With this background in maximum-likelihood estimation, we would like to compare the approach to another class of popular methods known as Bayesian estimation. As we saw in the previous sections, in the maximum-likelihood approach, we maximize the probability of the *measurements* over the model parameter θ

$$\hat{\theta}_{\text{MLE}} = \arg\max_{\theta} p(y;\theta) \tag{4.69}$$

Although in the MLE sections we wrote $p(y|\theta)$ to indicate that θ was a parameter, here we use instead $p(y;\theta)$ to emphasize that θ is an unknown parameter, *not* a random variable. In the MLE approach, $\hat{\theta}$ is a random variable, not θ, and we assess the confidence intervals for $\hat{\theta}$.

In Bayesian estimation, on the other hand, θ itself is modeled as a random variable. The information that we have about θ before the experiment is denoted by $p(\theta)$. In the experiment, we imagine drawing a value of θ as well as the measurement errors to create the data $y_i = x_i^T\theta + e_i$, $i = 1, \ldots, n$. With the measured y available, we then maximize $p(\theta|y)$ over θ to obtain the estimate

$$\hat{\theta}_{\text{BE}} = \arg\max_{\theta} p(\theta|y)$$

The conditional density $p(\theta|y)$ is known as the POSTERIOR density, i.e., the density for θ after the experiment, and the density $p(\theta)$ is known as the PRIOR, i.e., the density before experiment. In Bayesian estimation, we assess how much the measurement of y has changed our knowledge about θ. From Bayes's theorem we can express the posterior as

$$p(\theta|y) = \frac{p(y|\theta)p(\theta)}{p(y)}$$

Notice that $p(y|\theta)$ is *exactly the same functional form* as $p(y;\theta)$ in the MLE approach. Since the denominator does not depend on θ, in Bayesian estimation we estimate θ by the following equivalent maximization

$$\widehat{\theta}_{\mathrm{BE}} = \arg\max_{\theta} p(y|\theta)p(\theta) \tag{4.70}$$

The only difference in the estimators (4.69) and (4.70) is the presence of the prior $p(\theta)$ in the Bayesian approach. In the absence of knowledge about θ, we often assume that $p(\theta)$ is a uniform distribution. This is called the noninformative prior. Since $p(\theta)$ does not depend on θ with the noninformative prior, the MLE and BE estimates are identical in this case.

The posterior density of Bayesian estimation is a useful way to summarize the state of knowledge about the parameter θ given the available experiments. Since one has available the posterior density, confidence levels on random variable θ are determined directly from $p(\theta|y)$. Box and Tiao (1973) provide further discussion of Bayesian estimation. In Chapter 5 when we address the problem of state estimation, we will use the Bayesian approach.

4.8 PCA and PLS regression

Principal components analysis (PCA) and projection onto latent structures (also known as partial least squares) (PLS) are two methods used to develop empirical *linear* models between a vector of predictor or environmental variables x, and a vector of responses y. This is the same linear model discussed in Sections 4.7.3 and 4.7.4, so we can view these methods as alternatives to the maximum-likelihood estimation approach presented in those sections. The focus of these methods is on determining estimates of the linear model that can handle situations with possible collinearities in the x variables, and missing or erroneous information, such as unknown error structure. Collinearities in the data can make the maximum-likelihood estimator highly sensitive to outliers and nonnormal errors. Because the measurement error structure is regarded as unknown or at least unreliable, *robustness* of the estimated model to unmodeled effects is the goal, rather than statistical optimality as in the maximum-likelihood methods.

As in Section 4.7.3, let p-vector y and q-vector x be related by the linear model $y = \Theta x + e$, and we wish to determine the parameter matrix $\Theta \in \mathbb{R}^{p\times q}$ given data on y and x. We use $x_i, y_i, i = 1, 2, \ldots, n$ to denote the available samples. We assume $n > q$ (often $n \gg q$) so that we have more equations than unknowns, which is necessary for a well-conditioned estimation problem. It is customary to define data *matrices*

$$Y = \begin{bmatrix} y_1^T \\ y_2^T \\ \vdots \\ y_n^T \end{bmatrix} \qquad X = \begin{bmatrix} x_1^T \\ x_2^T \\ \vdots \\ x_n^T \end{bmatrix}$$

in which $Y \in \mathbb{R}^{n \times p}, X \in \mathbb{R}^{n \times q}$, and the model is $Y = X\Theta^T + E$. In order to use a more standard notation we let $B = \Theta^T \in \mathbb{R}^{q \times p}$, and we have the linear model

$$Y = XB + E$$

We wish to estimate parameters B from measurements X and Y without knowledge of the statistical structure of E. Given what we already know about least squares from Chapter 1, a natural approach would be to minimize some measure of the size of the residual matrix E over all choices of B. If we choose the sum of the squares of all the elements of matrix E as our measure, we have (the square of) the so-called Frobenius norm of the matrix

$$\|E\|_F = \left(\sum_{i=1}^{n} \sum_{j=1}^{p} E_{ij}^2 \right)^{1/2}$$

So our first candidate for estimating matrix B is

$$\min_B \|Y - XB\|_F^2$$

It is not difficult to show that the solution to this problem is the following

$$B_{\text{ls}} = (X^T X)^{-1} X^T Y = X^\dagger Y$$

with the usual pseudoinverse that we have seen in the standard vector least-squares problem in Chapter 1. Notice that by taking the transpose, this is also the maximum-likelihood estimate given in (4.58) for the case in which the measurement error in y is assumed normally distributed with covariance R, whether the covariance is known, or unknown and must be estimated from the data.

Also, we already know that $X^T X$ has an inverse if and only if the columns of X are linearly independent; see Proposition 1.19. Since we may not have control over the experimental conditions, we often must contend with datasets in which X has dependent or nearly dependent columns, i.e., we have near collinearity in the columns of X. In such cases, the maximum-likelihood estimate B_{ls} is unreliable and sensitive to small changes in the data or small errors in the assumed model structure.

SVD. But we also have a clear idea what to do about this issue given our background with singular value decomposition (SVD). We first replace X with its (real) SVD $X = USV^T$, and since X has more rows than columns, we obtain

$$X = \begin{bmatrix} U_1 & U_2 \end{bmatrix} \begin{bmatrix} \Sigma \\ 0 \end{bmatrix} V^T \qquad \Sigma = \mathrm{diag}(\sigma_1, \cdots, \sigma_q), \quad \sigma_1 \geq \cdots \geq \sigma_q > 0$$

in which U_1 contains the first q columns of U, and U_2 contains the remaining $n - q$ columns. Multiplying the partitioned matrices gives

$$X = U_1 \Sigma V^T$$

Next to handle the case in which Σ has several small singular values, corresponding to matrix X with columns that are nearly collinear, we approximate X by setting any small singular values to zero. Assume we have ℓ large singular values, and $q - \ell$ small singular values that are nearly zero. In this case, the rank of X may be q, but with small perturbations to the data in matrix X, it can easily drop to rank ℓ. We have

$$\begin{aligned} X &= \begin{bmatrix} U_\ell & U_q \end{bmatrix} \begin{bmatrix} \Sigma_\ell & 0 \\ 0 & \Sigma_q \end{bmatrix} \begin{bmatrix} V_\ell^T \\ V_q^T \end{bmatrix} \\ &= U_\ell \Sigma_\ell V_\ell^T + U_q \Sigma_q V_q^T \\ X &\approx U_\ell \Sigma_\ell V_\ell^T \end{aligned} \tag{4.71}$$

Using this lower-rank SVD in place of X then gives the following more robust least-squares estimate

$$B_{\mathrm{SVD}} = V_\ell \Sigma_\ell^{-1} U_\ell^T Y \tag{4.72}$$

The ill-conditioning caused by inverting Σ with all q singular values is overcome by inverting only the largest ℓ singular values. Thus the SVD estimate is less sensitive to errors in the data than the least-squares or maximum-likelihood estimate. Realize also that only the maximum-likelihood estimate is *unbiased*. By suppressing the small singular values, we introduce a small bias in B_{SVD}, but greatly reduce the variance in the estimate.

PCR. Given this background in the SVD approach, we are in an excellent position to summarize the principal component regression (PCR) method. In PCR, the X data matrix is decomposed as follows

$$X = TP^T$$

with orthogonal matrices T, known as the scores, and P, known as the loadings. Only the first ℓ principal components are retained, and the matrix X is approximated by $X = T_\ell P_\ell^T$ in which T_ℓ and P_ℓ are the first ℓ columns of T and P, respectively. The principal component regression for B is given by the following

$$B_{\text{PCR}} = P_\ell^T (T_\ell^T T_\ell)^{-1} T_\ell^T Y$$

So the correspondence with the SVD approach is as follows. The scores in PCR are the product of the singular values and the left singular vectors $T_\ell = U_\ell \Sigma_\ell$. The loadings are the right singular vectors, $P_\ell = V_\ell$. Substituting these relationships into the formula for B_{PCR} shows that

$$B_{\text{PCR}} = B_{\text{SVD}}$$

and the two approaches are equivalent. So one advantage of learning the SVD as part of linear algebra is that you have also learned PCR.

PLSR. A potential drawback of the PCR approach is that only the predictor variables are evaluated. The principal components are selected to maximize the information about matrix X. But there is no guarantee that these components can represent the responses Y. To improve the predictive capability of the model, the PLS regression (PLSR) adds a very interesting wrinkle. In this approach, one does not start with the SVD of X but with the SVD of $X^T Y$, which includes information about both X and Y and the correlation between them. Note that $X^T Y \in \mathbb{R}^{q \times p}$, which is a small matrix regardless of the number of samples, n. So computing the SVD of a matrix of the dimension of $X^T Y$, which is done repeatedly in PLSR, is a *fast* computation. The components, called latent variables, are obtained recursively as follows (Mevik and Wehrens, 2007). The first left and right singular vectors u_1 and v_1, are used to obtain the scores t_1 and w_1, respectively, via

$$t_1 = Xu_1 = E_1 u_1 \qquad w_1 = Yv_1 = F_1 v_1$$

in which the matrices E_1 and F_1 are initialized as X and Y, respectively. The X scores are then usually normalized $t_1 = t_1 / \sqrt{t_1^T t_1}$. We now define the two loadings, p_1 and q_1 using the same score t_1

$$p_1 = E_1^T t_1 \qquad q_1 = F_1^T t_1$$

Next the data matrices are deflated by subtracting the information in the current latent variable via

$$E_{i+1} = E_i - t_i p_i^T \qquad F_{i+1} = F_i - t_i q_i^T$$

The next iterate starts with the SVD of $E_{i+1}^T F_{i+1}$ in place of $X^T Y$ and the process is repeated. As in PCR, the number of latent variables $\ell \leq q$ is chosen as the number of iterations of the algorithm. The left singular vectors u_i, the scores t_i, and the loadings p_i and q_i for $i = 1, 2, \ldots, \ell$ are stored as the columns of the four matrices U, T, P, and Q. We do not require the right singular vectors v_i. Finally we compute the R matrix from

$$R = U(P^T U)^{-1}$$

and use the low-rank approximation, $X \approx TR^T$. The PLS solution is then the least-squares solution of $Y = TR^T B$ giving

$$B_{\text{PLS}} = R(T^T T)^{-1} T^T Y = RQ^T$$

Cross validation. In both PCR and PLS we need to decide how many principal components or latent variables to retain in the model. The most widely accepted method to make this decision is known as cross validation. In cross validation the dataset is divided into two or more sets; one set is used for fitting the parameters, and the other set is used to evaluate the predictive power of the model using the remaining data that have not been used in the fitting process. The validation error, defined as $E_v = Y_v - X_v \hat{B}_\ell$, in which $\hat{B}_\ell$ is the estimated model parameter matrix using the fitting dataset (X, Y) and the chosen number of principal components or latent variables, ℓ. To determine the best value of ℓ to use for estimating B, one finds the ℓ that minimizes $\|E_v\|_F^2$. This value of ℓ is large enough that the model fits the data accurately, but not so large that the model has been fit to the noise in the data. We demonstrate the cross validation technique with the following example.

Example 4.23: Comparing PCR and PLSR

Consider a dataset with five predictor variables, $x \in \mathbb{R}^5$, to model a vector of two responses, $y \in \mathbb{R}^2$. The dataset has 200 samples. The

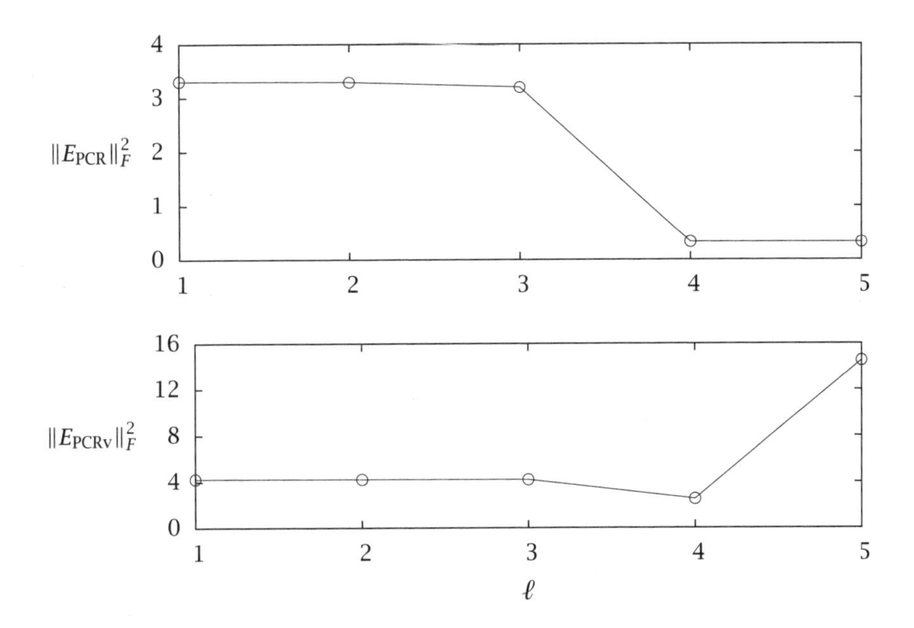

Figure 4.11: The sum of squares fitting error (top) and validation error (bottom) for PCR versus the number of principal components ℓ; cross validation indicates that four principal components are best.

data are available in file `pca_pls_data.dat` on the website `www.che.wisc.edu/~jbraw/principles`.

We would like to estimate the coefficient B in the model $Y = XB$. Compare the results using PCR and PLSR for the regression. Show the prediction error in Y for the number of principal components or latent variables ranging from one to five (full least squares). Which regression method provides the best fit with the smallest number of principal components/latent variables?

Solution

First we divide the 200 samples into two sets, and use the first 100 samples for estimating the parameter matrix B, and the second 100 samples for cross validation. For principal component analysis, we compute the SVD of the 5×100 X matrix. The five singular values are

$$\Sigma = \text{diag}(15.1,\ 3.26,\ 2.72,\ 2.67,\ 0.0226)$$

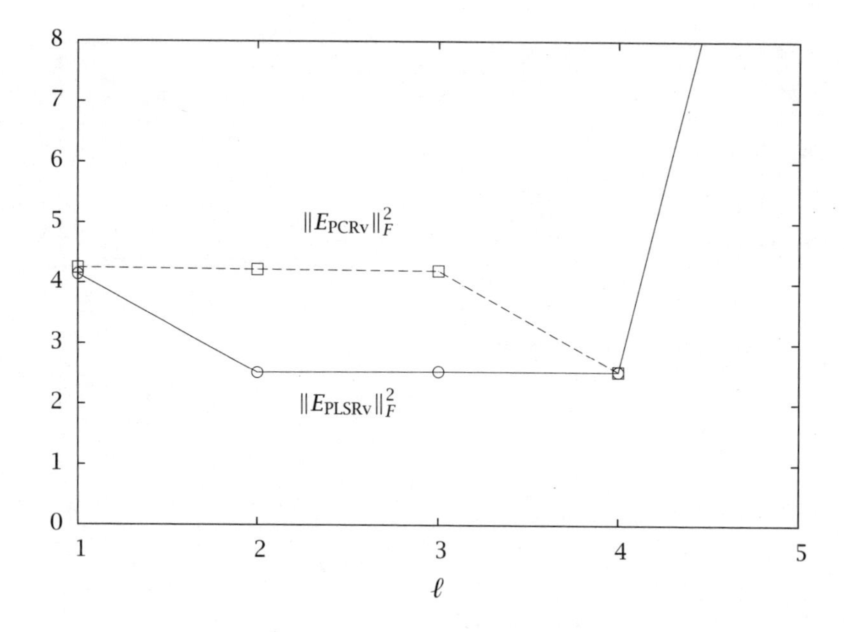

Figure 4.12: The sum of squares validation error for PCR and PLSR versus the number of principal components/latent variables ℓ; note that only two latent variables are required versus four principal components.

We see that X has four large singular values and one near zero, indicating that the rank of X is nearly four. Next we estimate B_{PCR} using (4.72) for $\ell = 1, 2, 3, 4, 5$ and calculate the sum of squares of the fitting error, $\|Y - XB_{\text{PCR}}\|_F^2$. The results are shown in the top of Figure 4.11. It is not surprising that the fitting error decreases with increasing number of principal components. As we see, the fitting error contains little information about how many principal components to use. After estimating the parameters, we then compute the output responses for the validation data and compute $\|Y_v - X_v B_{\text{PCR}}\|_F^2$ in which X_v, Y_v are the predictor and response variables in the validation dataset. This validation error is plotted in the bottom of Figure 4.11. Here we see that we should use four principal components in the model, in agreement with the SVD analysis of X. Using the unreliable smallest singular value in the regression causes a large error when trying to predict response data that have not been used in the fitting process.

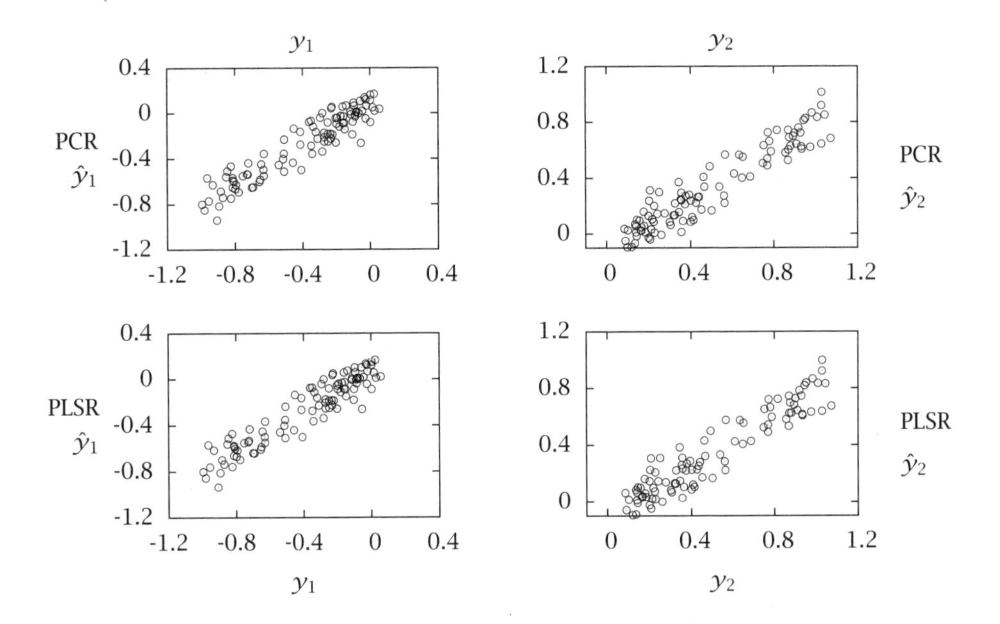

Figure 4.13: Predicted versus measured outputs for the validation dataset. Top: PCR using *four* principal components. Bottom: PLSR using *two* latent variables. Left: first output. Right: second output.

Next we implement the PLS regression algorithm as described above for $\ell = 1, 2, 3, 4, 5$ latent variables. The validation error is shown in Figure 4.12 along with the validation error of PCR. Notice that only *two* latent variables are required to obtain the same error as *four* principal components. This reduction in model order is the primary benefit of the PLSR approach. By evaluating the SVD of $X^T Y$ instead of only X, we obtain the latent variables that can explain the responses Y, not just the variables with independent information in X, which is what PCR provides.

Next, in Figure 4.13 we present the predicted responses versus the measured responses for the validation dataset. A perfect prediction would be a straight line with a slope of 45 degrees. Note that these data were *not* used in the fitting process, so this plot displays the predictive capability of the model. We see that the PLS model with two latent variables has roughly the same predictive capability as the PCR model

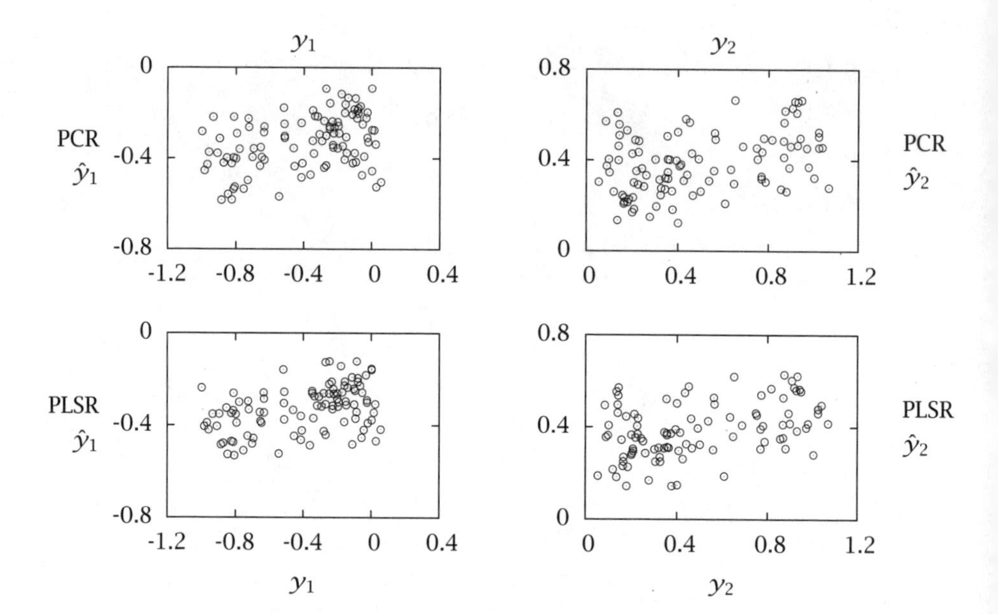

Figure 4.14: Effect of undermodeling. Top: PCR using *three* principal components. Bottom: PLSR using *one* latent variable.

with four principal components. Finally, in Figure 4.14 we make the same comparison if we use only *three* principal components and *one* latent variable. Notice that we obtain significantly worse predictions of the validation dataset, indicating that we have undermodeled the data by choosing too few variables for the regression. □

By now there is an extensive literature including many books and research monographs on model regression with PCR and PLSR. Many researchers have documented the usefulness and robustness of these techniques to identify linear empirical models in numerous applications. The understanding of PCR is reasonably complete, since it is based on the SVD of the single matrix, X. By contrast, the understanding of PLSR is not as complete. PLS was introduced by H. Wold in the 1960s in the field of econometrics (Wold, 1966). The use of PLS in the fields of analytical chemistry and chemometrics was pioneered by S. Wold, Martens, and Kowalski. The tutorial by Geladi and Kowalski (1986) and historical reviews by S. Wold (2001) and Martens (2001) summarize the approach and early contributions. Its use in process monitoring and control was developed by MacGregor, Marlin, Kresta, and Skagerberg (1991). Kaspar and Ray (1993) discovered the connection between the early PLS algorithms and the singular value decomposition, which we exploited here to compactly express the PLS algorithm. An efficient recursive formulation was developed by Qin (1998). Efficient numerical implementations of PCR and PLSR are available in several high-level computing languages such as R, Octave, and MATLAB, which make it easy for the user to try out these approaches (Mevik and Wehrens, 2007).

As demonstrated in the example, starting with the SVD of $X^T Y$ rather than X is useful for finding the smallest number of latent variables that have the most predictive capability. But PLSR research has not yet provided a complete analysis of the method and its salient properties. We do not know, for example, in what sense PLSR is an optimal estimator or whether there might be, as yet undiscovered, better methods. Adding to the complexity, several different alternative PLSR algorithms have been developed. The appearance of many different algorithms has in turn generated some confusion and controversy. To clarify matters, connections between the properties of several of the different algorithms have been established. But until some optimality properties of PLSR are uncovered, research on the PLSR approach will likely continue. In any field, a *valuable* technique that also defies easy *explanation* is a prime target for further research.

4.9 Appendix — Proof of the Central Limit Theorem

In this appendix we provide a complete proof of Theorem 4.16. We follow the basic approach outlined in the stimulating papers by Le Cam (1986) and Pollard (1986). Moreover, in this version of the central limit theorem we not only establish convergence to the normal distribution as $n \to \infty$, but also develop an approach that leads to *bounds* valid for finite n on the distance of the sum's distribution from the normal distribution. This version of the central limit theorem and, more importantly, the techniques used to establish it are wide ranging and worth knowing for researchers making extensive use of random variables. As you will see, the proof is elementary, by which we mean that none of the steps require any advanced techniques that are not already familiar to the reader. But the proof is rather long. Note also that this material can be skipped without affecting the understanding of any other section in the text.

Proof. We start by considering two sums of independent random variables; let $S_n = X_i + X_2 + \cdots + X_n$ and $T_n = Y_1 + Y_2 + \cdots + Y_n$, in which $\mathcal{E}(X_k) = \mathcal{E}(Y_k) = 0$ and $\text{var}(X_k) = \text{var}(Y_k) = \sigma_k^2$. The zero mean assumption is not restrictive. If the original X_k have nonzero mean μ_k, consider instead the zero mean, shifted variables $\tilde{X}_k = X_k - \mu_k$. Next define R_k as follows

$$R_k = \sum_{j<k}^{n} X_j + \sum_{j>k}^{n} Y_j, \quad k = 1, 2, \ldots, n$$

so that

$$\begin{aligned}
R_1 &= Y_2 + Y_3 + \cdots + Y_n \\
R_2 &= X_1 + Y_3 + Y_4 + \cdots + Y_n \\
R_3 &= X_1 + X_2 + Y_4 + \cdots + Y_n \\
&\cdots \\
R_n &= X_1 + X_2 + \cdots + X_{n-1}
\end{aligned}$$

Notice from this definition that R_k and X_k as well as R_k and Y_k are also independent for $k = 1, 2, \ldots, n$. We see shortly why the R_k variables are useful.

We also require an approximation theorem; the form we choose here is motivated by a nice, unpublished note of F.W. Scholz (2011).

Theorem 4.24 (Taylor's theorem with bound on remainder). *Let f be a bounded function on $\mathbb{R}$ with three continuous, bounded derivatives. Consider the second-order Taylor series with remainder*

$$f(x+h) = f(x) + f'(x)h + \frac{f''(x)}{2}h^2 + r(x,h)$$

The remainder satisfies the following bound for all $h \in \mathbb{R}$

$$\sup_{x \in \mathbb{R}} |r(x,h)| \leq K_f \min(h^2, |h|^3) \tag{4.73}$$

The term $|h|^3$ is expected from the standard Taylor expansion, but including the term h^2 gives a better bound for large h, which we shall find useful subsequently. Exercise 4.54 discusses how to prove this theorem, which is not difficult.

So we assume that f has three continuous, bounded derivatives, and express $f(X_k + R_k)$ as

$$f(X_k + R_k) = f(R_k) + X_k f'(R_k) + \frac{X_k^2}{2} f''(R_k) + r(R_k, X_k)$$

Performing a similar expansion for $f(Y_k + R_k)$, taking expectations, and subtracting gives

$$\mathcal{E}(f(X_k + R_k) - f(Y_k + R_k)) = (\mathcal{E}(X_k) - \mathcal{E}(Y_k))\mathcal{E}(f'(R_k)) +$$
$$(1/2)(\mathcal{E}(X_k^2) - \mathcal{E}(Y_k^2))\mathcal{E}(f''(R_k)) + \mathcal{E}(r(R_k, X_k) - r(R_k, Y_k))$$

where we have used the fact that $\mathcal{E}(AB) = \mathcal{E}(A)\mathcal{E}(B)$ for A and B independent random variables. Noting that the first two terms cancel, taking absolute values, and using (4.73) gives

$$\left|\mathcal{E}(f(X_k + R_k) - f(Y_k + R_k))\right| \leq K_f \mathcal{E}(g(X_k) + g(Y_k)) \tag{4.74}$$

where we used the fact[4] that $\left|\mathcal{E}(f(X))\right| \leq E(\left|f(X)\right|)$ and defined $g(X) = \min(X^2, |X|^3)$ to compress the notation. Next comes the reason for introducing the R_k variables. Notice that differencing the sum of $f(R_k + X_k)$ and $f(R_k + Y_k)$ leaves only two terms

$$\sum_{k=1}^{n} f(R_k + X_k) - f(R_k + Y_k) = f(R_n + X_n) - f(R_1 + Y_1) = f(S_n) - f(T_n)$$

Taking expectations and then absolute values and using (4.74) then gives

$$\left|\mathcal{E}(f(S_n) - f(T_n))\right| \leq K_f \sum_{k=1}^{n} \mathcal{E}(g(X_k) + g(Y_k)) \tag{4.75}$$

Establishing this inequality is the first major step.

But we wish to bound the distance between the two cumulative distributions F_{S_n} and F_{T_n}, so we next choose an appropriate function $f(\cdot)$ to achieve this goal. Consider the step function $f_1(w; x)$ depicted in Figure 4.15, in which w is the argument to the function and x is considered a fixed parameter. Using $f_1(w; x)$ we have immediately

$$\mathcal{E}(f_1(S_n)) = \int_{-\infty}^{\infty} f_1(w; x) p_{S_n}(w) dw = \int_{-\infty}^{x} p_{S_n}(w) dw = F_{S_n}(x)$$

The function $f_1(\cdot)$ is known as an indicator function, because $f_1(S_n)$ *indicates* when the random variable S_n satisfies $S_n \leq x$. So this is

[4]Since $f(x) \leq |f(x)|$ for all x, multiply by the density $p_X(x)$ and integrate.

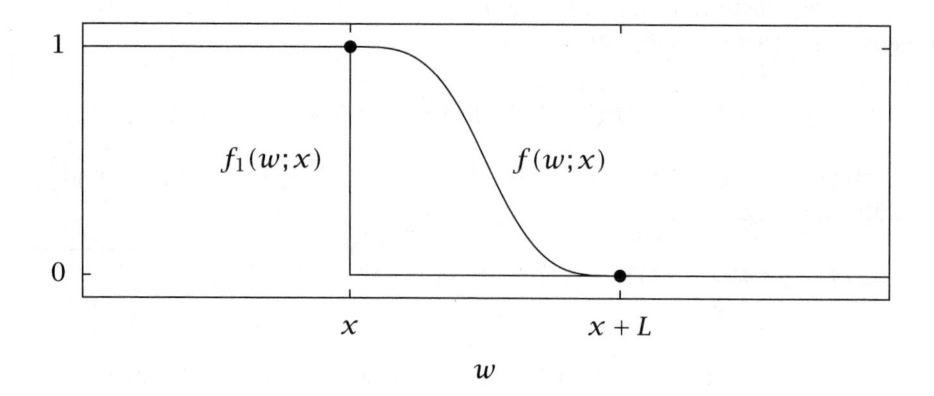

Figure 4.15: The indicator (step) function $f_1(w;x)$ and its smooth approximation, $f(w;x)$. A piecewise fifth-order polynomial gives continuous derivatives up to third order; see Exercise 4.57 for details.

the kind of function we seek, but, of course, f_1 does not have even a bounded first derivative, let alone three bounded derivatives as required in our development. So we first smooth out this function as depicted in Figure 4.15. Exercise 4.57 gives an example of a piecewise polynomial function f with the required smoothness. Moreover, there exists an $L_0 > 0$ such that $K_f = 20L^{-3}$ is a valid upper bound in (4.73) for every L satisfying $0 < L \leq L_0$; see (4.97). We will require this bound shortly.

Computing $\mathcal{E}f(S_n)$ gives

$$\int_{-\infty}^{\infty} p_{S_n}(w)f(w;x)dw = \int_{-\infty}^{x} p_{S_n}(w)dw + \int_{x}^{x+L} p_{S_n}(w)f(w;x)dw$$

$$\mathcal{E}(f(S_n)) = F_{S_n}(x) + \int_{x}^{x+L} p_{S_n}(w)f(w;x)dw$$

and, subtracting the analogous expression for T_n and rearranging gives

$$\begin{aligned}
F_{S_n}(x) - F_{T_n}(x) &= \mathcal{E}(f(S_n) - f(T_n)) - \int_{x}^{x+L} (p_{S_n}(w) - p_{T_n}(w))f(w;x)dw \\
&\leq \mathcal{E}(f(S_n) - f(T_n)) + \int_{x}^{x+L} p_{T_n}(w)f(w;x)dw \\
&\leq \mathcal{E}(f(S_n) - f(T_n)) + b_nL
\end{aligned}$$

with $b_n = \max_w p_{T_n}(w)$. Since we will later choose the Y_k variables, we have some control over the constant b_n. Taking absolute values and substituting (4.75) then gives the following bound

$$|F_{S_n}(x) - F_{T_n}(x)| \le K_f \sum_{k=1}^{n} \mathcal{E}(g(X_k) + g(Y_k)) + b_n L \tag{4.76}$$

Establishing this inequality is the second major step. Note that if we choose L small, we make K_f large, so making the sum on the right-hand side small will require a judicious choice of L.

Next we choose the Y_k to be $N(0, \sigma_k^2)$, and the scaled sum $T_n/s_n = \sum_{k=0}^{n} Y_k/s_n$ has zero mean and unit variance for all n, i.e, it is a standard normal, denoted Z with distribution function $\Phi(x)$. This gives immediately for (4.76) the value $b_n = \max_w p_Z(w) = 1/\sqrt{2\pi}$, which is also independent of n for this choice of Y_k. The variable $Z_n = S_n/s_n = \sum_{k=1}^{n} X_k/s_n$ is a sum of scaled X_k, and also has zero mean and unit variance for all n. Applying (4.76) to these variables gives

$$\sup_x |F_{Z_n}(x) - \Phi(x)| \le K_f \sum_{k=1}^{n} \mathcal{E}(g(X_k/s_n) + g(Y_k/s_n)) + b_n L \tag{4.77}$$

To evaluate the right-hand side, we partition the interval of integration as discussed before

$$\mathcal{E}(g(X_k/s_n)) = E(g(X_k/s_n); |X_k| \le \epsilon s_n) + E(g(X_k/s_n); |X_k| > \epsilon s_n)$$

Next we use the fact that $g(X_k/s_n) \le |X_k/s_n|^3$ in the first term and that $g(X_k/s_n) \le (X_k/s_n)^2$ in the second term to obtain the bound

$$\begin{aligned}
\mathcal{E}(g(X_k/s_n)) &\le \mathcal{E}(|X_k/s_n|^3; |X_k| \le \epsilon s_n) + \mathcal{E}((X_k/s_n)^2; |X_k| > \epsilon s_n) \\
&\le \frac{\epsilon}{s_n^2}\mathcal{E}(|X_k|^2; |X_k| \le \epsilon s_n) + \frac{1}{s_n^2}\mathcal{E}(X_k^2; |X_k| > \epsilon s_n) \\
&\le \frac{\epsilon \sigma_k^2}{s_n^2} + \frac{1}{s_n^2}\mathcal{E}(X_k^2; |X_k| > \epsilon s_n)
\end{aligned}$$

Performing the sum gives

$$\sum_{k=1}^{n} \mathcal{E}(g(X_k/s_n)) \le \epsilon + \frac{1}{s_n^2} \sum_{k=1}^{n} E(X_k^2; |X_k| > \epsilon s_n)$$

The second term goes to zero as $n \to \infty$ by the Lindeberg condition. So for large enough n it is smaller than the first term, which is independent of n. So as $n \to \infty$ we have that

$$\sum_{k=1}^{n} \mathcal{E}(g(X_k/s_n)) \leq 2\epsilon$$

We also can show that the normally distributed Y_k variables satisfy the Lindeberg conditions if the X_k do. See Exercise 4.56 for the steps. So we have that as $n \to \infty$

$$\sum_{k=1}^{n} \mathcal{E}(g(X_k/s_n) + g(Y_k/s_n)) \leq 4\epsilon \tag{4.78}$$

Next we choose L, and therefore K_f, as follows

$$L = \left(\sum_{k=1}^{n} \mathcal{E}(g(X_k/s_n) + g(Y_k/s_n))\right)^{1/4} \leq (4\epsilon)^{1/4}$$

To use the bound in (4.97), we require $L \leq L_0$. Therefore setting $\epsilon_0 = L_0^4/4 > 0$, we have from the previous inequality that for every $\epsilon < \epsilon_0$,

$$K_f = 20L^{-3} = \left(20\sum_{k=1}^{n} \mathcal{E}(g(X_k/s_n) + g(Y_k/s_n))\right)^{-3/4}$$

Substituting these values for K_f and L into (4.77) and using (4.78) gives

$$\sup_x \left|F_{Z_n}(x) - \Phi(x)\right| \leq c\epsilon^{1/4}$$

with $c = (1/2)5^{-3/4} + 1/\sqrt{\pi} \approx 0.71$. Since this bound holds for all $\epsilon \leq \epsilon_0$, we have established that

$$\lim_{n\to\infty} \sup_x \left|F_{Z_n}(x) - \Phi(x)\right| = 0$$

and the proof is complete. ∎

4.10 Exercises

Exercise 4.1: Consequences of the axioms of probability

(a) If $B \subseteq A$, show that $\Pr(A \setminus B) = \Pr(A) - \Pr(B)$.

(b) From the definition, the events $\mathcal{A}$ and $\mathcal{B}$ are independent if $\Pr(\mathcal{A} \cap \mathcal{B}) = \Pr(\mathcal{A})\Pr(\mathcal{B})$. If $\mathcal{A}$ and $\mathcal{B}$ are independent, show that $\overline{\mathcal{A}}$ and $\overline{\mathcal{B}}$ are independent.

Exercise 4.2: Statistical independence condition in densities

Show that two random variables ξ and η are statistically independent if and only if

$$p_{\xi,\eta}(x,y) = p_\xi(x)p_\eta(y), \qquad \text{all } x, y \tag{4.79}$$

Exercise 4.3: Statistical indpendence of functions of random variables

Consider statistically independent random variables, $\xi \in \mathbb{R}^m$ and $\eta \in \mathbb{R}^n$. Define random variable $\alpha \in \mathbb{R}^p$ and $\beta \in \mathbb{R}^q$ as $\alpha = f(\xi), \beta = g(\eta)$. Show that α and β are statistically independent for all functions $f(\cdot)$ and $g(\cdot)$. Summarizing

> Statistical independence of random variables (ξ, η) implies statistical independence of random variables $(f(\xi), g(\eta))$ for all $f(\cdot)$ and $g(\cdot)$.

Note that $f(\cdot)$ and $g(\cdot)$ are not required to be invertible.

Exercise 4.4: Trace of a matrix function

Derive the following formula for differentiating the trace of a function of a square matrix

$$\frac{d\operatorname{tr}(f(A))}{dA} = g(A^T) \qquad g(x) = \frac{df(x)}{dx} \tag{4.80}$$

in which g is the usual scalar derivative of the scalar function f.

Exercise 4.5: Derivatives of determinants

For $A \in \mathbb{R}^{n\times n}$ nonsingular, derive the following formulas

$$\frac{d\det A}{dA} = (A^{-1})^T \det A \qquad \frac{d\ln\det A}{dA} = (A^{-1})^T$$

Exercise 4.6: Transposing the maximum-likelihood problem statement

Consider again the estimation problem for the model given in (4.56), but this time express it in transposed form

$$y_i^T = x_i^T\Theta + e_i^T \qquad e_i \sim N(0,R) \tag{4.81}$$

(a) Derive the maximum-likelihood estimate for this case. Show all steps in the derivation. Arrange the data in matrices

$$\widetilde{Y} = \begin{bmatrix} y_1^T \\ \vdots \\ y_n^T \end{bmatrix} \qquad \widetilde{X} = \begin{bmatrix} x_1^T \\ \vdots \\ x_n^T \end{bmatrix}$$

and show the maximum-likelihood estimate can be expressed as

$$\widehat{\widetilde{\Theta}} = \left(\widetilde{X}^T\widetilde{X}\right)^{-1}\widetilde{X}^T\widetilde{Y}$$

Expressing the model this way gives an estimate formula that is analogous to what other problem?

(b) Find the resulting probability density for the estimate and give the analogous result corresponding to (4.59).

(c) Which form of the model do you prefer and why?

Exercise 4.7: Joint, marginal, mean, and covariance

We consider two discrete-valued random variables, ξ and η. Calculate the joint density, $p_{\xi,\eta}(x,y)$, both marginal densities, $p_\xi(x), p_\eta(y)$, the means, $\mathcal{E}(\xi), \mathcal{E}(\eta)$, and covariance, $\text{cov}(\xi,\eta)$, for the following two cases

(a) We throw two dice, and ξ and η are the values on each die.

(b) We throw two dice, ξ is the value on one die and η is the sum of the two values.

Exercise 4.8: Probability density of the inverse function

Consider a scalar random variable $\xi \in \mathbb{R}$ and let the random variable η be defined by the inverse function

$$\eta = \xi^{-1}$$

(a) If ξ is distributed uniformly on $[a, 1]$ with $0 < a < 1$, what is the density of η?

(b) Is η's density well defined if we allow $a = 0$? Explain your answer.

Exercise 4.9: Expectation as a linear operator

(a) Consider the random variable x to be defined as a linear combination of the random variables a and b

$$x = a + b$$

Show

$$\mathcal{E}(x) = \mathcal{E}(a) + \mathcal{E}(b)$$

Do a and b need to statistically independent for this statement to be true?

(b) Next consider the random variable x to be defined as a scalar multiple of the random variable a

$$x = \alpha a$$

in which α is a scalar. Show

$$\mathcal{E}(x) = \alpha \mathcal{E}(a)$$

(c) What can you conclude about $\mathcal{E}(x)$ if x is given by the linear combination

$$x = \sum_i \alpha_i v_i$$

in which v_i are random variables and α_i are scalars.

Exercise 4.10: Calculating mean and variance from data

We are sampling a real-valued, scalar random variable $x(k) \in \mathbb{R}$ at time k. Assume the random variable comes from a distribution with mean $\overline{x}$ and variance P, and the samples at different times are statistically independent.

A colleague has suggested the following formulas for estimating the mean and variance from N samples

$$\hat{x}_N = \frac{1}{N}\sum_{j=1}^{N} x(j) \qquad \hat{P}_N = \frac{1}{N}\sum_{j=1}^{N}(x(j) - \hat{x}_N)^2$$

(a) Prove the estimate of the mean is unbiased for all N, i.e., show

$$\mathcal{E}(\hat{x}_N) = \overline{x}, \qquad \text{all } N$$

(b) Prove the estimate of the variance is not unbiased for any N, i.e., show

$$\mathcal{E}(\hat{P}_N) \neq P, \qquad \text{any } N$$

(c) Using the result above, provide an improved formula for the variance estimate that is unbiased for all N. How large does N have to be before these two estimates of P are within 1%?

Exercise 4.11: The sum of throwing two dice

Using (4.23), what is the probability density for the sum of throwing two dice? On what number do you want to place your bet? How often do you expect to win if you bet on this outcome?

Make the standard assumptions: the probability density for each die is uniform over the integer values from one to six, and the outcome of each die is independent of the other die.

Exercise 4.12: The product of throwing two dice

Using (4.23), what is the probability density for the product of throwing two dice? On what number do you want to place your bet? How often do you expect to win if you bet on this outcome?

Make the standard assumptions: the probability density for each die is uniform over the integer values from one to six, and the outcome of each die is independent of the other die.

Exercise 4.13: Expected sum of squares

Given random variable x has mean m and covariance P, show that the expected sum of squares is given by the formula (Selby, 1973, p.138)

$$\mathcal{E}(x^T Q x) = m^T Q m + \operatorname{tr}(QP)$$

Recall that the trace of a square matrix A, written $\operatorname{tr}(A)$, is defined to be the sum of the diagonal elements

$$\operatorname{tr}(A) = \sum_i A_{ii}$$

Exercise 4.14: Normal distribution

Given a normal distribution with scalar parameters m and σ

$$p_\xi(x) = \sqrt{\frac{1}{2\pi\sigma^2}} \exp\left[-\frac{1}{2}\left(\frac{x-m}{\sigma}\right)^2\right] \tag{4.82}$$

By direct calculation, show that

(a)

$$\mathcal{E}(\xi) = m$$
$$\operatorname{var}(\xi) = \sigma^2$$

(b) Show that the mean and the maximum likelihood are equal for the normal distribution. Draw a sketch of this result. The maximum-likelihood estimate, $\hat{x}$, is defined as

$$\hat{x} = \arg\max_{x} p_{\xi}(x)$$

in which arg returns the solution to the optimization problem.

Exercise 4.15: The size of an ellipse's bounding box

Here we derive the size of the bounding box depicted in Figure 4.3. Consider a real, positive definite, symmetric matrix $A \in \mathbb{R}^{n \times n}$ and a real vector $x \in \mathbb{R}^n$. The set of x for which the scalar $x^T Ax$ is constant are n-dimensional ellipsoids. Find the length of the sides of the smallest box that contains the ellipsoid defined by

$$x^T Ax = b$$

Hint: consider the equivalent optimization problem to minimize the value of $x^T Ax$ such that the ith component of x is given by $x_i = c$. This problem defines the ellipsoid that is tangent to the plane $x_i = c$, and can be used to answer the original question.

Exercise 4.16: Conditional densities are positive definite

We showed in Example 4.19 that if ξ and η are jointly normally distributed as

$$\begin{aligned} \begin{bmatrix} \xi \\ \eta \end{bmatrix} &\sim N(m, P) \\ &\sim N\left(\begin{bmatrix} m_x \\ m_y \end{bmatrix}, \begin{bmatrix} P_x & P_{xy} \\ P_{yx} & P_y \end{bmatrix}\right) \end{aligned}$$

then the conditional density of ξ given η is also normal

$$(\xi|\eta) \sim N(m_{x|y}, P_{x|y})$$

in which the conditional mean is

$$m_{x|y} = m_x + P_{xy}P_y^{-1}(y - m_y)$$

and the conditional covariance is

$$P_{x|y} = P_x - P_{xy}P_y^{-1}P_{yx}$$

Given the joint density is well defined, prove the marginal densities and the conditional densities are also well defined, i.e., given $P > 0$, prove $P_x > 0$, $P_y > 0$, $P_{x|y} > 0$, $P_{y|x} > 0$.

Exercise 4.17: Transform of the multivariate normal density

Show the Fourier transform of the multivariate normal density given in (4.12) is

$$\varphi(u) = \exp\left[iu^T m - \frac{1}{2}u^T Pu\right]$$

Exercise 4.18: The difference of two exponentially distributed random variables

The random variables τ_1 and τ_2 are statistically independent and identically distributed with the exponential density

$$p_\tau(t) = e^{-t} \qquad t \geq 0$$

Define the new random variable y to be the difference

$$y = \tau_1 - \tau_2$$

We wish to calculate y's probability density p_y.

(a) First introduce a new random variable $z = \tau_2$ and define the transformation from (τ_1, τ_2) to (y, z). Find the inverse transformation from (y, z) to (τ_1, τ_2). What is the determinant of the Jacobian of the inverse transformation?

(b) What is the joint density $p_{\tau_1,\tau_2}(t_1, t_2)$? Sketch the region in (y, z) that corresponds to the region of nonzero probability of the joint density p_{τ_1,τ_2} in (τ_1, τ_2).

(c) Apply the formula given in (4.23) to obtain the transformed joint density $p_{y,z}$.

(d) Integrate over z in this joint density to obtain p_y.

(e) Generate 1000 samples of τ_1 and τ_2, calculate y, and plot y's histogram. Does your histogram of the y samples agree with your result from (d)? Explain why or why not.

Exercise 4.19: Surface area and volume of a sphere in n dimensions

In three-dimensional space, $n = 3$, the surface area and volume of the sphere are given by

$$S_3(r) = 4\pi r^2 \qquad V_3(r) = 4/3\pi r^3$$

You are also familiar with the formulas for $n = 2$, in which case "surface area" is the circumference of the circle and "volume" is the area of the circle

$$S_2(r) = 2\pi r \qquad V_2(r) = \pi r^2$$

If we define s_n and v_n as the constants such that

$$S_n(r) = s_n r^{n-1} \qquad V_n(r) = v_n r^n$$

we have

$$s_2 = 2\pi \qquad v_2 = \pi$$
$$s_3 = 4\pi \qquad v_3 = 4/3\pi$$

We seek the generalization of these results to the n-dimensional case. Compute the formulas for s_n and v_n and show

$$s_n = \frac{2(\pi)^{n/2}}{\Gamma(n/2)} \qquad v_n = \frac{\pi^{n/2}}{\Gamma(n/2+1)} \tag{4.83}$$

Exercise 4.20: Surface area and volume of an ellipsoid in n dimensions

The results for surface area and volume of a sphere in n dimensions can be extended to obtain the surface area and volume of an ellipse (ellipsoid, hyperellipsoid) in n dimensions. Let x be an n-vector. The surface of an ellipse is defined by the equation

$$x^T A x = R^2$$

in which $A \in \mathbb{R}^{n \times n}$ is a symmetric, positive definite matrix and R^2 is the square of the ellipse "radius." Let the interior of the ellipse of size R be denoted by the set σ_R

$$\sigma_R = \{x \mid x^T A x \leq R^2\}$$

We wish to compute the volume of the ellipse, which is defined by the following integral

$$V_n^e(R) = \int_{\sigma_R} dx$$

The surface area, $S_n^e(R)$, is defined to have the following relationship with the volume

$$V_n^e(R) = \int_0^R S_n^e(r)dr \qquad \frac{dV_n^e(r)}{dr} = S_n^e(r)$$

(a) Derive formulas for s_n^e and v_n^e such that

$$S_n^e(R) = s_n^e R^{n-1} \qquad V_n^e(R) = v_n^e R^n$$

for the ellipse.

(b) Show that your result subsumes the formula for the volume of the 3-dimensional ellipse given by

$$\left(\frac{x}{a}\right)^2 + \left(\frac{y}{b}\right)^2 + \left(\frac{z}{c}\right)^2 = 1 \qquad V = \frac{4}{3}\pi abc$$

Exercise 4.21: Definite integrals of the multivariate normal and χ^2

(a) Derive the following n-dimensional integral over an elliptical region

$$\int_{\sigma_b} e^{-x^T A x} dx = \frac{\pi^{n/2}}{(\det A)^{1/2}} \frac{\gamma(n/2, b)}{\Gamma(n/2)} \qquad \sigma_b = \{x \mid x^T A x \leq b\}$$

(b) Let ξ be distributed as a multivariate normal with mean m and covariance P

$$\xi \sim N(m, P)$$

and let α denote the total probability that ξ takes on a value x inside the ellipse $(x - m)^T P^{-1}(x - m) \leq b$. Use the integral in the previous part to show

$$\alpha = \frac{\gamma(n/2, b/2)}{\Gamma(n/2)} \tag{4.84}$$

(c) The $\chi^2(n, \alpha)$ function is defined to invert this relationship and give the size of the ellipse that contains total probability α

$$\chi^2(n, \alpha) = b \tag{4.85}$$

Plot $\frac{\gamma(n/2, x/2)}{\Gamma(n/2)}$ and $\chi^2(n, x)$ versus x for various n (try $n = 1, 4$), and display the inverse relationship given by (4.84) and (4.85).

Exercise 4.22: Normal distributions under linear transformations

Given the normally distributed random variable, $\xi \in \mathbb{R}^n$, consider the random variable, $\eta \in \mathbb{R}^n$, obtained by the linear transformation

$$\eta = A\xi$$

in which A is a nonsingular matrix. Using the result on transforming probability densities, show that if $\xi \sim N(m, P)$, then $\eta \sim N(Am, APA^T)$. This result establishes that (invertible) linear transformations of (nonsingular) normal random variables are normal.

Exercise 4.23: Normal with singular covariance

Consider the random variable $\xi \in \mathbb{R}^n$ and an arbitrary positive semidefinite covariance matrix P_x with rank $r < n$. Starting with the definition of a singular normal, Definition 4.10, show that the density for $\xi \sim N(m_x, P_x)$ is given by

$$p_\xi(x) = \frac{1}{(2\pi)^{r/2}(\det \Lambda_1)^{1/2}} \exp\left[-\frac{1}{2}(x - m_x)^T Q_1 \Lambda_1^{-1} Q_1^T (x - m_x)\right] \delta(Q_2^T(x - m_x))$$

in which matrices $\Lambda \in \mathbb{R}^{r\times r}$ and orthonormal $Q \in \mathbb{R}^{n\times n}$ are obtained from the eigenvalue decomposition of P_x

$$P_x = Q\Lambda Q^T = \begin{bmatrix} Q_1 & Q_2 \end{bmatrix} \begin{bmatrix} \Lambda_1 & 0 \\ 0 & 0 \end{bmatrix} \begin{bmatrix} Q_1^T \\ Q_2^T \end{bmatrix}$$

and $\Lambda_1 > 0 \in \mathbb{R}^{r\times r}$, $Q_1 \in \mathbb{R}^{n\times r}$, $Q_2 \in \mathbb{R}^{n\times(n-r)}$. On what set of x is the density nonzero?

Exercise 4.24: Linear transformation and singular normals

Prove Theorem 4.12, which generalizes the result of Exercise 4.22 to establish that *any* linear transformation of a normal is normal. And for this statement to hold, we must expand the meaning of normal to include the singular case.

Exercise 4.25: Useful identities in least-squares estimation

Establish the following two useful results using the matrix inversion formula

$$\left(A^{-1} + C^T B^{-1} C\right)^{-1} = A - AC^T \left(B + CAC^T\right)^{-1} CA$$
$$\left(A^{-1} + C^T B^{-1} C\right)^{-1} C^T B^{-1} = AC^T \left(B + CAC^T\right)^{-1} \tag{4.86}$$

Exercise 4.26: Least-squares parameter estimation and Bayesian estimation

Consider a model linear in the parameters

$$y = X\theta + e \tag{4.87}$$

in which $y \in \mathbb{R}^p$ is a vector of measurements, $\theta \in \mathbb{R}^m$ is a vector of parameters, $X \in \mathbb{R}^{p\times m}$ is a matrix of known constants, and $e \in \mathbb{R}^p$ is a random variable modeling the measurement error. The standard parameter estimation problem is to find the best estimate of θ given the measurements y corrupted with measurement error e, which we assume is distributed as

$$e \sim N(0, R)$$

(a) Consider the case in which the measurement errors are independently and identically distributed with variance σ^2, $R = \sigma^2 I$. For this case, the classic least-squares problem and solution are

$$\min_{\theta} \|y - X\theta\|^2 \qquad \hat{\theta} = \left(X^T X\right)^{-1} X^T y$$

Consider the measurements to be sampled from (4.87) with true parameter value θ_0. Show that using the least-squares formula, the parameter estimate is distributed as

$$\hat{\theta} \sim N(\theta_0, P_{\hat{\theta}}) \qquad P_{\hat{\theta}} = \sigma^2 \left(X^T X\right)^{-1}$$

(b) Now consider again the model of (4.87) and a Bayesian estimation problem. Assume a prior distribution for the random variable θ

$$\theta \sim N(\overline{\theta}, \overline{P})$$

Compute the conditional density of θ given measurement y, show this density is a normal, and find its mean and covariance

$$(\theta|y) \sim N(m, P)$$

Show that Bayesian estimation and least-squares estimation give the same result in the limit of a noninformative prior. In other words, if the covariance of the prior is large compared to the covariance of the measurement error, show

$$m \approx (X^T X)^{-1} X^T y \qquad P \approx P_{\hat{\theta}}$$

(c) What (weighted) least-squares minimization problem is solved for the general measurement error covariance

$$e \sim N(0, R)$$

Derive the least-squares estimate formula for this case.

(d) Again consider the measurements to be sampled from (4.87) with true parameter value θ_0. Show that the weighted least-squares formula gives parameter estimates that are distributed as

$$\hat{\theta} \sim N(\theta_0, P_{\hat{\theta}})$$

and find $P_{\hat{\theta}}$ for this case.

(e) Show again that Bayesian estimation and least-squares estimation give the same result in the limit of a noninformative prior.

Exercise 4.27: Least-squares and minimum-variance estimation

Consider again the model linear in the parameters and the least-squares estimator from Exercise 4.26

$$y = X\theta + e \qquad e \sim N(0, R)$$

$$\hat{\theta} = \left(X^T R^{-1} X\right)^{-1} X^T R^{-1} y$$

Show that the covariance of the least-squares estimator is the smallest covariance of all linear, unbiased estimators.

Exercise 4.28: Two stages are not better than one

We can often decompose an estimation problem into stages. Consider the following case in which we wish to estimate x from measurements of z, but we have the model between x and an intermediate variable, y, and the model between y and z

$$\begin{aligned} y &= Ax + e_1 \qquad \operatorname{cov}(e_1) = Q_1 \\ z &= By + e_2 \qquad \operatorname{cov}(e_2) = Q_2 \end{aligned}$$

(a) Write down the optimal least-squares problem to solve for $\hat{y}$ given the z measurements and the second model. Given $\hat{y}$, write down the optimal least-squares problem for $\hat{x}$ in terms of $\hat{y}$. Combine these two results together and write the resulting estimate of $\hat{x}$ given measurements of z. Call this the two-stage estimate of x.

(b) Combine the two models together into a single model and show the relationship between z and x is

$$z = BAx + e_3 \qquad \operatorname{cov}(e_3) = Q_3$$

Express Q_3 in terms of Q_1, Q_2 and the models A, B. What is the optimal least-squares estimate of $\hat{x}$ given measurements of z and the one-stage model? Call this the one-stage estimate of x.

(c) Are the one-stage and two-stage estimates of x the same? If yes, prove it. If no, provide a counterexample. Do you have to make any assumptions about the models A, B?

Exercise 4.29: Let's make a deal!

Consider the following contest of the American television game show of the 1960s, Let's Make a Deal. In the show's grand finale, a contestant is presented with three doors. Behind one of the doors is a valuable prize such as an all-expenses-paid vacation to Hawaii or a new car. Behind the other two doors are goats and donkeys. The contestant selects a door, say door number one. The game show host, Monty Hall, then says,

"Before I show you what is behind your door, let's reveal what is behind door number three!" Monty always chooses a door that has one of the booby prizes behind it. As the goat or donkey is revealed, the audience howls with laughter. Then Monty asks innocently,

"Before I show you what is behind your door, I will allow you one chance to change your mind. Do you want to change doors?" While the contestant considers this option, the audience starts screaming out things like,

"Stay with your door! No, switch, switch!" Finally the contestant chooses again, and then Monty shows them what is behind their chosen door.

Let's analyze this contest to see how to *maximize* the chance of winning. Define

$$p(i, j, y), \qquad i, j, y = 1, 2, 3$$

to be the probability that you chose door i, the prize is behind door j and Monty showed you door y (named after the data!) after your initial guess. Then you would want to

$$\max_j p(j|i, y) \tag{4.88}$$

for your optimal choice after Monty shows you a door.

(a) Calculate this conditional density and give the probability that the prize is behind door i, your original choice, and door $j \neq i$.

(b) You need to specify a model of Monty's behavior. Please state the one that is appropriate to Let's Make a Deal.

(c) For what other model of Monty's behavior is the answer that it does not matter if you switch doors. Why is this a poor model for the game show?

Exercise 4.30: A nonlinear transformation and conditional density

Consider the following relationship between the random variable y, and x and w

$$y = f(x) + w$$

The author of a famous textbook wants us to believe that

$$p_{y|x}(Y|X) = p_w(Y - f(X))$$

Derive this result and state what additional assumptions on the random variables x and w are required for this result to be correct.

Exercise 4.31: Least squares and confidence intervals

A common model for the temperature dependence of the reaction rate is the Arrhenius model. In this model the reaction rate (rate constant, k) is given by

$$k = k_0 \exp(-E/T) \tag{4.89}$$

in which the parameters k_0 is the preexponential factor and E is the activation energy, scaled by the gas constant, and T is the temperature in Kelvin. We wish to estimate k_0 and E from measurements of the reaction rate (rate constant), k, at different temperatures, T. In order to use linear least squares we first take the logarithm of (4.89) to obtain

$$\ln(k) = \ln(k_0) - E/T$$

Assume you have made measurements of the rate constant at 10 temperatures evenly distributed between 300 and 500 K. Model the measurement process as the true value plus measurement error e, which is distributed normally with zero mean and 0.001 variance

$$\ln(k) = \ln(k_0) - E/T + e \qquad e \sim N(0, 0.001)$$

Choose true values of the parameters to be

$$\ln(k_0) = 1 \qquad E = 100$$

(a) Generate a set of experimental data for this problem. Estimate the parameters from these data using least squares. Plot the data and the model fit using both (T, k) and $(1/T, \ln k)$ as the (x, y) axes.

(b) Calculate the 95% confidence intervals for your parameter estimates. What are the coordinates of the semimajor axes of the ellipse corresponding to the 95% confidence interval?

(c) What are the coordinates of the corners of the box corresponding to the 95% confidence interval?

(d) Plot your result by showing the parameter estimate, ellipse, and box. Are the parameter estimates highly correlated? Why or why not?

Exercise 4.32: A fourth moment of the normal distribution

You have established the following matrix integral result involving the second moment of the normal distribution

$$\int_{-\infty}^{\infty} xx^T \exp\left(-\frac{1}{2}x^T P^{-1} x\right) dx = (2\pi)^{n/2} (\det P)^{1/2} P$$

Establish the following matrix result involving a fourth moment

$$\int_{-\infty}^{\infty} xx^T xx^T \exp\left(-\frac{1}{2}x^T P^{-1} x\right) dx = (2\pi)^{n/2} (\det P)^{1/2} \left[2PP + \mathrm{tr}(P)P\right]$$

First you may want to establish the following result for scalar x

$$\begin{aligned} i_p &= \int_{-\infty}^{\infty} x^p \exp\left(-\frac{1}{2}\frac{x^2}{\sigma^2}\right) dx \\ &= \begin{cases} 0 & p \text{ odd} \\ 2^{\frac{p+1}{2}} \sigma^{p+1} \Gamma(\frac{p+1}{2}) & p \text{ even} \end{cases} \end{aligned}$$

Exercise 4.33: The χ^2 and χ densities

Let $X_i, i = 1, 2, \ldots, n$, be statistically independent, normally distributed random variables with zero mean and unit variance. Consider the random variable Y to be the sum of squares

$$Y = X_1^2 + X_2^2 + \cdots + X_n^2$$

(a) Find Y's probability density. This density is known as the χ^2 density with n degrees of freedom, and we say $Y \sim \chi_n^2$. Show that the mean of this density is n.

(b) Repeat for the random variable

$$Z = \sqrt{X_1^2 + X_2^2 + \cdots X_n^2}$$

This density is known as the χ density with n degrees of freedom, and we say $Z \sim \chi_n$.

Exercise 4.34: The t-distribution

Assume that the random variables X and Y are statistically independent, and X is distributed as a normal with zero mean and unit variance and Y is distributed as χ^2 with n degrees of freedom. Show that the density of random variable t defined as

$$t = \frac{X}{\sqrt{Y/n}}$$

is given by

$$p_t(z; n) = \frac{1}{\sqrt{n\pi}} \frac{\Gamma(\frac{n+1}{2})}{\Gamma(\frac{n}{2})} \left(\frac{z^2}{n} + 1\right)^{-\frac{n+1}{2}} \qquad t\text{-distribution (density)} \qquad (4.90)$$

This distribution is known as Student's t-distribution after its discoverer, the chemist W. S. Gosset (Gosset, 1908), writing under the name Student.

Exercise 4.35: The F-distribution

Given random variables X and Y are independently distributed as χ^2 with n and m degrees of freedom, respectively. Define the random variable F as the ratio,

$$F = \frac{X/n}{Y/m}$$

Show that F's probability density is

$$p_F(z; n, m) = \frac{\sqrt{\dfrac{(zn)^n m^m}{(zn+m)^{n+m}}}}{zB(\frac{n}{2}, \frac{m}{2})} \qquad z \geq 0 \quad n, m \geq 1$$

in which B is the complete Beta function (Abramowitz and Stegun, 1970, p. 258) defined by

$$B(n, m) = \frac{\Gamma(n)\Gamma(m)}{\Gamma(n+m)}$$

This density is known as the F-distribution (density).

Exercise 4.36: Relation between t- and F-distributions

Given the random variable F is distributed as $p_F(z; 1, m)$ distribution with parameters $n = 1$ and m, consider the transformation

$$\mathcal{T} = \pm\sqrt{F}$$

Show that the random variable $\mathcal{T}$ is distributed as a t-distribution with parameter m

$$p_{\mathcal{T}}(z; m) = p_t(z; m)$$

Exercise 4.37: Independence and conditional density

Consider two random variables A, B with joint density $p_{AB}(a, b)$, and well-defined marginals $p_A(a)$ and $p_B(b)$ and conditional $p_{A|B}(a|b)$. Show that A and B are statistically independent if and only if the conditional of A given B is independent of b

$$p_{A|B}(a|b) \neq f(b)$$

Exercise 4.38: Independent estimates of parameter and variance

(a) Show that $\hat{\theta}$ and $\widehat{\sigma^2}$ given in (4.48) and (4.49) are statistically independent.

(b) Are the random variables $\hat{\theta}$ and $y - X\hat{\theta}$ statistically independent as well? Explain why or why not.

Exercise 4.39: Many samples of the vector least-squares problem

We showed for the model

$$y = X\theta + e \qquad e \sim N(0, R)$$

that the maximum-likelihood estimate is given by (4.64)

$$\hat{\theta} = \left(X^T R^{-1} X\right)^{-1} X^T R^{-1} y$$

Use this result to solve the n-sample problem given by the following model

$$\begin{bmatrix} y_1 \\ y_2 \\ \vdots \\ y_p \end{bmatrix}_i = \begin{bmatrix} x_1^T \\ x_2^T \\ \vdots \\ x_p^T \end{bmatrix} \begin{bmatrix} \theta \end{bmatrix} + \begin{bmatrix} e_1 \\ e_2 \\ \vdots \\ e_p \end{bmatrix}_i$$

$$y_i = X_i \qquad \theta + e_i \qquad e_i \sim N(0, R)$$

First stack the y_i samples in an enlarged vector $\widetilde{y}$, and define the corresponding stacked $\widetilde{X}$ matrix and $\widetilde{e}$ measurement error

$$\begin{bmatrix} y_1 \\ y_2 \\ \vdots \\ y_n \end{bmatrix} = \begin{bmatrix} X_1 \\ X_2 \\ \vdots \\ X_n \end{bmatrix} \begin{bmatrix} \theta \end{bmatrix} + \begin{bmatrix} e_1 \\ e_2 \\ \vdots \\ e_n \end{bmatrix} \qquad e_i \sim N(0, R)$$

$$\widetilde{y} = \widetilde{X} \qquad \theta + \widetilde{e} \qquad \widetilde{e} \sim N(0, \widetilde{R})$$

(a) What is the covariance matrix $\widetilde{R}$ for the new $\widetilde{e}$ measurement error vector?

(b) What is the corresponding formula for $\widehat{\theta}$ in terms of $\widetilde{y}$ for this problem?

(c) What is the probability density for this $\widehat{\theta}$?

(d) Does this result agree with (4.65)? Discuss why or why not.

Exercise 4.40: Vector and matrix least-squares problems

A colleague has an old but good piece of software that solves the traditional vector least-squares problem with constraints on the parameters

$$y = A\theta + e \qquad e \sim N(0, R)$$

in which y, θ, e are vectors and A, R are matrices. If the constraints are not active, the code produces the well-known solution

$$\widehat{\theta} = \left(A^T R^{-1} A\right)^{-1} A^T R^{-1} y \tag{4.91}$$

You would like to use this code to solve your matrix model problem

$$y_i = \Theta x_i + e_i \qquad e_i \sim N(0, R)$$

in which y_i, x_i, e_i are vectors, Θ is a matrix, i is the sample number, $i = 1, \ldots, n$, and you have n statistically independent samples. Your colleague suggests you stack your problem into a vector and find the solution with the existing code. So you arrange your measurements as

$$Y = \begin{bmatrix} y_1 & \cdots & y_n \end{bmatrix} \qquad X = \begin{bmatrix} x_1 & \cdots & x_n \end{bmatrix} \qquad E = \begin{bmatrix} e_1 & \cdots & e_n \end{bmatrix}$$

and your model becomes the matrix equation

$$Y = \Theta X + E \tag{4.92}$$

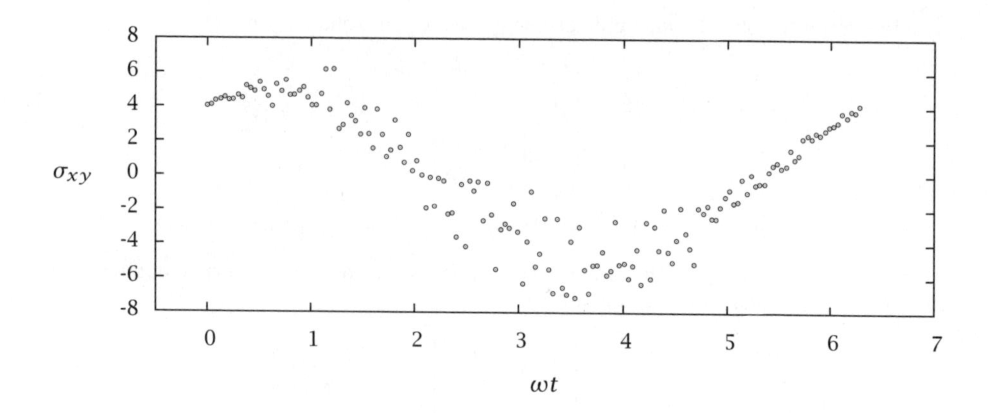

Figure 4.16: Typical strain versus time data from a molecular dynamics simulation from data file `rohit.dat` on the website `www.che.wisc.edu/~jbraw/principles`.

You looked up the answer to your estimation problem when the constraints are not active and find the formula

$$\hat{\Theta} = YX^T \left(XX^T\right)^{-1} \tag{4.93}$$

You do not see how this answer can come from your colleague's code because the answer in (4.91) obviously depends on R but your answer above clearly does not depend on R. Let's get to the bottom of this apparent contradiction, and see if we can use vector least-squares codes to solve matrix least-squares problems.

(a) What vector equation do you obtain if you apply the vec operator to both sides of the matrix model equation, (4.92)?

(b) What is the covariance of the vector $\text{vec}E$ appearing in your answer above?

(c) Apply (4.91) to your result in (a) and obtain the estimate $\text{vec}\hat{\Theta}$.

(d) Apply the vec operator to the matrix solution, (4.93), and obtain another expression for $\text{vec}\hat{\Theta}$.

(e) Compare your two results for $\text{vec}\hat{\Theta}$. Are they identical or different? Explain any differences. Does the parameter estimate depend on R? Explain why or why not.

Exercise 4.41: Estimating a material's storage and loss moduli from molecular simulation

Consider the following strain response model[5]

$$\sigma_{xy}(\omega t) = G_1 \sin \omega t + G_2 \cos \omega t$$

[5]This problem was motivated by Rohit Malshe's preliminary exam on May 7, 2007.

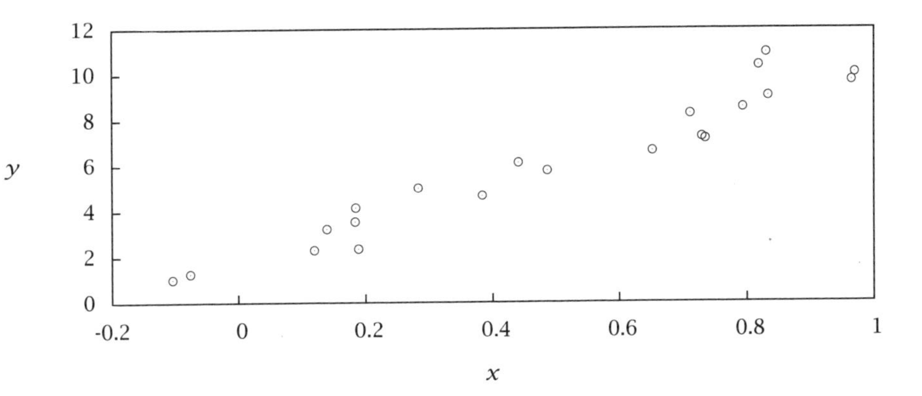

Figure 4.17: Plot of y versus x from data file errvbls.dat on the website www.che.wisc.edu/~jbraw/principles.

in which σ_{xy} is the strain, G_1 is the storage modulus, and G_2 is the loss modulus (G_1 and G_2 are positive scalars). We wish to estimate G_1 and G_2 from measurements of σ_{xy} at different times t for a given forcing frequency ω.

The strain "measurement" in this case actually comes from a molecular dynamics simulation. The simulation computes a noisy realization of $\sigma_{xy}(\omega t)$ for the given material of interest. A representative simulation data set is provided in Figure 4.16. These data are given in file `rohit.dat` on the website www.che.wisc.edu/~jbraw/principles so you can download them.

(a) Without knowing any details of the molecular dynamics simulation, suggest a reasonable least-squares estimation procedure for G_1 and G_2.

Find the optimal estimates and 95% confidence intervals for your recommended estimation procedure.

Plot your best-fit model as a smooth time function along with the data.

Are the confidence intervals approximate or exact in this case? Why?

(b) Examining the data shown in Figure 4.16, suggest an improved estimation procedure. What traditional least-squares assumption is violated by these data?

How would you implement your improved procedure if you had access to the molecular dynamics simulation so you could generate as many replicate "measurements" as you would like at almost no cost.

Exercise 4.42: Who has the error?

You are fitting some n laboratory measurements to a linear model

$$y_i = mx_i + b + e_{yi} \qquad i = 1, 2, \ldots, n$$

in which you have been told that the x variable is known with high accuracy and the y variable has measurement error e_y distributed as

$$e_y \sim N(0, 0.03)$$

The data are shown in Figure 4.17 and are given in file `errvbls.dat` on the website `www.che.wisc.edu/~jbraw/principles`.

(a) Given these assumptions, write the model as

$$y = X\theta + e_y$$

find the best estimate of the slope and intercept

$$\hat{\theta} = \begin{bmatrix} \hat{m} \\ \hat{b} \end{bmatrix}$$

and the 95% probability confidence ellipse, and also the plus/minus bounds on the parameter estimates.

(b) Plot the data, and the line of best fit to these data.

(c) Due to some confusion in the lab, you are told later that actually y is known with high accuracy and the x variable has measurement error e_x distributed as

$$e_x \sim N(0, 0.01)$$

Transform the model so that it is linear in a transformed parameter vector ϕ

$$x_i = f(y_i, \phi_1, \phi_2) + e_{xi} \qquad i = 1, 2, \dots, n$$

What are f and ϕ for the transformed model?

(d) Given these assumptions, write the model as

$$x = Y\phi + e_x$$

find the best estimate $\hat{\phi}$ for this model. Add this line of best fit to the plot of the data and the line of best fit from the previous model. Clearly label which line corresponds to which model.

(e) Compute the 95% confidence ellipse and plus/minus bounds for $\hat{\phi}$.

(f) Can you tell from the estimates and the fitted lines which of these two proposed models is more appropriate for these data? Discuss why or why not.

Exercise 4.43: Independence of transformed normals

Consider n independent samples of a scalar, zero-mean normal random variable with variance σ^2 arranged in a vector $e = \begin{bmatrix} e_1 & e_2 & \cdots & e_n \end{bmatrix}^T$ so that

$$e \sim N(0, \sigma^2 I_n)$$

Consider random variables x and y to be linear transformations of e, $x = Ae$ and $y = Be$.

(a) Provide necessary and sufficient conditions for matrices A and B so that x and y are independent.

(b) Given that the conditions on A and B are satisfied, what can you conclude about x and y if e has variance $\sigma^2 I_n$ but is not necessarily normally distributed.

Exercise 4.44: The multivariate t-distribution

Assume that the random variables $X \in \mathbb{R}^p$ and $Y \in \mathbb{R}_{\geq 0}$ are statistically independent, $X \sim N(0, \Sigma)$ and $Y \sim \chi_n^2$.

(a) Show that the density of random variable t defined as

$$t = \frac{X}{\sqrt{Y/n}} + m$$

with $m \in \mathbb{R}^p$ a constant is given by

$$p_t(z; p, n, m, \Sigma) = \frac{\Gamma(\frac{n+p}{2})}{\Gamma(\frac{n}{2})(n\pi)^{p/2}(\det \Sigma)^{1/2}} \left(1 + \frac{1}{n}(z-m)^T \Sigma^{-1}(z-m)\right)^{-(n+p)/2} \tag{4.94}$$

This distribution is known as the **multivariate t-distribution**, and was discovered by Cornish (1954), and Dunnett and Sobel (1954).

(b) Show that the ratio of the least-squares estimates of the parameters and variance for the case of the linear model with unknown measurement variance are distributed as

$$\frac{\hat{\theta} - \theta_0}{\sqrt{\frac{n}{n-n_p}\widehat{\sigma^2}}} + \theta_0 \sim t(n_p, n - n_p, \theta_0, (X^T X)^{-1})$$

Exercise 4.45: Integrals of the multivariate t-distribution and the F-statistic

Given the random variable t is distributed as a multivariate t defined in Exercise 4.44, consider the p-dimensional hyperellipse σ_b, of size $b \in \mathbb{R}_{\geq 0}$, centered at $m \in \mathbb{R}^p$

$$\sigma_b = \{z \mid (z-m)^T \Sigma^{-1}(z-m) \leq b\}$$

Show that the value of b that gives probability α in the multivariate t-distribution is given by the following F-statistic

$$b = pF(n, p, \alpha)$$

in which $F(n, p, \alpha)$ is defined in (4.54).

Exercise 4.46: Confidence interval for unknown variance

Consider again $\hat{\theta}$ and $\widehat{\sigma^2}$ from (4.48) and (4.49) and define the new random variable Z as the ratio

$$Z = \frac{\hat{\theta} - \theta_0}{\sqrt{\frac{n}{n-n_p}\widehat{\sigma^2}}} + \theta_0$$

in which $\hat{\theta}$ and $\widehat{\sigma^2}$ are statistically independent as shown in Exercise 4.38.

(a) Show Z is distributed as a multivariate t-distribution as defined in Exercise 4.34.

(b) Show that lines of constant probability of the multivariate t-distribution are ellipses in $\hat{\theta}$ as in the normal distribution.

(c) Define an α-level confidence interval using the multivariate t-distribution in place of the normal distribution and show that

$$(\hat{\theta} - \theta_0)^T \left(\frac{X^T X}{\widehat{\sigma^2}} \right) (\hat{\theta} - \theta_0) \leq \frac{n_p n}{n - n_p} F(n_p, n - n_p, \alpha)$$

in agreement with (4.55).

Exercise 4.47: Adding two uniformly distributed random variables

Given two independent, uniformly distributed random variables, $X \sim U[0,1]$ and $Y \sim U[4,5]$, find the density for $Z = X + Y$. Note that the transformation from (X, Y) to Z is not an invertible transformation.

Exercise 4.48: Product of two unit variance normals

Let X and Y be independent scalar random variables distributed identically as $N(0,1)$. Find and plot the density for $Z = XY$. Is $p_Z(z)$ well defined for all z? If not, explain why not.

Exercise 4.49: A useful integral in Fourier transforms of normals

Derive the definite integral used in taking the Fourier transform of the normal density

$$\int_0^\infty e^{-a^2x^2} \cos bx dx = \frac{\sqrt{\pi}}{2a} e^{-b^2/(4a^2)} \qquad a \neq 0$$

Hint: consider first the exponential version of the integral on $(-\infty, \infty)$. We wish to show that

$$\int_{-\infty}^\infty e^{-a^2x^2} e^{ibx} dx = \frac{\sqrt{\pi}}{a} e^{-b^2/(4a^2)} \qquad a \neq 0 \tag{4.95}$$

which gives the integral of interest as well as a second result

$$\int_0^\infty e^{-a^2x^2} \sin bx dx = 0 \qquad a \neq 0$$

To proceed, complete the square on the argument of the exponential and show that

$$-a^2x^2 + ibx = -a^2 \left((x - \frac{ib}{2a^2})^2 + \frac{b^2}{4a^4} \right)$$

Then perform the integral by noticing that integrating the normal distribution gives

$$\int_{-\infty}^\infty e^{-(1/2)(x-m')^2/\sigma^2} dx = \sqrt{2\pi}\sigma$$

even when $m' = im$ is complex valued instead of real valued. This last statement can be established by a simple contour integration in the complex plane and noting that the exponential function is an entire function, i.e., has no singularities in the complex plane.

Exercise 4.50: Orthogonal transformation of normal samples

Let vectors $x_1, x_2, \ldots, x_n \in \mathbb{R}^p$ be n independent samples of a normally distributed random variable with possibly different means but identical variance, $x_i \sim N(m_i, R)$. Consider the transformation

$$y_i = \sum_{j=1}^{n} c_{ij} x_j \qquad i = 1, 2, \ldots, n$$

in which matrix C is orthogonal.

Show that the y_i are independently distributed as $y_i \sim N(v_i, R)$ in which $v_i = \sum_{j=1}^{n} C_{ij} m_j$ for $i = 1, 2, \ldots, n$.

Hint: to reduce the algebra, you may wish to start off by arranging the x_i and y_i samples in the following matrices

$$X = \begin{bmatrix} x_1 & x_2 & \cdots & x_n \end{bmatrix} \qquad Y = \begin{bmatrix} y_1 & y_2 & \cdots & y_n \end{bmatrix}$$

and deduce the relationship between X, Y, and C given in the problem statement.

Exercise 4.51: Estimated variance and the Wishart distribution

Let vectors $e_1, e_2, \ldots, e_n \in \mathbb{R}^p$ be n independent samples of a normally distributed random variable with zero mean and identical variance, $e_i \sim N(0, R)$. Define the matrix

$$S = \sum_{i=1}^{n} e_i e_i^T$$

The distribution for random matrix S is known as the Wishart distribution, written

$$S \sim W(R, n)$$

and integer n is known as the number of degrees of freedom. Sometimes the fact that the e_i have p components (R is a $p \times p$ matrix) is also indicated using the notation $S \sim W_p(R, n)$ or $S \sim W(R, p, n)$.

Consider the estimation problem of Section 4.7.4 written in the form

$$Y = \Theta_0 X + E \qquad E = \begin{bmatrix} e_1 & \cdots & e_n \end{bmatrix}$$

(a) Show that the $EE^T \sim W_p(R, n)$.

(b) Define $\hat{E} = Y - \hat{\Theta} X$ and show that $n\hat{R} = \hat{E}\hat{E}^T$.

(c) Show that $\hat{E}\hat{E}^T \sim W(R, n - q)$, and therefore that $n\hat{R} \sim W_p(R, n - q)$.

Hint: take the SVD of the $q \times n$ matrix X for $q < n$

$$X = U \begin{bmatrix} \Sigma & 0 \end{bmatrix} \begin{bmatrix} V_1^T \\ V_2^T \end{bmatrix}$$

Define $Z = EV$, which can be partitioned as $\begin{bmatrix} Z_1 & Z_2 \end{bmatrix} = E \begin{bmatrix} V_1 & V_2 \end{bmatrix}$, and show that $\hat{E}\hat{E}^T = Z_2 Z_2^T$. Work out the distribution of $Z_2 Z_2^T$ from the definition of the Wishart distribution and the result of Exercise 4.50.

Exercise 4.52: Singular normal distribution as a delta sequence

Two generalized functions $f(\cdot)$ and $g(\cdot)$ are defined to be equal (in the sense of distributions) if they produce the same integral for all test functions $\phi(\cdot) \in \Phi$

$$\langle f, \phi \rangle = \langle g, \phi \rangle$$

$$\int_{-\infty}^{\infty} f(x)\phi(x)dx = \int_{-\infty}^{\infty} g(x)\phi(x)dx$$

The space of test functions Φ is defined to be the set of all smooth (nongeneralized) functions that vanish outside of a compact set $\mathbb{C} = [-c, c]$ for some $c > 0$.

Show that the zero mean normal density $n(x, \sigma)$

$$n(x, \sigma) = \frac{1}{\sqrt{2\pi}\sigma} e^{-\frac{1}{2}(x/\sigma)^2}$$

is equal to the delta function $\delta(x)$ in the limit $\sigma \to 0$.

Exercise 4.53: Error bound for the Taylor series of the exponential

Derive the bound (4.31) used in establishing the central limit theorem for sums of identically distributed random variables

$$\left| e^{ix} - \sum_{m=0}^{n} \frac{(ix)^m}{m!} \right| \leq \frac{|x|^{n+1}}{(n+1)!}$$

Hint: expand e^{ix} in a Taylor series with remainder term at $x = 0$ and take magnitudes. Note that for this particular function, the inequality above turns out to be an equality.

Exercise 4.54: Error bound for the remainder term in Taylor series

Derive the bound (4.73) for a second-order Taylor series of a bounded function f having three continuous, bounded derivatives

$$r(x,h) = f(x+h) - \left(f(x) + f^{(1)}(x)h + \frac{f^{(2)}(x)}{2}h^2 \right)$$

$$\sup_{x \in \mathbb{R}} |r(x,h)| \leq K_f \min(h^2, |h|^3)$$

Show that the following K_f is valid for any $b > 0$

$$K_f = \max\left([(2/b^2)M_f^{(0)} + (1/b)M_f^{(1)} + (1/2)M_f^{(2)}],\ (b/6)M_f^{(3)},\ (1/6)M_f^{(3)} \right) \tag{4.96}$$

with $M_f^{(i)} = \sup_{x \in \mathbb{R}} f^{(i)}(x)$.

Hints: first expand $f(x+h)$ about $f(x)$ to second order using the standard Taylor theorem with remainder. This gives the $|h|^3$ bound. For the second-order bound, first take absolute values of the definition of $r(x,h)$ and use the triangle inequality. Choose a constant $b > 0$ and consider two cases: $|h| \leq b$ and $|h| > b$. Develop second-order bounds for both cases and then combine them to obtain a second-order bound for all h. Finally, combine the second-order and third-order bounds by taking the smaller.

Exercise 4.55: Lindeberg conditions

Show that the following are special cases of the Lindeberg conditions given in Assumption 4.15.

(a) The de Moivre-Laplace central limit theorem assumption that the X_i are independent and identically distributed with mean zero and variance σ^2.

(b) The Lyapunov central limit theorem assumption that there exists $\delta > 0$ such that as $n \to \infty$

$$\frac{1}{s_n^{2+\delta}} \sum_{k=1}^{n} \mathcal{E}(|X_k|^{2+\delta}) \to 0$$

Note that the Lyapunov assumption implies only part (b) of Assumption 4.15.

(c) The bounded random variable assumption, i.e., there exists $B > 0$ such that $|X_i| \leq B, i = 1, 2, \ldots$.

Therefore, by proving Theorem 4.16, we have also proved the de Moivre-Laplace and the Lyapunov versions of the central limit theorem. We have also shown that the central limit theorem holds for bounded random variables, provided that $s_n \to \infty$.

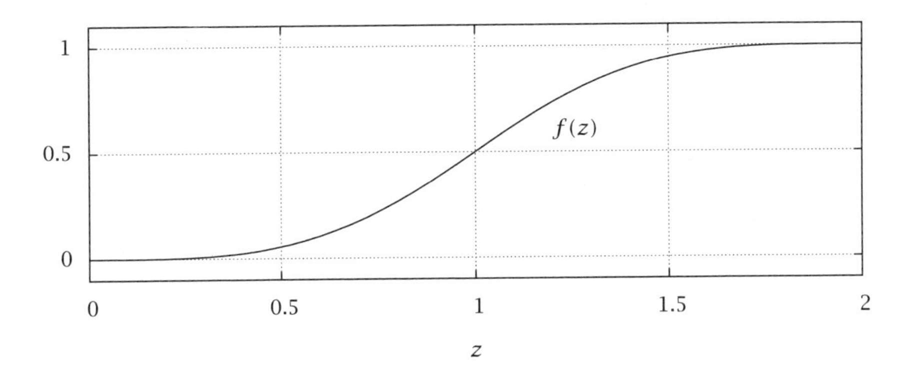

Figure 4.18: Smooth approximation to a unit step function, $H(z-1)$.

Exercise 4.56: Normal random variables satisfy Lindeberg conditions

Let $X_i, i = 1, 2, \ldots, n$ be independent with mean zero and variance σ_i^2. Let $Y_i, i = 1, 2, \ldots, n$ be independent *normals* with mean zero and variance σ_i^2. Show that if the X_i satisfy the Lindeberg conditions listed in Assumption 4.15, then so do the Y_i.

Hint: using the X_i variables, show that for n sufficiently large and any $\epsilon > 0$, $\sigma_i / s_n \leq \sqrt{2}\epsilon$ for all i. This result shows that no single random variable can account for a significant fraction of the sum's variance as n becomes large. Next evaluate the Lindeberg condition for the Y_i variables, and use the fact that $\sum_i \sigma_i^3 \leq (\max_i \sigma_i) \sum_i \sigma_i^2 = (\max_i \sigma_i) s_n^2$.

Exercise 4.57: Smoothing a step (indicator) function

We construct a suitably smooth indicator function as shown in Figure 4.15. To simplify the presentation, first consider the set up in Figure 4.18. We seek a monotone function $f(z)$ with three continuous derivatives that increases from zero at $z = 0$ to one at $z = 2$. We shall then rescale the z-axis to make this function as sharp as we please.

(a) Divide the interval in half and consider a fifth-order polynomial on $z \in [0, 1]$.

$$p(z) = a_0 + a_1 z + a_2 z^2 + a_3 z^3 + a_4 z^4 + a_5 z^5$$

To have $p(z)$ and its first three derivatives vanish at $z = 0$, we require $a_0 = a_1 = a_2 = a_3 = 0$. We will reflect this function about the $y = 1/2$ and $z = 1$ lines to provide the matching function $q(z)$ on $z \in [1, 2]$, or, in equations, $q(z) = -p(2-z) + 1$. Note that the symmetry implies $p^{(i)}(1) = (-1)^{i+1} q(1)$, so that all odd derivatives are automatically continuous at $z = 1$, and the even derivatives are negatives of each other at $z = 1$. So we require that the even derivatives at $z = 1$ are zero. We therefore have two conditions, $p(1) = 1/2$ and $p''(1) = 0$, to find the remaining two coefficients

$$p(1) = a_4 + a_4 = 1/2 \qquad p''(1) = 12a_4 + 20a_5 = 0$$

Solve these equations and show that $a_4 = 5/4$ and $a_5 = -3/4$.

(b) The candidate function $f(z)$ is therefore

$$f(z) = \begin{cases} 0, & z \leq 0 \\ (5/4)z^4 - (3/4)z^5, & 0 < z \leq 1 \\ 1 - (5/4)(2-z)^4 + (3/4)(2-z)^5, & 1 \leq z < 2 \\ 0, & 2 \leq z \end{cases}$$

Plot this function and its first three derivatives, and check that they are continuous at $z = 0, 1, 2$. Show that the maxima in magnitude of the derivatives are given by

$$M_f^{(0)} = 1 \qquad M_f^{(1)} = 5/4 \qquad M_f^{(2)} = 20/9 \qquad M_f^{(3)} = 15$$

and check these values also on your plots.

(c) Next we rescale. Let $w = (1 - z/2)L + x$ and $f(z) = f(2(1-(w-x)/L)) = \tilde{f}(w)$. The function $\tilde{f}(w)$ now has the required properties of Figure 4.15. Show that the derivative bounds are scaled by

$$M_{\tilde{f}}^{(i)} = (2/L)^i M_f^{(i)}$$

(d) Show finally that because of this scaling with L, there exists $L_0 > 0$ such that the bound in (4.96) is given by

$$K_{\tilde{f}} = 20L^{-3} \qquad \text{for every } L \text{ satisfying } 0 < L \leq L_0 \tag{4.97}$$

For an even smoother, seventh-order polynomial, with a smaller third derivative, see Thomasian (1969, p.486).

Exercise 4.58: Properties of PLSR algorithm

Given the PLSR algorithm described in Section 4.8, show the following properties.

(a) $T = XR$

(b) $T^T T = I_q$

(c) Q minimizes $\left\| Y - TQ^T \right\|_F^2$ for given Y and T.

Exercise 4.59: Using PCR and PLSR

Write your own PCR and PLSR algorithm and apply it to the data given in Example 4.23. The data are available in file `pca_pls_data.dat` on the website `www.che.wisc.edu/~jbraw/principles`.

(a) Reproduce the results given in Example 4.23.

(b) Estimate parameter B using both PCR and PLSR using the number of principal components/latent variables $\ell = 1, 2, 3, 4, 5$. Compare your estimates to the value that was used to generate the data.

$$B^T = \begin{bmatrix} 0 & 0 & 0 & -1 & 0 \\ 0 & 0 & 0 & 1 & 0 \end{bmatrix}$$

Bibliography

M. Abramowitz and I. A. Stegun. *Handbook of Mathematical Functions.* National Bureau of Standards, Washington, D.C., 1970.

T. W. Anderson. *An Introduction to Multivariate Statistical Analysis.* John Wiley & Sons, New York, third edition, 2003.

T. Bayes. An essay towards solving a problem in the doctrine of chances. *Phil. Trans. Roy. Soc.*, 53:370–418, 1763. Reprinted in *Biometrika*, 35:293–315, 1958.

G. E. P. Box and G. C. Tiao. *Bayesian Inference in Statistical Analysis.* Addison-Wesley, Reading, Massachusetts, first edition, 1973.

E. A. Cornish. The multivariate t-distribution associated with a set of normal sample deviates. *Aust. J. Phys.*, 7:531, 1954.

C. W. Dunnett and M. Sobel. A bivariate generalization of student's t-distribution with tables for certain special cases. *Biometrika*, 41:153, 1954.

R. Durrett. *Probability: Theory and Examples.* Cambridge University Press, fourth edition, 2010.

W. Feller. Über den zentralen Grenzwertsatz der Warscheinlichkeitsrechnung. *Math. Z.*, 40:512–559, 1935.

P. Geladi and B. R. Kowalski. Partial least-squares regression: A tutorial. *Anal. Chim. Acta*, 185:1–17, 1986.

W. S. Gosset. The probable error of a mean. *Biometrika*, 6:1–25, 1908.

M. H. Kaspar and W. H. Ray. Partial least squares modelling as successive singular value decompositions. *Comput. Chem. Eng.*, 17(10):985–989, 1993.

A. N. Kolmogorov. *Foundations of Probability.* Chelsea Publishing Company, New York, 1950. Translation of "Grundbegriffe der Wahrscheinlichkeitrechnung, Ergebnisse der Mathematik," 1933.

L. Le Cam. The central limit theorem around 1935. *Statist. Sci.*, 1(1):78–96, 1986.

P. Lévy. Propriétés asymptotiques des sommes de variables indépendantes on enchainées. *J. Math. Pures Appl.*, pages 347–402, 1935.

J. W. Lindeberg. Eine neue Herleitung des Exponentialgesetzes in der Wahrscheinlichkeitsrechnung. *Math. Z.*, 15:211–225, 1922.

J. F. MacGregor, T. F. Marlin, J. Kresta, and B. Skagerberg. Multivariate statistical methods in process analysis and control. In Y. Arkun and W. H. Ray, editors, *Chemical Process Control-CPCIV.* CACHE, 1991.

H. Martens. Reliable and relevant modelling of real world data: a personal account of the development of PLS regression. *Chemom. Intell. Lab. Syst.*, 58:85–95, 2001.

B.-H. Mevik and R. Wehrens. The pls package: Principal component and partial least squares regresion in R. *J. Stat. Softw.*, 18:1–24, 2007.

A. Papoulis. *Probability, Random Variables, and Stochastic Processes.* McGraw-Hill, Inc., second edition, 1984.

D. Pollard. Comment on: The central limit theorem around 1935. *Statist. Sci.*, 1(1):94–95, 1986.

G. Polyá. Über den zentralen Grenzwertsatz der Warscheinlichkeitsrechnung und das Momentproblem. *Math. Z.*, 8:171–180, 1920.

S. J. Qin. Recursive PLS algorithms for adaptive data modeling. *Comput. Chem. Eng.*, 22(4/5):503–514, 1998.

S. M. Selby. *CRC Standard Mathematical Tables.* CRC Press, twenty-first edition, 1973.

Student. The probable error of a mean. *Biometrika*, 6:1–25, 1908.

A. J. Thomasian. *The structure of probability theory with applications.* McGraw-Hill, 1969.

A. W. van der Vaart. *Asymptotic Statistics.* Cambridge University Press, 1998.

J. Wishart. The generalised product moment distribution in samples from a normal multivariate population. *Biometrika*, 20A:32–52, 1928.

H. Wold. Estimation of principal components and related models by iterative least squares. In P. R. Krishnaiah, editor, *Multivariate Analysis*, pages 391–420. Academic Press, 1966.

S. Wold. Personal memories of the early PLS development. *Chemom. Intell. Lab. Syst.*, 58:83–84, 2001.

5

Stochastic Models and Processes

5.1 Introduction

We are by now expert in using (deterministic) differential and partial differential equations as models of chemical and biological systems. These equations capture equations of motion, conservation of mass and energy, and many of the fundamental principles useful in analysis and design of chemically reacting systems. Chapters 2 and 3 were mainly devoted to developing this program. The motivation for *stochastic* processes and differential equations is to incorporate into the model the *random* effects of the internal system (discrete molecules) and the external environment on the system of interest. In some applications at fine length scales, the random effects are mainly due to the internal random behavior of the molecules. But even in applications at large scales, the random effects of the external environment are often quite important to understand and interpret the (noisy) measurements coming from a system.

In this chapter, we illustrate the usefulness of random variables and random processes in the modeling and analysis of systems of interest to chemical and biological engineers. We find the basic probability and statistics that we covered in Chapter 4 indispensable tools in carrying out this program. We study three main examples: (i) the Wiener process as a model of diffusion in transport phenomena, (ii) the Poisson process as a model of chemical reactions and kinetics at the small scale, and (iii) the Kalman filter for reducing the effects of noise in process measurements, a fundamental task in systems engineering. By covering representative examples from transport phenomena, chemical kinetics, and systems engineering, we hope to both introduce random models and processes, as well as demonstrate their wide range of applicability in modern chemical and biological engineering.

5.2 Stochastic Processes for Continuous Random Variables

5.2.1 Discrete Time Stochastic Processes

Our target in this part of the chapter is an understanding of the structure and dynamics of continuous time stochastic processes: the stochastic analogs of deterministic differential equations. In building up to these, it is instructive to start with the conceptually simpler stochastic difference equation. Consider the following example

$$x(k+1) = Ax(k) + G\xi(k) \tag{5.1}$$

in which $k \in \mathbb{I}_{\geq 0}$ is the sample number in discrete time, ξ is a random variable, assumed to have some fixed and known probability density, and $\xi(k), k = 0, 1, 2, \ldots$ are independent, identically distributed samples of ξ. If we define a sampling interval Δt, then $t = k\Delta t$. Because of the influence of the random variable ξ, the variable x is also a random variable. In general it can take any value, so we call it a continuous random variable in contrast to the integer-valued or discrete random variables we encounter in Section 5.3.

We wish to study the statistical properties of the process $x(k)$ due to the random disturbance ξ. Because the process is linear, an explicit solution is simply calculated

$$x(k) = A^k x(0) + \sum_{j=0}^{k-1} A^{k-j-1} G\xi(j) \qquad k \geq 0 \tag{5.2}$$

There is no difficulty expressing the solution to the stochastic difference equation; in fact we cannot determine by looking at the form of the solution if $\xi(k)$ is a random variable or simply a deterministic function of time. This is the perfect place to start because everything is well defined regardless of whether or not ξ is a random variable. We build some simple intuition with stochastic difference equations and then proceed to continuous time systems. We shall also see that difference equations arise whenever we wish to *numerically approximate* the solution to stochastic differential equations, so some facility with the difference equations is highly useful.

The INTEGRATED WHITE-NOISE process provides a starting point for understanding many important aspects of stochastic processes. Consider a system with scalar x, $A = 1$, zero initial condition and $\xi(k) =$

$w(k)$, where the $w(k)$ are independent and unit normals $w \sim N(0,1)$

$$x(k+1) = x(k) + Gw(k) \qquad x(0) = 0 \tag{5.3}$$

We wish to find the probability density of $x(k)$ versus time for this process.

We have $x(1) = x(0) + Gw(0) = Gw(0)$, so $x(1) \sim N(0, G^2)$. Since the $w(k)$ sequence is independent of x, we have for $k = 2$

$$x(2) = x(1) + Gw(1) \qquad \begin{bmatrix} x(1) \\ w(1) \end{bmatrix} \sim N\left(\begin{bmatrix} 0 \\ 0 \end{bmatrix}, \begin{bmatrix} G^2 & 0 \\ 0 & 1 \end{bmatrix}\right)$$

Noting that

$$x(2) = \begin{bmatrix} 1 & G \end{bmatrix} \begin{bmatrix} x(1) \\ w(1) \end{bmatrix}$$

and using Theorem 4.12 on the linear transformation of a normal we have that

$$x(2) \sim N(0, 2G^2)$$

Continuing this process gives

$$x(k) \sim N(0, kG^2) \qquad k \geq 0$$

and we have that the variance of $x(k)$ increases linearly with time and the mean remains zero for the integrated white-noise process. If we choose $G = \sqrt{\Delta t}$, then $x(k) \sim N(0, k\Delta t)$ and the system satisfies

$$x(t) \sim N(0, t) \qquad t \geq 0 \quad G = \sqrt{\Delta t}$$

or equivalently its probability density $p(x)$ satisfies

$$p(x,t) = \frac{1}{\sqrt{2\pi t}} \exp\left(-\frac{1}{2}\frac{x^2}{t}\right)$$

Similarly, if we let $G = \sqrt{2D\Delta t}$ where D is a constant, then

$$x(t) \sim N(0, 2Dt) \qquad t \geq 0$$

or

$$p(x,t) = \frac{1}{2\sqrt{\pi D t}} \exp\left(-\frac{1}{4}\frac{x^2}{Dt}\right)$$

This is precisely (3.70) from Chapter 3, which describes the transient spread by diffusion of a delta-function initial condition. Thus we see already the first sign of what turns out to be a deep and important connection between diffusion and stochastic processes: (5.3) is a model for a particle undergoing Brownian motion in one dimension. Continuing with the diffusion analogy, if we consider $x(k\Delta t) = x(t)$ to be a position variable, then the mean square displacement is given by

$$\mathcal{E}\left(x^2(t)\right) = \text{var}(x(t)) = kG^2 = 2Dt$$

For diffusion processes, the mean square displacement increases linearly in time.

The analysis above can be extended to the case where the random term has nonzero mean: $\xi \sim N(m, 1)$, which we can write $\xi = m + w$ with w defined as above. Now

$$x(k+1) = x(k) + Gm + Gw(k)$$

Defining $v = Gm/\Delta t$ this becomes

$$x(k+1) = x(k) + v\Delta t + Gw(k)$$

Again, if we interpret x as a particle position, then the particle travels or "drifts" a distance $v\Delta t$ in one time interval as well as diffusing. Letting $G = \sqrt{2D}$

$$x(t) \sim N(vt, 2Dt)$$

The particle drifts with a velocity v so its mean position changes linearly with time, while also diffusing.

Finally, we return to the case where ξ is drawn from an arbitrary distribution rather than a normal. With $A = 1$ and $x(0) = 0$, (5.2) becomes

$$x(k) = G \sum_{j=0}^{k-1} \xi(j)$$

That is, the solution becomes a sum of independent identically distributed (IID) random variables. In Section 4.5 we learned the remarkable fact that sums of IID random variables converge to a normal distribution. Thus as $k \to \infty$, $x(k)$ becomes normally distributed even if the noise that drives it is not. So, for example, if we can only observe the process $x(t)$ at time intervals that are infrequent compared to Δt, it will be virtually impossible to know whether the underlying noise was Gaussian or not—the resulting process $x(k)$ will be. This result is one reason why, in the absence of further information, taking the noise in a system to be normally distributed is often a good approximation.

5.2.2 Wiener Process and Brownian Motion

We now wish to define the continuous time version of the discrete time integrated white noise or Brownian motion just presented. This process, denoted $W(t)$, is known as a Wiener process in honor of the mathematician Norbert Wiener. The property that we retain in taking the limit as $\Delta t \to 0$ is that $W(t)$ is normally distributed with zero mean and linearly increasing variance or

$$W(t) \sim N(0, t) \qquad t \geq 0$$

By analogy with the results above, a diffusion process $x(t)$ with diffusivity D and $x(0) = 0$ would simply be

$$x(t) = \sqrt{2D}W(t) \tag{5.4}$$

Note that the linear increase in variance with time should hold for any starting time s, giving

$$W(t) - W(s) \sim N(0, t - s) \qquad s \geq 0, \quad t \geq s \tag{5.5}$$

The increment of the Wiener process is denoted

$$\Delta W(t - s) = W(t) - W(s)$$

Considering distinct time instants t_i, with $t_i > t_{i-1}$, we define $\Delta t_i = t_i - t_{i-1}$ and $\Delta W(t_i) = W(t_i) - W(t_{i-1})$. Increments involving non-overlapping time intervals are independent. The Wiener increments have a number of important properties that follow from their definitions

$$\mathcal{E}(\Delta W(t_i)) = 0 \tag{5.6}$$

$$\mathcal{E}((B\Delta W(t_i))(B\Delta W(t_j))) = B^2 \Delta t_i \delta_{ij} \tag{5.7}$$

$$\mathcal{E}\left(\Delta W(t_i)^n\right) = 0 \quad \text{for } n \text{ odd} \tag{5.8}$$

$$\mathcal{E}\left(\Delta W(t_i)^{2m}\right) \propto \Delta t_i^m \quad \text{for integer } m \tag{5.9}$$

In Theorem 4.12 we saw that the distribution of a sum of normally distributed random variables is also normally distributed. A number of important results for Wiener processes follow from this fact. A Wiener process can be written as a sum of N Wiener increments for any N

$$W(t - t_0) = \sum_{i=1}^{N} \Delta W(t_i) \tag{5.10}$$

where $t_N = t$ and the only restriction on t_i is that $t_i > t_{i-1}$. Accordingly, a diffusion (Brownian motion) process can be written as a sum of Wiener increments multiplied by $\sqrt{2D}$

$$x(t - t_0) = \sum_{i=1}^{N} \sqrt{2D}\Delta W(t_i) \tag{5.11}$$

Furthermore, for separate Wiener processes W_1, W_2, W_3

$$\sqrt{2D_1}\Delta W_1(t_i) + \sqrt{2D_2}\Delta W_2(t_i) = \sqrt{2(D_1 + D_2)}\Delta W_3(t_i) \tag{5.12}$$

In other words, the sum of two diffusion processes is equivalent to a different diffusion process whose diffusivity is the sum of the first two.

To visualize a trajectory of a Brownian motion process $x(t)$, we can use (5.11), generating points $x(t)$ at constant time intervals Δt. Observing that now $\Delta W \sim N(0, \Delta t)$, this is equivalent to evaluating the discrete time process

$$x((k+1)\Delta t) = x(k\Delta t) + \sqrt{2D\Delta t}\, w(k) \qquad x(0) = 0$$

with $w(k) \sim N(0, 1)$ defined as above. Figure 5.1 shows a trajectory of this process for sample time $\Delta t = 10^{-6}$ and diffusivity $D = 5 \times 10^5$. Notice that the roughness is quite apparent in the top row of Figure 5.1. But by looking at finer time scales, we can see the effect of the finite step size in the discrete time approximation. The continuous time Wiener process defined in (5.5) maintains its roughness at all time scales; Figure 5.2 shows how the path should appear between the samples if we chose the step size properly for this magnification. Unlike more familiar functions, the Wiener process is very irregular. Thus it is important to address its continuity and smoothness properties.

The Wiener process is continuous. A crude argument for this statement is that $|\Delta W| \propto \sqrt{\Delta t}$, which approaches zero as $\Delta t \to 0$. A more refined one is presented in Exercise 5.4. On the other hand, because of the $\Delta t^{1/2}$ behavior of ΔW, we arrive at the perhaps surprising fact that

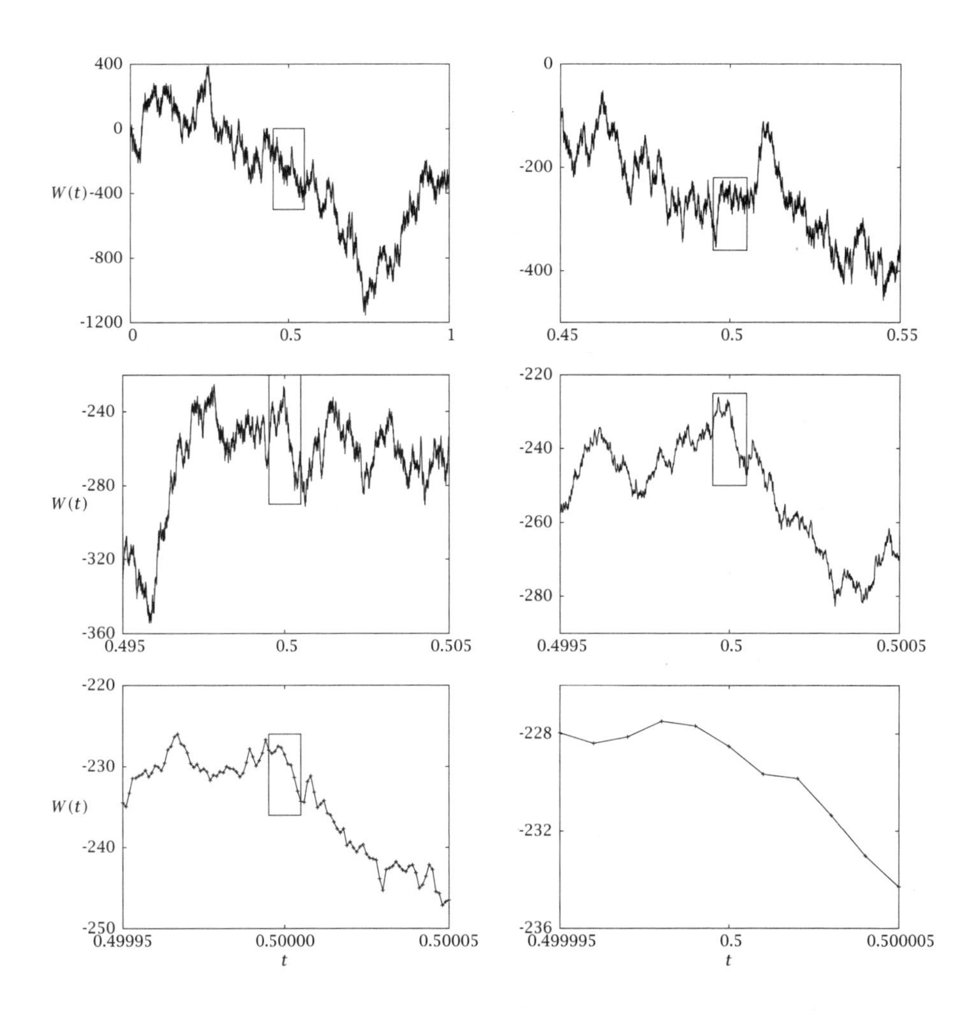

Figure 5.1: A simulation of the Wiener process with fixed sample time $\Delta t = 10^{-6}$ and $D = 5 \times 10^5$. The boxed region in each figure is expanded in the next plot to display a decreasing time scale of interest. The true Wiener process is rough at *all* time scales and therefore $dW(t)/dt$ does not exist. The top row shows an adequate sampling rate to display the roughness of the Wiener process. The middle row shows the time scale of interest starting to become too small for the given sample time. The bottom row shows a time scale of interest much too small for the given sample time; one can see the samples and the straight lines drawn between them.

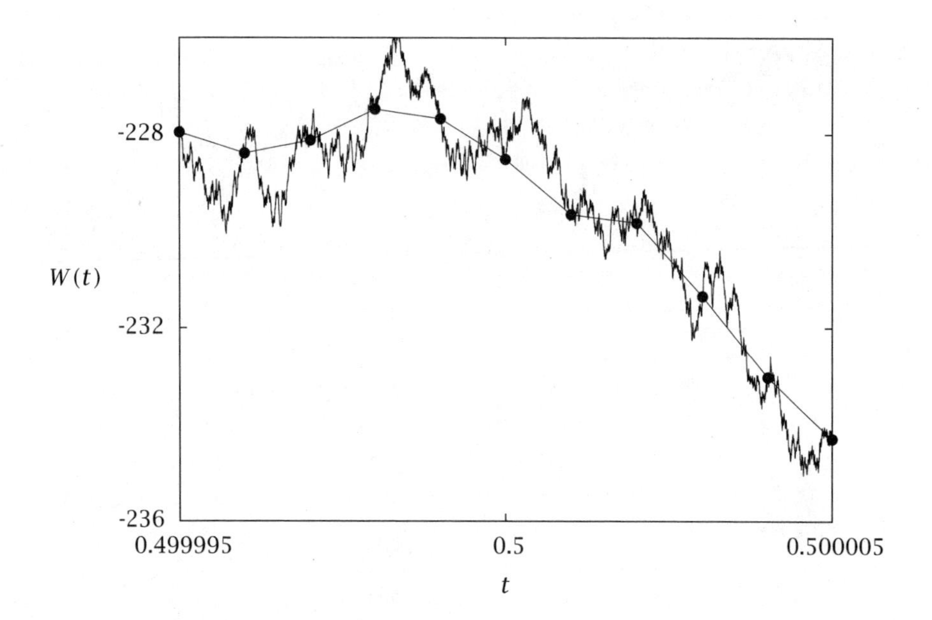

Figure 5.2: Sampling faster on the last plot in Figure 5.1; the sample time is decreased to $\Delta t = 10^{-9}$ and the roughness is restored on this time scale. Thought question: how did we generate a random walk that passes exactly through the solid sample points taken from Figure 5.1? Hint: certainly not by trial and error! Such a process is called a Brownian bridge (Bhattacharya and Waymire, 2009).

the Wiener process is *not* differentiable[1]

$$\begin{aligned}
\mathcal{E}\left(\left|\frac{\Delta W}{\Delta t}\right|\right) &= \frac{1}{\Delta t}\mathcal{E}\left(|\Delta w|\right) \\
&= \frac{1}{\Delta t}\frac{1}{\sqrt{2\pi\Delta t}}\int_{-\infty}^{\infty}|x|\exp\left(\frac{-x^2}{2\Delta t}\right)dx \\
&= \frac{\sqrt{2\Delta t/\pi}}{\Delta t} \\
&= \sqrt{\frac{2}{\pi}\frac{1}{\sqrt{\Delta t}}}
\end{aligned}$$

This diverges as $\Delta t^{-1/2}$ as $\Delta t \to 0$.

[1]The results of Exercise 5.8 were applied in this derivation.

Now let us return for the moment to the discrete time integrated white-noise process, (5.3). Considering a sampling interval Δt and letting $x(k+1) - x(k) = \Delta x$ and $Gw(k) = B\Delta W$, we can rewrite this as

$$\Delta x = B\Delta W \tag{5.13}$$

Under other circumstances we could divide by Δt and let it shrink to zero, yielding

$$\frac{dx}{dt} = \sqrt{2D}\frac{dW}{dt}$$

We have just found, however, that dW/dt does not exist. Nevertheless, we can define a differential of the Wiener process as the Wiener increment $W(t + \Delta ts) - W(t)$ when Δt becomes the infinitesimal dt

$$dW(t) = W(t + dt) - W(t) \sim N(0, dt)$$

This is also known as the white-noise process. It is *not* continuous. Now we can write (5.13) in differential form

$$dx = B\,dW \tag{5.14}$$

This is the most elementary STOCHASTIC DIFFERENTIAL EQUATION. With initial condition $x(0) = 0$, its solution is (5.4).

5.2.3 Stochastic Differential Equations

Basic ideas

To motivate and introduce stochastic differential equations, consider first the deterministic differential equation

$$\frac{dx}{dt} = f(x, t) \tag{5.15}$$

When we wish to augment this model to include some random effects, one might try

$$\frac{dx}{dt} = f(x, t) + g(x, t)\eta(t)$$

in which $\eta(t)$ is a random variable, often a normally distributed, zero mean random variable, as discussed in Chapter 4.

We have already run into problems with this formulation. Even to model a "well-behaved" (e.g., continuous) stochastic process like diffusion, we have seen that the random term would have to take on the form

$$g(x, t)\eta(t) = B\frac{dW}{dt}$$

But we also saw that $\frac{dW}{dt}$ does not exist. Extending what we did above for Brownian motion, we thus consider differentials instead of derivatives and write a general stochastic differential equation (SDE) in the form

$$dx = \mathcal{A}(x,t)\,dt + \mathcal{B}(x,t)\,dW \tag{5.16}$$

Formally, we can integrate this to yield

$$x(t) = x(0) + \int_0^t \mathcal{A}(x(t'),t')\,dt' + \int_0^t \mathcal{B}(x(t'),t')\,dW(t') \tag{5.17}$$

The first integral is classical. The second would be as well if $\frac{dW}{dt}$ existed, in which case would just write that

$$\int_0^t \mathcal{B}(x(t'),t')\,dW(t') = \int_0^t \mathcal{B}(x(t'),t')\frac{dW}{dt}\,dt'$$

This integral is nontrivial and to understand it we need to understand a little bit about the calculus of stochastic processes.

Elementary Stochastic Calculus

Stochastic integrals of the form

$$S = \int_{t_0}^t G(t')dW(t')$$

are more complex than conventional integrals because both G and dW can vary stochastically (think of the case $G(t) = W(t)$). Nevertheless, as with conventional integrals, we can divide the interval $[t_0, t]$ into n subintervals $t_0 \leq t_1 \leq t_2 ... \leq t_{n-1} \leq t$, and choose intermediate time points τ_i such that $t_{i-1} \leq \tau_i \leq t_i$. Now the integral S is approximated by the sum

$$S_n = \sum_{i=1}^{n} G(\tau_i)(W(t_i) - W(t_{i-1}))$$

In normal calculus this sum converges to the same value independent of the choice of the τ_i; in stochastic calculus this is not the case. We will choose $\tau_i = t_{i-1}$, yielding the ITÔ STOCHASTIC INTEGRAL[2]. Thus (5.16) is an Itô stochastic differential equation.

[2]Other choices are used in various situations—for example the STRATONOVICH stochastic integral takes $\tau_i = (t_{i-1} + t_i)/2$. Stochastic calculus is complex and technical; Gardiner (1990) provides a detailed discussion that is accessible to the non-mathematician.

The Itô stochastic integral corresponds to a stochastic "rectangle rule," with the function value chosen at the left side of the subinterval. One practical reason for this choice is that it is the one most straightforwardly applied in numerical solutions of stochastic differential equations. The EULER-MARUYAMA scheme generalizes the explicit Euler method to the stochastic case, using this rectangle rule approximation

$$x(t + \Delta t) = x(t) + \mathcal{A}(x(t), t)\Delta t + \mathcal{B}(x(t), t)\Delta W(t + \Delta t) \tag{5.18}$$

where $\Delta W(t + \Delta t) \sim N(0, \Delta t)$. This is the standard method for finding trajectories of SDEs and is often referred to as BROWNIAN DYNAMICS SIMULATION; it is not highly accurate, but higher-order schemes for SDEs are very complex to implement (Kloeden and Platen, 1992).

A more fundamental reason for working with the Itô integral is that, when applied to (5.17), it corresponds to a noise term that does not change the mean of $x(t)$, because its expected value is zero

$$\mathcal{E}\left(\int_{t_0}^{t} G(t')\, dW(t')\right) = 0 \tag{5.19}$$

This is easily seen by taking the expected value of the discrete sum and using the fact that for the Itô integral, $G(\tau_i)$ and $(W(t_i) - W(t_{i-1}))$ are independent

$$\begin{aligned}
\mathcal{E}\,(S_n) &= \sum_{i=1}^{n} \mathcal{E}\,(G(t_{i-1})(W(t_i) - W(t_{i-1}))) \\
&= \sum_{i=1}^{n} \mathcal{E}(G(t_{i-1}))\mathcal{E}\,(W(t_i) - W(t_{i-1})) \\
&= 0
\end{aligned}$$

because $\mathcal{E}\,(W(t_i) - W(t_{i-1})) = 0$. This calculation makes clear that the choice of τ_i matters: if τ_i were not taken to be t_{i-1}, then $G(\tau_i)$ and $(W(t_i) - W(t_{i-1}))$ would not be independent and $\mathcal{E}(S_n)$ would not necessarily be zero.

By considering integrals of the form

$$\int G(t')[dW(t')]^{2+N}$$

and using the Itô expression for S_n one can show that

$$\int_{t_0}^{t} G(t')[dW(t')]^{2+N} = \begin{cases} \int_{t_0}^{t} G(t')dt' & N = 0 \\ 0 & N > 0 \end{cases}$$

This result tells us how to treat higher differentials involving dW and dt in general

$$[dW(t)]^2 = dt, \qquad [dW(t)]^{2+N} = 0, \qquad dW\,dt = 0$$

and so on. If dW_i and dW_j are different white-noise processes, e.g., corresponding to different components of a vector of such processes, then

$$dW_i dW_j = \delta_{ij} dt \tag{5.20}$$

Unlike in regular calculus, in working with differentials of W, one must keep terms up to dW^2. To understand why, simply recall that $\mathcal{E}(|\Delta W|) \propto \sqrt{\Delta t}$.

We can use the above observations about stochastic differentials to derive the ITÔ STOCHASTIC CHAIN RULE. Let F be a function of t and $W(t)$. Then

$$dF(t, W(t)) = \left(\frac{\partial F}{\partial t} + \frac{1}{2}\frac{\partial^2 F}{\partial W^2}\right) dt + \frac{\partial F}{\partial W}\, dW(t)$$

For example, if we let $F = x(t, W(t)) = \mathcal{A}(t - t_0) + \mathcal{B}(W(t) - W(t_0))$, where $\mathcal{A}$ and $\mathcal{B}$ are constants, then application of the chain rule gives us back the constant coefficient SDE $dx = \mathcal{A}\,dt + \mathcal{B}\,dW$.

Now consider a function $f(x(t))$, where $x(t)$ evolves according to (5.16). The differential of f can be written

$$\begin{aligned} df(x(t)) &= f(x(t+dt) - f(x(t))) \\ &= f'(x(t))dx(t) + \frac{1}{2}f''(x(t))(dx(t))^2 \\ &= f'(x(t))(\mathcal{A}\,dt + \mathcal{B}\,dW) + \frac{1}{2}f''(x(t))(\mathcal{A}\,dt + \mathcal{B}\,dW)^2 \end{aligned}$$

Noting that $dt^2 = 0$ and $dW^2 = dt$, we have ITÔ'S FORMULA

$$df(x(t)) = (\mathcal{A}f' + \frac{1}{2}\mathcal{B}^2 f'')\,dt + \mathcal{B}f'\,dW \tag{5.21}$$

Example 5.1: Diffusion on a plane in Cartesian and polar coordinate systems

We can write two-dimensional Brownian motion in Cartesian coordinates as

$$dx = B\,dW_x \tag{5.22}$$
$$dy = B\,dW_y \tag{5.23}$$

where W_x and W_y are independent Wiener processes and $B = \sqrt{2D}$. How would we write the same process in polar coordinates?

Solution

As a brief prelude, observe that for a particle starting at the origin, the mean square displacement satisfies

$$\begin{aligned}\mathcal{E}\left(r^2(t)\right) &= B^2\left(\mathcal{E}\left(W_x^2\right) + \mathcal{E}\left(W_y^2\right)\right)\\ &= B^2\,(t+t)\\ &= 4Dt\end{aligned}$$

This result easily extends to Brownian motion in any number d of dimensions, giving the result $\mathcal{E}\left(r^2(t)\right) = 2dDt$.

Returning to the specific question at hand, consider the radial coordinate first and keep in mind that we may need to keep terms up to quadratic in dx and dy

$$dr = \frac{\partial r}{\partial x}dx + \frac{\partial r}{\partial y}dy + \frac{1}{2}\frac{\partial^2 r}{\partial x^2}dx^2 + \frac{1}{2}\frac{\partial^2 r}{\partial x \partial y}dxdy + \frac{1}{2}\frac{\partial^2 r}{\partial y^2}dy^2$$

Here all the partials can be evaluated from the formulas $r = \sqrt{x^2+y^2}$ and $\theta = \tan^{-1}\left(\frac{y}{x}\right)$. Now using the SDEs and noting that $dx^2 = dy^2 = B^2dt$, $dxdy = 0$, we have that

$$dr = \cos\theta B\; dW_x + \sin\theta B\; dW_y + \frac{1}{2r}B^2\;dt$$

Now, using (5.12) we see that $\cos\theta\; dW_x + \sin\theta B\; dW_y$ is a diffusion process with variance dt. We will denote this process as dW_r, so

$$dr = \frac{B^2}{2r}\;dt + B\;dW_r \tag{5.24}$$

Consider a particle that starts at $r = 0$. Applying Itô's formula with $f = r^2$ and taking the expected value we find that

$$\mathcal{E}(d(r^2)) = 2B^2\;dt$$

Letting $B^2 = 2D$ we find that $\mathcal{E}(r^2) = 4Dt$ in two dimensions, as we should.

Now we turn to the equation for θ

$$d\theta = \frac{\partial \theta}{\partial x}dx + \frac{\partial \theta}{\partial y}dy + \frac{1}{2}\frac{\partial^2 \theta}{\partial x^2}dx^2 + \frac{1}{2}\frac{\partial^2 \theta}{\partial x \partial y}dxdy + \frac{1}{2}\frac{\partial^2 \theta}{\partial y^2}dy^2$$

By symmetry there cannot be any drift term—positive and negative changes in θ must be equally likely. Evaluating derivatives we find

$$d\theta = \frac{B}{r^2}(-y dW_x + x dW_y)$$

Using (5.12) again we can replace $-y dW_x + x dW_y$ with $r\, dW_\theta$

$$d\theta = \frac{B}{r}\, dW_\theta \tag{5.25}$$

□

Example 5.2: Average properties from sampling

Often we are interested in an "average" property of the model rather than a single realization of the stochastic equation. Consider again the random walk model of the diffusion process on the plane, (5.22)-(5.23). Simulate the process and compute an estimate of the mean square displacement versus time.

Solution

We approximate this process for simulation with the discrete process

$$X(k+1) = X(k) + V\Delta t + \sqrt{2D\Delta t}P \tag{5.26}$$

where $X = (x, y)^T$, k is the sample number, Δt is the sample time, and time is $t = k\Delta t$. The velocity of the particles is $V = (v_x, v_y)^T$ and the random two-vector P is the two-dimensional normal distribution with zero mean and covariance equal to a 2×2 identity matrix

$$P \sim N(0, I)$$

This choice provides uncorrelated steps in the x and y directions. In the ensuing discussion we choose $\Delta t = 1$ so $k = t$. We also take $v_x = v_y = 0$ here so there is no drift, only diffusion. A representative simulation of (5.26) is given in Figure 5.3.

We can approximate average properties by simulating many trajectories or equivalently many independent particles, and then taking the average. Let $X_i(k)$ be the position of the ith particle at sample time k, which follows the evolution

$$X_i(k+1) = X_i(k) + \sqrt{2D}n_i \tag{5.27}$$

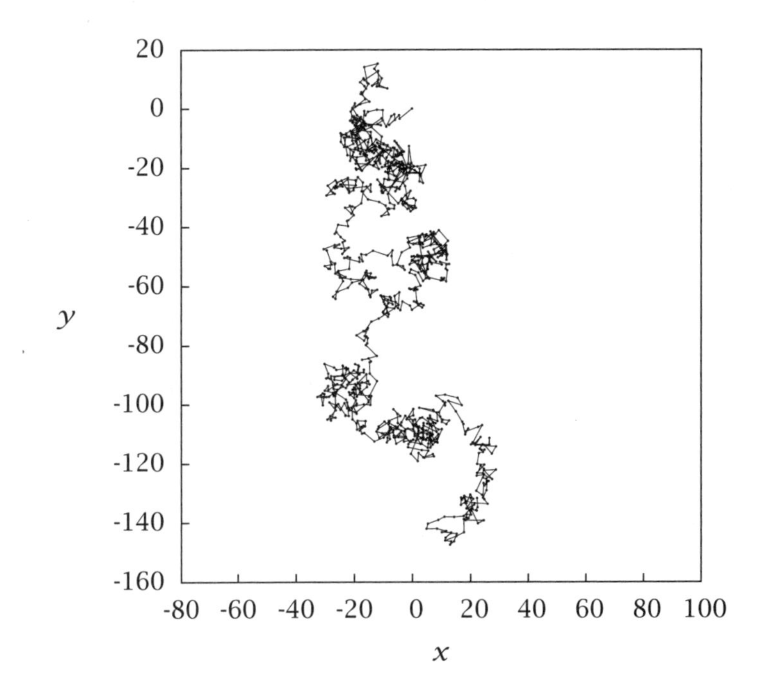

Figure 5.3: A representative trajectory of the discretely sampled Brownian motion; $D = 2$, $V = 0$, $n = 500$.

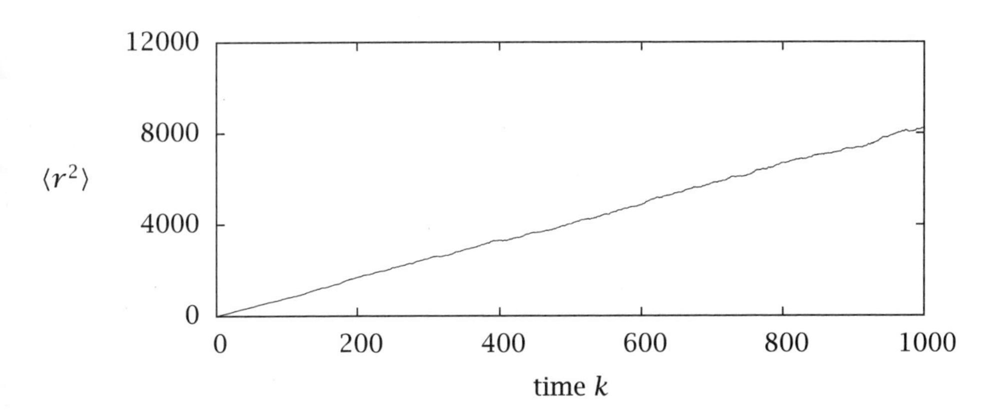

Figure 5.4: The mean square displacement versus time; $D = 2$, $V = 0$, $n = 500$.

The squared displacement of the ith particle is given by

$$r_i^2(k) = X_i^T(k)X_i(k) \tag{5.28}$$

and the mean square displacement is given by the average over many particles

$$\langle r^2 \rangle(k) \approx \frac{1}{n}\sum_{i=1}^{n} r_i^2(k) \qquad n \text{ large}$$

Figure 5.4 shows the mean square displacement for the random walk with no drift and $D = 2$ for the diffusion coefficient. We use $n = 500$ particles for this simulation. Notice that the mean square displacement grows linearly with time. The simulation agrees with Einstein's analysis of diffusion (Einstein, 1905). See also (Gardiner, 1990, pp.3–5) as well as our analyses above

$$\langle r^2 \rangle(k) = 4Dk \tag{5.29}$$

□

5.2.4 Fokker-Planck Equation

There are two ways to think about solving an SDE. We can find particular trajectories—this is what the Euler-Maruyama scheme above will do. We can also consider the evolution of the probability density $p(x,t)$. In considering the integrated white-noise and Wiener processes, we observed the connection between the evolution of $p(x,t)$ and the diffusion equation. The Wiener process is the solution to $dx = dW$. Because its trajectories $x(t) - x(0) \sim N(0,t)$, the density $p(x,t)$ for a trajectory starting at $x = x_0$ is a solution to the transient diffusion equation

$$\frac{\partial p}{\partial t} = D\frac{\partial^2 p}{\partial x^2}, \qquad p(0,t) = \delta(x - x_0) \tag{5.30}$$

with $D = \frac{1}{2}$. To generalize this result, consider the time evolution of the expected value of an arbitrary function $f(x(t))$, where $x(t)$ evolves according to the Itô SDE (5.16). Using Itô's formula and the result $\mathcal{E}(\mathcal{B}f'\,dW) = 0$, which is the infinitesimal version of (5.19)

$$\begin{aligned}
\mathcal{E}(df(x(t))) &= \mathcal{E}\left((\mathcal{A}f' + \frac{1}{2}\mathcal{B}^2 f'')\,dt + \mathcal{B}f'\,dW\right) \\
&= \mathcal{E}\left((\mathcal{A}f' + \frac{1}{2}\mathcal{B}^2 f'')\,dt\right)
\end{aligned}$$

This can be rewritten as

$$\frac{d}{dt}\int f(x)p(x,t)\,dx = \int (\mathcal{A}f' + \frac{1}{2}\mathcal{B}^2 f'')p(x,t)\,dx$$

Rearranging and integrating by parts yields

$$\int f(x)\frac{\partial p(x,t)}{\partial t}\,dx = \int f(x)\left(\frac{\partial}{\partial x}(-\mathcal{A}p(x,t)) + \frac{1}{2}\frac{\partial}{\partial x^2}\left(\mathcal{B}^2 p(x,t)\right)\right)dx$$

Finally, since f is arbitrary, this result can only hold in general if

$$\frac{\partial p(x,t)}{\partial t} = \frac{\partial}{\partial x}(-\mathcal{A}(x,t)p(x,t)) + \frac{\partial^2}{\partial x^2}\left(\frac{1}{2}\mathcal{B}^2(x,t)p(x,t)\right) \tag{5.31}$$

This is the evolution equation for $p(x,t)$, often called the FOKKER-PLANCK EQUATION (FPE). For a trajectory starting at $x = x_0$, the initial condition for this equation is again $p(0,t) = \delta(x - x_0)$. The equation can be put into conservation form

$$\frac{\partial p(x,t)}{\partial t} = -\frac{\partial}{\partial x}\left\{\mathcal{A}(x,t)p(x,t) - \frac{\partial}{\partial x}\left(\frac{1}{2}\mathcal{B}^2(x,t)p(x,t)\right)\right\} \tag{5.32}$$

The term inside the curly brackets is the flux of probability density and this equation bears obvious similarities to equations we are familiar with from transport phenomena. It shows us that trajectories of an Itô SDE have a drift coefficient $\mathcal{A}(x,t)$ and a diffusion coefficient $\mathcal{D}(x,t) = \frac{1}{2}\mathcal{B}^2(x,t)$. This is sometimes called the "short-time" diffusivity, because one can show using Itô's formula (Exercise 5.5) that for particle at position x' at time t'

$$\mathcal{E}\left(\frac{d(x-x')^2}{dt}\right)\bigg|_{t=t'} = 2\mathcal{D}(x',t') \tag{5.33}$$

Similarly, the instantaneous drift velocity of the trajectory is (as in the deterministic case)

$$\mathcal{E}\left(\frac{d(x-x')}{dt}\right)\bigg|_{t=t'} = \mathcal{A}(x',t') \tag{5.34}$$

The probability density must integrate to unity

$$\int p(x,t)\,dx = 1 \tag{5.35}$$

Elaborating on the connection between the FPE and transport equations, we recall that the convection-diffusion equation in one dimension is

$$\frac{\partial c}{\partial t} = -\frac{\partial}{\partial x}\left\{v(x)c - D\frac{\partial}{\partial x}c\right\}$$

The first term in the flux expression inside the curly brackets is analogous to that for the FPE, with $\mathcal{A} = v$, but there is an important difference in the second term. When the (short-time) diffusivity $\mathcal{D} = \frac{1}{2}\mathcal{B}^2$ in the FPE varies with position, it is not equivalent to the (gradient) diffusivity D that appears in the transport equation. Exercise 5.2 explores these differences in further detail.

We also can generalize the analysis to an n-vector random process x, with components x_i, $i = 1, 2, \ldots, n$. The SDE and FPEs for this case are

$$dx_i = \mathcal{A}_i(x,t)dt + \mathcal{B}_{ij}(x,t)\, dW_j \tag{5.36}$$

$$\frac{\partial}{\partial t}p = -\sum_{i=1}^{n}\frac{\partial}{\partial x_i}(\mathcal{A}_i(x,t)p) + \sum_{i=1}^{n}\sum_{j=1}^{n}\frac{\partial^2}{\partial x_i \partial x_j}(\mathcal{D}_{ij}(x,t)p) \tag{5.37}$$

Here p is a function of all components x_i and time, $p = p(x_1, \ldots, x_n, t)$, and $\mathcal{D}_{ij} = \frac{1}{2}B_{ik}B_{jk}$ are the elements of the diffusion coefficient *matrix*. The derivation of (5.37) from (5.36) makes use of the MULTIDIMENSIONAL ITÔ FORMULA

$$df(x) = \left(\mathcal{A}_i\frac{\partial f}{\partial x_i} + \frac{1}{2}\mathcal{B}_{ik}\mathcal{B}_{jk}\frac{\partial}{\partial x_i}\frac{\partial}{\partial x_j}f\right) dt + B_{ij}\frac{\partial f}{\partial x_i}dW_j \tag{5.38}$$

As in the scalar case, probability is conserved

$$\int p(x_1, x_2, \ldots, x_n)\, dx_1\, dx_2\, \ldots\, dx_n = 1 \tag{5.39}$$

In vector/matrix notation the equations are written

$$d\boldsymbol{x} = \boldsymbol{\mathcal{A}}\cdot\boldsymbol{x}(t)dt + \boldsymbol{\mathcal{B}}(\boldsymbol{x},t)\cdot d\boldsymbol{W} \tag{5.40}$$

$$\frac{\partial}{\partial t}p(\boldsymbol{x},t) = -\boldsymbol{\nabla}\cdot(\boldsymbol{\mathcal{A}}(\mathbf{x},t)p(\boldsymbol{x},t)) + \boldsymbol{\nabla\nabla}\colon(\boldsymbol{\mathcal{D}}(\boldsymbol{x},t)p(\boldsymbol{x},t)) \tag{5.41}$$

with

$$\boldsymbol{\mathcal{D}}(\boldsymbol{x},t) = \frac{1}{2}\boldsymbol{\mathcal{B}}\cdot\boldsymbol{\mathcal{B}}^T \tag{5.42}$$

This result indicates that $\boldsymbol{\mathcal{D}}$ is symmetric positive semidefinite. For numerical integration of multidimensional SDEs, the Euler-Maruyama scheme extends straightforwardly.

Example 5.3: Transport of many particles suspended in a fluid

A large number of particles, each obeying the equation

$$dx = v\,dt + \sqrt{2D}\,dW \tag{5.43}$$

are moving in a fluid. How do we describe the evolution of the concentration field in the fluid?

Solution

The probability density for an individual particle evolves as

$$p_t = -v p_x + D p_{xx}$$

For the many-particle system, we define an n-particle joint density function (n is on the order of Avogadro's number) as

$p(x_1, x_2, \ldots, x_n, t)dx_1 dx_2 \cdots dx_n$
probability density that particles 1 through n are located at x_1 through x_n, respectively, at time t

The concentration of particles at x, $c(x,t)$, is then

$$c(x,t) = \sum_{j=1}^{n} \int_{\Omega} p(x_1, \ldots, x_j, \ldots, x_n)\delta(x_i - x_j) \prod_{i=1}^{n} dx_i \tag{5.44}$$

The jth term in the sum represents the probability that the jth particle is located at x at time t, and the sum over all particles gives the total concentration. If the particle motions are independent

$$p(x_1, \ldots, x_n; t) = \prod_{i=1}^{n} p_i(x_i; t)$$

Performing the integral in (5.44) gives

$$c(x,t) = \sum_{j=1}^{n} p_j(x,t)$$

which indicates that the linear superposition of each particle's probability of being at location x produces the total concentration at x. If the particles are identical, $p_j(x,t) = p(x,t), j = 1, \ldots, n$, this reduces to

$$c(x,t) = np(x,t)$$

The evolution equation for c is therefore

$$c_t(x,t) = -v c_x(x,t) + D c_{xx}(x,t)$$

The conclusion is that the concentration profile created by many non-interacting, identical particles obeys the same evolution equation as the probability density of a single particle. Averaging the behavior of many particles does not "average away" the diffusion term in the evolution equation of the total concentration $c(x,t)$. See Deen (1998, pp. 59-63) for further discussion of this case. □

Example 5.4: Fokker-Planck equations for diffusion on a plane

Example 5.1 introduced the stochastic differential equations for diffusion on a plane in Cartesian and polar coordinate representations. For the Cartesian representation, (5.22) and (5.23) have probability density $p(x,y)$ that satisfies the diffusion equation

$$\frac{\partial p(x,y)}{\partial t} = D\left(\frac{\partial^2}{\partial x^2} + \frac{\partial^2}{\partial x^y}\right) p(x,y) = D\nabla^2 p(x,y)$$

with normalization (conservation of probability) condition

$$\iint_{-\infty}^{\infty} p(x,y)\, dx\, dy = 1$$

If we rewrite this equation in polar coordinates we get

$$\frac{\partial p(r,\theta)}{\partial t} = D\left(\frac{1}{r}\frac{\partial}{\partial r}\left(r\frac{\partial}{\partial r}\right) + \frac{1}{r^2}\frac{\partial^2}{\partial \theta^2}\right) p(r,\theta) = D\nabla^2 p(r,\theta) \quad (5.45)$$

and

$$\int_0^{2\pi}\int_0^{\infty} p(r,\theta) r\, dr\, d\theta = 1$$

Do we get the same result if we start with the polar coordinate form of the stochastic differential equations, (5.24) and (5.25)? Why or why not?

Solution

Equations (5.24) and (5.25) can be written as the system

$$\begin{bmatrix} dr \\ d\theta \end{bmatrix} = \begin{bmatrix} \frac{D}{r} \\ 0 \end{bmatrix} dt + \begin{bmatrix} B & 0 \\ 0 & \frac{B}{r} \end{bmatrix} \begin{bmatrix} dW_r \\ dW_\theta \end{bmatrix}$$

With regard to (5.36) and (5.37), $x_1 = r, x_2 = \theta$ and

$$\mathcal{A} = \begin{bmatrix} \frac{D}{r} \\ 0 \end{bmatrix}$$

$$\mathcal{D} = \frac{1}{2}\begin{bmatrix} B & 0 \\ 0 & \frac{B}{r} \end{bmatrix}\begin{bmatrix} B & 0 \\ 0 & \frac{B}{r} \end{bmatrix}^T = \begin{bmatrix} D & 0 \\ 0 & \frac{D}{r^2} \end{bmatrix}$$

Inserting these expressions into (5.37) and denoting the probability density as $p_P(r, \theta)$ yields

$$\frac{\partial p_P(r,\theta)}{\partial t} = -D\frac{\partial}{\partial r}\frac{p_P(r,\theta)}{r} + D\left(\frac{\partial^2}{\partial r^2} + \frac{1}{r^2}\frac{\partial^2}{\partial \theta^2}\right) p_P(r,\theta) \tag{5.46}$$

This is *not* the transient diffusion equation in polar coordinates.

We begin to understand this difference by writing the normalization condition, (5.39)

$$\int_0^{2\pi}\int_0^{\infty} p_P(r,\theta)\, dr\, d\theta$$

This differs by a factor of r in the integrand from the conventional area integral in polar coordinates. The reason is simple: in going from the SDE to the FPE, we did not tell Itô's formula about the geometry of area elements on the plane, but only to take an SDE written with variables $x_1 = r, x_2 = \theta$ and write the corresponding FPE. There is no paradox here, only a message to be careful about coordinate transformations.

Finally, we wish to understand the relationship between p and p_P. Motivated by the factor of r difference in the normalization conditions, we might guess that $p_P(r,\theta) = crp(r,\theta)$ where c is a constant. Indeed, making this substitution into (5.46), we recover the transient diffusion equation in polar coordinates, (5.45). For a process starting at the origin at $t = 0$, the normalized solutions (Exercise 5.9) are

$$p(r,\theta,t) = \frac{1}{4\pi Dt}\, e^{-r^2/(4Dt)}$$

and

$$p_P(r,\theta,t) = rp(r,\theta,t)$$

□

5.3 Stochastic Kinetics

5.3.1 Introduction, and Length and Time Scales

Our next application of interest is reaction networks and chemical kinetics taking place at small numbers of molecules. First we start with a continuum kinetics example to define some useful nomenclature. Consider the following two-step series reaction

$$\mathrm{A} \xrightarrow{k_1} \mathrm{B} \qquad \mathrm{B} \xrightarrow{k_2} \mathrm{C}$$

We define the species vector of concentrations $c = \begin{bmatrix} c_A & c_B & c_C \end{bmatrix}^T$, and denote the stoichiometry for the reaction network with the stoichiometric matrix

$$\nu = \begin{bmatrix} -1 & 1 & 0 \\ 0 & -1 & 1 \end{bmatrix}$$

We let ν_i, $i = 1, 2, \ldots, n_r$ denote the rows of the stoichiometric matrix, written as column vectors

$$\nu_1 = \begin{bmatrix} -1 \\ 1 \\ 0 \end{bmatrix} \qquad \nu_2 = \begin{bmatrix} 0 \\ -1 \\ 1 \end{bmatrix}$$

We assume the reaction takes place in a well-mixed reactor and assume some rate law for the reaction kinetics, such as

$$r_1 = k_1 c_A \qquad r_2 = k_2 c_B \qquad r = \begin{bmatrix} r_1 \\ r_2 \end{bmatrix}$$

As taught in every undergraduate chemical engineering curriculum, the material balances for the three species is then given by

$$\frac{d}{dt} c = \nu^T r(c) = \sum_{i=1}^{n_r} \nu_i r_i(c) \tag{5.47}$$

The solution of this model with a pure reactant A initial condition is shown in Figure 5.5.

Next we consider reactions taking place at small concentrations. Instead of the common case in which we have on the order of Avogadro's number of reacting molecules, assume we have only tens or hundreds of molecules moving randomly in a constant-volume, well-mixed, reactor. At such low concentrations, the deterministic concentration assumption makes no sense, and we have to consider the random behavior of the molecules. But we still have to choose an appropriate length and time scale of interest. Indeed, if we move down to the length scale of the atoms, we can model the electron bonds deforming continuously in time from reactants through transition states to products. We choose instead a larger time and length scale so that each reaction that

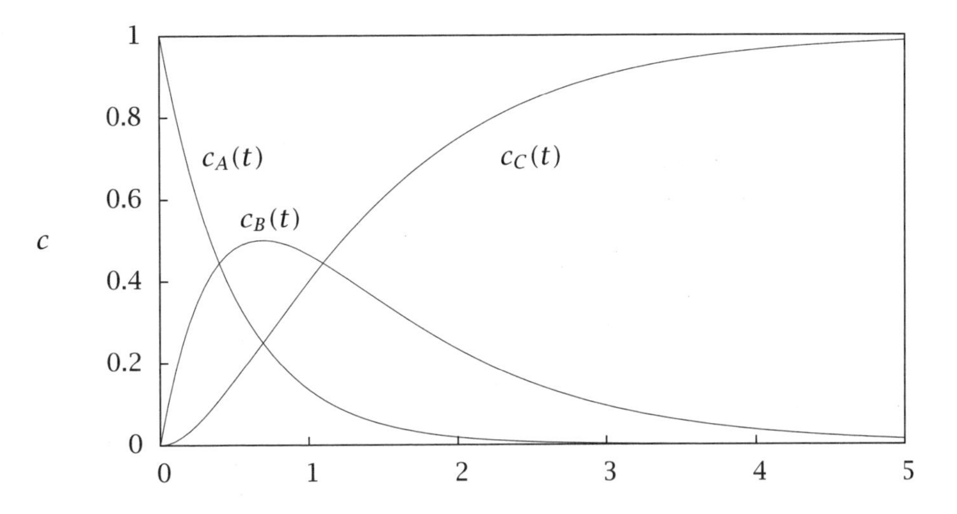

Figure 5.5: Two first-order reactions in series in a batch reactor, $c_{A0} = 1, c_{B0} = c_{C0} = 0, k_1 = 2, k_2 = 1$.

takes place can be regarded as a single instantaneous event causing a discrete change in the number of reactants and products. At this scale, we track the integer-valued numbers of reactant and products, and we treat the reaction events as random jump processes. This choice of length and time scale makes the discrete Poisson process the natural description for stochastic kinetics.

5.3.2 Poisson Process

Just as the Wiener process $W(t)$ is the simplest mathematical process appropriate for modeling diffusion, the Poisson process $Y(t)$ is the simplest mathematical process appropriate for modeling stochastic chemical kinetics. The POISSON PROCESS is an integer-valued counting process. Time is modeled as a continuous variable, but the value of the Poisson process is discrete. The Poisson process is characterized by a rate parameter, $\lambda > 0$, and for small time interval Δt, the probability of an event taking place in this time interval is proportional to $\lambda \Delta t$. To start off, we assume that parameter λ is constant. The probability that an event does *not* take place in the interval $[0, \Delta t]$ is therefore approximately $1 - \lambda \Delta t$. Let random variable τ be the time of the first event of the Poisson process starting from $t = 0$. We then have for small Δt

$$\Pr(\tau > \Delta t) \approx 1 - \lambda \Delta t$$

Like the Wiener process, the Poisson process has independent increments, which means that the number of events in disjoint time intervals are independent. The independent increment assumption coupled with the fact that λ does not change implies that the probability that an event does not take place in two consecutive time intervals $[0, 2\Delta t]$ is $\Pr(\tau > 2\Delta t) \approx (1 - \lambda\Delta t)^2$. Continuing this argument to n intervals gives for $t = n\Delta t$

$$\Pr(\tau > t) \approx (1 - \lambda\Delta t)^n \approx (1 - \lambda\Delta t)^{t/\Delta t}$$

Taking the limit as $\Delta t \to 0$ gives

$$\Pr(\tau > t) = e^{-\lambda t}$$

From the probability axioms and the definition of τ's probability distribution, we than have

$$\Pr(\tau \leq t) = F_\tau(t) = 1 - e^{-\lambda t}$$

Differentiating to obtain the density gives the exponential density

$$p_\tau(t) = \lambda e^{-\lambda t} \tag{5.48}$$

The exponential distribution should be familiar to chemical and biological engineers because of the residence-time distribution of a well-mixed tank. The residence-time distribution of the CSTR with volume V and volumetric flowrate Q satisfies (5.48) with λ being the dilution rate or inverse mean residence time, $\lambda = V/Q$.

Figure 5.6 shows a simulation of the *unit* Poisson process, i.e., the Poisson process with $\lambda = 1$. If we count many events, the sample path looks like the top of Figure 5.7, which resembles a "bumpy" line with slope equal to λ, unity in this case. The frequency count of the times to next event, τ, are shown in the bottom of Figure 5.7, and we can clearly see the exponential distribution with this many events. Note that to generate a sample of the exponential distribution for the purposes of simulation, one can simply take the negative of the logarithm of a uniformly distributed variable on $[0, 1]$. Most computational languages provide functions to give pseudorandom numbers following a uniform distribution, so it is easy to produce samples from the exponential distribution as well. See Exercise 5.14 for further discussion.

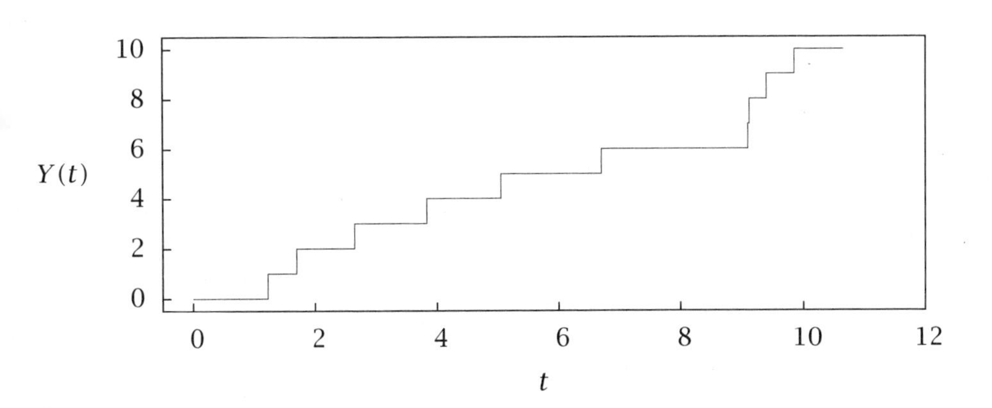

Figure 5.6: A sample path of the unit Poisson process.

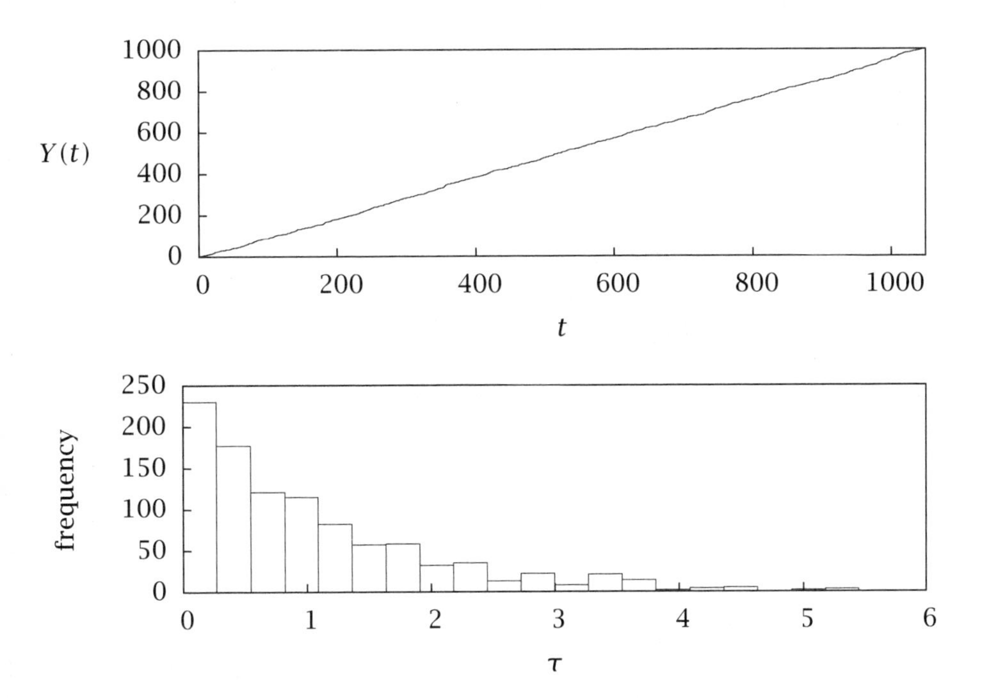

Figure 5.7: A unit Poisson process with more events; sample path (top) and frequency distribution of event times τ.

The time of the first event also characterizes the probability $\Pr(Y(t) = 0)$ for $t \geq 0$. The probability that Y is still zero at time t is the same as the probability that the first event has occurred at some time greater than t, or $\Pr(Y(t) = 0) = \Pr(\tau > t) = 1 - \Pr(\tau \leq t)$. Therefore we have the relationships

$$\Pr(Y(t) = 0) = 1 - F_\tau(t) = e^{-\lambda t}$$

We next generalize the discussion to find the probability density for the time of the second and subsequent events. Let random variable τ_2 denote the time of the second event. We wish to compute the density $p_{\tau_2,\tau}(t_2, t)$. Because of the independent increments property and the fact that λ is constant, we have for the joint density

$$p_{\tau_2,\tau}(t_2,t) = \begin{cases} p_\tau(t_2 - t)p_\tau(t), & t_2 > t \\ 0 & t_2 \leq t_1 \end{cases}$$

Integrating the joint density gives the marginal

$$\begin{aligned} p_{\tau_2}(t_2) &= \int_0^{t_2} p_\tau(t_2 - t)p_\tau(t)dt \\ &= \int_0^{t_2} \lambda e^{-\lambda(t_2-t)}\lambda e^{-\lambda t}dt \\ p_{\tau_2}(t_2) &= \lambda^2 t_2 e^{-\lambda t_2} \end{aligned}$$

or $p_{\tau_2}(t) = \lambda^2 t e^{-\lambda t}$. We can then use induction to obtain the density of the time for the nth event, $n > 2$. Assuming that τ_{n-1} has density $\lambda^{n-1}t^{n-2}e^{-\lambda t}/(n-2)!$, we have for τ_n

$$\begin{aligned} p_{\tau_n}(t_n) &= \int_0^{t_n} p_{\tau_n,\tau_{n-1}}(t_n,t)p_{\tau_{n-1}}(t)dt \\ &= \int_0^{t_n} p_\tau(t_n - t)\frac{\lambda^{n-1}}{(n-2)!}t^{n-2}e^{-\lambda t}dt \\ &= \int_0^{t_n} \lambda e^{-\lambda(t_n-t)}\frac{\lambda^{n-1}}{(n-2)!}t^{n-2}e^{-\lambda t}dt \\ p_{\tau_n}(t_n) &= \frac{\lambda^n}{(n-1)!}t_n^{n-1}e^{-\lambda t_n} \end{aligned} \tag{5.49}$$

From here we can work out $\Pr(Y(t) = n)$ for any n. For $Y(t)$ to be n at time t, we must have time $\tau_n \leq t$ and time $\tau_{n+1} > t$, i.e., n events have occurred by time t but $n + 1$ have not. In terms of the joint density, we have

$$\Pr(Y(t) = n) = \int_t^\infty \int_0^t p_{\tau_{n+1},\tau_n}(t',t)dtdt'$$

As before, the independent increments property allows us to express the joint density as $p_{T_{n+1},T_n}(t',t) = p_T(t'-t)p_{T_n}(t)$ for $t' \geq t$. Substituting this and (5.49) into the previous equation gives

$$\Pr(Y(t)=n) = \int_t^\infty \int_0^t \lambda e^{-\lambda(t'-t)} \frac{\lambda^n}{(n-1)!} t^{n-1} e^{-\lambda t} dt dt'$$

$$\Pr(Y(t)=n) = \frac{(\lambda t)^n}{n!} e^{-\lambda t} \tag{5.50}$$

See Exercise 5.13 for an alternative derivation. The discrete density appearing on the right-hand side of (5.50), i.e, $p(n) = e^{-a}a^n/n!$ with parameter $a = \lambda t$, is known as the Poisson density. Its mean and variance are equal to a (see Exercise 5.12). So we have that $\mathcal{E}(Y(t)) = \lambda t$, which is consistent with Figure 5.7.

Because λ and t appear only as the product λt, the Poisson process with intensity λ, now denoted $Y_\lambda(t)$, can be expressed in terms of the unit Poisson process, denoted $Y(t)$, with the relation

$$Y_\lambda(t) = Y(\lambda t) \qquad t \geq 0$$

The justification is as follows. We have just shown

$$\Pr(Y_\lambda(t) = n) = \frac{(\lambda t)^n}{n!} e^{-\lambda t}$$

and, for the unit Poisson process, we have $\Pr(Y(t) = n) = t^n e^{-t}/n!$, which is equivalent on the substitution of λt for t. Because the increments are independent, we also have the property for all $n \geq 0$

$$\Pr(Y(t) - Y(s) = n) = \Pr(Y(t-s) = n) \qquad t \geq s$$

which is similar to (5.5) for the Wiener process.

Nonhomogeneous Poisson process. Next we consider the *nonhomogeneous* Poisson process in which the intensity $\lambda(t)$ is time varying. We define the Poisson process for this more general case so that the probability of an event during time interval $[t, t+\Delta t]$ is proportional to $\lambda(t)\Delta t$ for Δt small. We can express the nonhomogeneous process also in terms of a unit Poisson process with the relation

$$Y_\lambda(t) = Y\Big(\int_0^t \lambda(s)ds\Big) \qquad t \geq 0$$

To see that the right-hand side has the required property, we compute the probability that an event occurs in the interval $[t, t+\Delta t]$. Let $z(t) =$

$\int_0^t \lambda(s)ds$. We have

$$\begin{aligned}
\Pr(t \leq \tau \leq t+\Delta t) &= \Pr(Y_\lambda(t+\Delta t) - Y_\lambda(t) > 0) \\
&= \Pr(Y(z(t+\Delta t) - Y(z(t))) > 0) \\
&= \Pr(Y(z(t+\Delta t) - z(t)) > 0) \\
&= 1 - \Pr(Y(\int_t^{t+\Delta t} \lambda(s)ds) = 0) \\
&= 1 - e^{-\int_t^{t+\Delta t} \lambda(s)ds}
\end{aligned}$$

For Δt small, we can approximate the integral as $\int_t^{t+\Delta t} \lambda(s)ds \approx \lambda(t)\Delta t$ giving

$$\Pr(t \leq \tau \leq t+\Delta t) \approx 1 - (1 - \lambda(t)\Delta t) = \lambda(t)\Delta t$$

and we have the stipulated probability.

Random time change representation of stochastic kinetics. With these results, we can now express the stochastic kinetics problem in terms of the Poisson process. Assume n_r reactions take place between n_s chemical species with stoichiometric matrix $\nu \in \mathbb{R}^{n_r \times n_s}$, and denote its row vectors, written as columns, by $\nu_i, i = 1, 2, \ldots, r$. Let $X(t) \in \mathbb{I}^{n_s}$ be an integer-valued random variable vector of the chemical species numbers, and let $r_i(X), i = 1, 2, \ldots, n_r$ be the kinetic rate expressions for the n_r reactions. We assign to each reaction an independent Poisson process Y_i with intensity r_i. Note that this assignment gives n_r nonhomogeneous Poisson processes because the species numbers change with time, i.e., $r_i = r_i(X(t))$. The Poisson processes then count the number of times that each reaction fires as a function of time. Thus the Poisson process provides the *extents* of the reactions versus time. From these extents, it is a simple matter to compute the species numbers from the stoichiometry. We have that

$$\boxed{X(t) = X(0) + \sum_{i=1}^{n_r} \nu_i Y_i(\int_0^t r_i(X(s))ds)} \tag{5.51}$$

This is the celebrated random time change representation of stochastic kinetics due to Kurtz (1972).

Notice that this representation of the species numbers has $X(t)$ appearing on both sides of the equation. This integral equation representation of the solution leads to many useful solution properties and simulation algorithms. We can express the analogous integral equation for the deterministic continuum mass balance given in (5.47)

$$c(t) = c(0) + \sum_{i=1}^{n_r} \nu_i \int_0^t r_i(c(s))ds$$

Comparing the two results, we see the obvious similarities; the key differences are that the species number vector X is an integer-valued random variable, and the Poisson process Y fires the reactions at random times.

5.3.3 Stochastic Simulation

The random time change representation suggests a natural simulation or sampling strategy for the species numbers $X(t)$. We start with a chosen or known initial condition, $X(0)$. We then select based on each reaction, n_r exponentially distributed proposed times for the next reactions, $\tau_i, i = 1, 2, \ldots, n_r$. These exponential distributions have intensities equal to the different reaction rates, $r_i(X(0))$. As mentioned previously, we obtain a sample of an exponential $F_{\tau_i}(t) = 1 - e^{-r_i t}$ by drawing a sample of a uniformly distributed RV on $[0, 1]$, u, and rescaling the logarithm

$$\tau_i = -(1/r_i) \ln u_i \quad i = 1, 2, \ldots, n_r$$

We then select the reaction with the smallest event time as the reaction to fire, giving

$$t_1 = \min_{i \in [1, n_r]} \tau_i \qquad i_1 = \arg \min_{i \in [1, n_r]} \tau_i$$

We then update the species numbers at the chosen reaction time with the stoichiometric coefficients of the reaction that fires

$$X(t_1) = X(0) + \nu_{i_1}$$

This process is then repeated to provide a simulation over the time interval of interest. This simulation strategy is known as the FIRST REACTION METHOD (Gillespie, 1977). We summarize the first reaction method with the following algorithm.

Algorithm 5.5 (First reaction method).

Require: Stoichiometric matrix and reaction-rate expressions, $\nu_i, r_i(X)$, $i = 1, 2, \ldots, n_r$; initial species numbers, X_0; stopping time T.

1: Initialize time $t = 0$, time index $k = 1$, and species numbers $X(t) = X_0$.

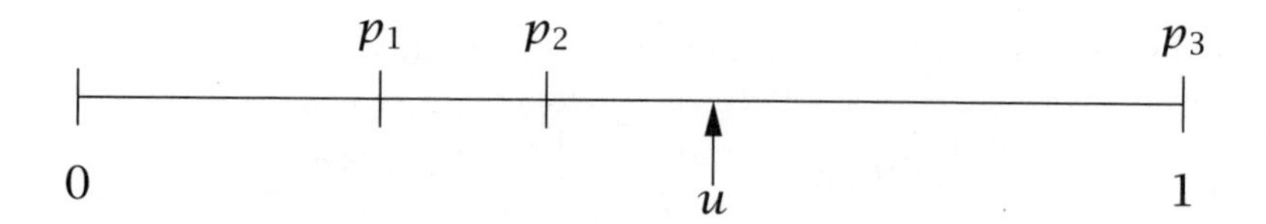

Figure 5.8: Randomly choosing a reaction with appropriate probability. The interval is partitioned according to the relative sizes of the reaction rates. A uniform random number u is generated to determine the reaction. In this case, since $p_2 \le u \le p_3$, $m = 3$ and the third reaction is selected.

2: Evaluate rates $r_i = r_i(X(t))$. If $r_i = 0$, all i, exit (system is at steady state.)
3: Choose n_r independent samples of a uniformly distributed RV, u_i. Compute random times for each reaction $\tau_i = (1/r_i)\ln u_i$.
4: Select smallest time and corresponding reaction, $\overline{\tau}_k = \min_{i\in[1,n_r]} \tau_i$, $i_k = \arg\min_{i\in[1,n_r]} \tau_i$.
5: Update time and species numbers: $t_k = t + \overline{\tau}_k$, $X(t_k) = X(t) + \nu_{i_k}$.
6: Set $t = t_k$, replace $k \leftarrow k + 1$. If $t < T$, go to Step 2. Else exit.

Gibson and Bruck (2000) show how to conserve random numbers in this approach by saving the $n_r - 1$ random numbers that were *not* selected at the current iteration, and reusing them at the next iteration. With this modification, the method is termed the NEXT REACTION METHOD.

An alternative, and probably the most popular, simulation method was proposed also by Gillespie (1977, p. 2345). In this method, the reaction rates are added together to determine a total reaction rate $r = \sum_{i=1}^{n_r} r_i(X(t))$. The time to the next reaction is distributed as $p_\tau(t) = re^{-rt}$. So sampling this density provides the *time* of the next reaction, which we denote τ. To determine *which* reaction fires, the following cumulative sum is computed

$$p_i = \sum_{i=1}^{n_r} r_i/r, \qquad i = 0, 1, 2, \ldots n_r$$

Note that $0 = p_0 \le p_1 \le p_2 \le \cdots \le p_{n_r} = 1$, so the set of p_i are a partition of $[0, 1]$ as shown in Figure 5.8 for $n_r = 3$ reactions. The length of each interval indicates the relative rate of each of the n_r reactions. So to determine which reaction m fires, let u be a sample from the uniform distribution on $[0, 1]$, and determine the interval m in which u falls by the condition

$$p_{m-1} \leq u \leq p_m$$

Given the reaction that fires m and the time of the reaction τ, we then update the species numbers in the standard way

$$X(t + \tau) = X(t) + \nu_m$$

This method is known as Gillespie's DIRECT METHOD or simply the STOCHASTIC SIMULATION ALGORITHM (SSA). We summarize this method with the following algorithm.

Algorithm 5.6 (Gillespie's direct method or SSA).

Require: Stoichiometric matrix and reaction-rate expressions, $\nu_i, r_i(X)$, $i = 1, 2, \ldots, n_r$; initial species numbers, X_0; stopping time T.

1: Initialize time $t = 0$, time index $k = 1$, and species numbers $X(t) = X_0$.
2: Evaluate rates $r_i = r_i(X(t))$ and total rate $r = \sum_i r_i$. If $r = 0$, exit (system is at steady state.)
3: Choose two independent samples, u_1, u_2, of a uniformly distributed RV on $[0, 1]$. Compute time of next reaction $\tau = (1/r) \ln u_1$.
4: Select which reaction, i_k, as follows. Compute the cumulative sum, $p_i = \sum_{j=1}^{i} r_j / r$ for $i \in [0, n_r]$. Note $p_0 = 0$. Find index i_k such that $p_{i_k - 1} \leq u_2 \leq p_{i_k}$.
5: Update time and species numbers: $t_k = t + \tau$, $X(t_k) = X(t) + \nu_{i_k}$.
6: Set $t = t_k$, replace $k \leftarrow k + 1$. If $t < T$, go to Step 2. Else exit.

Figure 5.9 shows the results when starting with $n_A = 100$ molecules. Notice the random aspect of the simulation gives a rough appearance to the number of molecules versus time, which is quite unlike the deterministic simulation presented in Figure 5.5. Because the *number* of molecules is an integer, the simulation is discontinuous with jumps at the reaction event times. But in spite of the roughness, we can already make out the classic behavior of the series reaction: loss of starting material A, appearance and then disappearance of the intermediate species B, slow increase in final product C. Note also that Figure 5.9 is only *one* simulation or sample of the random process. Unlike the deterministic models, if we repeat this simulation, we obtain a different sequence of random numbers and a different simulation. To compute accurate expected or average behavior of the system, we perform many of these random simulations and then compute the sample averages of quantities we wish to report.

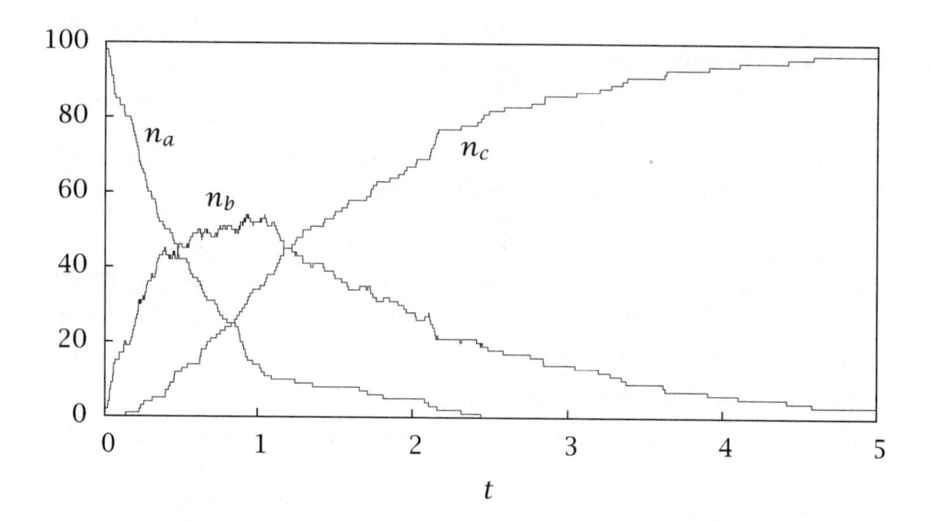

Figure 5.9: Stochastic simulation of the first-order series reaction $A \longrightarrow B \longrightarrow C$ starting with 100 A molecules.

5.3.4 Master Equation of Chemical Kinetics

The simulations in the previous section allow us to envision many possible simulation trajectories depending on the particular sequence of random numbers we have chosen. Some reflection leads us to consider instead modeling the evolution of the probability density of the state. We shall see that we can either solve this evolution equation directly, or average over many randomly chosen simulation trajectories to construct the probability density by brute force. Both approaches have merit, but here we focus on expressing and solving the evolution equation for the probability density.

Consider the reversible reaction

$$A + B \underset{k_{-1}}{\overset{k_1}{\rightleftharpoons}} C \tag{5.52}$$

taking place in a constant-volume, well-stirred reactor. Let $p(a, b, c, t)$ denote the probability density for the system to have a molecules of species A, b molecules of species B, and c molecules of species C at time t. We seek an evolution equation governing $p(a, b, c, t)$. The probability density evolves due to the chemical reactions given in (5.52). Consider the system state (a, b, c, t); if the forward event takes place, the system moves from state (a, b, c, t) to $(a-1, b-1, c+1, t+dt)$. If the reverse reaction event takes place the system moves from state (a, b, c, t) to $(a+1, b+1, c-1, t+dt)$. We have expressions for the rates of these two events

$$r_1 = k_1 ab \qquad r_{-1} = k_{-1} c$$

These are the rates required for the trajectory simulations of the previous sections. But here we are asking for more. Here we want to know how these reaction events occurring at these rates change the probability density that the system is in state (a, b, c, t). This evolution equation for the probability density is known as the master equation for chemical kinetics. The master equation for this chemical example system is

$$\begin{aligned} \frac{\partial p(a,b,c,t)}{\partial t} &= -(k_1 ab + k_{-1} c) \cdot p(a,b,c,t) \\ &\quad + k_1 (a+1)(b+1) \cdot p(a+1, b+1, c-1, t) \\ &\quad + k_{-1}(c+1) \cdot p(a-1, b-1, c+1, t) \end{aligned} \tag{5.53}$$

We see that the reaction rate for each event is multiplied by the probability density that the system is in that state.

Because we have a single reaction, we can simplify matters by defining ε to be the extent of the reaction. The numbers of molecules of each species are calculated from the initial numbers and reaction extent given the reaction stoichiometry

$$a = a_0 - \varepsilon \qquad b = b_0 - \varepsilon \qquad c = c_0 + \varepsilon$$

We see that $\varepsilon = 0$ corresponds to the initial state of the system. Using the reaction extent, we define $p(\varepsilon, t)$ to be the probability density that the system has reaction extent ε at time t. Converting (5.53) we obtain

$$\begin{aligned} \frac{\partial p(\varepsilon,t)}{\partial t} &= -(k_1(a_0-\varepsilon)(b_0-\varepsilon) - k_{-1}(c_0+\varepsilon)) \cdot p(\varepsilon,t) \\ &\quad + k_1(a_0-\varepsilon+1)(b_0-\varepsilon+1) \cdot p(\varepsilon-1,t) \\ &\quad + k_{-1}(c_0+\varepsilon+1) \cdot p(\varepsilon+1,t) \end{aligned} \tag{5.54}$$

The four terms in the master equation are depicted in Figure 5.10. Given a_0, b_0, c_0 we can calculate the range of possible extents. For simplicity, assume we start with only reactants A and B so $c_0 = 0$. Then the minimum extent is $\varepsilon = 0$ because we cannot fire the reverse reaction from this starting condition. If we fire the forward reaction $n = \min(n_{A0}, n_{B0})$ times, the limiting reactant A or B is completely consumed and no further forward reactions are possible. Therefore the range of possible extents is $0 \leq \varepsilon \leq n$. We now can write $n + 1$ equations stemming from the master equation and place them in the matrix form

$$\frac{d}{dt}\begin{bmatrix} p_0 \\ p_1 \\ p_2 \\ \vdots \\ p_{n-1} \\ p_n \end{bmatrix} = \begin{bmatrix} \beta_0 & \gamma_0 & & & & \\ \alpha_1 & \beta_1 & \gamma_1 & & & \\ & \alpha_2 & \beta_2 & \gamma_2 & & \\ & & \ddots & \ddots & \ddots & \\ & & & & & \gamma_{n-1} \\ & & & & \alpha_n & \beta_n \end{bmatrix}\begin{bmatrix} p_0 \\ p_1 \\ p_2 \\ \vdots \\ p_{n-1} \\ p_n \end{bmatrix} \tag{5.55}$$

in which $p_j(t)$ is shorthand for $p(j,t)$, and α_j, β_j, and γ_j are the following rate expressions evaluated at different extents of the reaction

$$\begin{aligned} \alpha_j &= k_1(a_0 - j + 1)(b_0 - j + 1) \\ \beta_j &= -k_1(a_0 - j)(b_0 - j) - k_{-1}(c_0 + j) \\ \gamma_j &= k_{-1}(c_0 + j + 1) \end{aligned}$$

We can also write this model as

$$\begin{aligned} \frac{dP}{dt} &= AP \\ P(0) &= P_0 \end{aligned} \tag{5.56}$$

in which P is the column vector of probabilities for the different reaction extents

$$P = \begin{bmatrix} p_0 & p_1 & \cdots & p_n \end{bmatrix}^T$$

and the A matrix contains all the model parameters.

The essential connection between the stochastic and deterministic approaches to the well-mixed chemical kinetics problem is that the stochastic model's probability density becomes arbitrarily sharp at the solution to the deterministic problem as the number of molecules increases. Figure 5.11 displays the solution to (5.55) starting with 20 A molecules, 100 B molecules and 0 C molecules. The extent of reaction is scaled by the initial number of A molecules. Notice that the probability density spreads out rapidly as time increases and there is significant uncertainty in the equilibrium state.

If we increase the starting number of molecules by a factor of 10, we obtain the results depicted in Figure 5.12. Notice the sharpening

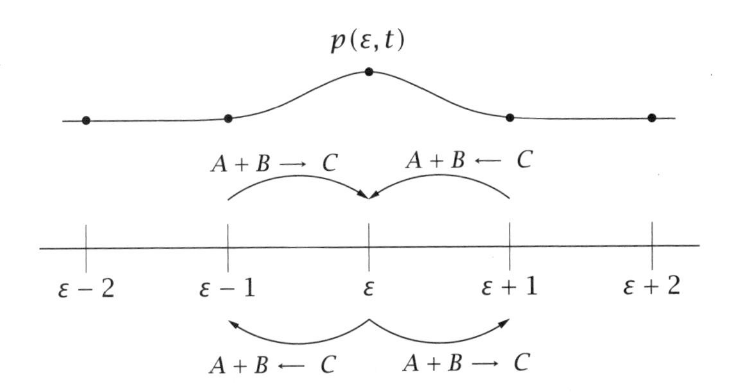

Figure 5.10: Master equation for chemical reaction A + B $\rightleftharpoons$ C. The probability density at state ε changes due to forward and reverse reaction events. The rate of change is proportional to the reaction rate times the probability density of being in that state.

in the probability density. We can see that the extent versus time is traced out by the peak in probability density is approaching the mass action kinetics limit. You can imagine the sharpness in the density if we started out with on the order of Avogadro's number of A molecules. As stressed earlier, however, if we are not operating near that limit, the random fluctuation may be an important physical behavior to include in the model. To describe this behavior, the stochastic approach is essential and the deterministic approach cannot be substituted.

The master equation, (5.56), is a simple *linear*, constant-coefficient differential equation, and the solution is

$$P(t) = e^{At} P_0$$

The challenge in solving the master equation directly is its *high dimension*. The dimension of P is the number of different species values that the system can reach by reaction. If we have a single reaction, the extent can range from zero, its initial value, to a value that exhausts some limiting species. Denote this limiting species's initial number by n_0, the dimension of the state vector P is then n_0. But if we have multiple reactions, we multiply n_r by the limiting species corresponding to all the combinations of reactions. The scaling is on the order of the product $n_0 n_r$. If we have 1000 initial molecules with 10 reactions, the dimension of the master equation P vector is already on the order of

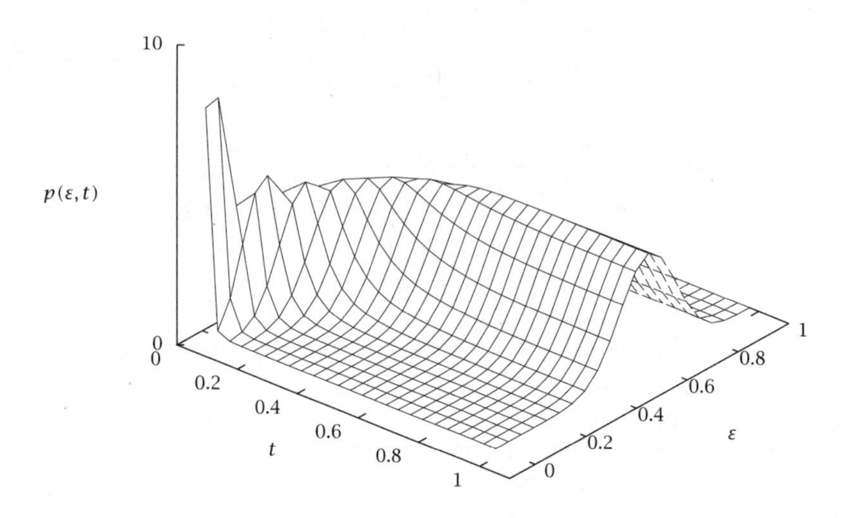

Figure 5.11: Solution to master equation for $A + B \rightleftharpoons C$ starting with 20 A molecules, 100 B molecules and 0 C molecules, $k_1 = 1/20, k_{-1} = 3$. Congratulations, you now understand what is displayed on the cover of the text.

10^4. The A matrix already contains 10^8 elements, although it would be quite sparse. Thus solving the master equation becomes computationally intractable for problems of even modest size. The best we can hope for with these larger models is to sample the master equation with simulations. Even simulating enough trajectories to obtain reliable sample averages can be quite time consuming, which motivates research efforts to develop *efficient* simulation algorithms and sampling strategies.

Given this basic understanding, we now express the general master equation for n_r reactions with the random variable X (species numbers) as the state of the system rather than the reaction extents. Given a system in state $x \in \mathbb{I}^{n_s}$, reaction i with stoichiometric vector ν_i can reach state x from only state $x - \nu_i$, and can leave this state to reach state $x + \nu_i$. We then have for the evolution of the probability density

$$\frac{d}{dt} p_X(x,t) = \sum_i r_i(x - \nu_i) p_X(x - \nu_i, t) - \Big(\sum_i r_i(x)\Big) p_X(x,t) \quad (5.57)$$

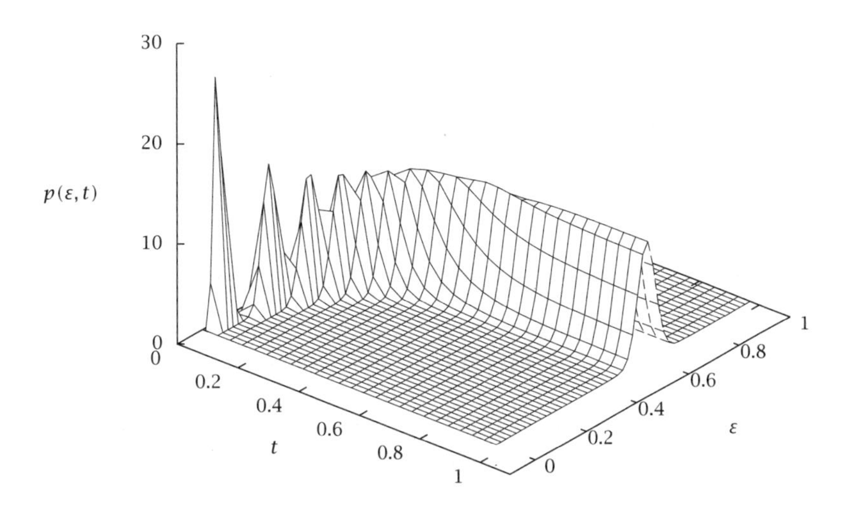

Figure 5.12: Solution to master equation for A + B $\rightleftharpoons$ C starting with 200 A molecules, 1000 B molecules and 0 C molecules, $k_1 = 1/200, k_{-1} = 3$.

with initial condition $p_X(x, 0) = p_0(x)$. Equation (5.57) is the CHEMICAL MASTER EQUATION for a general reaction network. It is also known as the forward Kolmogorov equation in the mathematics literature.

Applying (5.57) to the previous example we have

$$x = \begin{bmatrix} n_A \\ n_B \\ n_C \end{bmatrix} \quad \nu_1 = \begin{bmatrix} -1 \\ -1 \\ 1 \end{bmatrix} \quad \nu_{-1} = \begin{bmatrix} 1 \\ 1 \\ -1 \end{bmatrix} \quad r_1(x) = k_1 n_A \quad r_{-1}(x) = k_{-1} n_C$$

and master equation

$$\frac{d}{dt} p_X(x, t) = r_1(x - \nu_1, t) p_X(x - \nu_1, t) + r_{-1}(x - \nu_{-1}, t) p_X(x - \nu_{-1}, t) \\ - (r_1(x, t) + r_{-1}(x, t)) p_X(x, t)$$

or, written to show the species numbers

$$\frac{d}{dt}p_X\left(\begin{bmatrix} n_A \\ n_B \\ n_C \end{bmatrix}, t\right) = k_1(n_A+1)p_X\left(\begin{bmatrix} n_A+1 \\ n_B+1 \\ n_C-1 \end{bmatrix}, t\right) +$$

$$k_{-1}(n_C+1)p_X\left(\begin{bmatrix} n_A-1 \\ n_B-1 \\ n_C+1 \end{bmatrix}, t\right) - (k_1 n_A + k_{-1} n_C)p_X\left(\begin{bmatrix} n_A \\ n_B \\ n_C \end{bmatrix}, t\right)$$

5.3.5 Microscopic, Mesoscopic, and Macroscopic Kinetic Models

Next we would like to explore how the discrete stochastic kinetic model of a microscopic system transforms into the deterministic kinetic model of a macroscopic system that is familiar to undergraduate chemical and biological engineers. Along the way, we derive a model for the the regime bridging the microscopic and macroscopic levels, which is sometimes called the mesoscopic regime. Our goal is to start with the microscopic chemical master equation and take the limit as the system size becomes large. We use the system volume Ω for the size parameter. The procedure we follow is given by van Kampen (1992, pp. 244-263) and is known as the OMEGA EXPANSION. The essentials of the approach are perhaps best explained by taking a concrete (and nonlinear) example. Consider the bimolecular reaction

$$2A \longrightarrow B$$

In the deterministic macroscopic description, we have a reaction-rate expression $r = \widetilde{k}c^2$, in which c is molar concentration of A, an intensive variable, and the rate constant $\widetilde{k}$ has units of $l^3/(\text{mol}\cdot t)$, so the rate has units of $\text{mol}/(t \cdot l^3)$, a rate of reaction per volume, which is also intensive. The mole balance for species A in a well-mixed system is the familiar

$$\frac{dc}{dt} = -2\widetilde{k}c^2 \qquad c(0) = c_0 \tag{5.58}$$

For these same kinetics, at the small scale, we have the microscopic chemical master equation

$$\frac{d}{dt}P(n,t) = -k\frac{n(n-1)}{2\Omega}P(n,t) + k\frac{(n+2)(n+1)}{2\Omega}P(n+2,t) \qquad n \geq 0 \tag{5.59}$$

in which n is the number of A molecules in the well-mixed system of volume Ω. Here n is a discrete (nonnegative, integer-valued) random

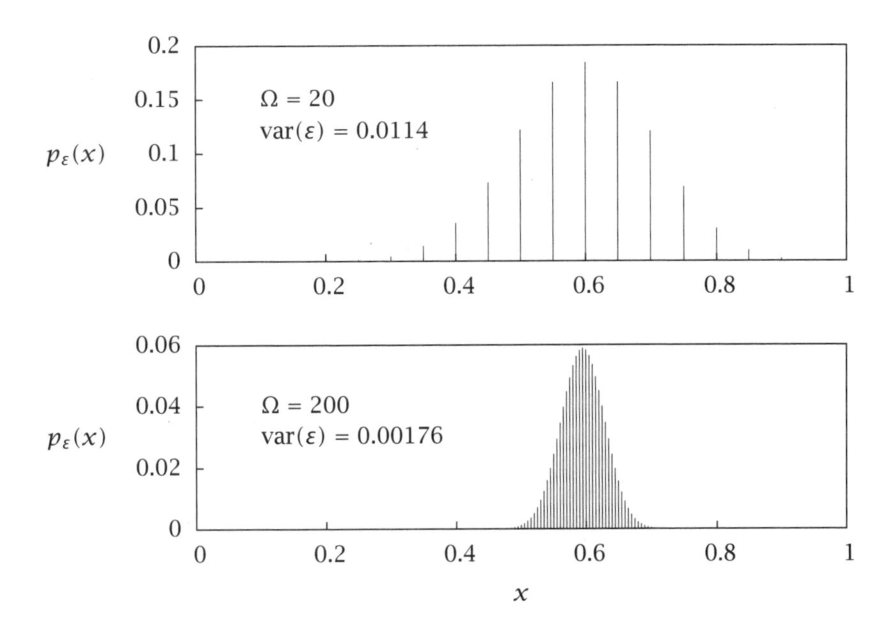

Figure 5.13: The equilibrium reaction extent's probability density for Reactions 5.52 at system volume $\Omega = 20$ (top) and $\Omega = 200$ (bottom). Notice the decrease in variance in the reaction extent as system volume increases.

variable. As Ω becomes large, we expect the concentration $c = n/\Omega$ to be well described by the ODE (5.58). It is initially far from clear how we take this limit to make this transition from a discrete-valued random variable n to a continuous-valued deterministic variable c.

To motivate the appropriate analysis, we first look at solutions to the master equation for increasing values of Ω. Figure 5.13 shows the final equilibrium distributions of the scaled reaction extent, ξ, from Figures 5.11 and 5.12. We have increased the system size from $\Omega = 20$ in the top figure to $\Omega = 200$ in the bottom figure. We also show the variance in random variable ε in the two simulations. Notice that for a ten-fold *increase* in Ω, the variance has *decreased* by almost this same ten-fold amount. From these solutions to the master equation we have some idea what to expect. For a large system, the integer increments in the number of molecules n become so fine that we can approximately replace them by a continuous variable c. But we also see randomness in the concentration, and although the (relative) magnitude of the concentration fluctuations decreases as the system size increases, it is not zero. In fact, we see that the familiar normal distribution appears to describe the probability distribution of the fluctuations and the variance scales with Ω^{-1}.

Therefore we are led to hypothesize that we can approximate n as a combination of the deterministic concentration c and a *continuous* random variable ξ to capture the fluctuations. Based on our numerical experiments, the form we choose is

$$n = c\Omega + \xi\Omega^{1/2} \tag{5.60}$$

so that the variance in n/Ω scales with Ω^{-1}, i.e., $\mathrm{var}(n/\Omega) = \Omega^{-1}\mathrm{var}(\xi)$. We are neglecting terms of order Ω^0 and lower in the expansion of n in (5.60). Thus we are expressing n/Ω as a perturbation solution in increasing powers of small parameter $\Omega^{-1/2}$. The additional complication in this case compared to our previous perturbation examples in Chapters 2 and 3, is that we are also changing from a discrete variable n to continuous variables c and ξ.

The master equation describes the density of random variable n, $P(n,t)$, and we wish to deduce an evolution equation for the density of random variable ξ, which we denote $\Pi(\xi,t)$. And we also expect the analysis to show that the familiar differential equation (5.58) describes the deterministic variable c. As a transformation of random variables, we are considering the two densities to be related by

$$P(n,t) = P(c\Omega + \xi\Omega^{1/2}, t) = \Pi(\xi, t)$$

in which we suppress the dependence of n on c. Consider c to be some known function of time when expressing the transformation between the two random variables n and ξ.

Given this transformation, the partial derivatives are related by $P_t = \Pi_t + \Pi_\xi \xi_t$ and ξ_t is found by differentiating (5.60) holding n constant, which yields

$$\xi_t = -\dot{c}\Omega^{1/2}$$

in which $\dot{c}$ represents the time derivative of $c(t)$. Substituting this into the relation for the partial derivatives gives

$$P_t = \Pi_t - \dot{c}\Omega^{1/2}\Pi_\xi$$

This is the first step. We have the left-hand side of the master equation evaluated in terms of the new density Π. Next we work on the right-hand side.

The term $P(n,t)$ is simply the transformed $\Pi(\xi,t)$, but we also require $P(n+2,t)$. First we solve (5.60) for ξ so that we know what $n+2$ corresponds to in variable ξ

$$\xi(n) = n\Omega^{-1/2} - c\Omega^{1/2} \tag{5.61}$$

Using (5.61) gives $\xi(n+2) - \xi(n) = 2\Omega^{-1/2}$. Next we use a Taylor series to represent $\Pi(\xi(n+2),t)$ in terms of $\Pi(\xi(n),t)$, denoted simply as Π, and its derivatives

$$\Pi(\xi(n+2),t) = \Pi + 2\Omega^{-1/2}\Pi_\xi + \frac{4\Omega^{-1}}{2!}\Pi_{\xi\xi} + \frac{8\Omega^{-3/2}}{3!}\Pi_{\xi\xi\xi} + \cdots \tag{5.62}$$

The number of terms retained in the Taylor series determines the order of the approximation for the density Π. We now can easily transform the remaining terms in n using (5.60)

$$\frac{n(n-1)}{\Omega} = c^2\Omega + 2c\xi\Omega^{1/2} + (-c+\xi^2) - \xi\Omega^{-1/2}$$
$$\frac{(n+2)(n+1)}{\Omega} = c^2\Omega + 2c\xi\Omega^{1/2} + (3c+\xi^2) + 3\xi\Omega^{-1/2} + 2\Omega^{-1}$$

Now we combine all of these ingredients by substituting them into the master equation (5.59) giving

$$\begin{aligned}\Pi_t - \dot{c}\Omega^{1/2}\Pi_\xi = {} & \frac{k}{2}[4c + 4\xi\Omega^{-1/2} + 2\Omega^{-1}]\,\Pi + \\ & k[c^2\Omega^{1/2} + 2c\xi + (3c+\xi^2)\Omega^{-1/2} + 3\xi\Omega^{-1} + 2\Omega^{-3/2}]\,\Pi_\xi + \\ & \quad k[c^2 + 2c\xi\Omega^{-1/2} + (3c+\xi^2)\Omega^{-1} + 3\xi\Omega^{-3/2} + 2\Omega^{-2}]\,\Pi_{\xi\xi}\end{aligned}$$

in which we have kept up to the second-order term in (5.62).

The third and final step is to extract from this large equation the information provided at the different orders of the expansion parameter Ω.

Order $\Omega^{1/2}$. Collecting the terms of order $\Omega^{1/2}$ gives $(\dot{c} + kc^2)\Pi_\xi = 0$ and, since $\Pi_\xi \neq 0$, we deduce

$$\frac{dc}{dt} = -kc^2 \tag{5.63}$$

which is the macroscopic equation (5.58) after noting that the usual macroscopic convention absorbs a factor of one-half into the definition of the rate constant, i.e., $\tilde{k} = k/2$.

Order Ω^0. Collecting the terms of order Ω^0 gives

$$\Pi_t = 2kc\,\Pi + 2kc\xi\,\Pi_\xi + kc^2\,\Pi_{\xi\xi}$$

which can be rearranged into

$$\frac{\partial}{\partial t}\Pi = -\frac{\partial}{\partial \xi}(-2kc\xi\,\Pi) + \frac{\partial^2}{\partial \xi^2}(kc^2\,\Pi)$$

This is the familiar Fokker-Planck equation, (5.41), which we can write as an equivalent SDE

$$d\xi = -2kc\xi\,dt + \sqrt{2kc^2}\,dW$$

Because this is a linear Fokker-Planck equation (the drift term is linear in ξ and the diffusivity is independent of ξ), this equation is sometimes referred to as the linear noise approximation.

To simulate the model at this level of approximation, we first solve (5.63) for $c(t)$, and then perform a random walk simulation for the fluctuation term $\xi(t)$, which depends on $c(t)$. We combine these two parts for $n(t)$ using (5.60). This description in which $c(t)$ is deterministic and $\xi(t)$ is a continuous random walk is the mesoscopic description. We see the results in Figure 5.14. The top figure shows the discrete simulation using KMC for volume $\Omega = 500$ and initial condition of 500 A molecules $n_0 = 500$. Note that the plot has a log scale on the time axis to more clearly show the evolution at early times. These two simulations display quite similar character. To compare them more quantitatively, we could compute several low-order moments of the densities by computing sample averages over many simulations.

As a more comprehensive alternative, we compute the corresponding cumulative probability distributions at the selected time $t = 1$, shown as the dashed line in Figure 5.14. We obtain the cumulative distribution for the discrete model by solving the master equation and summing

$$F(n,t) = \sum_{n'=0}^{n} P(n',t) \qquad 0 \le n \le n_0$$

We can obtain the density for the omega expansion by solving the PDE for $\Pi(\xi,t)$, shifting the mean by the deterministic $c(t)$, and integrating for the cumulative distribution. Or we can instead derive a corresponding evolution equation for ξ's cumulative density

$$F_\xi(x,t) = \int_{-\infty}^{x} \Pi(\xi,t)\,d\xi$$

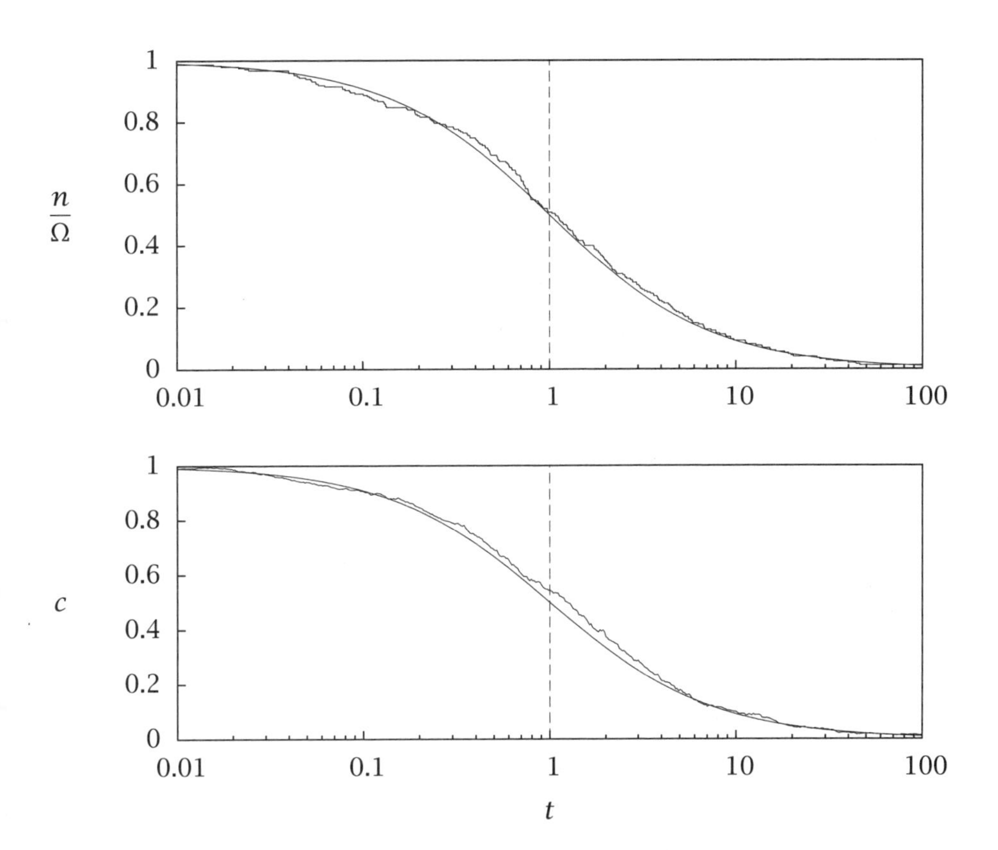

Figure 5.14: Simulation of 2 A ⟶ B for $n_0 = 500$, $\Omega = 500$. Top: discrete simulation; bottom: SDE simulation.

and shift its mean by $c(t)$, $F_c(x,t) = F_\xi(x - c(t), t)$. Exercise 5.22 discusses this approach in more detail. The results are shown in Figure 5.15. The staircase function is the solution to the discrete master equation at time $t = 1$, at which time the deterministic concentration is one-half, i.e., $c(t) = 1/2$ at $t = 1$. The steps in $x = n/\Omega$ are caused by the zero probability at all the odd integer values of n in the discrete model. The smooth function is the omega expansion, which we can see is in reasonably close agreement with the discrete model for $\Omega = 500$.

Finally, in the limit as $\Omega \to \infty$, the fluctuation ξ becomes negligible compared to c, and we have the familiar deterministic macroscopic description, (5.63) or (5.58). In Figure 5.15, this limit would be observed

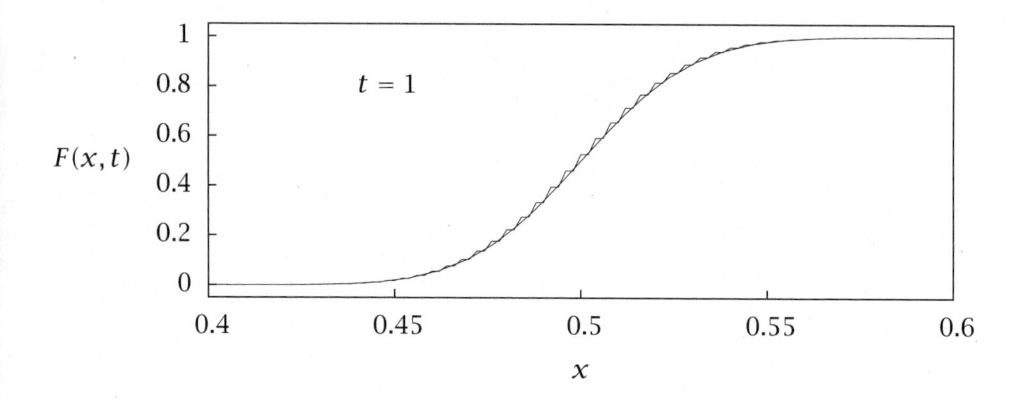

Figure 5.15: Cumulative distribution for 2 A ⟶ B at $t = 1$ with $n_0 = 500$, $\Omega = 500$. Discrete master equation (steps) versus omega expansion (smooth).

by the two functions converging to a unit step function at the value of $x = c(t)$. See also Exercise 5.22.

There is now an extensive and rapidly growing literature on stochastic kinetics. The book chapter by Anderson and Kurtz (2011) is highly recommended for those interested in a current and comprehensive overview of most of the topics covered here as well as more advanced topics on: relevant central limit theorems for Poisson processes, martingales, and scaling and model reduction.

5.4 Optimal Linear State Estimation

5.4.1 Introduction

Sensors are how we learn about the world. Our five natural senses provide us with our first exposure to sensors, i.e., the type built in by nature. Since humans are very curious about the world, people have been hard at work for a long time augmenting the natural senses by constructing artificial or man-made sensors. Some of mankind's biggest advances in science and engineering were precipitated by a breakthrough in sensor technology, e.g., the telescope, the microscope, detectors for electromagnetic radiation outside the visible range, etc.

One of the important things that we know about sensors is that they are limited and imperfect indicators of the world around us. They often are affected by nature in ways that the user does not intend or desire; they often conflate many different physical effects into a single signal, which makes it challenging for us to interpret what the sensor is telling us. Finally, all sensors, as well as the systems that we are trying to measure, are subject to uncontrolled and random effects.

One of the fundamental problems in systems engineering is to devise methods for taking these imperfect measurements of imperfectly controlled systems and deciding on the *best* estimate of the state of the system. We are inherently trading off *two* sources of error: the sensor's random error or noise, and the system's random fluctuation or disturbance. We may decide in some situation that a change in a sensor signal indicates that the system has changed. But we may decide in a different situation that a change in a sensor signal is caused by a random effect or disturbance to the *sensor* itself, and the system is completely *unchanged.* Optimally combining these two sources of information: what the sensor tells us and the other knowledge that we have about the system's behavior, is the task of state estimation.

To make these concepts precise, we consider a linear system. Let $x \in \mathbb{R}^n$ be an n-vector containing all the relevant information about a system of interest

$$x^+ = Ax + Bu$$

The u variables are the input variables that also affect the evolution of the system. If we control the inputs, they are called actuators, i.e., the valves in a chemical plant. If the inputs are not controlled by us, they are regarded as *disturbances,* and often given another letter to indicate this difference. We use $w \in \mathbb{R}^n$ to represent the disturbances. Because of the central limit theorem, these will be considered *normally distributed* random variables with zero mean and variance Q. The dynamic model is then

$$x^+ = Ax + Bu + w$$

The initial state of the system x_0 is also generally unknown and will be considered a normally distributed random variable with mean $\overline{x}_0$ and variance $P(0)$. Now we consider the sensors. Let $y \in \mathbb{R}^p$ be the p-vector of available measurements. Normally $p < n$ indicating that we are not measuring every relevant property of the system. Because sensors are expensive, often $p \ll n$ indicating that we have a complex system with many states, but are information poor with few measurements. The sensor is also affected by random disturbances, which we denote by v. Because the input u is considered known, we can remove it from the model for simplicity without changing any important features of the state estimation problem. The linear model of interest is then

$$x^+ = Ax + w$$
$$y = Cx + v$$

and the disturbances and unknown initial condition satisfy

$$w \sim N(0, Q_w) \qquad v \sim N(0, R_v) \qquad x(0) \sim N(\overline{x}_0, P(0))$$

If the measurement process is quite noisy, then R is large. If the measurements are highly accurate, then R is small. Similar considerations apply for the process noise, Q. If the state is subjected to large disturbances, then Q is large, and if the disturbances are small, Q is small. Again we choose zero mean for w because the nonzero mean disturbances should have been accounted for in the system model. The variance $P(0)$ reflects our confidence in the initial state. If we know how the system starts off, $P(0)$ is small. If we have little knowledge, we take $P(0)$ large. Recall the noninformative prior is a uniform distribution, which we can approximate by taking $P(0)$ very large. In industrial applications, the initial condition may be known with high accuracy for batch processes. But the initial condition is usually considered largely unknown when analyzing a dataset taken from a continuous process.

We require three main results concerning normals, conditional normals, and linear transformation. These follow directly from the properties of the normal established in Chapter 4, but see Exercise 5.24 for some hints if you have difficulty deriving any of these. Recall also the normal function notation (4.13)

$$n(x, m, P) = \frac{1}{(2\pi)^{n/2}(\det P)^{1/2}} \exp\left[-\frac{1}{2}(x-m)^T P^{-1}(x-m)\right]$$

which was introduced in Chapter 4, and will be used frequently in the following discussion.

Joint independent normals. If $p_{x|z}(x|z)$ is normal, and y is statistically independent of x and z and normally distributed

$$p_{x|z}(x|z) = n(x, m_x, P_x)$$
$$y \sim N(m_y, P_y) \qquad y \text{ independent of } x \text{ and } z$$

then the conditional joint density of (x, y) given z is

$$p_{x,y|z}(x, y|z) = n(x, m_x, P_x)\, n(y, m_y, P_y)$$

$$p_{x,y|z}\left(\begin{bmatrix} x \\ y \end{bmatrix} \middle| z\right) = n\left(\begin{bmatrix} x \\ y \end{bmatrix}, \begin{bmatrix} m_x \\ m_y \end{bmatrix}, \begin{bmatrix} P_x & 0 \\ 0 & P_y \end{bmatrix}\right) \tag{5.64}$$

Linear transformation of a normal. If x and z are jointly normally distributed with conditional density $p_{x|z}(x|z)$ having mean m and variance P, and y is a linear transformation of x, $y = Ax$, then $p_{y|z}(y|z)$ is normal with mean Am and variance APA^T

$$p_{x|z}(x|z) = n(x, m, P) \qquad y = Ax$$

$$p_{y|z}(y|z) = n(y, Am, APA^T) \tag{5.65}$$

Conditional of a joint normal. If the joint conditional density of (x, y) given z is normal

$$p_{x,y|z}\left(\begin{bmatrix} x \\ y \end{bmatrix} \middle| z\right) = n\left(\begin{bmatrix} x \\ y \end{bmatrix}, \begin{bmatrix} m_x \\ m_y \end{bmatrix}, \begin{bmatrix} P_x & P_{xy} \\ P_{yx} & P_y \end{bmatrix}\right)$$

then the conditional density of x given (y, z) is also normal

$$p_{x|y,z}(x|y, z) = n(x, m, P) \tag{5.66}$$

in which

$$m = m_x + P_{xy}P_y^{-1}(y - m_y) \qquad P = P_x - P_{xy}P_y^{-1}P_{yx}$$

Note that the conditional mean m is itself a random variable because it depends on the random variable y.

5.4.2 Optimal Dynamic Estimator

We have specified the random process of interest

$$x^+ = Ax + w \tag{5.67}$$

$$y = Cx + v \tag{5.68}$$

with known densities

$$w \sim N(0, Q_w) \qquad v \sim N(0, R_v) \qquad x(0) \sim N(\overline{x}_0, P(0))$$

We will next derive the optimal estimator for this process. As part of this derivation, we will derive the probability densities of the state as a function of time. This is the same pattern that we followed in the first two sections on Brownian motion and stochastic kinetics. We started with the random process (Wiener and Poisson processes), and then we derived their probability density equations (Fokker-Plank and chemical master equations).

Because we have assumed a prior, the density of $x(0)$, we are using Bayesian estimation. The overall game plan is as follows. The initial state $x(0)$ is assumed normal. Our optimal estimate before measurement is denoted $\hat{x}^-(0)$. The minus sign indicates estimate *before* measurement. We obtain from the sensor measurement $y(0)$. We then compute the conditional density of $x(0)|y(0)$. We show that is also normal. The maximum of that conditional density is our optimal estimate *after* measurement, denoted $\hat{x}(0)$. We are combining the measurement with the prior to calculate the posterior. Then we use the random process (5.68) to forecast the state forward one time step to obtain $x(1)$. We show that the density of $x(1)$ (conditioned on $y(0)$) is also normal,[3] and the maximum of that density is our estimate at $k = 1$ before measurement, $\hat{x}^-(1)$. Then we add measurement $y(1)$ and compute the conditional density of $x(1)|y(0), y(1)$; its maximum gives $\hat{x}(1)$, and we continue the iteration. So now we fill in the details.

Combining the measurement. We start off at $k = 0$ with estimate $\hat{x}^-(0) = \overline{x}_0$ and consider the effect of adding the first measurement. We obtain noisy measurement $y(0)$ satisfying

$$y(0) = Cx(0) + v(0)$$

in which $v(0) \sim N(0, R)$ is the measurement noise. Given the measurement $y(0)$, we next obtain the conditional density $p_{x(0)|y(0)}(x(0)|y(0))$. This conditional density describes the change in our knowledge about $x(0)$ after we obtain measurement $y(0)$. This step is the essence of state estimation. To derive this conditional density, first consider the pair of variables $(x(0), y(0))$ given as

$$\begin{bmatrix} x(0) \\ y(0) \end{bmatrix} = \begin{bmatrix} I & 0 \\ C & I \end{bmatrix} \begin{bmatrix} x(0) \\ v(0) \end{bmatrix}$$

We assume that the noise $v(0)$ is statistically independent of $x(0)$, and use the independent joint normal result (5.64) to express the joint

[3] Because we have linear transformations of normals at each step of the procedure, *every density in sight* will be normal.

density of $(x(0), v(0))$

$$\begin{bmatrix} x(0) \\ v(0) \end{bmatrix} \sim N\left(\begin{bmatrix} \overline{x}_0 \\ 0 \end{bmatrix}, \begin{bmatrix} Q(0) & 0 \\ 0 & R \end{bmatrix} \right)$$

From the previous equation, the pair $(x(0), y(0))$ is a linear transformation of the pair $(x(0), v(0))$. Therefore, using the linear transformation of normal result (5.65), and the density of $(x(0), v(0))$ gives the density of $(x(0), y(0))$

$$\begin{bmatrix} x(0) \\ y(0) \end{bmatrix} \sim N\left(\begin{bmatrix} \overline{x}_0 \\ C\overline{x}_0 \end{bmatrix}, \begin{bmatrix} Q(0) & Q(0)C^T \\ CQ(0) & CQ(0)C^T + R \end{bmatrix} \right)$$

Given this joint density, we then use the conditional of a joint normal result (5.66) to obtain

$$p_{x(0)|y(0)}(x(0)|y(0)) = n(x(0), m, P)$$

in which

$$\begin{aligned} m &= \overline{x}_0 + L(0)\left(y(0) - C\overline{x}_0\right) \\ L(0) &= Q(0)C^T(CQ(0)C^T + R)^{-1} \\ P &= Q(0) - Q(0)C^T(CQ(0)C^T + R)^{-1}CQ(0) \end{aligned}$$

We see that the conditional density $p_{x(0)|y(0)}$ is normal. The *optimal* state estimate is the value of $x(0)$ that maximizes this conditional density. For a normal, that is the mean, and we choose $\hat{x}(0) = m$. We also denote the variance in this conditional after measurement $y(0)$ by $P(0) = P$ with P given in the previous equation. The change in variance after measurement ($Q(0)$ to $P(0)$) quantifies the information increase by obtaining measurement $y(0)$. The variance after measurement, $P(0)$, is always less than or equal to $Q(0)$, which implies that we can only gain information by measurement; but the information gain may be small if the measurement device is poor and the measurement noise variance R is large.

Forecasting the state evolution. Next we consider the state evolution from $k = 0$ to $k = 1$, which satisfies

$$x(1) = \begin{bmatrix} A & I \end{bmatrix} \begin{bmatrix} x(0) \\ w(0) \end{bmatrix}$$

in which $w(0) \sim N(0, Q)$ is the process noise. We next calculate the conditional density $p_{x(1)|y(0)}$. Now we require the conditional version of the joint density $(x(0), w(0))$. We assume that the process noise $w(0)$ is statistically independent of both $x(0)$ and $v(0)$, hence it is also independent of $y(0)$, which is a linear combination of $x(0)$ and $v(0)$. Therefore we use (5.64) to obtain

$$\begin{bmatrix} x(0) \\ w(0) \end{bmatrix} \sim N\left(\begin{bmatrix} \hat{x}(0) \\ 0 \end{bmatrix}, \begin{bmatrix} P(0) & 0 \\ 0 & Q \end{bmatrix}\right)$$

We then use the conditional version of the linear transformation of a normal (5.65) to obtain

$$p_{x(1)|y(0)}(x(1)|y(0)) = n(x(1), \hat{x}^-(1), P^-(1))$$

in which the mean and variance are

$$\hat{x}^-(1) = A\hat{x}(0) \qquad P^-(1) = AP(0)A^T + Q$$

We see that forecasting forward one time step may increase or decrease the conditional variance of the state. The term $AP(0)A^T$ may be smaller or larger than $P(0)$, but the process noise Q always makes a positive contribution.

Given that $p_{x(1)|y(0)}$ is also a normal, we are situated to add measurement $y(1)$ and continue the process of adding measurements followed by forecasting forward one time step until we have processed all the available data. Because this process is recursive, the storage requirements are small. We need to store only the current state estimate and variance, and can discard the measurements as they are processed. The required online calculation is minor. These features make the optimal linear estimator an ideal candidate for rapid online application. We next summarize the state estimation recursion.

General time step k. Denote the measurement trajectory by

$$\mathbf{y}(k) = \{y(0), y(1), \dots y(k)\}$$

At time k the conditional density with data $\mathbf{y}(k-1)$ is normal

$$p_{x(k)|\mathbf{y}(k-1)}(x(k)|\mathbf{y}(k-1)) = n(x(k), \hat{x}^-(k), P^-(k))$$

and we denote the mean and variance with a superscript minus to indicate these are the statistics *before* measurement $y(k)$. At $k = 0$, the recursion starts with $\hat{x}^-(0) = \overline{x}_0$ and $P^-(0) = Q(0)$ as discussed previously. We obtain measurement $y(k)$, which satisfies

$$\begin{bmatrix} x(k) \\ y(k) \end{bmatrix} = \begin{bmatrix} I & 0 \\ C & I \end{bmatrix} \begin{bmatrix} x(k) \\ v(k) \end{bmatrix}$$

The density of $(x(k), v(k))$ follows from (5.64) since measurement noise $v(k)$ is independent of $x(k)$ and $\mathbf{y}(k-1)$

$$\begin{bmatrix} x(k) \\ v(k) \end{bmatrix} \sim N\left(\begin{bmatrix} \hat{x}^-(k) \\ 0 \end{bmatrix}, \begin{bmatrix} P^-(k) & 0 \\ 0 & R \end{bmatrix}\right)$$

Equation (5.65) then gives the joint density

$$\begin{bmatrix} x(k) \\ y(k) \end{bmatrix} \sim N\left(\begin{bmatrix} \hat{x}^-(k) \\ C\hat{x}^-(k) \end{bmatrix}, \begin{bmatrix} P^-(k) & P^-(k)C^T \\ CP^-(k) & CP^-(k)C^T + R \end{bmatrix}\right)$$

We note $\{\mathbf{y}(k-1), y(k)\} = \mathbf{y}(k)$, and using the conditional density result (5.66) gives

$$p_{x(k)|\mathbf{y}(k)}\left(x(k)|\mathbf{y}(k)\right) = n\left(x(k), \hat{x}(k), P(k)\right)$$

in which

$$\begin{aligned}
\hat{x}(k) &= \hat{x}^-(k) + L(k)\left(y(k) - C\hat{x}^-(k)\right) \\
L(k) &= P^-(k)C^T(CP^-(k)C^T + R)^{-1} \\
P(k) &= P^-(k) - P^-(k)C^T(CP^-(k)C^T + R)^{-1}CP^-(k)
\end{aligned}$$

We forecast from k to $k+1$ using the model

$$x(k+1) = \begin{bmatrix} A & I \end{bmatrix} \begin{bmatrix} x(k) \\ w(k) \end{bmatrix}$$

Because $w(k)$ is independent of $x(k)$ and $\mathbf{y}(k)$, the joint density of $(x(k), w(k))$ follows from a second use of (5.64)

$$\begin{bmatrix} x(k) \\ w(k) \end{bmatrix} \sim N\left(\begin{bmatrix} \hat{x}(k) \\ 0 \end{bmatrix}, \begin{bmatrix} P(k) & 0 \\ 0 & Q \end{bmatrix}\right)$$

and a second use of the linear transformation result (5.65) gives

$$p_{x(k+1)|\mathbf{y}(k)}(x(k+1)|\mathbf{y}(k)) = n(x(k+1), \hat{x}^-(k+1), P^-(k+1))$$

in which

$$\begin{aligned}
\hat{x}^-(k+1) &= A\hat{x}(k) \\
P^-(k+1) &= AP(k)A^T + Q
\end{aligned}$$

and the recursion is complete.

Summary. We place all the required formulas for implementing the optimal estimator in one place for easy reference. The initial conditions for $k = 0$ are

$$\boxed{\hat{x}^-(0) = \overline{x}_0 \qquad P^-(0) = Q(0)}$$

The update equations for time $k \geq 0$ are

$$\hat{x}(k) = \hat{x}^-(k) + L(k)(y(k) - C\hat{x}^-(k)) \tag{5.69}$$

$$L(k) = P^-(k)C^T(CP^-(k)C^T + R)^{-1} \tag{5.70}$$

$$P(k) = P^-(k) - P^-(k)C^T(CP^-(k)C^T + R)^{-1}CP^-(k) \tag{5.71}$$

$$\hat{x}^-(k+1) = A\hat{x}(k) \tag{5.72}$$

$$P^-(k+1) = AP(k)A^T + Q \tag{5.73}$$

The full densities of the state before and after measurement are

$$\begin{aligned} p_{x(k)|\mathbf{y}(k-1)}(x(k+1)|\mathbf{y}(k)) &= n(x(k), \hat{x}^-(k), P^-(k)) \\ p_{x(k)|\mathbf{y}(k)}(x(k)|\mathbf{y}(k)) &= n(x(k), \hat{x}(k), P(k)) \end{aligned}$$

These formulas provide the celebrated Kalman filter (Kalman, 1960). One of Kalman's key contributions was to use the state-space model to describe the system dynamics. As we see here, after that step, the solution of the optimal filtering problem reduces to a few well-known results about normals, linear transformation, and conditional density. One of the main practical advantages of the Kalman filter is the extremely efficient implementation. One can update and store the conditional mean and variance with only a few matrix multiplications and finding one matrix inverse. This efficient recursion makes the Kalman filter ideal for *online* state estimation where one would like to find the optimal estimate in real time as the sensor measurements become available.

5.4.3 Optimal Steady-State Estimator

Notice from (5.70) that the *optimal* estimator has a time-varying gain, $L(k)$, coming from the time-varying recursion for $P(k)$ and $P^-(k)$, given by (5.71) and (5.73). If we are willing to give up a small amount of performance during small initial times, we can obtain an even simpler filter. Assume for the moment that these recursions converge to a steady state. The steady state then satisfies

$$\begin{aligned} P_s &= P_s^- - P_s^- C^T(CP_s^- C^T + R)^{-1}CP_s^- \\ P_s^- &= AP_sA^T + Q \end{aligned}$$

Substituting P_s from the first equation into the second equation and eliminating P_s give the steady-state covariance before measurement as the solution to the following algebraic Riccati equation

$$P_s^- = Q + AP_s^- A^T - AP_s^- C^T (CP_s^- C^T + R)^{-1} CP_s^- A^T$$

The steady-state filter gain then follows from (5.70)

$$L_s = P_s^- C^T (CP_s^- C^T + R)^{-1}$$

and the optimal steady-state estimate before measurement, i.e., the conditional mean of the state given measurements, is obtained by combining (5.69) and (5.72) giving

$$\hat{x}^-(k+1) = A\hat{x}^-(k) + AL_s(y(k) - C\hat{x}^-(k))$$

Implementing this filter as data $y(k)$ become available is extremely efficient. Offline one solves the steady-state Riccati equation, P_s^-, and computes the steady-state filter gain, L_s. Online one has to store only L_s and current estimate $\hat{x}^-(k)$, and implement a few matrix-vector multiplications and vector additions after $y(k+1)$ is measured to obtain the next estimate, $\hat{x}^-(k+1)$. We have an ideal algorithm that combines extremely small storage requirements and extremely fast computation making the steady-state Kalman filter ideal for many applications in many engineering disciplines.

In any design problem, including state estimator design, we usually have many, sometimes conflicting, design objectives. Optimality is certainly one desirable objective. But we would also like some performance guarantees on the estimator. For example, if the disturbances to the system are small does the estimate error become small as we collect more measurements? We formulate this objective as a stability question in the final section. To motivate that discussion, consider the following case: $A = I, C = 0$, i.e., the system is an integrator and we are not making any measurements. Even without disturbances, the system evolution is $x^+ = x$, and therefore $x(k) = x_0$ for all $k \geq 0$. But (5.70) gives that $L_s = 0$, so the estimator equation is $\hat{x}^+ = \hat{x}$ and therefore $\hat{x}(k) = \overline{x}_0$ for all $k \geq 0$. Since the RV $x(0)$ is not necessarily at its mean, $x(0) \neq \overline{x}_0$, and we see that the state estimate does not converge to the system state no matter how many "measurements" we make. This *system* needs to be redesigned before we can obtain a state estimator that converges to the system state. It is clear what is wrong with this system since $C = 0$ provides no information from the sensor, but to detect *all* such badly designed systems, we introduce the concept of observability.

5.4.4 Observability of a Linear System

The basic idea of observability is that any two distinct states can be *distinguished* by applying some input and observing the two system outputs over some finite time interval (Sontag, 1998, p.262–263). The general definition for nonlinear systems can be quite complex, but observability for linear systems is much simpler. First of all, the applied input is irrelevant and we can set it to zero. Therefore consider the linear time-invariant system (A, C) with zero input and disturbances

$$x^+ = Ax$$
$$y = Cx$$

with initial condition $x(0) = x_0$. The solution for the state is $x(k) = A^k x_0$, and the output is therefore

$$y(k) = CA^k x_0 \tag{5.74}$$

The system is observable if there exists a finite N, such that for every x_0, N measurements $\{y(0), y(1), \ldots, y(N-1)\}$ distinguish uniquely the initial state x_0. As shown in Exercise 5.26, if we cannot determine the initial state using n measurements, we cannot determine it using $N > n$ measurements. Therefore we can develop a convenient test for observability as follows. For n measurements, the system model gives

$$\begin{bmatrix} y(0) \\ y(1) \\ \vdots \\ y(n-1) \end{bmatrix} = \begin{bmatrix} C \\ CA \\ \vdots \\ CA^{n-1} \end{bmatrix} x_0 \tag{5.75}$$

The question of *observability* is therefore a question of *uniqueness* of solutions to these linear equations. The matrix appearing in this equation is known as the *observability matrix* $\mathcal{O}$

$$\mathcal{O} = \begin{bmatrix} C \\ CA \\ \vdots \\ CA^{n-1} \end{bmatrix} \tag{5.76}$$

From Section 1.3.6 of Chapter 1, we know that the solution to (5.75) is unique if and only if the *columns* of the $np \times n$ observability matrix are linearly independent. Therefore, we have that the system (A, C) is observable if and only if

$$\text{rank}(\mathcal{O}) = n$$

We see in the next section that observability is a sufficient condition for estimator stability.

To illustrate this observability analysis in a chemical engineering context, we present the following example (Ray, 1981, p.58).

Example 5.7: Observability of a chemical reactor

Consider an isothermal, continuous well-stirred tank reactor (CSTR) with first-order liquid-phase reactions

$$\text{A} \xrightarrow{k_1} \text{B} \qquad \text{B} \xrightarrow{k_2} \text{C}$$

The volumetric flowrate Q_f and tank volume V_R are constant. The concentration of A in the feed c_{Af} is the manipulated variable, and $c_{Bf} = 0$. Let $x = \begin{bmatrix} c_A & c_B \end{bmatrix}^T$.

(a) Write down the mass balances for species A and B and show that

$$\dot{x} = A_c x + B_c u$$

What are matrices A_c and B_c for this problem?

(b) Consider measuring only species A reactor concentration with sample time $\Delta t > 0$. What is matrix C_c in this case? Is the system with this sampled measurement observable?

(c) Consider measuring only species B reactor concentration. What is matrix C_c in this case? Is the system with this sampled measurement observable? Provide a physical explanation if this answer differs from the answer to the previous part.

Solution

(a) Assuming constant density, the mass balances for A and B are

$$\frac{d}{dt}\begin{bmatrix} c_A \\ c_B \end{bmatrix} = \begin{bmatrix} -(F/V + k_1) & 0 \\ k_1 & -(F/V + k_2) \end{bmatrix} \begin{bmatrix} c_A \\ c_B \end{bmatrix} + \begin{bmatrix} k/V \\ 0 \end{bmatrix} c_{Af}$$

$$A_c = \begin{bmatrix} -(F/V + k_1) & 0 \\ k_1 & -(F/V + k_2) \end{bmatrix} \qquad B_c = \begin{bmatrix} k/V \\ 0 \end{bmatrix}$$

We can convert this continuous time system into a discrete time system by approximating the time derivative with an explicit Euler method[4]

$$\frac{dx}{dt} \approx \frac{x(k+1) - x(k)}{\Delta t}$$

giving

$$x^+ = Ax \qquad y = C_c x$$

$$A = \begin{bmatrix} 1 - (\Delta t)(F/V + k_1) & 0 \\ (\Delta t)k_1 & 1 - (\Delta t)(F/V + k_2) \end{bmatrix}$$

(b) For measuring only species A we have $C_c = \begin{bmatrix} 1 & 0 \end{bmatrix}$. We then check the observability matrix for the DT system, giving

$$\mathcal{O}(A, C_c) = \begin{bmatrix} 1 & 0 \\ 1 - (\Delta t)(F/V + k_1) & 0 \end{bmatrix}$$

which has rank one. Since $\text{rank}(\mathcal{O}) < n$, the system is *not* observable.

(c) For measuring only species B we have $C_c = \begin{bmatrix} 0 & 1 \end{bmatrix}$. This gives the observability matrix

$$\mathcal{O}(A, C_c) = \begin{bmatrix} 0 & 1 \\ (\Delta t)k_1 & 1 - (\Delta t)(F/V + k_2) \end{bmatrix}$$

which has rank two for all sample times $\Delta t > 0$. Since $\text{rank}(\mathcal{O}) = n$, the system *is* observable.

The answers are different because measuring A tells us how much total B we have *produced*, but we have no information about how much B was present initially nor how much was consumed to produce C. Therefore we cannot reconstruct the B concentration from the model and the A concentration. Measuring species B, however, provides information about how much A is in the reactor, because the A concentration affects the production rate of B. The B measurement information plus the mass balances enable us to reconstruct the A concentration. The value of the rank condition of the observability matrix is that it makes rigorous this kind of physical intuition and reasoning. □

[4]Improving the numerical approximation does not change the observability analysis that follows.

5.4.5 Stability of an Optimal Estimator

Optimality of a filter is one desirable characteristic, but systems engineers often care about other characteristics such as stability. Stability in this situation means that the state estimate "gets close" (in some sense that we make precise shortly) to the true state as more measurements become available. As shown previously with an unobservable system ($A = I, C = 0$), we can have situations in which optimal estimators are not stable. In these situations, optimality is small consolation, and the estimator is not useful.

We define estimate error as the difference between the true system state and our estimate of the state. We shall use the estimate before measurement to illustrate

$$\widetilde{x}(k) = x(k) - \hat{x}^-(k)$$

The evolution of the estimate error can be given by substituting the

$$\widetilde{x}(k+1) = Ax(k) + w(k) - A\hat{x}^-(k) - AL_s(y(k) - C\hat{x}^-(k))$$

Substituting the system measurement $y(k) = Cx(k) + v(k)$ and combining terms gives

$$\boxed{\widetilde{x}(k+1) = (A - AL_sC)\widetilde{x}(k) + w(k) - AL_sv(k)}$$

Estimator stability is the question of whether $(A - AL_sC)$ is a stable matrix, i.e., has all its eigenvalues inside the unit circle.

We have the following theorem covering the stability of the steady-state estimator.

Theorem 5.8 (Riccati iteration and estimator stability). *Given (A, C) observable, $Q > 0$, $R > 0$, $P^-(0) \geq 0$, and the discrete Riccati equation*

$$P^-(l+1) = Q + AP^-(l)A' - AP^-(l)C'(CP^-(l)C' + R)^{-1}CP^-(l)A', \quad l = 0, 1, \ldots$$

Then

(a) There exists $P_s^- \geq 0$ such that for every $P^-(0) \geq 0$

$$\lim_{l\to\infty} P^-(l) = P_s^-$$

and P_s^- is the unique solution of the steady-state Riccati equation

$$P_s^- = Q_w + AP_s^-A' - AP_s^-C'(CP_s^-C' + R)^{-1}CP_s^-A'$$

among the class of positive semidefinite matrices.

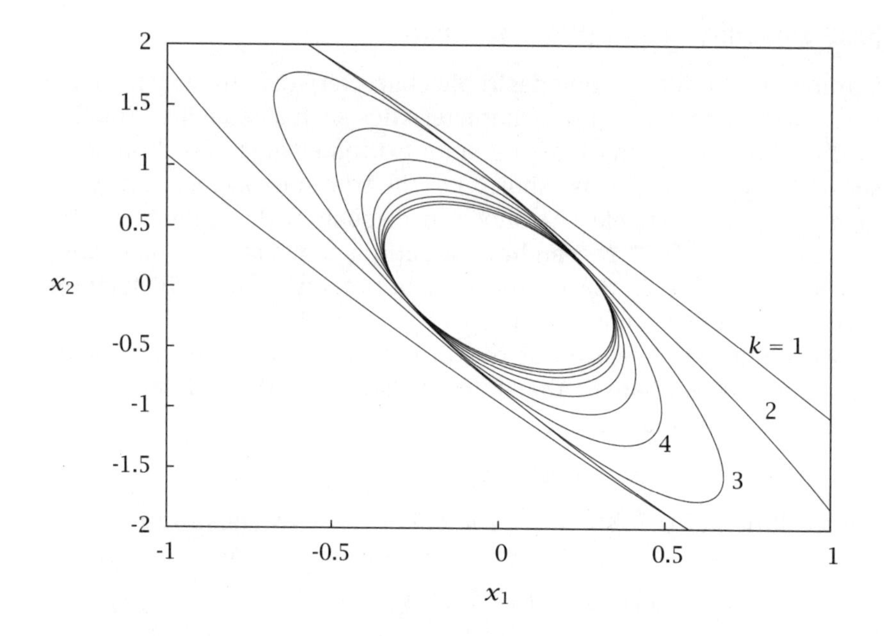

Figure 5.16: The change in 95% confidence intervals for $\hat{x}(k|k)$ versus time for a stable, optimal estimator. We start at $k = 0$ with a noninformative prior, which has an infinite confidence interval.

(b) The matrix $A - AL_sC$ in which

$$L_s = P_s^- C'(CP_s^- C' + R)^{-1}$$

is a stable matrix.

Bertsekas (1987, pp. 59-64) provides a proof of the "dual" of this theorem, which can be readily translated to this case.

So what is the payoff for knowing how to design a stable, optimal estimator? Assume we have developed a linear empirical model for a chemical process describing its normal operation around some nominal steady state. After some significant unmeasured process disturbance, we have little knowledge of the state. So we take initial variance $P^-(0)$ to be large (the noninformative prior). Figure 5.16 shows the evolution of our 95% confidence intervals for the state as time increases and we obtain more measurements. We see that the optimal estimator's confidence interval returns to its steady-state value after about only

10 measurements. Recall that the conditional variances of state given measurement do *not* require the measurements. Only the optimal estimates $\hat{x}(k)$ depend on the data. So we can assess the information quality of our sensor system before we even examine the data. But realize that the noise parameters Q and R almost always need to be determined from process data before we can perform this analysis. Moreover, if we plan to use feedback control to move the disturbed process back to its optimal operating point, the better our estimate of the state, the better our control and therefore process performance.

State estimation is a fundamental topic appearing in many branches of science and engineering, and has a large literature. A nice and brief annotated bibliography describing the early contributions to optimal state estimation of the *linear Gaussian* system is provided by Åström (1970, pp. 252-255). Kailath (1974) provides a comprehensive and historical review of *linear* filtering theory including the historical development of Wiener-Kolmogorov theory for filtering and prediction that preceded Kalman filtering (Wiener, 1949; Kolmogorov, 1941). Jazwinski (1970) provides an early and comprehensive treatment of the optimal stochastic state estimation problem for linear and *nonlinear* systems. Many optimal control texts contain discussions of the nonlinear state estimation problem (Bryson and Ho, 1975; Stengel, 1994). The moving horizon estimation (MHE) method, which uses online optimization to address system nonlinearity and constraints, is presented by Rawlings and Mayne (2009, Ch. 4).

5.5 Exercises

Exercise 5.1: Random walk with the uniform distribution

Consider again a discrete-time random walk simulation

$$x(k+1) = x(k) + v\Delta t + \sqrt{2D\Delta t}\, w(k) \tag{5.77}$$

in which $x, w \in \mathbb{R}^2$, k is sample number, Δt is the sample time with $t = k\Delta t$. Instead of using normally distributed steps as in Figure 5.3, let $w = 2\sqrt{3}(u - 1/2)$ in which $u \sim U(0,1)$

$$p_u(a_1, a_2) = \begin{cases} 1 & 0 \leq a_1, a_2 \leq 1 \\ 0 & \text{otherwise} \end{cases}$$

We then have that $w \sim U(-\sqrt{3}, \sqrt{3})$ with zero mean and unit variance. The Octave or MATLAB function `rand` generates samples of u, from which we can generate samples of w with the given transformation.

(a) Calculate a trajectory for this random walk in the plane and compare to Figure 5.3 for the normal distribution.

(b) Calculate the mean square displacement for 500 trajectories and compare to Figure 5.4 for the normal distribution.

(c) Derive the evolution equation for the probability density $p_x(x,t)$ for this process in the limit as Δt goes to zero. How is this model different from the usual diffusion equation given by (5.30) for the random walk with normally distributed steps?

Exercise 5.2: The different diffusion coefficients D and $\mathcal{D}$

In the chapter we compared two models for the evolution of concentration undergoing convection and diffusion processes

$$\frac{\partial c}{\partial t} = -\frac{\partial}{\partial x}(v(x,t)c) + \frac{\partial}{\partial x}(D(x,t)\frac{\partial c}{\partial x})$$

and

$$\frac{\partial c}{\partial t} = -\frac{\partial}{\partial x}(v(x,t)c) + \frac{\partial^2}{\partial x^2}(\mathcal{D}(x,t)c)$$

in which we consider x, v, and D scalars. The first is derived from conservation of mass with a flux law defined by $N = -D\partial c/\partial x$. The second is the Fokker-Planck equation corresponding to the following random walk model of diffusion

$$dx = v(x,t)dt + \sqrt{2\mathcal{D}(x,t)}\, dW$$

(a) Show that when the diffusivity $D(x,t)$ does not depend on x, these two models are equivalent and $\mathcal{D}(t) = D(t)$.

(b) Show that the Fokker-Planck equation can always be written in the following convection-diffusion form with a modified drift term

$$\frac{\partial c}{\partial t} = -\frac{\partial}{\partial x}(\tilde{v}(x,t)c) + \frac{\partial}{\partial x}(\mathcal{D}(x,t)\frac{\partial c}{\partial x})$$

and find the expression for $\tilde{v}(x,t)$.

Exercise 5.3: The diffusion coefficient matrices D and $\mathcal{D}$

Repeat Exercise 5.2 but for the case in which $\mathbf{x}$ and $\mathbf{v}$ are n-vectors and $\mathbf{D}$ and $\boldsymbol{\mathcal{D}}$ are $n \times n$ diffusion coefficient matrices

$$\frac{\partial c}{\partial t} = -\nabla \cdot (\mathbf{v}(\mathbf{x},t)c) + \nabla \cdot (\mathbf{D}(\mathbf{x},t) \cdot \nabla c)$$

and

$$\frac{\partial c}{\partial t} = -\nabla \cdot (\mathbf{v}(\mathbf{x},t)c) + \nabla\nabla\colon (\boldsymbol{\mathcal{D}}(\mathbf{x},t)c)$$

Exercise 5.4: Continuity of random processes

We know that the Wiener process is too rough to be differentiated, but is it even continuous? To answer this question, we first have to extend the definition of continuity to cover random processes such as $W(t)$. We use the following definition.

Definition 5.9 (Continuity (with probability one)). A scalar random process $r(t)$ is continuous (with probability one) if for all $\varepsilon > 0$ and $\alpha < 1$, there exists $\delta > 0$, which generally depends on ε and α, such that

$$\Pr(|r(t) - r(s)| \leq \varepsilon) \geq \alpha \qquad \text{for all } t, s \text{ satisfying } |t - s| \leq \delta$$

In other words, a random process is continuous if squeezing the times close enough together squeezes the values of the random process together, with probability as close to one as desired.

Using this definition, prove that the Wiener process $W(t)$ is continuous (with probability one) and that the white-noise process $dW(t)$ is discontinuous, i.e., not continuous. Hence, establish that integrating the discontinuous white-noise process smooths it enough to creates a continuous Wiener process.

Exercise 5.5: Multidimensional Itô Formula and Moments of Multidimensional SDEs

(a) Use Itô's formula to derive (5.33) and (5.34).

(b) Derive the multidimensional form of Itô's formula, (5.38), for an SDE in the form

$$dx_i = \mathcal{A}_i(x,t)dt + \mathcal{B}_{ij}(x,t)dW_j$$

Recall (5.20).

(c) Use this formula to derive the multidimensional versions of (5.33) and (5.34):

$$\mathcal{E}\left(\frac{d(x_i - x_i')(x_j - x_j')}{dt}\right)\bigg|_{t=t'} = 2\mathcal{D}_{ij}(x',t')$$

$$\mathcal{E}\left(\frac{d(x_i - x_i')}{dt}\right)\bigg|_{t=t'} = \mathcal{A}_i(x',t')$$

Exercise 5.6: Diffusion equation in one dimension with Laplace transform

Consider the diffusion equation on the line

$$\frac{\partial c}{\partial t} = D\nabla^2 c \qquad 0 < t, \quad -\infty < x < \infty$$

We wish to calculate the response $c(x,t)$ to an impulse source term at $t = 0$, $c(x,0) = \delta(x)$.

(a) In Chapter 3, we already solved this problem using the Fourier transform. Here we try the Laplace transform. Take the Laplace transform of the one-dimensional diffusion equation with this initial condition and show

$$D\frac{d^2\overline{c}(x,s)}{dx^2} - s\overline{c}(x,s) = -\delta(x) \tag{5.78}$$

(b) What are the two linearly independent solutions to the homogeneous equation? Break the problem into two parts and solve the differential equation for $x > 0$ and $x < 0$. You have four unknown constants at this point.

(c) Which of the two linearly independent solutions is bounded for $x \to \infty$? Which of these two solutions is bounded for $x \to -\infty$? Use this reasoning to find two of the unknown constants.

(d) Use continuity of $\overline{c}(x, s)$ at $x = 0$ to find one more unknown constant. Integrate (5.78) across a small interval containing zero to obtain a condition on the change in the first derivative

$$\frac{d\overline{c}(x = 0^+, s)}{dx} - \frac{d\overline{c}(x = 0^-, s)}{dx}$$

(e) Use this jump condition to find the last constant and obtain the full transform $\overline{c}(x, s)$ valid for all x.

(f) Invert this transform and show

$$c(x, t) = \frac{1}{2\sqrt{\pi D t}} e^{-x^2/(4Dt)} \qquad 0 < t, \quad -\infty < x < \infty \tag{5.79}$$

State which inversion formula you are using.

(g) Compute the mean square displacement for this concentration profile.

$$p(x, t) = c(x, t)$$

$$\langle x^2 \rangle = \int_{-\infty}^{\infty} p(x, t) x^2 dx$$

Exercise 5.7: Random walk in one dimension

Prepare a simulation of a random walk in one dimension for $D = 2$. Start the particles at $x = 0$ at $t = 0$ and simulate until $t = 1000$.

(a) Show the trajectories of the random walks for five particles on the same plot.

(b) Plot the mean square displacement versus time for 1000 particles. Compare this result to the analytical solution given in Exercise 5.6(g). Describe any differences.

(c) Plot the histogram of particle locations at $t = 1000$ for 1000 particles. On the same plot, compare this histogram to the analytical result given in (5.79). Describe any differences.

Exercise 5.8: More useful integrals

Use the definition of the complete gamma function and establish the following integral relationship

$$\int_0^{\infty} x^p e^{-ax^n} dx = \frac{\Gamma(\frac{p+1}{n})}{n a^{(p+1)/n}} \qquad a > 0$$

For the case $n = 2$, this relation reduces to

$$\int_0^{\infty} x^p e^{-ax^2} dx = \frac{\Gamma(\frac{p+1}{2})}{2 a^{(p+1)/2}} \qquad a > 0 \tag{5.80}$$

which proves useful in the next exercises.

Exercise 5.9: Diffusion equation in cylindrical coordinates with Laplace transform

Consider the diffusion equation in cylindrical coordinates with symmetry in the θ coordinate

$$\frac{\partial c}{\partial t} = \frac{1}{r} \frac{\partial}{\partial r} r D \frac{\partial c}{\partial r} \qquad 0 < t, \quad 0 < r < \infty$$

We wish to calculate the response $c(r, t)$ to an impulse source term at $t = 0$, $c(r, 0) = \frac{1}{2\pi r}\delta(r)$.

(a) Take the Laplace transform of the diffusion equation with this initial condition and show

$$D\frac{1}{r}\frac{d}{dr}r\frac{d\overline{c}(r,s)}{dr} - s\overline{c}(r,s) = -\frac{1}{2\pi r}\delta(r) \tag{5.81}$$

(b) What are the two linearly independent solutions to the homogeneous equation?

(c) Which of the two linearly independent solutions is bounded for $r \to \infty$? Use this reasoning to determine one of the unknown constants.

(d) Integrate (5.81) across a small interval containing zero to obtain a condition on the change in the first derivative

$$\lim_{r\to 0^+} r\frac{d\overline{c}(r,s)}{dr} - \lim_{r\to 0^-} r\frac{d\overline{c}(r,s)}{dr}$$

(e) Use this jump condition to find the second constant and obtain the transform

$$\overline{c}(r,s) = \frac{1}{2\pi D}K_0\left(\sqrt{\frac{s}{D}}r\right)$$

(f) Invert this transform and show

$$c(r,t) = \frac{1}{4\pi D t}e^{-r^2/(4Dt)} \qquad 0 < t, \quad 0 < r < \infty$$

State which inversion formula you are using.

(g) Compute the mean square displacement for this concentration profile

$$\langle r^2 \rangle = \int_0^{2\pi}\int_0^{\infty} r^2 c(r,t) r\, dr\, d\theta$$

Exercise 5.10: Diffusion equation in spherical coordinates with Laplace transform

Consider the diffusion equation in spherical coordinates with symmetry in the θ and ϕ coordinates

$$\frac{\partial c}{\partial t} = \frac{1}{r^2}\frac{\partial}{\partial r}r^2 D\frac{\partial c}{\partial r} \qquad 0 < t, \quad 0 < r < \infty$$

We wish to calculate the response $c(r,t)$ to an impulse source term at $t = 0$, $c(r,0) = \frac{1}{4\pi r^2}\delta(r)$.

(a) Take the Laplace transform of the diffusion equation with this initial condition and show

$$D\frac{1}{r^2}\frac{d}{dr}r^2\frac{d\overline{c}(r,s)}{dr} - s\overline{c}(r,s) = -\frac{1}{4\pi r^2}\delta(r) \tag{5.82}$$

(b) What are the two linearly independent solutions to the homogeneous equation?

(c) Which of the two linearly independent solutions is bounded for $r \to \infty$? Use this reasoning to find one of the unknown constants.

(d) Integrate (5.82) across a small interval containing zero to obtain a condition on the change in the first derivative

$$\lim_{r\to 0^+} r^2\frac{d\overline{c}(r,s)}{dr} - \lim_{r\to 0^-} r^2\frac{d\overline{c}(r,s)}{dr}$$

(e) Use this jump condition to find the second constant and obtain the full transform $\bar{c}(r,s)$ valid for all r.

(f) Invert this transform and show

$$c(r,t) = \frac{1}{8(\pi D t)^{\frac{3}{2}}} e^{-r^2/(4Dt)} \qquad 0 < t, \quad 0 < r < \infty$$

State which inversion formula you are using.

(g) Compute the mean square displacement for this concentration profile

$$\langle r^2 \rangle = 4\pi \int_0^\infty r^2 c(r,t)\, r^2\, dr$$

Exercise 5.11: Probabilty distributions for diffusion on the plane

This exercise provides another view of the issues raised in Example 5.4. Consider again the diffusion equation, (5.30), repeated here

$$\frac{\partial p}{\partial t} = D \nabla^2 p$$

subject to a unit impulse at the origin at $t = 0$.

We consider solving this equation in the plane using both rectangular coordinates (x, y) and polar coordinates (r, θ).

(a) Using rectangular coordinates, let $p(x, y, t)$ satisfy (5.30)

$$\frac{\partial p}{\partial t} = D \left(\frac{\partial^2 p}{\partial x^2} + \frac{\partial^2 p}{\partial y^2} \right)$$

with initial condition

$$p(x, y, t) = \delta(x)\delta(y) \qquad t = 0$$

Solve this equation and show

$$p(x, y, t) = \frac{1}{4\pi D t} e^{-(x^2+y^2)/(4Dt)} \tag{5.83}$$

Notice this $p(x, y, t)$ is a valid probability density (positive, normalized).

(b) If we consider the two components (x, y) as time-varying random variables with the probability density given by (5.83)

$$\xi = \begin{bmatrix} x \\ y \end{bmatrix}$$

$$p_\xi(x, y, t) = \frac{1}{4\pi D t} e^{-(x^2+y^2)/(4Dt)}$$

then we say ξ is distributed as follows

$$\xi \sim N(0, (2Dt)I)$$

in which I is a 2×2 identity matrix. The position random variable in rectangular coordinates is normally distributed with zero mean and covariance $(2Dt)I$.

(c) Next define a new random variable,

$$\eta = \begin{bmatrix} r \\ \theta \end{bmatrix}$$

as the (invertible) transformation from rectangular to polar coordinates. We have for the transformation, inverse transformation and Jacobian

$$\eta = f(\xi) \qquad \begin{bmatrix} r \\ \theta \end{bmatrix} = \begin{bmatrix} \sqrt{x^2+y^2} \\ \tan^{-1}(y/x) \end{bmatrix}$$

$$\xi = f^{-1}(\eta) \qquad \begin{bmatrix} x \\ y \end{bmatrix} = \begin{bmatrix} r\cos\theta \\ r\sin\theta \end{bmatrix}$$

$$\frac{df^{-1}(\eta)}{d\eta} = \begin{bmatrix} \cos\theta & -r\sin\theta \\ \sin\theta & r\cos\theta \end{bmatrix} \qquad \left|\frac{\partial f^{-1}(\eta)}{\partial \eta}\right| = r$$

Use the rule for finding the probability density of a transformed random variable[5] and show

$$p_\eta(r,\theta,t) = \frac{1}{4\pi Dt} r e^{-r^2/(4Dt)}$$

This is the quantity denoted p_P in Example 5.4. Calculate the marginal p_r by integration and show

$$p_r(r,t) = \frac{1}{2Dt} r e^{-r^2/(4Dt)}$$

Note that these are both well-defined probability densities (positive, normalized). The first is the probability density of the pair of random variables (r,θ), and the second is the marginal density of the random variable r for particles undergoing the Brownian motion.

Exercise 5.12: Mean and variance of the Poisson distribution

Given that discrete random variable Y has the Poisson density

$$p_Y(n) = \frac{a^n}{n!} e^{-a}$$

for $n = 0, 1, \ldots$, and parameter $a \in \mathbb{R} \geq 0$, show that

$$\mathcal{E}(Y) = a \qquad \text{var}(Y) = a$$

Exercise 5.13: Alternate derivation of Poisson process density

Consider the Poisson process probability $\Pr(Y(t) = n)$ for $n \geq 0$. Show that

$$\Pr(Y(t) = n) = \Pr(Y(t) \geq n) - \Pr(Y(t) \geq n+1)$$

You may want to review Exercise 4.1(a). Using the definition of the event time τ_n, show that

$$\Pr(Y(t) = n) = \int_0^t p_{\tau_n}(t)dt - \int_0^t p_{\tau_{n+1}}(t)dt$$

Substitute (5.49) and use integration by parts to show (5.50)

$$\Pr(Y(t) = n) = \frac{(\lambda t)^n}{n!} e^{-\lambda t}$$

[5] $p_\eta(y) = p_\xi(f^{-1}(y)) \left|\frac{\partial f^{-1}(y)}{\partial y}\right|$. See (4.23).

Exercise 5.14: Generating samples from an exponential distribution

Let random variable u be distributed uniformly on $[0,1]$. Define random variable τ by the transformation

$$\tau = -\frac{1}{\lambda} \ln u$$

Show that τ has density (see Section 4.3.2)

$$p_\tau(t) = \lambda e^{-\lambda t}$$

Thus uniformly distributed random samples can easily be transformed to give exponentially distributed random samples as required for simulating Poisson processes and stochastic kinetics.

Exercise 5.15: State-space form for master equation

Write the linear state-space model for the master equation in the extent of the reaction describing the single reaction

$$\mathrm{A} + \mathrm{B} \underset{k_{-1}}{\overset{k_1}{\rightleftharpoons}} \mathrm{C} \tag{5.84}$$

Assume we are not measuring anything.

(a) What are x, A, B, C, D for this model?

(b) What is the dimension of the state vector in terms of the initial numbers of molecules in the system.

Exercise 5.16: Properties of the kinetic matrix

(a) Show that for a valid master equation the row sum is zero for each column of the A matrix in (5.56).

(b) Show that this result holds for the A given in (5.55) for the reaction $A + B \rightleftharpoons C$.

(c) What is the row sum for each column of the A_θ matrix in the sensitivity equation? Show this result.

Exercise 5.17: Reaction probabilities in stochastic kinetics

Consider a stochastic simulation of the following reaction

$$a\,\mathrm{A} + b\,\mathrm{B} \underset{k_{-1}}{\overset{k_1}{\rightleftharpoons}} c\,\mathrm{C} + d\,\mathrm{D}$$

(a) Write out the two reaction probabilities $h_i(n_j), i = 1, -1$ considering the forward and reverse reactions as separate events.

(b) Compare these to the deterministic rate laws $r_i(c_j), i = 1, -1$ for the forward and reverse reactions considered as elementary reactions. Why are these expressions different? When do they become close to being the same?

Exercise 5.18: Sampling to solve the master equation

Consider the stochastic simulation method described in the chapter. We may view the trajectory of each simulation using this method as a sample of the probability density of the system. Show that the expectation of this sampling process satisfies the chemical master equation, (5.53).

Exercise 5.19: The evolution of the mean concentration

Consider the simple irreversible reaction

$$A \xrightarrow{k} B \qquad r = kn_A$$

in which n_A is the number of A molecules and k is a rate constant. The reactor volume starts with n_{A0} A molecules. Consider $p(n_A, t)$, the probability that the reactor volume contains n_A molecules at time t.

(a) Over what range of n_A is $p(n_A, t)$ defined? Call this set N. Write the evolution equation for $p(n_A, t), n_A \in N$.

(b) Define the mean of A's probability density by

$$\langle n_A(t) \rangle = \sum_{n_A \in N} n_A p(n_A, t)$$

From this definition and the evolution of the probability density, write an evolution equation for $\langle n_A(t) \rangle$. The probability density itself should not appear in the evolution equation for the mean.

(c) How is the mean's evolution equation related to the usual mass action kinetics governing the macroscopic concentration $c_A(t)$?

Exercise 5.20: Stochastic simulation for nonlinear kinetics[6]

Consider the reversible, second-order reaction

$$A + B \underset{k_{-1}}{\overset{k_1}{\rightleftharpoons}} C \qquad r = k_1 c_A c_B - k_{-1} c_C$$

(a) Solve the deterministic material balance for a constant-volume batch reactor with

$$k_1 = 1 \text{ L/mol·min} \qquad k_{-1} = 1 \text{ min}^{-1}$$
$$c_A(0) = 1 \text{ mol/L} \qquad c_B(0) = 0.9 \text{ mol/L} \qquad c_C(0) = 0 \text{ mol/L}$$

Plot the A, B, and C concentrations out to $t = 5$ min.

(b) Compare the result to a stochastic simulation using an initial condition of 400 A, 360 B and zero C molecules. Notice from the units of the rate constants that k_1 should be divided by 400 to compare simulations. Figure 5.17 is a representative comparison for one sequence of pseudorandom numbers.

(c) Repeat the stochastic simulation for an initial condition of 4000 A, 3600 B, zero C molecules. Remember to scale k_1 appropriately. Are the fluctuations noticeable with this many starting molecules?

---

[6] See also Exercise 4.17 in (Rawlings and Ekerdt, 2012)

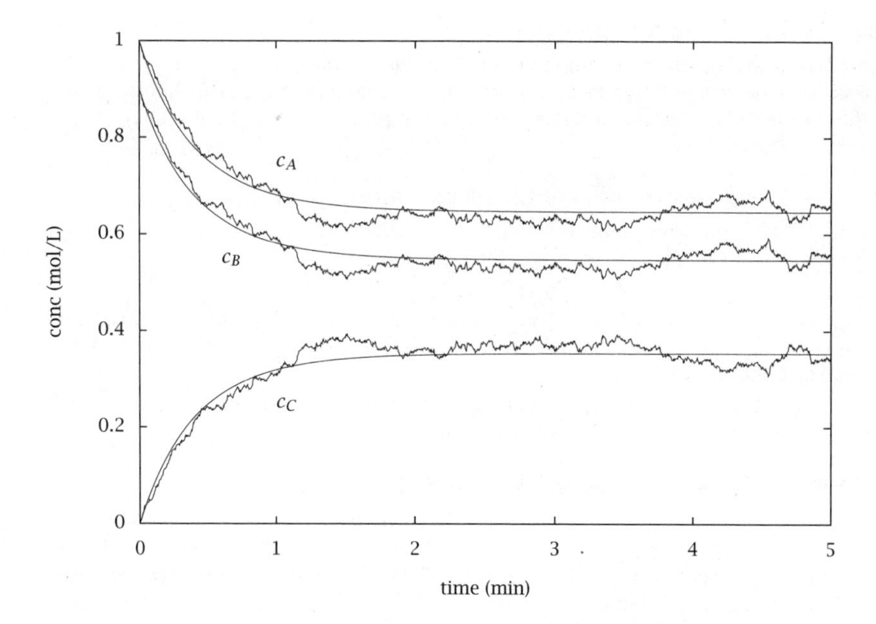

Figure 5.17: Deterministic simulation of reaction $A + B \rightleftharpoons C$ compared to stochastic simulation starting with 400 A molecules.

Exercise 5.21: What happened to my rate?

Consider a well-mixed continuum setting in which we have positive, real-valued concentrations of reacting molecules of two types, A and B, as depicted in Figure 5.18. Let the concentration of A and B molecules in the volume of interest be denoted c_{A0}, c_{B0}. Consider the three possible irreversible reactions between these species using the elementary rate expressions

$$\begin{aligned} \mathrm{A} + \mathrm{A} &\xrightarrow{k_1} \mathrm{C} \qquad r_1 = k_1 c_A^2 \\ \mathrm{A} + \mathrm{B} &\xrightarrow{k_2} \mathrm{D} \qquad r_2 = k_2 c_A c_B \\ \mathrm{B} + \mathrm{B} &\xrightarrow{k_3} \mathrm{E} \qquad r_3 = k_3 c_B^2 \end{aligned} \tag{5.85}$$

Consider also the total rate of reaction

$$r = r_1 + r_2 + r_3$$

(a) If the A and B species are chemically similar so the different reactions' rate constants are all similar, $k_1 = k_2 = k_3 = k$, and the concentrations of A and B are initially equal, the total rate is given by

$$r = 3kc_{A0}^2$$

But if we erase the distinctions between A and B completely and relabel the B molecules in Figure 5.18 as A molecules, we obtain the new concentrations of A

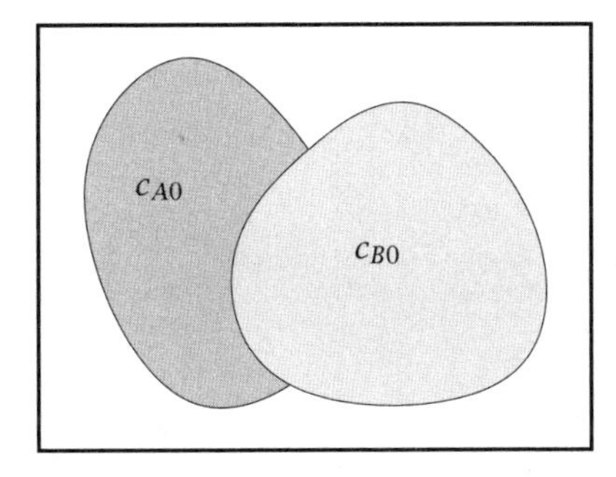

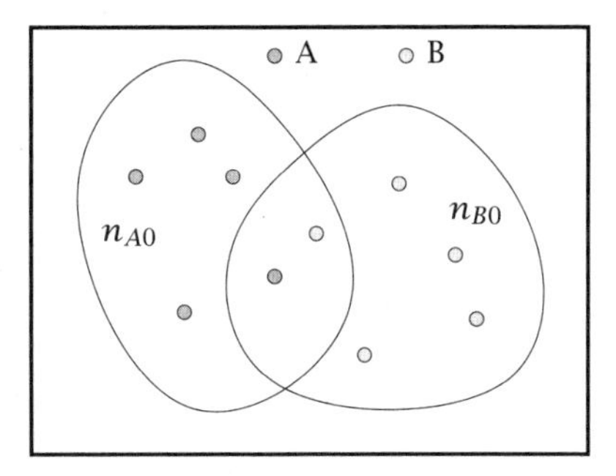

Figure 5.18: Species A and B in a well-mixed volume element. Continuum and molecular settings.

and B as $c_A = 2c_{A0}$, $c_B = 0$ and the total rate is then

$$r = r_1 + r_2 + r_3$$
$$r = k_1 c_A^2 + k_2 c_A c_B + k_3 c_B^2$$
$$r = k(2c_{A0})^2 + k(2c_{A0} \cdot 0) + k(0)^2$$
$$r = 4kc_{A0}^2$$

Why are these two total rates different and which one is correct?

(b) Repeat your analysis of the reaction rates if we reduce the length scale and consider the molecular kinetic setting in which we have integer-valued n_{A0}, n_{B0} molecules of A and B in the volume of interest.

(c) Perform a stochastic simulation of the molecular setting using the following parameters

$$n_{A0} = 50 \qquad n_{B0} = 60 \qquad n_{C0} = n_{D0} = n_{E0} = 0$$
$$k_1 = k_2 = k_3 = k = 10\text{sec}^{-1}$$

Make a plot of all species versus time. Print the plot and the simulation code.

Exercise 5.22: Cumulative distribution for the omega expansion

Given the governing equation for the fluctuation density in the omega expansion

$$\frac{\partial}{\partial t}\Pi = -\frac{\partial}{\partial \xi}(-2kc\xi\,\Pi) + \frac{\partial^2}{\partial \xi^2}(kc^2\,\Pi)$$

Define the cumulative distribution

$$F(x,t) = \int_{-\infty}^{x} \Pi(\xi,t)d\xi$$

(a) Derive the PDE governing F's evolution. What are the corresponding boundary conditions and initial condition?

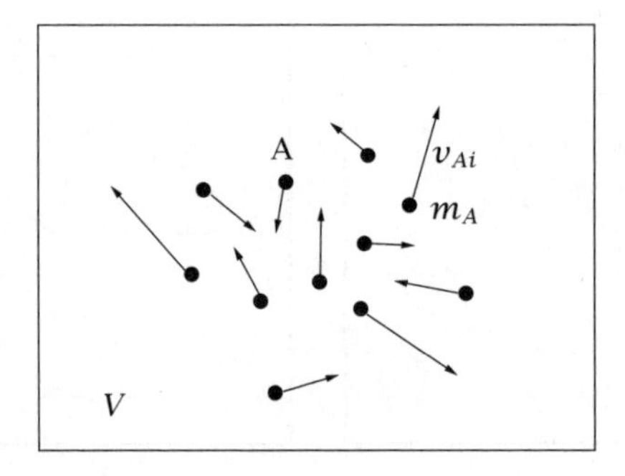

Figure 5.19: Molecular system of volume V containing molecules of mass m_A with velocity v_{Ai}.

(b) Solve this PDE numerically and compare to Figure 5.15 in the text. Increase Ω holding $c_0 = n_0/\Omega$ fixed and describe the effect on F.

Exercise 5.23: Properties of the Maxwell-Boltzmann distribution

Consider the simple molecular system depicted in Figure 5.19 with a large number of ideal gas molecules of species A with molecular weight m_A. The system volume is V. Molecule i has velocity v_{Ai}, $i = 1, 2, \ldots n$. A velocity vector is denoted $v = \begin{bmatrix} v_x & v_y & v_z \end{bmatrix}^T$ with corresponding x, y, z components. These velocities are considered samples of a random variable with fixed and known distribution.

The Maxwell-Boltzmann distribution for the zero mean fluctuation velocity in an ideal gas is

$$p_u(u_x, u_y, u_z) = \left(\frac{m}{2\pi k_B T}\right)^{3/2} e^{-\frac{1}{2}\frac{m}{k_B T}(u_x^2 + u_y^2 + u_z^2)}$$

in which m is the molecule mass, T is absolute temperature, and k_B is the Boltzmann constant, $k_B = R/N_{\text{Av}}$. This distribution is a multivariate normal with zero mean and variance matrix $\frac{k_B T}{m} I$, which we write as

$$u \sim N\left(0, \frac{k_B T}{m} I\right)$$

Denote the A species mean velocity (drift term) as v_A. The A molecule velocities are then distributed as

$$v_{Ai} \sim N\left(v_A, \frac{k_B T}{m_A} I\right) \qquad \text{all } i \tag{5.86}$$

Starting from the distribution (5.86), derive the following expectations in terms of the mean species velocity v_A and k_B, T, m_A.

1. $\mathcal{E}(v_{Ai})$
2. $\mathcal{E}(v_{Ai} v_{Ai}^T)$
3. $\mathcal{E}(v_{Ai}^2)$ in which $v_{Ai}^2 = v_{Ai} \cdot v_{Ai} = v_{Ai}^T v_{Ai}$
4. $\mathcal{E}(v_{Ai}^2 v_{Ai})$

Exercise 5.24: The normal's properties for optimal linear estimation

Establish the three properties (5.64)–(5.66) used in deriving the optimal linear estimator. Some hints follow.

(a) For (5.64), use the independence of y to establish that

$$p_{x,y,z}(x,y,z) = p_{x,z}(x,z)p_y(y)$$

and divide both sides by $p_z(z)$.

(b) For (5.65), we are given that (x,z) is jointly distributed as

$$p_{x,z}(x,z) = n\left(\begin{bmatrix} x \\ z \end{bmatrix}, \begin{bmatrix} m_x \\ m_z \end{bmatrix}, \begin{bmatrix} P_x & P_{xz} \\ P_{zx} & P_z \end{bmatrix}\right)$$

Consider the linear transformation

$$\begin{bmatrix} y \\ z \end{bmatrix} = \begin{bmatrix} A & 0 \\ 0 & I \end{bmatrix} \begin{bmatrix} x \\ z \end{bmatrix}$$

and show that

$$\begin{bmatrix} y \\ z \end{bmatrix} \sim N\left(\begin{bmatrix} Am_x \\ m_z \end{bmatrix} = \begin{bmatrix} AP_xA^T & AP_{xz} \\ P_{zx}A^T & P_z \end{bmatrix}\right)$$

Now use the conditional density formula to obtain $p_{y|z}$.

(c) For (5.66), note that this property is derived in Example 4.20.

Exercise 5.25: Observability, controllability, and duality

Review the concept of controllability presented in Exercise 1.26. Show that (A,C) is observable if and only if (A^T, C^T) is controllable. This result marks the beginning of the interesting story of the duality between regulation and estimation.

Exercise 5.26: Observability with N measurements

Consider the linear system

$$x^+ = Ax \qquad y = Cx$$

Prove the statement made in the text that if $x(0)$ cannot be uniquely determined by n measurements $\{y(0), y(1), \dots y(n-1)\}$, then it cannot be determined by N measurements for any N.

Bibliography

D. F. Anderson and T. G. Kurtz. Continuous time Markov chain models for chemical reaction networks. In H. Koeppl, editor, *Design and Analysis of Biomolecular Circuits: Engineering Approaches to Systems and Synthetic Biology*. Springer, 2011.

K. J. Åström. *Introduction to Stochastic Control Theory*. Academic Press, San Diego, California, 1970.

D. P. Bertsekas. *Dynamic Programming*. Prentice-Hall, Inc., Englewood Cliffs, New Jersey, 1987.

R. N. Bhattacharya and E. C. Waymire. *Stochastic Processes with Applications*. Society for Industrial and Applied Mathematics, Philadelphia, 2009.

A. E. Bryson and Y. Ho. *Applied Optimal Control*. Hemisphere Publishing, New York, 1975.

W. M. Deen. *Analysis of Transport Phenomena*. Topics in chemical engineering. Oxford University Press, Inc., New York, 1998.

A. Einstein. Über die von der molekular-kinetischen Theorie der Wärme geforderte Bewegung von in ruhenden Flüssigkeiten suspendierten Teilchen. *Annalen der Physik*, 17:549, 1905.

C. W. Gardiner. *Handbook of Stochastic Methods for Physics, Chemistry, and the Natural Sciences*. Springer-Verlag, Berlin, Germany, second edition, 1990.

M. A. Gibson and J. Bruck. Efficient exact stochastic simulation of chemical systems with many species and many channels. *J. Phys. Chem. A.*, 104: 1876–1889, 2000.

D. T. Gillespie. Exact stochastic simulation of coupled chemical reactions. *J. Phys. Chem.*, 81:2340–2361, 1977.

A. H. Jazwinski. *Stochastic Processes and Filtering Theory*. Academic Press, New York, 1970.

T. Kailath. A view of three decades of linear filtering theory. *IEEE Trans. Inform. Theory*, IT-20(2):146–181, March 1974.

R. E. Kalman. A new approach to linear filtering and prediction problems. *Trans. ASME, J. Basic Engineering*, pages 35–45, March 1960.

P. E. Kloeden and E. Platen. *Numerical solution of stochastic differential equations.* Springer Verlag, Berlin, 1992.

A. N. Kolmogorov. Interpolation and extrapolation of stationary random sequences. *Bull. Moscow Univ., USSR, Ser. Math. 5*, 1941.

T. G. Kurtz. The relationship between stochastic and deterministic models for chemical reactions. *J. Chem. Phys.*, 57(7):2976–2978, 1972.

A. Lasota and M. C. Mackey. *Chaos, Fractals and Noise: Stochastic Aspects of Dynamics.* Springer-Verlag, New York, second edition, 1994.

J. B. Rawlings and J. G. Ekerdt. *Chemical Reactor Analysis and Design Fundamentals.* Nob Hill Publishing, Madison, WI, second edition, 2012.

J. B. Rawlings and D. Q. Mayne. *Model Predictive Control: Theory and Design.* Nob Hill Publishing, Madison, WI, 2009.

W. H. Ray. *Advanced Process Control.* McGraw-Hill, New York, 1981.

E. D. Sontag. *Mathematical Control Theory.* Springer-Verlag, New York, second edition, 1998.

R. F. Stengel. *Optimal Control and Estimation.* Dover Publications, Inc., 1994.

N. G. van Kampen. *Stochastic Processes in Physics and Chemistry.* Elsevier Science Publishers, Amsterdam, The Netherlands, second edition, 1992.

N. Wiener. *The Extrapolation, Interpolation, and Smoothing of Stationary Time Series with Engineering Applications.* Wiley, New York, 1949. Originally issued as a classified MIT Rad. Lab. Report in February 1942.

A

Mathematical Tables

A.1 Laplace Transform Table

The Laplace transform pairs used in the text are collected in Table A.1 with a reference to the page in the text where they are derived or first stated.

	$f(t)$	$\overline{f}(s)$	Page
1	$\alpha f(t) + \beta g(t)$	$\alpha \overline{f}(s) + \beta \overline{g}(s)$	105
2	$\dfrac{df(t)}{dt}$	$s\overline{f}(s) - f(0)$	105
3	$\dfrac{d^2 f(t)}{dt^2}$	$s^2\overline{f}(s) - sf(0) - f'(0)$	105
4	$\dfrac{d^n f(t)}{dt^n}$	$s^n\overline{f}(s) - \sum_{i=1}^{n} s^{n-i} f^{(i-1)}(0)$	105
5	$\int_0^t f(t')dt'$	$\dfrac{1}{s}\overline{f}(s)$	105
6	$t^n f(t)$	$(-1)^n \dfrac{d^n \overline{f}(s)}{ds^n}$	105, 225
7	$f(t-a)H(t-a)$	$e^{-as}\overline{f}(s)$	106
8	$e^{at} f(t)$	$\overline{f}(s-a)$	106
9	$\int_0^t f(t')g(t-t')dt'$	$\overline{f}(s)\overline{g}(s)$	106, 223

continued on next page

continued from previous page

	$f(t)$	$\overline{f}(s)$	page
10	$\lim_{t\to 0^+} f(t)$ initial value theorem	$\lim_{s\to\infty} s\overline{f}(s)$	106, 224
11	$\lim_{t\to\infty} f(t)$ final value theorem	$\lim_{s\to 0} s\overline{f}(s)^\dagger$	106, 224
12	$H(t)$	$\frac{1}{s}$	107
13	$\delta(t)$	1	113
14	$\delta^{(n)}(t) \quad n \geq 0$	s^n	113
15	t	$\frac{1}{s^2}$	107
16	$t^n \quad n > -1$	$\frac{\Gamma(n+1)}{s^{n+1}}$	107
17	e^{at}	$\frac{1}{s-a}$	107
18	$e^{At} \quad A \in \mathbb{R}^{n\times n}$	$(sI - A)^{-1}$	109
19	te^{at}	$\frac{1}{(s-a)^2}$	107
20	$\sin \omega t$	$\frac{\omega}{s^2+\omega^2}$	107
21	$\cos \omega t$	$\frac{s}{s^2+\omega^2}$	107
22	$\sinh \omega t$	$\frac{\omega}{s^2-\omega^2}$	107
23	$\cosh \omega t$	$\frac{s}{s^2-\omega^2}$	107
24	$e^{at} \sin \omega t$	$\frac{\omega}{(s-a)^2+\omega^2}$	107
25	$e^{at} \cos \omega t$	$\frac{s-a}{(s-a)^2+\omega^2}$	107
26	$\sum_{n=1}^{m} \frac{\overline{p}(s_n)}{\overline{q}'(s_n)} e^{s_n t} \quad \overline{q}(s_n)$ simple zero	$\frac{\overline{p}(s)}{\overline{q}(s)}$	308
27	$\sum_{n=1}^{m} e^{s_n t} \sum_{i=1}^{r_n} a_{ni} t^{i-1} \quad \overline{q}(s_n)$ zero of order r_n	$\frac{\overline{p}(s)}{\overline{q}(s)}^*$	308
28	$\frac{k}{2\sqrt{\pi t^3}} e^{-\frac{k^2}{4t}}$	$e^{-k\sqrt{s}} \quad k > 0$	330

continued on next page

continued from previous page

	$f(t)$	$\overline{f}(s)$	page
29	$\frac{1}{\sqrt{\pi t}}\,e^{-\frac{k^2}{4t}}$	$\frac{e^{-k\sqrt{s}}}{\sqrt{s}}$ $k>0$	330
30	$\text{erfc}\left(\frac{k}{2\sqrt{t}}\right)$	$\frac{e^{-k\sqrt{s}}}{s}$ $k>0$	330
31	$\frac{1}{2\sqrt{\alpha}}e^{\alpha t}\left\{e^{-k\sqrt{\alpha}}\text{erfc}\left(\frac{k}{2\sqrt{t}}-\sqrt{\alpha t}\right)-e^{k\sqrt{\alpha}}\text{erfc}\left(\frac{k}{2\sqrt{t}}+\sqrt{\alpha t}\right)\right\}$	$\frac{e^{-k\sqrt{s}}}{(s-\alpha)\sqrt{s}}$ $k>0$	330
32	$\frac{1}{2t}\,e^{-\frac{k^2}{4t}}$	$K_0(k\sqrt{s})$ $k>0$	331
33	$\frac{1}{k}e^{-\frac{k^2}{4t}}$	$\frac{K_1(k\sqrt{s})}{\sqrt{s}}$ $k>0$	331
34	$\frac{\sinh(x\sqrt{k})}{\sinh\sqrt{k}}-2\sum_{n=1}^{\infty}\frac{(-1)^{n+1}\pi n}{n^2\pi^2+k}\sin(n\pi x)\,e^{-(n^2\pi^2+k)t}$	$\frac{\sinh(x\sqrt{s+k})}{s\sinh\sqrt{s+k}}$	333
35	$1-2\sum_{n=1}^{\infty}\frac{(-1)^{n+1}}{n\pi x}\sin(n\pi x)\,e^{-n^2\pi^2 t}$	$\frac{\sinh(x\sqrt{s})}{xs\sinh\sqrt{s}}$	335
36	$1-2\sum_{n=0}^{\infty}\frac{(-1)^n}{(n+1/2)\pi}\cos((n+1/2)\pi x)\,e^{-((n+1/2)\pi)^2 t}$	$\frac{\cosh(x\sqrt{s})}{s\cosh\sqrt{s}}$	333
37	$1-2\sum_{n=1}^{\infty}\frac{1}{\alpha_n J_1(\alpha_n)}J_0(\alpha_n x)\,e^{-\alpha_n^2 t}$ $J_0(\alpha_n)=0$	$\frac{I_0(x\sqrt{s})}{sI_0(\sqrt{s})}$	335
38	$2\sum_{n=1}^{\infty}(-1)^{n+1}\sin(n\pi a)\,\sin(n\pi b)\,\cos(n\pi t)$	$\frac{\sinh(as)\sinh(bs)}{\sinh s}$	314
39	$2\sum_{n=1}^{\infty}\frac{(-1)^{n+1}}{n\pi}\sin(n\pi a)\,\sin(n\pi b)\,\sin(n\pi t)$	$\frac{\sinh(as)\sinh(bs)}{s\sinh s}$	341

Table A.1: Larger table of Laplace transforms.

† Final value exists if and only if $s\overline{f}(s)$ is bounded for $\text{Re}(s)\geq 0$.

* $a_{ni}=\frac{\phi^{(r_n-i)}(s_n)}{(r_n-1)!(i-1)!}$ $\phi(s)=(s-s_n)^{r_n}\overline{p}(s)/\overline{q}(s)$.

A.2 Statistical Distributions

The different probability distributions that have been discussed in the text are summarized in Table A.2 with a reference to the page in the text where they are first mentioned.

Distribution	Density	Page
uniform	$p(x) = 1/(b-a) \quad x \in [a,b]$	382
normal	$p(x) = \frac{1}{\sqrt{2\pi\sigma^2}} \exp\left(-\frac{1}{2}\frac{(x-m)^2}{\sigma^2}\right)$	352
multivariate normal	$p(x) = \frac{1}{(2\pi)^{n/2}\lvert P\rvert^{1/2}} \exp\left[-\frac{1}{2}(x-m)^T P^{-1}(x-m)\right]$	358
exponential	$p(x) = \lambda e^{-\lambda x} \quad x \geq 0, \quad \lambda > 0$	478
Poisson	$p(n) = \frac{a^n}{n!} e^{-a} \quad n = 0, 1, 2, \ldots, \quad a > 0$	481
chi	$p(x) = 2x \frac{\left(\frac{1}{2}\right)^{n/2}}{\Gamma(n/2)} x^{n-2} e^{-\frac{1}{2}x^2}$	441
chi-squared	$p(x) = \frac{\left(\frac{1}{2}\right)^{n/2}}{\Gamma(n/2)} x^{n/2-1} e^{-x/2} \quad x \geq 0, \quad n \geq 1$	441
F	$p(x) = \frac{\sqrt{\frac{(xn)^n m^m}{(xn+m)^{n+m}}}}{xB(\frac{n}{2}, \frac{m}{2})} \quad x \geq 0, \quad n, m \geq 1$	442
Student's t	$p(x) = \frac{\Gamma\left(\frac{n+1}{2}\right)}{\sqrt{n\pi}\,\Gamma\left(\frac{n}{2}\right)} \left(1 + \frac{x^2}{n}\right)^{-\frac{n+1}{2}}$	441
multivariate t	$p(x) = \frac{\Gamma\left(\frac{n+p}{2}\right)}{(n\pi)^{p/2}\,\Gamma\left(\frac{n}{2}\right)\lvert\Sigma\rvert^{1/2}} \left(1 + \frac{1}{n}(x-m)^T \Sigma^{-1}(x-m)\right)^{-\frac{n+p}{2}}$	447
Wishart	$p(X) = \frac{\lvert X\rvert^{\frac{n-p-1}{2}}}{2^{\frac{np}{2}} \lvert R\rvert^{\frac{n}{2}} \Gamma_p\left(\frac{n}{2}\right)} e^{-\frac{1}{2}\mathrm{tr}(R^{-1}X)} \quad X > 0$	412
Maxwell	$p(x) = x^2 e^{-\frac{1}{2}x^2}$	524
Maxwell-Boltzmann	$p_u(u_x, u_y, u_z) = \left(\frac{m}{2\pi k_B T}\right)^{3/2} e^{-\frac{1}{2}\frac{m}{k_B T}(u_x^2 + u_y^2 + u_z^2)}$	524

Table A.2: Statistical distributions defined and used in the text and exercises.

A.3 Vector and Matrix Derivatives

Definition. First consider $s(t)$, a real-valued, scalar function[1] of a real-valued scalar, $s : \mathbb{R} \to \mathbb{R}$. Assume the derivative, $ds(t)/dt$ exists. We wish to extend the definition of the derivative to vector and matrix-valued functions of vector and matrix-valued arguments. Many of the derivative operations, such as derivative of scalar-valued functions with respect to scalars, vectors, and matrices, can be conveniently expressed using the rules of vector/matrix operations. Other derivative operations, such as the derivative of matrix-valued functions with respect to vectors and matrices, produce *tensors* having more than two indices. We summarize here the most important formulas that can be expressed in matrix/vector calculus. To state how the derivatives are arranged into vectors and matrices, we require a more precise notation than we used in the text. Moreover, several different and conflicting conventions are in use in different fields; these are briefly described in Section A.3.1. So we state here the main results in a descriptive notation, and expect the reader can translate these results into the conventions of other fields.

We require a few preliminaries. Now let $s(x)$ be a scalar-valued function of vector x, $s : \mathbb{R}^n \to R$. Assume that all partial derivatives, $\partial s/\partial x_i, i = 1, 2, \ldots, n$ exist. The derivative ds/dx is then defined as the *column* vector

$$\frac{ds}{dx} = \begin{bmatrix} \dfrac{\partial s}{\partial x_1} \\ \dfrac{\partial s}{\partial x_2} \\ \vdots \\ \dfrac{\partial s}{\partial x_n} \end{bmatrix} \qquad \text{scalar-vector derivative}$$

The derivative ds/dx^T is defined as the corresponding *row* vector

$$\frac{ds}{dx^T} = \begin{bmatrix} \dfrac{\partial s}{\partial x_1} & \dfrac{\partial s}{\partial x_2} & \cdots & \dfrac{\partial s}{\partial x_n} \end{bmatrix}$$

and note that $(ds/dx)^T = ds^T/dx^T = ds/dx^T$. Next let $s(A)$ be a scalar-valued function of matrix A, $s : \mathbb{R}^{m\times n} \to \mathbb{R}$. Again, assuming all

[1]All of the formulas in this section are readily extended to complex-valued functions of a complex variable.

partial derivatives exist, the derivative ds/dA is then defined as

$$\frac{ds}{dA} = \begin{bmatrix} \frac{\partial s}{\partial A_{11}} & \frac{\partial s}{\partial A_{12}} & \cdots & \frac{\partial s}{\partial A_{1n}} \\ \frac{\partial s}{\partial A_{21}} & \frac{\partial s}{\partial A_{22}} & \cdots & \frac{\partial s}{\partial A_{2n}} \\ \vdots & \vdots & \ddots & \vdots \\ \frac{\partial s}{\partial A_{m1}} & \frac{\partial s}{\partial A_{m2}} & \cdots & \frac{\partial s}{\partial A_{mn}} \end{bmatrix} \qquad \text{scalar-matrix derivative}$$

As in the vector case, we define ds/dA^T as the transpose of this result, or $ds/dA^T = (ds/dA)^T$. These more general matrix derivatives do specialize to the two vector derivatives previously defined.

Next up is the vector-valued function of a vector, $f(x)$. Let $f : \mathbb{R}^n \to \mathbb{R}^m$. The quantity of most interest is usually the Jacobian matrix, which we denote by df/dx^T, defined by

$$\frac{df}{dx^T} = \begin{bmatrix} \frac{\partial f_1}{\partial x_1} & \frac{\partial f_1}{\partial x_2} & \cdots & \frac{\partial f_1}{\partial x_n} \\ \frac{\partial f_2}{\partial x_1} & \frac{\partial f_2}{\partial x_2} & \cdots & \frac{\partial f_2}{\partial x_n} \\ \vdots & \vdots & \ddots & \vdots \\ \frac{\partial f_m}{\partial x_1} & \frac{\partial f_m}{\partial x_2} & \cdots & \frac{\partial f_m}{\partial x_n} \end{bmatrix} \qquad \begin{array}{l}\text{vector-vector derivative} \\ \text{(Jacobian matrix)}\end{array}$$

The notation df/dx^T serves as a convenient reminder that the *column* vector f is distributed down the *column* and the *row* vector x^T is distributed across the *row* in the entries in the Jacobian matrix. The transpose of the Jacobian is simply $df^T/dx = (df/dx^T)^T$, which is easy to remember. Note that df/dx is a long column vector with mn entries coming from stacking the columns of the Jacobian matrix. This is the vec operator, so we have

$$\frac{df}{dx} = \text{vec}\left(\frac{df}{dx^T}\right)$$

The transpose, denoted df^T/dx^T, is a long row vector. These vector arrangements of the derivatives are not usually of much interest compared to the Jacobian matrix, as we shall see when we discuss the chain rule.

Inner product. The inner product of two vectors was defined in Chapter 1

$$(a, b) = a^T b = \sum_{i=1}^{n} a_i b_i \qquad a, b \in \mathbb{R}^n$$

We can extend this definition to linear spaces of matrices as follows

$$(A, B) = \text{tr}(A^T B) = \sum_{i=1}^{m} \sum_{j=1}^{n} A_{ij} B_{ij} \qquad A, B \in \mathbb{R}^{m \times n}$$

Because $\text{tr}(C) = \text{tr}(C^T)$ for any square matrix C, the matrix inner product can also be expressed as $(A, B) = \text{tr}(B^T A)$, which is valid also in the vector case.

Chain rules. One of the most important uses of these derivative formulas is a convenient expression of the chain rule. For scalar-valued functions we have two common forms.

$$ds = (\frac{ds}{dx}, dx) = \frac{ds}{dx^T} dx \qquad \text{scalar-vector} \qquad (A.1)$$

$$ds = (\frac{ds}{dA}, dA) = \text{tr}\left(\frac{ds}{dA^T} dA\right) \qquad \text{scalar-matrix} \qquad (A.2)$$

Notice that when written with inner products, these two formulas are identical. The vector chain rule can be considered a special case of the matrix chain rule, but since the vector case arises frequently in applications and doesn't require the trace, we state it separately. For vector-valued functions we have one additional form of the chain rule

$$df = \frac{df}{dx^T} dx \qquad \text{vector-vector} \qquad (A.3)$$

which is a matrix-vector multiplication of the Jacobian matrix of f with respect to x with the column vector dx. Because df is a vector, this chain rule is *not* expressible by an inner product as in the scalar case. But notice the similarity of the vector chain rule with the second equalities of the two scalar chain rules. Because of this similarity, all three important versions of the chain rule are easy to remember using this notation. There is no chain rule for matrix-valued functions that does not involve tensors.

Finally, we collect here the different matrix and vector differentiation formulas that have been used in the text and exercises. These are summarized in Table A.3, with a reference to the page in the text where they are first mentioned or derived.

	Derivative Formula	Page
1	$ds = \dfrac{ds}{dx^T} dx$ (chain rule 1)	
2	$ds = \mathrm{tr}\left(\dfrac{ds}{dA^T} dA\right)$ (chain rule 2)	
3	$df = \dfrac{df}{dx^T} dx$ (chain rule 3)	
4	$\dfrac{d}{dx} f^T g = \dfrac{df^T}{dx} g + \dfrac{dg^T}{dx} f$ (product rule)	
5	$\dfrac{d}{dx} x^T b = b$	
6	$\dfrac{d}{dx^T} b^T x = b^T$	
7	$\dfrac{d}{dx} x^T A x = Ax + A^T x$	
8	$\dfrac{d}{dx^T} Bx = B$	
9	$\dfrac{d}{dx} x^T B^T = B^T$	
10	$\dfrac{d}{dt} p(A) = q(A) \dfrac{d}{dt} A, \quad q(\cdot) = \dfrac{d}{d(\cdot)} p(\cdot)$	328
11	$\dfrac{d}{dt} \det A = \det(A)\, \mathrm{tr}\left(A^{-1} \dfrac{d}{dt} A\right), \quad \det A \neq 0$	328
12	$\dfrac{d}{dA} \mathrm{tr}(p(A)) = q(A^T), \quad q(\cdot) = \dfrac{d}{d(\cdot)} p(\cdot)$	431
13	$\dfrac{d}{dA} \det A = (A^{-1})^T \det A, \quad \det A \neq 0$	431
14	$\dfrac{d}{dA} \ln(\det A) = (A^{-1})^T, \quad \det A \neq 0$	431
15	$\dfrac{d}{dA} \mathrm{tr}(AB) = \dfrac{d}{dA} \mathrm{tr}(BA) = B^T$	
16	$\dfrac{d}{dA} \mathrm{tr}(A^T B) = \dfrac{d}{dA} \mathrm{tr}(BA^T) = B$	

continued on next page

continued from previous page

	Derivative Formula	Page
17	$\frac{d}{dA}\text{tr}(ABA^T) = A(B^T + B)$	
18	$\frac{d}{dA}\text{tr}(A^T BA) = (B + B^T)A$	

Table A.3: Summary of vector and matrix derivatives defined and used in the text and exercises; $s, t \in \mathbb{R}$, $x, b \in \mathbb{R}^n$, $A \in \mathbb{R}^{m\times n}$, $B \in \mathbb{R}^{m\times n}$, $f(\cdot)$ and $g(\cdot)$ are any differentiable functions, and $p(\cdot)$ is any matrix function defined as a power series.

A.3.1 Derivatives: Other Conventions

Given the many scientific fields requiring vector/matrix derivatives, chain rules, and so on, a correspondingly large number of different and conflicting notations have also arisen. We point out here some of the other popular conventions and show how to translate them into the notation used in this section.

Optimization. The dominant convention in the optimization field is to define the scalar-vector derivative ds/dx as a *row* vector instead of a column vector. The nabla notation for gradient, ∇s, is then used to denote the corresponding *column* vector. The Jacobian matrix is then denoted df/dx. So the vector chain rule reads in the optimization literature

$$df = \frac{df}{dx}\,dx \qquad \text{optimization convention}$$

Given that ds/dx is a row vector in the optimization notation, the first scalar chain rule reads

$$ds = \left(\left(\frac{ds}{dx}\right)^T, dx\right) = \frac{ds}{dx}dx \qquad \text{optimization convention}$$

The biggest problem with adopting the optimization field's conventions arises when considering the scalar-matrix derivative. The derivative ds/dA has the *same* meaning in the optimization literature as that used in this text. So the second scalar chain rule is also the same as in this text

$$ds = (\frac{ds}{dA}, dA) \qquad = \text{tr}\left(\frac{ds}{dA^T}\, dA\right) \qquad \text{optimization convention}$$

Notice the inconsistency in the chain rules: the scalar-matrix version directly above contains a transpose and the scalar-vector and vector-vector versions do not. The burden rests on the reader to recall these different forms of the chain rule and remember which ones require the transpose. The advantage of the notation used in this section is that *all* chain rules appear with a transpose, which is what one might anticipate due to the chain rule's required summation over an index. Also, in the notation used in this section, the ∇ operator is *identical* to d/dx and neither implies a transpose should be taken. Finally, there is no hint in the optimization field's notation ds/dx and ∇s as to which should be a column vector and which a row vector. The notation used in this section, ds/dx and ds/dx^T, makes that distinction clear.

Field theories of physics (transport phenomena, electromagnetism) As noted in Chapter 3, the literature in these areas primarily uses Gibbs vector-tensor notation and index notation. For example, the derivative of a scalar function s with regard to a vector argument $\boldsymbol{x}$ is

$$\boldsymbol{\nabla} s \quad \text{or} \quad \frac{\partial s}{\partial \boldsymbol{x}}$$

In Cartesian coordinates

$$\left(\frac{\partial s}{\partial \boldsymbol{x}}\right)_i = \frac{\partial s}{\partial x_i}$$

The derivative of a scalar s with respect to a tensor argument is similar

$$\left(\frac{\partial s}{\partial \boldsymbol{A}}\right)_{ij} = \frac{\partial s}{\partial A_{ij}}$$

The derivative of a vector function $\boldsymbol{f}$ with respect to a vector $\boldsymbol{x}$ is

$$\boldsymbol{\nabla} \boldsymbol{f} \quad \text{or} \quad \frac{\partial \boldsymbol{f}}{\partial \boldsymbol{x}}$$

where

$$(\boldsymbol{\nabla} \boldsymbol{f})_{ij} = \left(\frac{\partial \boldsymbol{f}}{\partial \boldsymbol{x}}\right)_{ij} = \frac{\partial f_j}{\partial x_i}$$

so $\left(\frac{\partial \boldsymbol{f}}{\partial \boldsymbol{x}}\right)$ is the *transpose* of the Jacobian. Therefore, the chain rule becomes

$$d\boldsymbol{f} = d\boldsymbol{x} \cdot \nabla \boldsymbol{f} = (\nabla \boldsymbol{f})^T \cdot d\boldsymbol{x}$$

Consistent with this notation, one can write the Taylor-series expansion of a vector field $\boldsymbol{f}$ around the origin as

$$\boldsymbol{f}(\boldsymbol{x}) = \boldsymbol{f}(\boldsymbol{0}) + \boldsymbol{x} \cdot \nabla \boldsymbol{f} + \frac{1}{2}\boldsymbol{x}\boldsymbol{x} : \nabla\nabla \boldsymbol{f} + \ldots$$

where derivatives are evaluated at the origin and

$$(\nabla\nabla \boldsymbol{f})_{jki} = \frac{\partial^2 f_i}{\partial x_j \partial x_k}$$

One must beware, however, that this ordering of indices is not used universally, primarily because some authors write the Taylor expansion as

$$\boldsymbol{f}(\boldsymbol{x}) = \boldsymbol{f}(\boldsymbol{0}) + \boldsymbol{J} \cdot \boldsymbol{x} + \frac{1}{2}\boldsymbol{K} : \boldsymbol{x}\boldsymbol{x} + \ldots$$

where

$$J_{ij} = \frac{\partial f_i}{\partial x_j}$$

$$K_{ijk} = \frac{\partial^2 f_i}{\partial x_j \partial x_k}$$

A.4 Exercises

Exercise A.1: Simple and repeated zeros

Assume all the zeros of $\overline{q}(s)$ are first-order zeros, $r_n = 1, \quad n = 1, 2, \ldots, m$, in entry 27 of Table A.1, and show that it reduces to entry 26.

Exercise A.2: Deriving the Heaviside expansion theorem for repeated roots

Establish the Heaviside expansion theorem for repeated roots, entry(27) in Table A.1.

Hints: Close the contour of the inverse transform Bromwich integral in (2.7) to the left side of the complex plane. Show that the integral along the closed contour except for the Bromwich line goes to zero, leaving only the residues at the singularities, i.e., the poles $s = s_n$, $n = 1, 2, \ldots, m$. Since $\phi(s)$ has no singularities, expand it in a Taylor series about the root $s = s_n$. Find the Laurent series for $\overline{f}(s)$ and show that the residues are the coefficients a_{in} given in the expansion formula. Note that this procedure remains valid if there are an infinite number of poles, such as the case with a transcendental function for $\overline{q}(s)$.

Exercise A.3: Laplace transform relations

Take the limit $k \to 0$ in entry 34 of Table A.1 and show that it produces entry 35.

Exercise A.4: Some invalid derivative formulas

Notice that entry 5 in Table A.3 is a special case of entry 9 with $B^T = b$. Also, entry 6 is a special case of entry 8 with $B = b^T$. Consider the following derivatives that could be included in Table A.3 in place of entries 5 and 6

$$\frac{d}{dx} b^T x = b \qquad \frac{d}{dx^T} x^T b = b^T$$

These can be derived by transposing the scalar numerators in entries 5 and 6, respectively. But notice that we do *not* find companion forms for these listed in Table A.3 with general matrix B replacing column vector b. Compute the following matrix derivatives and show that simply replacing b with general matrix B above does not generate correct formulas

$$\frac{d}{dx} B^T x \neq B \qquad \frac{d}{dx^T} x^T B \neq B^T$$

Note that you may want to use the vec operator to express the correct formulas. Next show that the *correct* matrix versions of these derivatives do reduce to the above formulas for $B = b$, a column vector.

Exercise A.5: Companion trace derivatives

(a) Use the fact that $\mathrm{tr}(AB) = \mathrm{tr}(BA)$ to establish that Formulas 15 and 16 in Table A.3 are equivalent formulas, i.e., assuming one of them allows you to establish the other one.

(b) On the other hand, show that Formulas 17 and 18 are equivalent by taking transposes of one of them to produce the other one.

Author Index

Citation Index

Subject Index